Two Manuals Combined into One Book

Military Police

And

Security Force Assistance

Department of Defense

Pentagon
Publishing

FM 3-19.4 (FM 19-4)

Field Manual
No. 3-19.4

Headquarters
Department of the Army
Washington, DC, 4 March 2002

Military Police Leaders' Handbook

Contents

Distribution Restriction: Approved for public release; distribution is unlimited.

*This publication supersedes FM 19-4, 7 May 1993.

Page

Preface

This field manual (FM) addresses military police (MP) maneuver and mobility support (MMS), area security (AS), internment and resettlement (I/R), law and order (L&O), and police intelligence operations (PIO) across the full spectrum of Army operations. Although this manual includes a discussion of corps and division MP elements, it primarily focuses on the principles of platoon operations and the tactics, techniques, and procedures (TTP) the platoon uses to accomplish its mission.

This FM provides the capabilities and organization of the MP, demonstrates the flexibility and diversity of MP in adapting to any mission throughout the full spectrum of Army operations, and characterizes the MP as a combat-force multiplier. Additionally, this manual identifies the fact that the Army will not conduct operations alone and defines the role of the MP in support of joint, multinational, and interagency operations.

The MP TTP are organized by the MP functions of MMS, AS, I/R, LO, and PIO with supporting tasks, both individual and collective, to help illustrate the functions.

NOTE: United States (US) policy regarding the use and employment of antipersonnel land mines (APLs) outlined in this FM is subject to the convention on certain conventional weapons and executive orders (EOs). Current US policy limits the use of non-self-destructing APLs to (1) defending the US and its allies from armed aggression across the Korean demilitarized zone and (2) training personnel engaged in demining and countermine operations. The use of the M18A1 claymore in the command-detonation mode is not restricted under international law or EO.

Appendix A complies with current Army directives which state that the metric system will be incorporated into all new publications. *Appendix B* deals with media relations.

The proponent of this publication is Headquarters (HQ) United States Army Training and Doctrine Command (TRADOC). Send comments and recommendations on Department of the Army (DA) Form 2028 directly to Commandant, US Army Military Police School, ATTN: ATSJ-MP-TD, 401 MANSCEN Loop, Suite 2060, Fort Leonard Wood, Missouri 65473-8926.

Unless this publication states otherwise, masculine nouns and pronouns do not refer exclusively to men.

Chapter 1

Military Police Overview

This chapter provides information about how MP are organized, equipped, and trained to provide combat support (CS) across the full spectrum of Army operations.

INTRODUCTION

1-1. As a flexible economy-of-force organization, MP provide a wide range of diverse support because of their agility and versatility to adapt to any mission or environment. As a combat multiplier, they support the maneuver commander through the five MP functions. MP contribute to the commander's overall combat power by integrating efforts with those of other combat, CS, and combat service support (CSS) elements.

1-2. In addition to single-service operations, MP also support joint, multinational, and interagency activities. MP support air base defense in concert with Air Force Security Forces, operate joint and multinational checkpoints, conduct combined police patrols, and exchange police information and criminal intelligence with the host nation (HN), military, and civilian police agencies.

1-3. MP have the capability to expedite the movement of combat resources, provide critical asset security and protection, conduct I/R, contribute to force protection efforts through L&O operations, and gather and disseminate police information and intelligence.

MILITARY POLICE FUNCTIONAL AREAS

1-4. With the old battlefield missions, the term "operations" was used extensively and carried too broad of a meaning. To clarify the specific tasks of the MP, the battlefield missions have been redefined into the following five functional areas:

- MMS
- AS
- I/R
- L&O
- PIO

1-5. Each of these MP functions have task areas and tasks that support them. MP functions are the broadest areas for which tasks are placed. Some of these tasks will require groupings that might not be related to the entire function. Therefore, task areas were created to group specific tasks. Specific tasks consist of two types—collective and individual. Individual tasks are further divided into leader and soldier tasks *(Figure 1-1)*. The collective and individual tasks that support the MP task areas are found in the MP mission training plans (MTP) and MP soldier's manuals (SMs).

1-6. MP procedures are the lowest level of detail. They explain the "how to" at the task level. Procedures include the standing methods used by a unit to accomplish tasks, weapon and equipment operating steps, crew drills, and staff action and coordination. They are the building blocks of individual and collective task accomplishment and serve as the foundation of tactics and techniques. Procedures are explained in the unit standing operating procedures (SOPs), MTPs, SMs, and similar publications.

MANEUVER AND MOBILITY SUPPORT

1-7. The MMS function involves the measures necessary to enhance combat movement and the ability to conduct

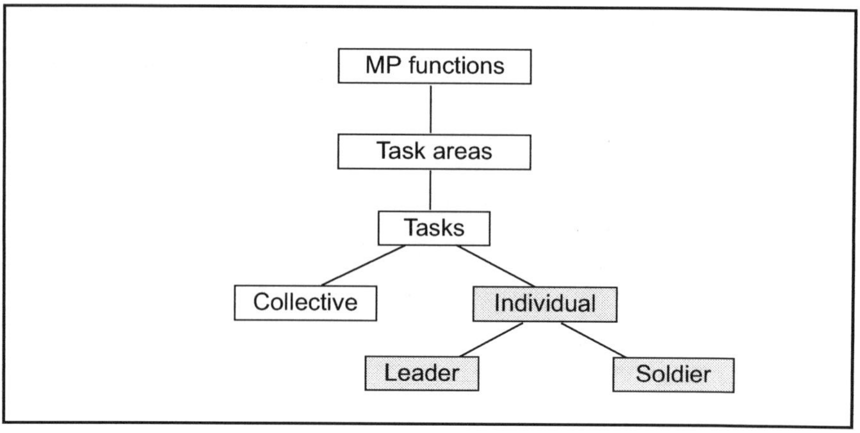

Figure 1-1. Relationship Between MP Functions and Tasks

movement of friendly resources in all environments. MP conduct MMS to ensure that the commanders receive personnel, equipment, and supplies when and where they are needed. The task areas that support the function of MMS include—

- MP support for river crossings, breaching, and passage-of-line operations.
- Straggler and dislocated civilian control.
- Route reconnaissance and surveillance.
- Main supply route (MSR) regulation enforcement.

1-8. The security and viability of the operational and tactical lines of communications (LOC) will be critical to continuous sustainment and recovery operations. MP ensure that logistics and supply operations are kept on time and arrive at the right place. Refer to *Chapter 5* for more information about MMS.

AREA SECURITY

1-9. The AS function consists of those security measures designed to give commanders freedom of maneuver and flexibility to conduct operations. The task areas that support AS include—

- Reconnaissance operations.
- Area damage control (ADC).
- Base and air base defense.
- Response force and tactical combat force (TCF) operations.
- Critical site, asset, and high-risk personnel security.
- Force protection and physical security.
- Antiterrorism.

1-10. MP performing AS contribute to securing and protecting the force and preserving combat power. Refer to *Chapter 6* for more information about AS.

INTERNMENT AND RESETTLEMENT

1-11. The I/R function consists of those measures necessary to provide shelter, sustain, guard, protect, and account for people (enemy prisoners of war [EPWs] and civilian internees [CIs], US military prisoners, and dislocated civilians [DC]). The task areas that support I/R include—

- EPW and CI handling.
- Populace and resource control.
- US military prisoner confinement.
- DCs control.

1-12. The international community, media, and public perceptions have increased sensitivity to the protection of human rights and the need for absolute accountability of interned, detained personnel, and refugees in military operations. Refer to *Chapter 7* for more information about I/R.

LAW AND ORDER

1-13. Task areas and tasks that minimize the effects of a criminal threat on friendly forces support the L&O function. MP conduct L&O to remove the conditions and opportunities that promote crime, thereby preventing diversion of military resources and maintaining military discipline. The task areas include—

- Law enforcement.
- Criminal investigations.
- US customs operations.
- Related L&O training.

1-14. Whether patrolling an installation's housing area, conducting counterdrug operations, enhancing security, or investigating war crimes, MP L&O capabilities are invaluable to the commander. Refer to *Chapter 8* for more information about L&O.

POLICE INTELLIGENCE OPERATIONS

1-15. The PIO function supports, enhances, and contributes to the commander's protection program, situational awareness, and battlefield visualization by portraying relevant threat information that may affect the operational and tactical environment. The task areas that support PIO include—

- Intelligence preparation of the battlefield (IPB).
- Active and passive roles.
- Police assessment process.

1-16. Whether in support of peacetime installation L&O or detecting threat forces in the battle space, the MP force employment provides the commander with substantial information and an intelligence source, particularly where the criminal element is the same as or closely aligned with the opposing forces (OPFOR) and government. Refer to *Chapter 9* for more information about PIO.

THREAT

1-17. Today, friendly forces encounter a broad range of traditional and nontraditional threats. No single dominating threat will be the undisputed focus of US security policy. Although overt attacks on the US and its strategic interests may be less common, stability operations and support operations will likely increase. Economic development and demographics, as well as the progression of social and cultural movements, will encompass an array of threat forces including nonnation forces (ethnic conflicts and terrorist activities) which challenge traditional nation and state environments. Additionally, nation and state forces (internal security forces and infantry-based and armor-mechanized-based armies) continue to present a global threat. These forces possess varying levels of military and advanced technology capabilities.

1-18. In recent military operations, a nontraditional criminal threat has emerged. The evolving criminal threat operates most often in the rear area, near ports, in built-up areas, and where troop populations are high. This threat is most likely to be detected at border crossings trying to disrupt the relocation efforts of DCs. They may commit crimes against particular ethnic groups or be at checkpoints and roadblocks trying to position weapons, explosives, or personnel in sustainment areas in order to disrupt military operations or kill friendly forces. Such a threat requires commanders to minimize its negative impact on friendly forces, resources, and operations. The MP continue to respond to nonmilitary threats including famine, health epidemics, illegal immigration, illegal drug traffic, and population dislocation.

MILITARY POLICE PLATOON ORGANIZATION AND LEADERSHIP

1-19. There are two basic MP platoon organizations, corps and division. Corps MP platoons are organized and equipped basically the same. Each division MP platoon supporting a different kind of division (such as heavy, light, airborne, or air assault) is designed under a different table(s) of organization and equipment (TOE).

CORPS MILITARY POLICE

1-20. Depending on the nature of the operation, corps MP are usually among the first forces deployed to support military operations around the world. They deploy early to areas devastated by natural or man-made disasters to assist disaster relief and damage assessment efforts. They provide security and force protection to friendly forces, critical facilities, and resources as units organize for military operations. In a developing theater, corps MP concentrate mission support to the main effort. Units whose assistance to the main effort is vital normally receive the highest priority for protection. Key facilities, such as traffic choke points, critical tunnels and bridges, and ammunition and fuel storage points may require special protection. As the theater matures, the focus may quickly change to other functions as MP adjust priorities to accommodate the change.

DIVISION MILITARY POLICE

1-21. Division MP are organized somewhat different depending on the type of division they are supporting. For example, a heavy division has one MP platoon providing direct support (DS) to each maneuver brigade and two MP platoons providing general support (GS) to the division's rear. Both airborne and air assault divisions have four MP platoons providing GS. A light infantry division is supported by three MP platoons.

1-22. In heavy divisions, where highly mobile forces are designed to move quickly over open ground, the overall need for MMS is significant. Division MP are likely to focus on expediting the forward movement of the critical combat resources into the division area. However, the priority could change quickly to removing EPWs from forward areas to freeing maneuver forces from guarding and caring for captives.

1-23. In airborne and air assault divisions, priority of MP support is most often needed for EPW operations and then for MMS to speed the movement of CS vehicles within the airhead.

1-24. For MP supporting any division, certain employment considerations remain constant. MP provide dedicated security for assets deemed critical by the division commander. This includes the division's main command post (CP) where MP operate outside the CP perimeter conducting screening missions designed to detect, disrupt, and delay enemy forces from disrupting the division's primary CP. Another consideration is MP accepting EPWs from capturing troops as far forward as possible.

SEPARATE BRIGADES

1-25. MP support to a separate brigade is normally provided by a four-squad MP platoon. The platoon is assigned to the brigade headquarters and headquarters company (HHC). The brigade has a separate provost marshal (PM) cell that serves as the command and control (C^2) element for the platoon. The brigade PM cell has operational control (OPCON) of all MP assets the same way the division PM has OPCON of the division MP assets. The brigade HHC provides sustainment support for both the PM cell and the MP platoon. The PM advises the separate brigade commander on matters pertaining to MP operations. The platoon leader directs the execution of the platoon's missions based on the priorities set forth by the PM and the supported commander.

1-26. The platoon can perform any of the five functions. The platoon leader may task organize the squads according to mission, enemy, troops, terrain, time available, and civilian considerations (METT-TC), and provide one squad to operate the EPW collecting point, one squad to provide a mobile security screen and occupy observation posts (OPs) around the brigade's CP, and two squads to conduct MMS and AS throughout the brigade's rear area.

INITIAL BRIGADE COMBAT TEAMS (IBCTs)

1-27. MP support to an IBCT may differ from that of other separate brigades. The IBCT is a preconfigured, ready-to-fight, combined-arms package. It is designed and optimized primarily for employment in small-scale contingencies operations (SSCO) in complex and urban terrain, confronting low-end and mid-range threats. The IBCT participates in war, with augmentation, as a subordinate maneuver component within a division or corps, in a variety of possible roles. It also participates in stability and/or support operations as an initial entry force. Civil unrest or complete turmoil normally characterizes these environments.

1-28. Organically, MP support to the IBCT is a two-person PM planning cell. The PM planning cell is located with the HHC's maneuver support cell and under the direct supervision of the brigade Operations and Training Officer (US Army) (S3). The role of the PM cell is significantly different from that of a traditional division PM or separate brigade PM. The main difference is the lack of organic or habitual MP assets in the IBCT. The absence of organic MP assets makes the job of the PM much more critical. The PM must—

- Understand the organization, capabilities, and limitations of the IBCT.

- Conduct effective liaison with higher HQ PM elements.
- Become an effective planner and anticipator of MP requirements.
- Task organize MP units effectively and efficiently.
- Assume C^2 of incoming MP forces or, if operating under a division, relinquish C^2 to a division or corps PM (if appropriate for effective employment of MP forces).

1-29. Depending on METT-TC, the IBCT may be augmented by MP elements ranging from a platoon to a battalion. Once the brigade receives MP augmentation, the PM then becomes a staff planner and coordinator for all MP activities.

1-30. Since the IBCT is a divisional brigade, the division PM and the IBCT PM play an important role in developing an optimum MP force package to support the brigade commander's concept of the operation. Despite the brigade's early time lines, the PM must consider and plan for MP augmentation forces as early as possible to free up valuable combat resources. SSCO that result in numerous EPWs, CIs, and refugees will hamper the maneuver force's freedom of movement.

1-31. Initially, MP priority of effort during the offense may be providing MMS for ground combat, CS, and CSS forces and taking control of EPWs and CIs. During the defense, the priority of MP support may shift to conducting AS and counterreconnaissance along the LOC, C^2 centers, and CSS bases. MP may also be required to conduct response force operations or become part of the TCF.

1-32. During stability and support operations, MP support may include—

- Order restoration.
- Crowd control.
- AS or force protection.
- DC operations.
- Noncombatant-evacuation operations.

PLATOON LEADER

1-33. The MP platoon leader is responsible to the company commander for the platoon's combat readiness, training, and discipline and the maintenance of its equipment. To be successful, the platoon leader must demonstrate an ability to lead soldiers and manage an organization, material, and time. He must be able to articulate the capabilities and limitations of the platoon to various non-MP staff sections. In combat, the platoon leader is responsible for accomplishing all the missions assigned to the platoon according to the commander's intent and preserving the platoon's fighting capability.

PLATOON SERGEANT (PSG)

1-34. The PSG leads elements of the platoon as directed by the platoon leader and assumes command of the platoon in the absence of the platoon leader. He directs the day-to-day activities of the platoon and ensures that the platoon has individual and team training and logistics needed to accomplish its mission. During tactical operations, he may assist in the control of the platoon.

TEAM AND SQUAD LEADER

1-35. The MP team leader is responsible to the squad leader for individual and team training and team discipline. He is responsible for the tactical employment and control of the team and the maintenance and operation of all vehicles and equipment organic to the team. During combat operations or anytime there is a threat, the team leader quickly assesses the situation, reports to his superiors, and takes appropriate action to protect the team according to the rules of engagement (ROE). A squad leader has the same responsibility for the squad as the team leader has for the team.

FORCE PROTECTION (FP) MEASURES

1-36. MP leaders at all levels must examine FP requirements and integrate FP measures throughout all the operations. Once higher HQ has established local FP policies, leaders set the example by complying with them. Leaders reduce the soldiers' exposure to hazards by strictly enforcing all the protective postures that may include—

- Traveling with at least two vehicles armed with at least one automatic weapon.
- Hardening of the vehicles.
- Wearing Kevlar® and body armor.
- Not driving off the road or cross-country.
- Placing off-post facilities off limits during nonduty hours.

1-37. MP leaders analyze and compensate for other threats such as disease, weather, crime, complacency, terrorism, morale, safety, and other considerations.

1-38. At the operational level, team and squad leaders establish a safety zone around their teams. The safety zone is the immediate area around the team where threat forces or events could harm the team or inflict casualties. In open terrain, the safety zone may extend out to the maximum effective range of the team's organic-weapon systems. However, when searching vehicles at a checkpoint or conducting crowd control, the safety zone may only be an arm's length from the team.

1-39. Team and squad leaders remain alert to threats that enter the team's safety zone. They must quickly assess any threat to the team and take appropriate action within the ROE to reduce the threat or move the team.

1-40. When required to operate in crowds, maintain eye contact with team members. Establish and maintain a safe distance between the team and the crowd. Never allow the team to become separated or surrounded.

1-41. When patrolling in built-up areas, the gunner scans the upper floors of the buildings and the streets to the vehicle's front, rear, and flanks and immediately reports any suspicious activity. The driver concentrates on the area directly in front of the vehicle looking for any unexploded munitions, scatterable mines, or other road hazards. All team members should stay awake, alert, and ready to react to danger.

MILITARY POLICE PLATOON MISSION, CAPABILITIES, AND LIMITATIONS

1-42. The platoon has one critical wartime mission which is to provide MP CS to an assigned area of operations (AO). MP CS consists of all five MP functions. The platoon performs its missions primarily mounted, taking full advantage of the high mobility multipurpose wheeled vehicle's (HMMWV's) versatility and the added protection and firepower of the armor security vehicle (ASV).

CAPABILITIES

1-43. The MP platoon is capable of operating day or night, in various terrain conditions, and under all weather and visibility conditions. Their mode of operation is possible through the deployment and employment of the three-person team throughout the battlefield. However, it is dependent on its parent unit for sustainment support. The platoon has self-protection capabilities such as nuclear, biological, chemical (NBC) detection equipment and a Platoon Early-Warning System (PEWS). The platoon's radio transmission range is increased with an OE-254 antenna. Because of extensive police training and law enforcement missions, the MP are highly skilled in the use of force and the employment of lethal and nonlethal technologies, information-collecting and dissemination, observation and surveillance, and crowd control. The MP platoon has a tremendous combat and noncombat information-collecting capability. This capability is the result of extensive area, zone, and route reconnaissance; daily contact with local nationals; conducting combined police patrols with HN military and civilian police agencies; and conducting field interviews. An MP platoon is capable of covering 500 square kilometers in rolling terrain; however, more severe terrain such as mountains, METT-TC, and mission objectives will affect this capability. For example, consider one mobile MP team per 10 kilometers of route coverage. For area coverage, begin with an estimate of one mobile MP team per 55 square kilometers.

1-44. Unlike most combat arms platoons, which maneuver together in formation, the MP platoon most often operates independently and dispersed over a large area. The platoon conducts combat operations, when required, through the employment of mobile combat systems containing three-man teams operating independently or in concert, and having vehicle crew-served and individual weapons capable of defeating a Level II threat and defending a position against dismounted threats.

1-45. Based on METT-TC, the platoon leader may task organize the platoon for certain missions. Normally, MP are employed as squads; however, individual teams may execute many MP tasks.

LIMITATIONS

1-46. During combat operations, the platoon is not organized and equipped to fight for extended periods unless it is augmented with indirect fire or close air support (CAS). Although the MP team is a lethal and highly mobile platform, it is not structured or equipped for prolonged autonomous missions. Leaders must use the MP team as a task organizational building block and avoid over tasking based solely on the number of teams available. The platoon has limited antiarmor capability and normally uses antiarmor weapons for self-protection and to break contact.

PEACETIME TRAINING

1-47. MP units train as they will fight. Peacetime training must replicate battlefield conditions and conform to Army doctrine. Leaders and soldiers must understand standardized doctrinal principles found in applicable manuals to ensure that training is conducted to standard. The following manuals provide the basic foundation for Army training:

- FMs.
- Training circulars (TCs).
- MTPs.
- Drill books.
- SMs.
- Army regulations (ARs)

1-48. *FM 25-100*, *FM 25-101*, and *TC 25-10* provide MP leaders with established training doctrine and assist the leaders in the development and execution of the training programs. These manuals introduce the concept of lane training and define it as a technique for training company, team, and smaller units on a series of selected soldier, leader, and collective tasks using specific terrain.

1-49. Lane training uses multiechelon techniques to maximize the efficient use of limited terrain and control conditions for formal or informal evaluations. Lane training is externally supported, resourced, and evaluated. It enables similar units to simultaneously or sequentially train on mission-related scenarios. Lane training is resource intensive, so commanders must maximize its benefit. Commanders narrow the focus and select only the most critical mission-essential task lists (METLs) items or collective tasks for training. Lane training is especially valuable for conducting specific METL tasks, situational training exercises (STXs), and other training events. It is often associated with training requiring movement over terrain; for example, movement to contact or conducting a route reconnaissance. Lane training can be modified to achieve benefits in L&O scenarios, such as special-reaction team (SRT) incidents, patrol incidents, traffic accidents, and so forth.

1-50. The lane training doctrine outlined in *FM 25-101* and *TC 25-10* can be tailored for small MP units by using the training execution model (TEM). The TEM follows the Army doctrine and training philosophy of hands-on METL training as outlined by *FMs 25-100* and *25-101*. Before the TEM can be implemented, the concept of the operation must be approved, evaluated, and directed from two levels up. For example, a squad leader must receive approval through the chain of command from his company commander to execute the training event; a platoon leader gets approval from the battalion commander and so forth.

1-51. The TEM incorporates the combined-arms training methodology and adjusts it to meet the MP training requirements. The TEM focuses the unit on the time available during the training cycle to train the most critical collective and individual tasks. The TEM consists of an eight-step training methodology that is based on leader certification of the lane expert and an observer/controller (OC) as well as subordinate unit leaders. For more information about TEM refer to *Appendix C*.

Chapter 2

Battle Command

This chapter provides the techniques and procedures used by MP leaders at company and platoon level to C^2 their organizations.

OVERVIEW

2-1. Battle command is the art of battle decision making, leading, and motivating soldiers and organizations into action to achieve victory with the least cost to the organization. Commanders must visualize the current and future state of both friendly and enemy forces. The commander positions himself where he can guide and motivate the soldiers and influence the outcome of the missions.

2-2. The company commander is responsible for all that the unit does or fails to do. He cannot delegate this responsibility, and the final decision and responsibility rest with him. He discharges his responsibility through an established chain of command and holds each subordinate leader responsible for the actions of the platoon or the section.

2-3. The commander must be proficient in the tactical employment of the unit. He must know the capabilities and limitations of the soldiers and the equipment. A commander does this through a continuous cycle of planning, executing, and assessing training. Through this training, the commander gets to know the soldiers.

2-4. MP commanders prioritize, assign missions, and allocate resources where they can best support the higher echelon commander's intent. The company commander makes most of the tactical decisions. Technological advances in today's operational environments have reduced the time available for decision making while increasing the possibilities that must be considered.

2-5. Thorough and sound operational planning is the key to successful combat and CS operations. Commanders must identify the opportunities and anticipate and avoid problems. They must analyze their options before making the decisions on which subordinate leaders will base their actions. Commanders balance competing risks and then identify and develop the best course of action (COA).

MILITARY DECISION-MAKING PROCESS (MDMP)

2-6. The MDMP is a single, established, and proven analytical process used at all the echelons of the commands. This is a seven-step process used when adequate planning time and enough staff support are available *(Table 2-1)*. This process is a detailed, deliberate, sequential, and time-consuming process that helps the commander and his staff examine a battlefield situation and reach logical decisions. The commander uses the entire staff during the process to explore the full range of probable and likely enemy and friendly COAs and to analyze and compare his own organization's capabilities with the enemy's.

Table 2-1. MDMP

Step	Action	Step	Action
1	Receipt of the mission	5	COA comparison
2	Mission analysis	6	COA approval
3	COA development	7	Orders production
4	COA analysis		

2-7. At company level, the commander normally uses the MDMP in a time-constrained environment without enough staff. A unit can shorten the process if it fully understands the role of each step of the process and the requirements to produce the necessary products. The application of the MDMP at company level and below is called the troop-leading procedures (TLP). *Figure 2-1, page 2-4* shows the relationship between MDMP and TLP.

2-8. MP commanders plan successful operations by anticipating possible future events and planning contingencies. MP leaders enhance both planning and execution of the operations when they—

- Use the military planning and decision-making process.
- Develop short- and long-range goals.
- Identify goals and objectives with a recognizable end-state.
- Coordinate goals and actions internally and externally.
- Base their plans on objective planning factors.
- Review their plans, continuously, in light of the METT-TC and updated information.
- Assign responsibilities and express expectations.
- Identify the options that may develop during an operation.
- Stand ready to accommodate the changes.

MILITARY PLANNING

2-9. Commanders select and carry out the developed COAs using military planning. Military planning guidelines include—

- Forecasting requirements by analyzing and evaluating facts and trends to predict what may occur.

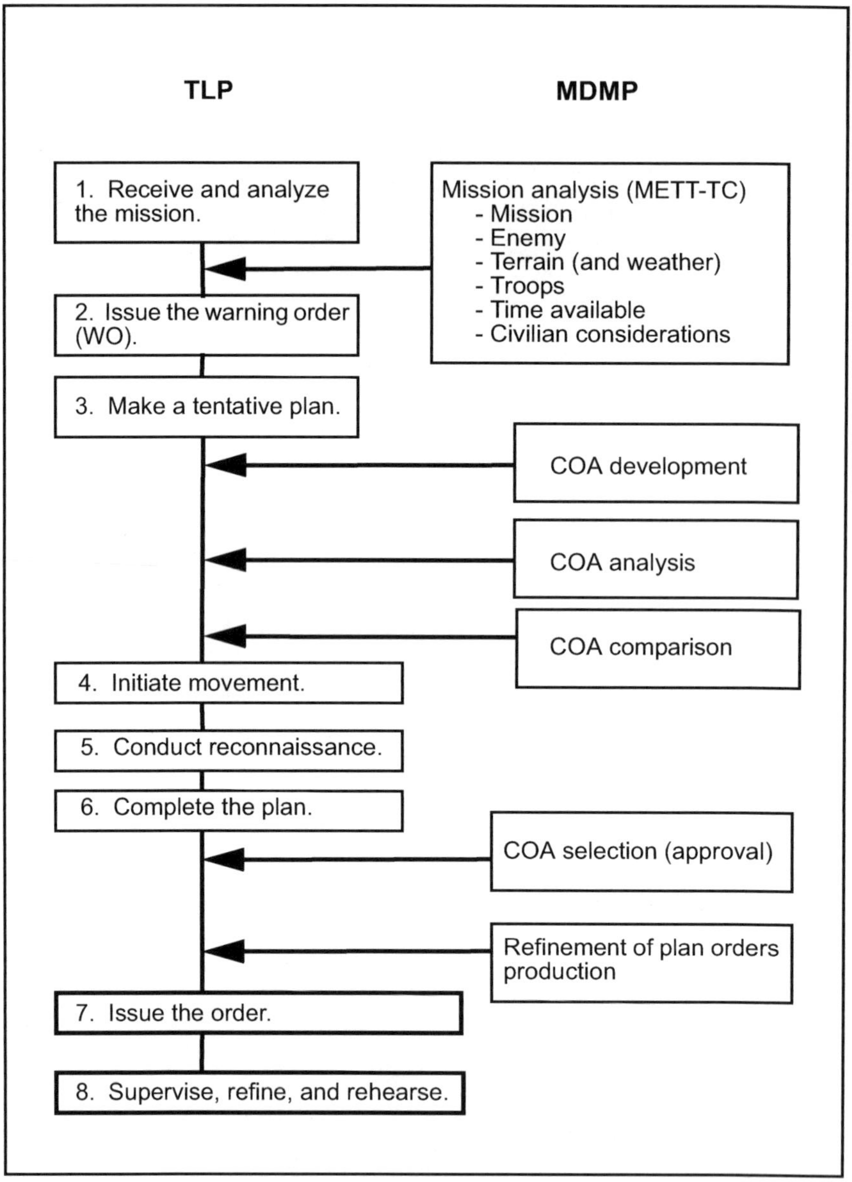

Figure 2-1. Relationship Between TLP and the MDMP

- Examining probable requirements and establishing priorities for further preparation.
- Studying implications and interrelationships of probable requirements.
- Analyzing the mission to determine tasks, their complexity, and their relative importance.
- Establishing guidance for further planning that will help keep all the elements focused on the commander's intent.
- Preparing studies and estimates to help formulate the COA and assess its feasibility.
- Selecting the COA, identifying the best course, and retaining other feasible courses for use in contingencies as alternate plans.
- Preparing the plan in detail and conducting rehearsals when time, resources, and security permit.

2-10. Use the following military planning guidelines to answer the three key questions of operational planning:

- What military condition must be produced to achieve the goal?
- What sequence of actions is most likely to produce that condition?
- How should resources be applied to accomplish that sequence of actions?

2-11. If the plan is not implemented immediately, retain it for later use. As events occur or new information becomes available, review and revise the plan accordingly.

ANALYSIS, FORECASTING, AND RISKS

2-12. Conducting a mission analysis is crucial to planning. The process begins by gathering facts and ascertaining current conditions, such as the—

- Higher-level mission and the commander's intent (one and two levels up).

- Current task organization (two levels down).
- Current unit status (locations, operation capabilities, and activities).
- Logistics situation (refer to *FM 101-5* for the logistic estimate format).

2-13. When facts are not available, the commander will need to develop assumptions. Assumptions must substitute for fact where information is not known. Keep in mind that as time passes between the receipt of a mission and the execution of a plan, facts are increasingly likely to have changed. Develop sound assumptions that can be used in place of facts.

2-14. Analyzing the higher-level mission and the commander's intent will help identify what tasks are required to accomplish the mission. As the mission is analyzed, identify both the *specified* and the *implied* tasks to be performed. *Specified tasks* are those stated in higher HQ orders and plans. *Implied tasks* (like crossing a river or passing through the lines of a unit lying between you and the objective) are not so stated, but must be accomplished to satisfy the overall operation. From among the specified and implied tasks, essential tasks that are crucial to the mission's success must be identified.

2-15. Integral to mission planning is the analysis of mission requirements in terms of time, space, and personnel. If MP are to balance the benefits of detailed planning against the need for immediate action, they must—

- Determine how much time there will be between receiving the mission and the deadline for having completed it.
- Know how long it will take to obtain and process information, make decisions, and issue orders.

- Know how long it will take subordinates to execute the orders, complete the mission, or carry out the operation.

2-16. Because each unit involved in an operation performs its planning based on the plans of the next higher level, allocation of adequate time for subordinate units to plan is a consideration at each level. Publishing SOPs reduces the number of details to be explained. It also promotes understanding and teamwork among commanders, staff, and troops.

2-17. When doing reverse planning, consider the classic allotment of one-third time for planning and two-thirds time for execution *(Table 2-2)*. Move backward from the time of execution to—

- Allocate time to accomplish each phase of an operation.
- Determine how much time is needed to rehearse.
- Determine how much time is needed for developing the plan.

Table 2-2. Reverse Planning Timetable

Time	Action
2230	Execute the mission.
2130-2215	Conduct inspection(s).
1845-2130	Conduct rehearsal(s).
1815-1845	Issue an operation order (OPORD).
1745-1815	Complete the plan.
1715-1745	Conduct the leaders' reconnaissance.
1630-1715	Issue a tentative plan.
1630	Issue a WO.
1600	Receive the mission.

2-18. When using terrain analysis, consider the layout of the battlefield. Appreciate the opportunities and limitations of the major terrain features, transportation networks, and built-up areas. Fit the operational concept and planning to that environment. Use the IPB to evaluate the area in terms of the military aspects of the terrain. Consider how to exploit the opportunities afforded by weather while minimizing its adverse effects.

2-19. Use current information on the threat to identify known enemy activities and threat capabilities that could affect this and future operations. Attempt to anticipate the enemy's objectives and intentions.

2-20. Consider available assets and determine acceptable levels of risk. At every echelon, MP disperse their assets and prioritize operations to meet the echelon commander's needs within the limits of the resources at hand. MP leaders must concentrate their efforts on key locations and accept risks elsewhere. When possible, recognize and moderate such risks in the choice of operations and in the contingency planning. The five steps to identify, analyze, and reduce risks are listed in *FM 100-14*.

COURSE OF ACTION IDENTIFICATION, DEVELOPMENT, AND SELECTION

2-21. The commander's selected COA will become the actual plan for accomplishing the mission. To ensure the best plan possible, identify several possible COAs, each significantly different from the others. In identifying COAs, do not overlook a check of the "basics" that include the—

- Commander's intent.
- Essential tasks.
- Effective use of C^2.
- Principles of war.
- Ethical considerations.
- Relative force ratio.

2-22. Consider preparing a COA statement (and sketch, if appropriate) for each option. The COA statement is the "how" of the operation. Ensure that it includes the following five elements:

- What – the type of action.
- When – the time the action will begin.
- Where – the assigned area.
- How – the use of available assets.
- Why – the purpose of the operation.

2-23. When developing the COA, analyze the relative combat power. Consider the initial array of the forces and develop the schemes of maneuver. Determine C^2 means and maneuver control measures.

2-24. Base doctrinal capabilities and planning on historical planning factors, and then relook the estimates in light of the available assets, the factors of METT-TC, the echelon commander's intent, and the mission's priorities. For example, when planning distribution of mobile assets for route coverage, begin with an estimate of one mobile MP team per 10 kilometers. For area coverage, begin with an estimate of one mobile MP team per 55 square kilometers.

2-25. When dispersing the assets into small combat elements, consider the classic ratios of friendly to enemy forces (3 to 1) to help ensure that the elements can concentrate enough combat power to accomplish the mission. Consider the speed and ease of reassembling the elements if dispersing them to distant sites.

2-26. Attempt to anticipate the enemy's likely moves. Consider ways to obstruct dangerous approaches to the area and avenues leading away from potential landing zones. Plan ways to combine the efforts of different resources, like enhancing the combat power for the base response and counterreconnaissance operations with fire support from field artillery or aviation. Attempt to determine the outcome of the operations by—

- Conserving unit strength through economy of force.
- Using terrain, weather, deception, and operations security (OPSEC) to your advantage.
- Focusing your efforts on enemy vulnerabilities.
- Ensuring unity of effort among subordinates and with your peers.

2-27. When planning for combat operations, whenever possible, develop a COA that avoids an enemy's strength and strikes at his weaknesses. Avoid head-on encounters with an enemy's forces. Seek to gain the element of surprise. When moving, plan to use indirect approaches and flank positions that do not attract immediate attention. Plan for fire support to increase MP combat power. Plan to operate on the enemy's flanks and rear, where direct fire is most effective, psychological shock is the greatest, and the enemy is least prepared to fight. Respond to and implement changes quickly and plan supplementary or alternative control measures to modify the plan as the situation dictates.

2-28. For a combat operation, the COA statement and sketches include the following:

- Allocated forces.
- Unit boundaries.
- Axes of advance.
- Routes for a forward or rearward passage of lines.
- Air axes for the maneuver of attack helicopters.
- Other control measures which may include—
  - Phase lines.
  - Assembly and holding areas.
  - Zones or sectors.
  - Battle positions.
  - Objectives.
  - Obstacles.
  - Routes.

2-29. Assess all of the feasible COAs after developing them. Consider constraints and restrictions on each COA. Weigh the available COA for the AO, for the level of responsibility, the commander's intent, and the mission's priorities. Decide on the best COA. For more information about the MDMP process, refer to *FM 101-5*.

2-30. Once the commander decides on a COA, he announces it in the form of orders that include his intent and concept of the operation. Based on these orders, the platoon leader uses TLP to organize his time during planning and preparation for the mission. Time management is the key. The platoon leader normally uses one-third of the available time to plan, prepare, and issue the order. The squad leaders have the remaining two-thirds of the time to prepare the squads for the mission. Whenever possible, TLP are integrated and accomplished concurrently rather than sequentially. Relationships between TLP and the MDMP are shown in *Figure 2-1, page 2-4*.

TROOP-LEADING PROCEDURES

2-31. TLP begin when the platoon leader is alerted for a mission and starts again when he receives a change or a new mission. Conducting TLP is an eight-step process *(Table 2-3, page 2-12)*. Steps 3 through 8 may not follow a rigid sequence. Several of the steps may be accomplished concurrently. In CS operations, platoon leaders rarely have enough time to go through each step in detail. However, the procedure must be followed, if only in abbreviated form. This ensures that nothing is left out of the planning and the preparation.

2-32. Risk management is an integral part of TLP and must be integrated into every aspect of mission

Table 2-3. The Eight Steps of TLP

Step	Action
1	Receive and analyze the mission.
2	Issue a WO.
3	Make a tentative plan.
4	Initiate movement.
5	Conduct a reconnaissance.
6	Complete the plan.
7	Issue the order.
8	Supervise, refine, and rehearse.

planning and execution. Identifying hazards and implementing control measures to mitigate those hazards will minimize operational accidents, thereby preserving combat power. Risk management is not a separate process, but rather infused into the entire TLP. For more information about risk management, refer to *FM 100-14*.

RECEIVE AND ANALYZE THE MISSION

2-33. TLP begin when the platoon leader is alerted for a mission and start again when he receives a change or a new mission. When the platoon leader receives the mission, he—

- Analyzes the mission.
- Considers the operations underway.
- Identifies the hazards associated with the mission, and considers the aspects of the current and future situations, environment, and known historical problems.
- Considers the time needed to plan and carry out the new mission. For example, will sleep plans be needed to ensure that all the teams are on a similar rest posture?

- Plans the use of available time. The most critical resource may be time, especially during daylight hours.
- Uses reverse planning to make a timetable. The timetable—
  - Identifies what must be done.
  - Works backwards from the time the soldiers have to be ready, allowing them time to do each task.
- Uses no more than one-third of the time for planning. The subordinate leaders need the remaining time to make preparations.

2-34. If time is too short to do the rest of the troop-leading steps in detail, at least do a fast mental review and—

- Make a quick map reconnaissance while sending for the subordinate leaders, depending on the level of the mission.
- Have the minimum control measures needed posted on their maps.
- Give an abbreviated order.
- Cite enemy and friendly situations.
- State the mission of the team, squad, or platoon and the concept of the operation.

2-35. If there is not enough time to do these actions, have the unit move out, then issue a fragmentary order (FRAGO) by radio or at the next scheduled halt. Continue the plan while moving.

ISSUE A WARNING ORDER

2-36. Issue an oral or written WO to the subordinate leaders as soon as possible. Give enough information for the unit to begin preparing for the mission. If needed, issue several WOs to keep subordinates informed. Refer to *Appendix D* for more information about WOs.

2-37. The unit SOP should detail what actions to take when a WO is received. Such actions may include drawing ammunition, rations, water, and communications gear and checking vehicles and equipment. Keep all the personnel informed of what they are to do and why they are to do it.

MAKE A TENTATIVE PLAN

2-38. Develop the plan based on the factors of METT-TC (using the OPORD format and the higher HQ order). The order may be specific about the tasks the unit is to do. The time available may be limited and the scheme of maneuver may be dictated. Nevertheless, the leader still must evaluate the mission in terms of METT-TC to see how MP elements can best carry out the commander's order. The leader must—

- Consider each factor and compare the COA to form a base for the plan.
- Include concepts for reconnaissance, coordination with adjacent and/or supporting units, and troop movement.
- Assess the identified hazards. Consider the impact of each hazard in terms of potential loss based on probability and severity.
- Identify control measures that will eliminate the hazards or mitigate them to an acceptable level. Make decisions on acceptable levels of risk based on potential benefits versus cost.
- Issue the plan, when firm, as an order. *Appendix D* contains the OPORD format.

INITIATE MOVEMENT

2-39. Instruct subordinate leaders to start moving to the assembly area (AA). Allow subordinate leaders enough time for their actions if the element has to move and reorganize for the mission.

CONDUCT A RECONNAISSANCE

2-40. Ensure that the terrain where the operation is to be conducted is reconnoitered. At a minimum, conduct a map reconnaissance. (A map reconnaissance is the easiest, but least reliable form of reconnaissance. It is a supplement to other types of reconnaissance.) Study the map for terrain features, natural barriers, and other characteristics. Have subordinate leaders help identify key terrain features. Refer to *FM 21-26* for map-reading skills. Follow up with a visual reconnaissance of the area to be used and the terrain over which you will operate. A visual reconnaissance can be done—

- On the ground. A ground reconnaissance is time consuming, but the most reliable type of reconnaissance. MP see terrain features up close and can note problems not easily seen using other reconnaissance methods.
- In the air. If available, air reconnaissance can cover terrain quickly. To do an air reconnaissance, show the pilot a map of the terrain to be reconnoitered. Specify the type of information to be gathered. Have one person in the plane track the patrol's route on a map. At critical points, if the aircraft can land, have part of the patrol dismount to make a ground reconnaissance while the rest of the patrol goes back into the air to provide overwatch security. If the aircraft cannot land, make a visual search for enemy activity or for the required information.

2-41. Use the information from the reconnaissance to verify or to change the plan and to modify the risk-assessment process. Adapt your tactics to the terrain and the abilities of the force. If the reconnaissance cannot be finished due to distance or enemy pressure, make the plan from what has been seen. Give instructions for later actions in general terms and confirm or change while moving over the terrain.

Provide a sketch that will aid with a sand table to help in the OPORD issue, if time permits.

COMPLETE THE PLAN

2-42. Add details or makes changes to the tentative plan (as a result of the reconnaissance and coordination with nearby and/or supporting agencies) and identify specific tasks for all the subordinate elements.

ISSUE THE ORDER

2-43. Issue an OPORD or FRAGO ensuring that—

- The soldiers know the plan.
- The instructions are stated clearly and concisely (use the OPORD format). Platoon and squad orders are usually issued orally. However, if time permits, they can be written. When the order is written, delete the service support and command and signal paragraphs if covered by the SOP.
- The soldiers are thoroughly briefed on the hazards associated with the mission and the control measures identified to mitigate the hazards.
- The subordinate leaders back brief the order and spot-check the soldiers.
- When possible, the order is given from the advantage point where soldiers can see the area in which they will operate. This lets the leader point out terrain features on the ground as well as on a map. If this cannot be done, use a terrain model (sand table) or a sketch to help explain the order.

SUPERVISE, REFINE, AND REHEARSE

2-44. Supervise, refine, and rehearse the preparation to ensure that the soldiers are ready for the mission. This includes—

- Using the feedback received from subordinates.

- Ensuring that every soldier knows the mission and understands the commander's intent for the operation. The unit must be able to carry out the mission in the absence of the leader that developed the plan.
- Continuing to prepare when the troop-leading steps have been completed.
- Having the soldiers rehearse their actions if there is enough time before an operation.

Rehearsals

2-45. Rehearsals build confidence and improve performance. They allow faults in a plan to surface. If possible, soldiers should rehearse on terrain and under conditions like those at the operation site. Give priority to actions to be taken in the objective area. A rehearsal is especially helpful when operating in reduced visibility.

2-46. Each type of rehearsal reflects an increase in mission realism and a corresponding increase in rehearsal benefit. Each technique increases the realism of the enemy, terrain, team and squad actions, and actual time and distance relationships. Rehearsal techniques fall into the following four categories:

- Back brief.
- Rock drill.
- Walk-through.
- Full-scale.

2-47. Regardless of the rehearsal technique, leaders must demand that soldiers demonstrate the known hazards associated with the mission and understand the control measures identified to counter them. Insist that rehearsals include prescribed control measures (either actions or equipment).

2-48. **Back Brief.** The back brief rehearsal is an event that occurs after an OPORD has been issued. The back brief is the quickest rehearsal technique. Subordinate leaders repeat back to the commander what he expects them to do and why, using a map or a sand table to explain their mission. The subordinate leader identifies all specified and implied tasks, determines their mission-essential tasks, and restates the mission. Items essential to the back brief are the—

- Commander's intent.
- Concept of the operation.
- Scheme of maneuver.
- Time to complete the tasks.

2-49. **Rock Drill.** A rock drill rehearsal is done by acting out the friendly and enemy actions based on the scheme of maneuver and the situation. Subordinate leaders rehearse their actions by moving objects, such as rocks, that represent them or the platoon. In acting out the plan, leaders can talk through their missions, critical tasks, actions, and decisions. All subordinate leaders act out their parts simultaneously so problems and disconnects in synchronization can be more clearly identified.

2-50. **Walk-Through.** A walk-through rehearsal is the acting out of actions that will occur during a mission using the actual vehicles and equipment that will be used to conduct the operation. Participants communicate with the same type of equipment they will use during the operation. During a walk-through, subordinate leaders rehearse—

- Movement techniques, both mounted and dismounted.
- Critical actions.
- Decision making.

2-51. Since the leaders are in a more realistic environment, they rehearse the finer aspects of synchronization, C^2, and squad and team actions. Aggressive portrayal of OPFOR is critical in walk-through rehearsals because it increases the rehearsal realism. This type of rehearsal is more difficult to orchestrate. It is the optimum balance between resource constraints and realism. Ensure that a walk-through rehearsal is the minimum goal for all the units.

2-52. **Full-Scale.** During a full-scale rehearsal, participants use real-time mounted and dismounted movements over the actual or similar terrain. It is normally conducted with all teams and squads to be used for the actual mission. At a minimum, one subunit must participate for a full-scale rehearsal. This type of rehearsal is the most resource intensive, but provides the most realistic training environment for the unit. It is often used to rehearse the operation plan (OPLAN) or OPORD when time is not an immediate constraint.

Inspection

2-53. The last action before an operation is inspecting. Allow ample time for subordinate leaders to correct problems. Inspect soldiers by checking their mental and physical readiness. Inspect their equipment and check for the following:

- Accountability.
- Serviceability.
- Weapons.
- Ammunition.
- Individual uniforms and equipment.
- Mission-essential equipment.
- Water and rations.
- Communications equipment.
- Vehicles.
- Camouflage.

2-54. The equipment used during a mission is based on the unit's SOP, a risk assessment, and special considerations. Ensure that the SOP specifies a combat load and a list of ammunition and equipment usually carried on missions. Base changes from the SOP combat load on the METT-TC.

2-55. *Appendix E* contains a sample precombat-inspection (PCI) checklist. Ensure that the soldiers have everything they need for the mission. Ensure that they—

- Know their duties.
- Have only the equipment needed.
- Are wearing their equipment correctly and securely.

ORDERS AND REPORTS

2-56. MP leaders translate their thoughts, evaluations, and decisions into understandable reports and orders. Battlefield communication requires standardized, streamlined procedures. Despite personal exhaustion or battle confusion, you must be able to rapidly report information or issue instructions that are simple, clear, and brief.

ORDERS

2-57. Combat orders are written or oral communications giving details of tactical operations and administration. The three most common types of combat orders at company level and below are—

- WO
- OPORD
- FRAGO

2-58. WOs and OPORDs generally have set formats. This helps ensure that the receiver understands the intent of the message and that all needed information is provided. Standardization helps save time in writing as well as interpreting orders. FRAGOs enhance what has been previously sent out in the OPORD, such as a change in the situation or mission. Refer to *Appendix D* for examples of orders.

Reports

2-59. MP report to higher HQ and provide information on which plans, decisions, and orders can be based. The information included in MP reports must be accurate and timely, and complete *negative* information ("There is no enemy at . . . . ") is often as important as positive information. Reports are the main record of operational events. The three broad categories of reports are—

- Administrative.
- Operational.
- Intelligence.

2-60. Commanders may specify report formats in their local tactical SOP. Treat friendly information, including administrative reports, as classified or sensitive in nature to keep information from falling into the enemy's hands. *FM 101-5-2* and *Appendix D* of this manual CONTAIN STANDARD REPORTING FORMATS.

SOPs

2-61. SOPs detail how forces will execute unit-specific techniques and procedures that commanders standardize to enhance effectiveness and flexibility. Commanders use the SOP to standardize routine or recurring actions not needing their personal involvement.

RULES OF ENGAGEMENT AND RULES OF INTERACTION (ROI)

2-62. MP usually are among the first CS forces deployed to troubled areas around the world. As a direct result of a peacetime L&O mission, MP continually train in the prudent use of force, crisis management, and operations requiring restrictive ROE and ROI.

2-63. ROE are the directives established by higher HQ that delineate the circumstances and limitations under which soldiers will initiate and/or continue engagement with belligerent forces. ROE may reflect the law of armed conflict and operational considerations, but are primarily concerned with the restraints on the use of force. ROE are the primary means by which commanders convey legal, political, diplomatic, and military guidance to soldiers. Leaders at every level must train their soldiers carefully and thoroughly concerning ROE and laws that govern armed conflict before deployment. During the conduct of the operation, leaders continue to train soldiers and stress firm, determined, and impartial execution of ROE to preclude inviting challenges from any of the belligerent parties.

2-64. ROI embody those human dimension skills needed to successfully interface with various categories of people. They spell out with whom, under what circumstances, and to what extent soldiers may interact with other forces and the civilian populace. ROI, when applied with good interpersonal communication (IPC) skills, improve the soldier's ability to accomplish the mission while reducing possible hostile confrontations. ROI and IPC, by enhancing the soldier's persuasion, negotiation, and communication skills, also improve his survivability. ROI founded on firm ROE provide the soldier with the tools to address nontraditional threats such as political friction, ideologies, cultural idiosyncrasies, and religious beliefs and rituals. ROI must be regionally and culturally specific. MP leaders must train soldiers on ROE and ROI using tactical vignettes or simulated events.

SITUATIONAL AWARENESS

2-65. Situational awareness is the ability to maintain a constant, clear mental picture of the tactical situation. This picture includes an understanding of both the friendly and enemy situations and of relevant terrain. It also includes relating events in time to form logical conclusions and make decisions that anticipate events. Since MP platoons normally operate dispersed, it is essential that all MP leaders maintain situational awareness so that they can make quick, sound tactical decisions. Situational awareness also permits MP leaders to anticipate events and relate separate pieces of information to form logical conclusions. One of the critical outcomes of situational awareness on the part of all MP is a reduction in fratricide incidents. Refer to *Appendix F* for fratricide avoidance.

BATTLEFIELD FRAMEWORK

2-66. The commander will structure the battlefield based on the conditions of METT-TC and his commander's intent. How he does this affects the MP platoon leader's mission planning and his ability to maintain situational awareness. Geographically, available assets, mission, and the AO influence the dispersion of MP assets. The framework of this AO can vary from an area that is dominated by several towns or large cities to an area that includes several bases and base clusters. MP can expect to operate in sustainment areas where there may be clear boundaries and closely tied adjacent units or in a decentralized structure with few secure areas and unit boundaries. Between these extremes are an unlimited number of possible variations. Maintaining situational awareness becomes more difficult as the battlefield becomes less structured. Modern, highly mobile operations with small forces lend themselves to a less rigid framework that challenges the MP ability to maintain an accurate *picture* of the battlefield. MP are the echelon commander's critical link to the battlefield, constantly gathering police, operational, and combat information.

BATTLEFIELD PICTURE

2-67. To have a clear picture of the battlefield, the MP leader must have virtually perfect knowledge of the friendly situation one level higher. This means the MP platoon leader must know the company situation and the location and mission of the adjacent company and platoons. It is also important that the platoon leader update his subordinate leaders periodically regarding the higher situation. The platoon leader must have a relatively complete knowledge of the terrain, and he must know as much as possible about the enemy. The requirement to maintain a real-time picture of the battlefield one level higher does not relieve the platoon leader of the requirement to understand the situation and the commander's intent two levels higher. The difference is that this understanding of the situation two levels higher does not have to be as specific or in real time.

2-68. Most of the information the platoon leader needs comes in the form of reports over communication channels. Subordinate leaders are required to periodically report their status. If an MP team is operating in an area that does not allow uninterrupted communications, the team leader coordinates with adjacent teams to relay his report. If an MP team does not report in a timely manner, the platoon leader must quickly determine the status of the overdue team.

2-69. If possible, the platoon leader monitors his platoon and company net. How effectively he can accomplish this is, to some degree, experience-dependent; however, there are techniques he can apply to relate the information he is receiving to the map and thereby track the tactical situation.

2-70. The platoon leader's map is the key to maintaining situational awareness. He plots all friendly position reports up to one level higher than his own and plots information from spot reports (SPOTREPs). He uses different colors for friendly and enemy elements to allow quick distinction. To avoid cluttering the map, he places a dot or symbol on the map where the element is located and labels the point with a number. The same number is then written in the map's margin (or beyond the AOs) with the complete SPOTREP or unit identification (ID) next to it. Include the time on this notation. As positions or reports are updated, the old symbol is crossed off and a new one with a corresponding notation is added; it is critical that updates to previous reports be clearly identified as such during transmission. This simple system can greatly increase the ability to track both friendly forces and enemy activity in a particular AO.

BATTLE SPACE

2-71. As mentioned earlier, an accurate picture of the battlefield provides the platoon leader with important tactical information, including friendly and enemy positions and relevant terrain. In turn, complete understanding of the military significance of this picture requires knowledge of the concept of battle space, the key element in the intellectual process of visualizing the battlefield.

2-72. At the most fundamental level, battle space is the three-dimensional *bubble* or area in which the platoon can acquire enemy forces and influence them with effective fires. This space is defined by the following numerous battlefield factors:

- The locations of the friendly forces, including the platoon's individual teams, OPs, and patrols.
- The effects of the terrain, weather, and movement.
- The ranges of all the available platoon weapons and sensing systems.

2-73. Each squad has its own battle space; the platoon battle space is the sum of the individual squads battle spaces. Platoon battle space is not restricted by boundaries; it can overlap with the battle space of adjacent units. For example, an enemy element that is spotted outside of the platoon's AO can still adversely affect the platoon's mission. Coordination is made with adjacent units to detect and destroy the threat.

2-74. Battle space has applications in all phases of mission planning and execution. During the planning process, it is a critical factor in selection of the routes and tentative positions. Once mission execution begins, the platoon leader's knowledge of the battle space is critical to issuing timely and effective orders as the situation changes.

2-75. The importance of battle space demands that the platoon leader direct most of his battle command effort toward managing and enhancing his space. He must be aware at every moment how battle space is changing as friendly and enemy forces move and terrain and visibility conditions change. He must evaluate how these changes affect his squads.

2-76. As the operation progresses, the platoon leader must take active measures to shape the battle space to his best advantage. One vital step in this process is to eliminate any gaps, or dead space, that exist within the *bubble*. The platoon leader can accomplish this in several ways, including maneuvering teams, repositioning OPs, and deploying patrols or remote sensors.

COMMAND POST OPERATIONS

2-77. Company- and larger-size elements have a tactical operations center (TOC) and platoons have CPs. The CP is wherever the platoon leader goes. It can be mobile or stationary. No matter the location, there must be communication with and command of the unit and a method for battle tracking. CP activities are a 24-hour operation. The leadership must ensure that there is a plan for continuous operations.

2-78. A mobile CP may be a HMMWV and an ASV. The platoon leader can make decisions on the move while having communication capabilities nearby. A leader's book can hold information on battle rosters, report formats, and a map of the battlefield.

2-79. A stationary CP may be the platoon leader's tent or office. Communications may be telephonic, by messenger, or by radio. Track the status of equipment and personnel by using charts. Use large maps to track the battle. Keep in mind that moving to another location requires taking charts and maps, so everything needs to be mobile.

2-80. Key personnel for CP operations are the platoon leader, the PSG, and the radio/telephone operator (RTO).

PLATOON LEADER

2-81. The platoon leader is responsible for C^2 of his organization and—

- Planning the missions according to the commander's guidance.
- Planning security to include the placement of crew-served weapons.
- Issuing orders and providing work priorities to subordinate leaders.
- Conducting PCI.
- Planning and conducting platoon rehearsals.
- Battle tracking (knowing) on a map the exact location of all the MP teams, to include the position of friendly platoons and known enemy, and a platoon-sector sketch.
- Adapting to new situations, making necessary adjustments, and issuing FRAGOs.

- Controlling the movement of the subordinate units.
- Ensuring that the communication net is established.
- Conducting risk assessment and continuously updating it.
- Reporting to higher HQ using the correct report format.

PLATOON SERGEANT

2-82. The PSG is responsible for the logistics of the operation and—

- Coordinating and providing all the needed class items.
- Ensuring equipment serviceability and accountability.
- Supervising the maintenance.
- Establishing and supervising a sleep plan.
- Supervising the security plan's execution.
- Supervising the maintenance of work priorities.
- Supervising PCIs.
- Supervising sanitation and hygiene.
- Adjusting, manning, and cross leveling the soldiers, weapons, and equipment.
- Coordinating and supervising morale services.
- Assisting the platoon leader in rehearsals.
- Supervising test firing.
- Assisting with battle tracking.
- Consolidating subordinate units' status reports for the platoon leader.
- Assisting with the reports.
- Providing technical and tactical advice to the platoon leader.
- Maintaining situational awareness.

RADIO/TELEPHONE OPERATOR

2-83. The RTO should be an experienced MP who is also the platoon leader's driver. The RTO—

- Communicates with higher HQ and separate units.
- Submits the required reports according to orders and SOPs.
- Maintains a record of communications.
- Maintains the radio and communication equipment.
- Assists with battle tracking.

Chapter 3

Shoot, Move, and Communicate

The ability of an MP unit to shoot, move, and communicate ensures its ability to detect, disrupt, and defend against the enemy and immeasurably adds to its survivability and maneuverability. MP are structured to be strategically, operationally, and tactically agile to respond to the increased range of worldwide MP requirements.

SHOOT

3-1. It is important that MP understand their shooting capabilities and limitations. Just as important is the understanding of firing techniques and associated fire distribution, reacting to air and armor attacks, calling for fire, and obtaining various fire support.

UNDERSTAND FIRE TECHNIQUES

3-2. Fire techniques include fire from or at a moving vehicle, fire distribution and control, and suppressive fire.

Fire From or at a Moving Vehicle

3-3. The key to forward maneuver is firing on the enemy. When maneuvering, the fire element—

- Attempts to destroy or suppress the enemy.
- Covers and protects the maneuver element as it advances.
- Moves, when possible, into its firing position undetected. Fire from an unexpected direction

has a greater effect than fire from a known position.

3-4. Firing on the move is less accurate than firing from a halt. However, to halt and fire takes more time and is more dangerous. A stationary vehicle is more likely to be hit than a moving vehicle. The team leader must decide whether to fire while moving or to fire from a short halt. He bases his decision on sound judgment and evaluation of the threat.

3-5. Crew-served weapons engage all targets on the move with free gunfire. To deliver this type of fire, the gunner removes the traversing and elevating (T&E) mechanism from the bottom of the receiver, allowing the gun to move freely in any direction. Accurate firing with crew-served weapons while moving is affected by—

- The terrain.
- The vehicle's speed.
- The team's proficiency.

3-6. When aiming from a moving vehicle or at a moving vehicle, or both, the gunner must lead the target. The speed of the firing vehicle, the time of flight, and the *angle of engagement* affect the amount of lead required. The time of flight is the required time it takes the projectile to move from the firing vehicle to the target. The angle of engagement is the angle found between the centerline of the vehicle and the gun when laid on the target. When a round is fired from the flank of a moving vehicle, the round drifts in the same direction and at the same speed as the vehicle. The longer the flight time and the larger the engagement angle, the greater the drift. Thus, the gunner must apply more lead to the shot. If a lead is required and the gunner is traversing left to keep on target, the gunner must lead left. If the gunner is traversing right to keep on the target, the gunner must lead right. This is true whether the firing vehicle is moving, the target is moving, or both are moving. *Table 3-1* shows the responsibilities of an MP team when firing on the move.

Table 3-1. Team Responsibilities When Firing While Moving

Position	Actions
Team leader	• Directs the driver. • Keeps the gunner oriented. • Senses the impact of the rounds–long, short, left, or right of the target. • Identifies additional targets. • Assists the gunner with reloading, if required. • Observes the surrounding terrain.
Gunner	• Develops a *feel* for the moving vehicle. • Tracks the position of the target with the MK19 grenade machine gun (GMG) despite the movement of the vehicle. • Remains alert to the sounds of the engine and transmission. These sounds indicate the type of terrain over which the vehicle is traveling and helps the gunner anticipate vehicle movements.
Driver	• Tries to maintain a steady gun platform while the gunner engages the targets. • Attempts to time the gear and direction changes so they occur immediately after firing and do not interfere with accuracy. • Informs the gunner of obstacles in the vehicle's path that might affect the gun's accuracy. • Announces "depression," "turn," and the like to warn the gunner of vehicle movements. • Announces, "steady" to let the gunner know when the vehicle is once again on a stable platform. The gunner assumes he has a stable platform unless the driver informs him otherwise.

Distribute Fire

3-7. MP leaders must distribute the fires of their organic weapons to destroy or suppress enemy positions. The following are the two methods to distribute fire on a target:

- Point fire. Point fire *(Figure 3-1, page 3-4)* is directed against one target (such as a machine gun position) with all the troops firing at the

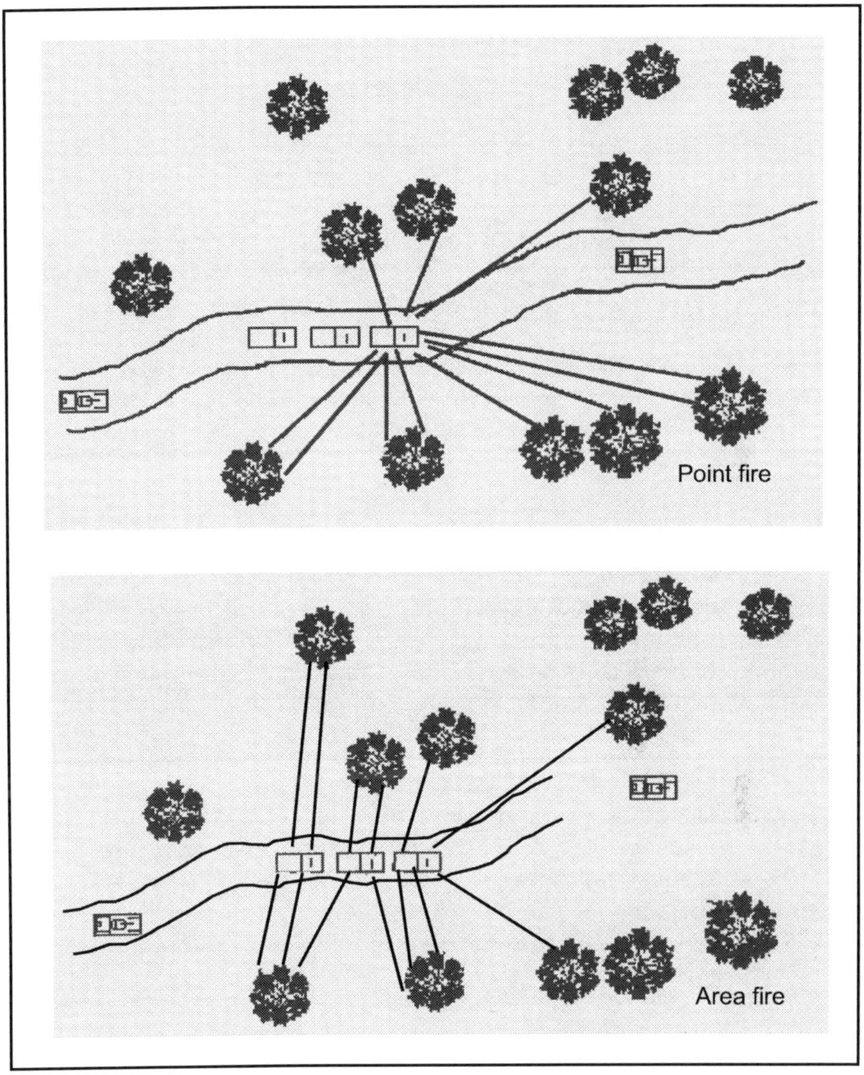

Figure 3-1. Fire Distribution

same target. Spreading out the base-of-fire element makes this type of fire particularly effective because the fire is directed from many sources.

- Area fire. Area fire *(Figure 3-1)* permits rapid cover of an entire area with fire from the left to the right and in depth, even if the enemy cannot be seen. This method is used without command and is the quickest and most effective way to bring all parts of a target under fire. Each member in the element is assigned a portion of the target. Fire is placed on likely locations for enemy positions rather than in a general area. If the leader wants fire on a wood line, he may shoot tracers to mark the center of the target. Soldiers to the left of the leader fire to the left of the tracers and soldiers to his right fire to the right of the tracers.

3-8. A rifleman fires his first shot on the part of the target that corresponds to his individual position. If he is left of the leader, he fires to the left of the leader's tracers. He then distributes his remaining shots over the part of the target extending a few meters right and left of his first shot. He covers the part of the target that he can hit without changing position.

3-9. A grenadier fires into the center of the target area of his team. He then distributes his shots over the remaining target area from the center to each side and from front to rear. A machine gunner covers part of the target depending on his position and how much of the target is in range. When possible, he covers the entire target of the team. When placing automatic suppressive fire on the enemy, the tendency is to shoot high. Therefore, he places the first bursts low and works up to the target. The squad leader tells the machine gunners where to shoot by assigning sectors of fire.

3-10. An MK19 gunner engages area targets with traversing and searching fire after the leader designates the width and depth of the target. If one MK19 GMG is being fired, the gunner engages the area target by adjusting his fire on the center of the mass, then traverses and searches to either flank. When he reaches the flank, he reverses direction and traverses and searches in the opposite direction. If two MK19 GMGs are being fired as a pair, the point of the initial lay and adjustment for both guns is on the midpoint of the target. After adjusting the fire on the center of the mass, fire is distributed by applying direction and elevation changes that give the most effective coverage of the target area. Usually, the right gun (number 1) fires on the right half, and the left gun (number 2) fires on the left half. *Appendix G* describes the MK19 qualification and familiarization tables and provides a sample scorecard.

Control Fire

3-11. Fire control is an essential component of fire distribution. A platoon leader must know what means he will direct the fire element to use when engaging the targets. He will communicate directly or use prearranged signals to identify the location of the target to the other units. He may use sound signals (such as voice, a horn, or a whistle), but must remember that they are only good for short distances and that their reliability and range are reduced by battle noise, weather, terrain, and vegetation. Use a radio to direct the base-of-fire element or adjust fires from reference points or landmarks, because a radio offers immediate voice communication. For example, he may say, "From the burning scout vehicle, northwest 50 meters, machine gun position." If portable radio equipment is not available, he uses prearranged visual signals, such as smoke or flares. A smoke round from a grenade launcher, unless it is being used for some other purpose, and a smoke canister can be used as a signal. Use these items during reduced visibility in addition to aiming stakes, illumination, night-vision devices, infrared chemical lights, and so forth.

Use Fire Commands

3-12. Leaders use fire commands to direct fire. A fire command has the following six parts:

- Alert. The leader alerts the soldiers to receive further instruction. He alerts the soldiers by name or unit designation, some type of visual or sound signal, personal contact, or any other practical way.
- Direction. The leader tells the soldiers the general direction to the target. In some cases, he pinpoints a target. The following are the three ways the leader can give the direction to the target:
  - Points with his armor rifle.
  - Fires tracer ammunition at a target.
  - Uses either target reference points (TRPs) or easily recognized man-made objects or terrain features. He gives the general direction just before giving the reference points.
- Description. The leader describes the target briefly but accurately and always gives the formation of the enemy soldiers.
- Range. The leader tells the soldiers the range to the target in meters.
- Method of fire. The leader tells the soldiers which weapons to fire, the type and amount of ammunition to fire, and the rate of fire.
- Command to fire. The leader tells the soldiers when to fire by using an oral command or visual signal. When he wants to control the exact moment of fire, he says, "At my command" (then pauses until ready to commence firing). When he wants to start firing on completion of the fire command, he just says, "Fire."

Use Subsequent Fire Commands

3-13. These commands adjust or change information given in the initial fire command. Only the elements that change are given.

Terminate Fire

3-14. Fire is terminated by the command or signal for *cease fire, end of mission*.

Suppress Fires

3-15. When the fire element is in position, it lays a heavy volume of fire on the enemy to suppress them. When the leader senses that the enemy is suppressed, he instructs the fire element to reduce its rate of fire as long as it keeps the enemy suppressed. As the movement element nears its objective, the fire element increases the rate of fire to keep the enemy down. This lets the movement element close enough to assault the enemy before the enemy can react. When the assault begins, or on a signal, the fire element stops firing, shifts its fire to another target, or walks its fire across the objective in front of the movement element, and then shifts or ceases fire.

3-16. Positions for fire elements are located so that movement of the maneuver element does not mask their fires. Fire element positions are often higher and usually to the flank of the maneuver element. The maneuver element neither masks the fire of the fire element nor moves outside the protective umbrella provided by the fire. A platoon or squad can point fire at one target or an area of several targets. In both cases, the leader must control the fire. He must ensure that the fire is directed on the enemy, not on the maneuver element.

Use Nonlethal Weapons (NLW)

3-17. The Department of Defense (DOD) defines NLW as weapons that are explicitly designed and primarily employed to incapacitate personnel or material while minimizing fatalities, permanent injury to personnel, and undesired damage to property and the environment. Unlike conventional weapons that destroy the targets principally through blast, penetration, and fragmentation, NLW employ means other than gross physical destruction to prevent the target from functioning.

3-18. NLW doctrine and concepts of operation are designed to reinforce deterrence and expand the range of options available to commanders. They enhance the capability of US forces to accomplish the following objectives:

- Discourage, delay, or prevent hostile actions.
- Limit escalation.
- Take military action in situations where use of lethal force is not the preferred option.
- Protect US forces.
- Disable equipment, facilities, and personnel temporarily.

NOTE: The zero probability of producing fatalities or permanent injuries is not a requirement of NLW. However, while complete avoidance of these effects is not guaranteed or expected, when properly employed, NLW significantly reduce them as compared with physically destroying the same target.

3-19. When drafting the ROE, it must be clearly articulated and understood that the role of NLW is an additional means of employing force for the particular purpose of limiting the probability of death or serious injury to noncombatants or belligerents. However, the use of deadly force must always remain an inherent right of individuals in instances when they, their fellow soldiers, or personnel in their charge are threatened with death or serious bodily harm. NLW add flexibility to the control of disturbances within the I/R facility and provide an environment where guard forces can permissively engage threatening targets *(Figure 3-2)* with limited risk of noncombatant casualties and collateral damage. Refer to *FM 90-40*.

3-20. The use of lethal force, employed under the standing ROE, will never be denied. At no time will forces be deployed without the ability to defend themselves against a lethal threat, nor will they forego normal training, arming, and equipping for combat. Nonlethal options are a complement to, not a replacement for, lethal force and seek to expand a proactive response across the range of military operations. Refer to *FM 90-40*.

3-21. The decision to use NLW against an adversary during a confrontation is delegated to the lowest possible level, preferably to the platoon or the squad. However, this requires that all personnel, not just the leaders, have a clear understanding of the ROE and the commander's intent. Refer to *FM 90-40*.

3-22. Commanders and public affairs officers must be prepared to address media questions and concerns regarding the use and role of NLW, and they must make it clear that the presence of NLW in no way indicates abandoning the option to employ deadly force in appropriate circumstances.

3-23. **Advantages of Employing Nonlethal Weapons**. NLW provide the commander with the flexibility to influence the situation favorably with reduced risk of noncombatant fatalities and collateral damage.

NLW	Meters												
	0	5	15	20	25	30	35	50	55	65	95	100	120
Modular crowd control munitions (MCCM)	Lethal zone			X	X	X	X	X	X	X	X	X	X
Stun hand grenade	Lethal zone				X	X	X	X	X	X	X	X	X
12-gauge point	Lethal zone	Lethal zone	Nonlethal zone				X	X	X	X	X	X	X
12-gauge area	Lethal zone	Lethal zone					X	X	X	X	X	X	X
40-millimeter point	Lethal zone	Lethal zone	Nonlethal zone						X	X	X	X	X
66-millimeter sting ball								Nonlethal zone					X
66-millimeter flash bang													X
66-millimeter CS grenade												X	X

Figure 3-2. Range of Munitions Contained in a Nonlethal Capability Set

3-24. NLW can be more humane, be consistent with the political and social implications implicit in humanitarian missions, be used during peacekeeping missions, and facilitate post-incident stabilization by reducing internee alienation and collateral damage.

3-25. The force that properly employs nonlethal options gains advantages over those who rely on lethal options alone, because the degree of provocation required to employ these options is substantially less. This advantage provides a more proactive posture and quicker response as well as a diminished likelihood of having a situation escalate to a point where deadly force is required to resolve a conflict within the I/R facility.

3-26. NLW options are less likely to provoke others and the use of NLW, in fact, may provoke a negative response. However, demonstrated restraint greatly diminishes feelings of anger and remorse when deadly force is required after nonlethal options fail.

3-27. **Military Police Nonlethal Weapons**. I/R facility commanders consider the use of force options discussed in *Chapter 2* and *AR 190-14* when dealing with disruptions within the compound. Facility commanders are encouraged by *AR 190-14* to substitute nonlethal devices for firearms when it is considered adequate for MP to safely perform their duties. Currently, MP have such nonlethal options as riot-control agents chlorobenzul-malononitrile (CS) and oleoresin capsicum (OC), military working dogs (MWD), an MP club, and a riot baton for crowd control. There are other nonlethal devices currently being tested and fielded that will be available to the I/R commander in the future.

3-28. **Nonlethal Training**. Soldiers and their leaders must be trained in the correct employment of NLW available to them. Soldiers and their leaders must understand the limited use of these systems in environments with restrictive ROE. Their training must be continuous at all levels to ensure that NLW are properly employed and that the leaders and the soldiers understand when and how to effectively employ them. They must understand that the incorrect application of NLW can have significant operational and political ramifications. Well-trained MP leaders, providing timely and clear guidance to MP soldiers using NLW, will ensure the mission's successful accomplishment.

3-29. Many NLW have both maximum effective and minimum safety ranges. Individuals struck short of the minimum safety range often suffer severe injuries or death while the effects of most nonlethal devices are greatly mitigated at longer ranges. In order to be effective, engage the threat within the *effective* zone, beyond the minimum safety range, and short of the maximum effective range.

3-30. When training with and planning for the use of NLW consider the following:

- Never apply NLW in a situation where deadly force is appropriate.
- Never apply NLW in a situation that will place troops in undue danger.
- Always cover NLW with deadly force.

3-31. **Nonlethal Tactics**. *FM 90-40* provides an in-depth discussion on the tactics associated with the employment of various NLW available to the commander, such as—

- Riot formations. Riot formations establish riot-control teams with a minimum response time. Because of the physical nature of riot control, individuals in riot control formations should not carry long rifles. Nonlethal attachments should follow closely behind the riot control formation. Lethal coverage should be provided for this entire formation. Refer to *FM 90-40*.

- Designated marksman. During a nonlethal engagement, the use of a designated marksman (DM) provides confidence and safety to those facing a riot. If a lethal threat is presented, the DM who is in an overwatch position and armed with a standard infantry rifle, mounted with a high-powered scope, can scan a crowd and identify agitators and riot leaders for apprehension as well as fire lethal rounds if warranted. Additionally, he is ideally suited for flank security and countersniper operations. Refer to *FM 90-40*.

React to An Air Attack

3-32. Passive and active air defense measures are used by MP to protect themselves from enemy air attacks during all missions.

3-33. **Passive Air Defense Measures.** MP employ passive air defense measures that include actions to avoid detection and air attack, and actions to limit damage if they are attacked. MP use active air defense measures to fight back against enemy aircraft. Enemy aircraft will attack and attempt to destroy any target seen. The passive air defense methods that limit enemy detection include—

- Concealment.
- Camouflage.
- Cover.

3-34. Dispersion is another passive air defense method. Its purpose varies from those of concealment, camouflage, and cover. While they are designed to hide MP and their vehicles and equipment from the view of the enemy, dispersion is employed to reduce the effects of an enemy air attack. Dispersion is important when MP are occupying a static position (such as a company TOC or platoon CP) or when they come under air attack. If MP come under air attack, they quickly disperse with their vehicles, move to a concealed position, if possible, and stop. Stationary vehicles are more difficult for the enemy to detect than moving vehicles.

3-35. Early warning (quick recognition of enemy aircraft) is a passive air defense method that affords MP an opportunity to take cover and one that may lead into active air defense measures. The warning may come through communication channels, OP and listening posts (LPs), or from convoy air guards. A whistle, a voice, a radio, or any other method can provide a warning.

3-36. All OP/LPs watch for enemy aircraft as a standard duty. When air sighting, the first person to see an enemy aircraft shouts, "Aircraft," then, "Front (Right, Left, or Rear)." In a convoy, air guards are given sectors of the sky to observe for enemy aircraft. When an enemy aircraft is spotted, the predetermined alarm (such as a horn or hand signal) is given until all vehicles are aware of the situation.

3-37. When an alarm is given, all dismounted troops take cover at once. They go below ground level, if possible. If the aircraft is not firing, MP withhold their fire to avoid disclosing their position and they allow the aircraft to pass. They stay concealed until the all clear is given. The MP leader initiates a size, activity, location, unit, time, and equipment (SALUTE) report (or refer to the SOP) for the sighting of hostile aircraft.

3-38. **Active Air Defense Measures**. Although passive measures are the first line of defense against air an attack, MP must be prepared to engage enemy aircraft. Low-flying hostile aircraft may appear suddenly from behind low hills, trees, or a haze. To gain surprise, they may attack with the sun behind them. Before MP fire at enemy aircraft, they must positively identify the aircraft as hostile. If the aircraft is making a firing run on the patrol, take cover and return fire. However, commanders may restrict active air defense when friendly aircraft are in the area.

3-39. In convoys, drivers alternately pull their vehicles off the road to the right and left, seeking concealment from air observation. Caution must be used if mines are a known threat. If the enemy aircraft is not attacking, the same actions stated earlier are taken. If the aircraft is attacking, MP dismount and seek cover away from the vehicle (the vehicle may be the aircraft's target) and return fire. All personnel remain under cover until the command is given to continue the mission.

3-40. Fire small arms at attacking aircraft during or after the first attack. MP direct fires to saturate the airspace through which the aircraft will fly without trapping the aircraft. For more detailed information refer to *FM 44-8*. When engaging hostile aircraft—

- Fire only on command unless under direct attack (being fired on by aircraft).
- Ensure that the direction of fire does not place rounds on friendly personnel, equipment, or positions.
- Deliver a large volume of fire.
- Lead a slow-moving aircraft and adjust fire by observing the flight of the rounds, especially if tracer rounds become available, using the free-gun technique of fire.
- Aim at the center mass of a grounded or hovering helicopter and a helicopter that is coming directly at your position. Cease-fire when the aircraft passes out of range.

3-41. Slow-moving rotary-winged aircraft that are on the ground, hovering, taking off, or landing are most successfully engaged by the MK19. The MK19 GMG's 40-millimeter ammunition is fired at a relatively slow speed and has a high trajectory at a long distance. The MK19 GMG rounds that do not hit the aircraft will detonate on impact with the ground. The location of friendly elements within range of the weapon must be considered at all times. *Figure 3-3* depicts the rules for selecting aiming points of various aircraft.

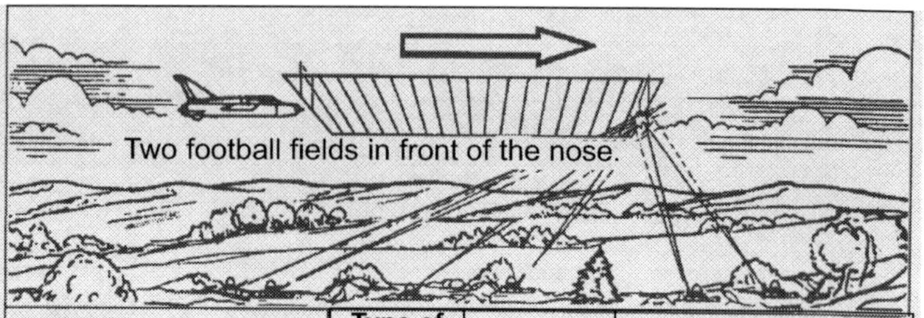

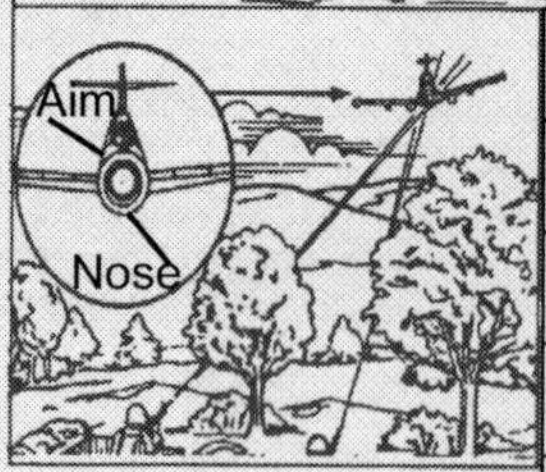

Type of Aircraft	Course	Aim Point
Jet	Crossing	Two football fields in front of the nose
Jet	Overhead	Two football fields in front of the nose
Jet	Directly at you	Slightly above the aircraft's nose
Helicopter	Crossing	One-half football field in front of the nose
Helicopter	Hovering	Slightly above the helicopter's body
Helicopter	Directly at you	Slightly above the helicopter's body

Figure 3-3. Rules for Selecting the Aim Point

REACT TO ARMOR

3-42. MP maneuver and operate in much of the battle space and can expect to encounter pockets of by-passed enemy armor within the rear area. MP engage enemy armor targets only for self-defense or when total surprise can be achieved. MP place antiarmor weapons on avenues of approach to defend against enemy armor. After initial contact, MP relocate immediately to avoid the fast-moving enemy armor and its firepower.

3-43. The (M136) antitank (AT) 4 provides antiarmor capability for MP teams. The AT4 is primarily employed against light armored vehicles, such as personnel carriers. It has a very limited capability against main battle tanks. The AT4 is issued as ammunition rather than as an individual weapon. It is carried and employed in addition to the basic weapon of MP.

3-44. The most stable firing positions for the AT4 are standing supported, prone, and prone supported. Whenever possible, use a supported position which provides more stability and aids in aiming.

Engage Armored Vehicles

3-45. The four methods to engage armored vehicles with the AT4 are—

- Single firing.
- Sequence firing.
- Pair firing.
- Volley firing.

3-46. Refer to *Table 3-2* for a description of the four methods. The best methods of engaging armored vehicles are pair and volley. Regardless of the method used, the closer the target, the better the chance for a first-round hit. Aim for the center mass of the target. The most vulnerable spots of armored vehicles are the top and the rear. The sides of the armored vehicles can also be penetrated.

3-47. An armored vehicle without the protection of dismounted infantry is vulnerable to a close attack by well-armed dismounted units. When an armored vehicle is buttoned up, visibility of the crew is restricted. This provides an opportunity for an armor-killer team to approach the vehicle with less risk of detection. The types of vehicles and the methods to engage enemy armored vehicles are—

- Stationary targets. Place the center post at the center of the visible mass for stationary targets. The firer does the same for vehicles that are moving toward or away from him.
- Slow-moving targets. Place the center post on the front leading edge of the vehicle (less than 10 miles per hour [mph]). This method is also applied to oblique moving targets.
- Fast-moving targets. These targets are moving more than 10 mph. If the target is moving to the left, place the right lead post at the center of the mass. If the target is moving toward the right, place the left lead post at the center of the mass.

Table 3-2. Methods of Engagement with the Light Antiarmor Weapon (LAW)/AT4

Method	Gunners	Firing Sequence	Probability of a Hit
Single firing	One	One gunner fires one AT4 at the target.	Low. Use only at ranges up to 200 meters for AT4s. Regardless of the method used, the closer the target the better the chances of a hit.
Sequence firing	One	One gunner fires two or more AT4s in turn. He prepares several weapons for firing before engaging the target. He gets the weapon, estimates the sight picture, and shoots the weapon in turn.	Good. If the first round misses, the gunner adjusts the range and the lead of the succeeding rounds until he gets a hit and fires until the target is destroyed.
Pair firing	Two or more	Each gunner fires one or more AT4s at a target, one at a time. They prepare several weapons for	Better. Two or more gunners track the target at one time, letting them get a target hit sooner. They can be ready to shoot as soon as an earlier

Table 3-2. Methods of Engagement with LAW/AT4 (Continued)

Method	Gunners	Firing Sequence	Probability of a Hit
		firing before engaging the target.	round hits. The first gunner sees a target, identifies it, and states the estimated range and lead to use. For example, the gunner, on spotting a fast-moving scout reconnaissance vehicle, says, "BMP 150 meters; fast target." He then fires at the target. If the first gunner misses, the second gunner gives a revised range and lead. This continues until one gets a hit. Once the correct range or lead has been found, all the gunners fire until the target is destroyed.
Volley firing	Two or more	Each gunner fires one or more AT4s on command or on signal until the target is destroyed. They prepare several weapons for firing before engaging the target.	This is the best method of engagement for an AT4 because the gunners shoot more rounds at a target at one time. This method is used only when the range and lead to the target have been determined. Range can be determined by using the map, pacing, or the results of pair firing after a target has been hit.

Estimate the Range of the Targets

3-48. A gunner has a better chance of hitting a target with the AT4 if he knows the range to the target. Determining the range is a learned skill and must be mastered by anyone who fires the AT4. Methods of range determination include—

- Using range finders.
- Measuring the distance on a map.

- Pacing.
- Firing the pair and sequence method.
- Using visual range estimation.

3-49. Visual range estimation is the least desirable method of range determination because of its inaccuracy. However, in the offense or in a hasty defense, it may be the only method available.

Disable Armored Vehicles

3-50. Armored vehicles are hard to destroy when firing at their front. Use the following three ways to disable them:

- Mobility kill. In this disabling method, the vehicle has stopped moving because a track or road wheel has been destroyed, or the vehicle has been hit in the engine compartment. The vehicle can no longer move but can return fire.
- Firepower kill. When a firepower kill has occurred, the main gun cannot return fire because of a hit in the turret, knocking out its capability to fire. The vehicle can still move, thus enabling it to get away.
- Catastrophic kill. In this kill, the vehicle is destroyed. To obtain a catastrophic kill, firers prepare to fire a second or third shot to destroy the vehicle.

CALL FOR FIRE

3-51. A call for fire is used to obtain fire support from other units. Fire support may be needed in the rear area if the enemy endangers key units or facilities. Fire support may come from mortars, artillery, Army aviation, and US Air Force aircraft. Before a mission, the commander will outline the TRPs and the priority of fires in the OPORD that will affect the call for fire response time. All MP must know how to call for and adjust fire. To call for fire the—

- Leader tells the RTO that a target has been seen.
- RTO initiates the call for fire while the target location is being determined.
- RTO sends the information as it is determined instead of waiting until a complete call for fire has been prepared.

3-52. MP may either go directly to the fire direction center (FDC) of the firing unit for artillery fire support or relay communications to MP leaders, the rear area CP fire support element, or when so directed, a TCF.

Use an Initial Call for Fire

3-53. Artillery fire support can provide the rear area with on-order fires to assist in countering threat incursions. A standard call-for-fire message is used to obtain artillery or other fire support. Regardless of the method of the target location used, the call for fire consists of six elements transmitted in three parts. There is a break and a read back after each part.

3-54. **First Transmission**. Send elements one and two during the first transmission.

- Element 1 includes the identification of the observer. This element tells the FDC who is calling and clears the net for the remainder of the call.
- Element 2 includes the WO. The type of fire support mission and the method of locating the target are identified in this element. See *Table 3-3* for the types of fire missions.

3-55. **Second Transmission**. Element 3 is sent during this transmission and includes the target's location. A target location may be provided by the grid coordinate

Table 3-3. Types of Artillery-Fire Missions

Mission	Description
Adjust fire	This mission is used when the observer is uncertain of the exact location of the target. The observer says, "Adjust fire."
Fire for effect	The observer uses this mission when he is certain of the location of the target. He is requesting a first-round fire for effect in this mission to validate the desired effect on the target with little or no adjustment. The observer says, "Fire for effect."
Suppress	This mission is used to quickly bring fire only on a preplanned target. A target description is not provided when requesting this mission. It is a simplified call for fire and is sent in one transmission. An example of what an observer may say is, "P25—this is P59—suppress AB2502—over."
Immediately suppress	This mission is similar to that of suppression. The difference is a planned target or a target of opportunity is firing at friendly soldiers or aircraft in this mission. The observer may say, "P25—this is P59—immediate suppression AB2503—over."

(normally six digits), a polar plot, or a shift from a known position method.

3-56. **Third Transmission**. Send elements 4, 5, and 6 during this transmission.

- Element 4 includes the target's description. A brief description of the target is given to the FDC using size, nature, activity, protection (SNAP). SNAP represents the size or shape, the nature or nomenclature, the activity, and the protection or posture of the target.
- Element 5 includes the method of engagement. This element consists of the type of adjustments, danger close, trajectory, ammunition, and distribution. The observer specifies how he wants to attack the target.

- Element 6 includes the methods of fire and control. The observer states who will give the command for fire to begin. If the observer wishes to control the time of firing, he will say, "At my command." If the observer does not say this, the FDC will fire as soon as the element is ready.

3-57. At a minimum, a call for fire must include the first four elements. Untrained observers need to send only the first four elements, and the FDC will decide the methods of engagement, fire, and control. Every MP must know that in order to put indirect fire on a target, the following information must be given to the FDC:

- Who he is.
- Where and what the target is.
- How close the target is to friendly troops.
- Where the target is in relation to his or other known positions.
- The direction from himself to the target. Determine the direction during—
  - Daylight with the mini-eye safe-laser infrared observation set (MELIOS), AN/PVS-6. The lightweight laser rangefinder is capable of determining ranges 50 through 9,995 meters. Refer to *TM 11-5860-202-10*.
  - Hours of darkness with the infrared (IR) illuminator, AN/PEQ-2A. The AN/PEQ-2A is for use with night vision devices (NVD) and can be used as either a handheld illuminator or pointer or can be weapon-mounted. In the weapon-mounted mode, the AN/PEQ-2A can be used to accurately direct fire as well as illuminate and designate targets. This item is fully waterproof and can be taken down to extended depths without risk of leakage.

3-58. Determining the direction to a target is an essential skill for the observer. Direction is an integral part of terrain map association, adjustment of fire, and target location. Use the following paragraphs to manually determine the direction to a target.

- Use a compress. Using an M2 or lensatic compress, the forward observer (FO) can measure the direction. The FO will add or subtract the grid of magnetic (GM) angle to determine the grid direction to send to the FDC.
- Scale from a map. Using a protractor or an observed fire (OF) fan, the FO can scale the direction from a map to an accuracy of 10 mils.
- Measure from a reference point. Using a reference point with a known distance, the FO can measure the angle between the reference point and the target and add or subtract the measured angle to or from the known direction to determine the direction to the target. The angle between the reference point and the target can be measured with binoculars or with the hand measurement technique depicted in *Figure 3-4.*
- Estimate. With a thorough terrain-map analysis, the FO can estimate the direction by visualizing the 8-cardinal directions (north [N], northeast [NE], east [E], southeast [SE], south [S], southwest [SW], west [W], and northwest [NW]).

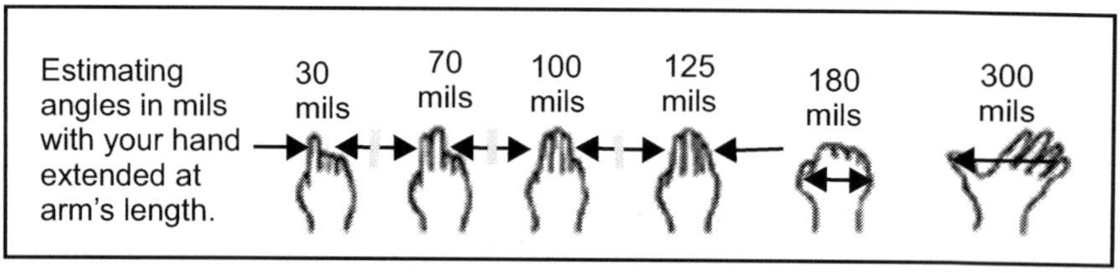

Figure 3-4. Determine the Direction to a Target

NOTE: The observer tries to be as accurate as possible, and the use of mils is preferred. All measured directions sent to the FDC will be rounded to the nearest 10 mils.

Identify the Elements of a Call for Fire

3-59. FDC personnel will help in the call for fire and subsequent adjustments by asking leading questions to obtain the information needed. Refer to *FM 6-30* for call for fire. The elements of a call for fire include the following:

- Observer identification. The observer identification tells who you are. Use the call signs from the signal operating instructions (SOI).
- Warning order. The WO alerts the firing units of the—
  - Type of mission. This includes adjusting fire, firing for effect, suppressing fire, and immediately suppressing fire.
  - Size of element to fire. Omission indicates a request for one field artillery (FA) battery. If the fire mission requires an element larger than a FA battery, state the size needed, such as 2 FA batteries (battalions).
  - Method of the target's location. The grid has no announcement. Announce the word "polar" for the polar plot. Shift from a known point by announcing the word "shift" followed immediately by the designation (target number) of the known point.
- Target location. The target location enables the FDC to plot the target.
  - Grid: Two-character, 6-digit grid, such as NA123456.
  - Polar: Direction (grid azimuth) and distance (meters) to the target from the observer's position. Give the difference in elevation if there is a vertical shift of over 35 meters between the observer and the target.
  - Shift from a known point or the direction to the target (grid azimuth). The three types of shifts are the lateral shift (left or right) in

meters, the flange shift (add or drop) in meters, and the vertical shift (up or down) over 35 meters from the known point and target.

- Target description. The target description helps the FDC to select the type and the amount of ammunition. A word picture of the target (for example, the number and type of vehicles or personnel observed).
- Method of engagement. The method of engagement tells the FDC how to attack the target, which includes the following:
    - Type of engagement. Area fire is standard without a request. Request precision fire only to destroy a point target.
    - Trajectory. A low-angle trajectory is standard without a request. An high-angle trajectory is at the request of the observer or when required due to masking terrain.
    - Danger close. Danger close is announced when applicable.
    - Ammunition. Ammunition is the type of projectile, the type of fuse action, and the volume of fire desired in the fire-for-effect phase stated in rounds per howitzer.
    - Distribution. Distribution is the type of sheaf desired and parallel is standard without request.
- Method of fire and control. The method of fire and control tells the FDC how you want to control the delivery and adjustment of the fire.
    - Method of fire. In area fire, the adjustment normally is conducted with one howitzer or with the center gun of a mortar platoon or section. If for any reason the observer

determines that *platoon right (left)* will be more appropriate, he may request it. Adjusting at extreme distances may be easier with two guns firing. The normal interval of time between rounds fired by a platoon or battery right (left) is 5 seconds. If the observer wants some other interval, he may so specify.

- Method of control. Use the following methods of control: *Fire when ready* is standard and no request is required; use *at my command* when weapons fire at the observer's command; use *cannot observe* when fire will not be observed; use *time on target* when rounds land at a specified time; use *continuous illumination* when the FDC will determine when to fire; use *coordinated illumination* when illumination rounds are fired only when the target is engaged; use *cease loading* when missions have two or more rounds in effect, causing the firing unit to stop loading rounds; and use *check firing* when there is a temporary halt in firing.

- Danger close. Include the term *danger close* in the method-of-engagement portion of the call for fire when the target is within 400 meters of any friendly troops or mortars and 600 meters for field artillery. When adjusting naval gunfire, announce "danger close" when the target is located within 750 meters when using naval guns that are 5 inches or smaller. For naval guns larger than 5 inches, announce "danger close" when the target is within 1,000 meters. The creeping method of adjustment will be used exclusively during danger close missions. The FO makes range changes by creeping the rounds to the target using corrections of less than 100 meters.

Plot a Target Location

3-60. Selecting a targeting method includes giving the directions in mils, degrees, or cardinal points of the compress (N, NE, S, SW, E, SE, W, and NW). Give a deviation of left or right and the distance in meters. Use the following paragraphs when plotting a target location:

3-61. Grid. Determine a 2-character, 6-digit grid for the target. Then, determine a grid direction to the target, and send it after the call for fire and before any subsequent corrections.

3-62. Polar. Determine the grid direction to the target. Determine the distance from the observer to the target. Determine if any significant vertical interval exists.

3-63. Shift. Refer to *Figure 3-5* and determine—

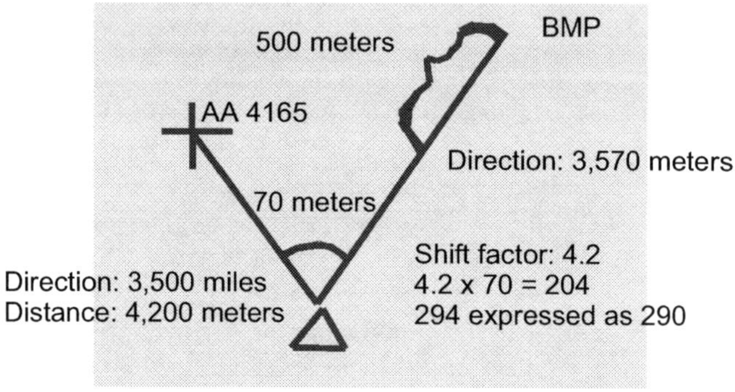

Figure 3-5. Call for Fire Elements

- The grid direction to the target.
- The lateral shift to the target from the known point. W = R times m (mil relation formula), when—

$W = R \times m$

W = width of the lateral shift (the unknown)

R = shift factor, the distance to the known point divided by 1,000 and expressed to one decimal place

m = measured angle in mils from the known point to the target

- The range shift from the known point to the target.
- Any significant vertical interval that may exist.

Example of plotting a target location: Complete target location – direction 3,570, right 290, add 500

Adjust Fire

3-64. The objective in adjusting fire is to move the center of the impact to within 50 meters of the center of the target. The observer accomplishes this by sending the FDC subsequent corrections, which are deviation (lateral) and range corrections. The FDC can talk the observer through the adjustments if they are necessary. The burst is moved to, and kept on, the observer target line in order to get a positive range spotting. The observer target line is the line of sight (an imaginary line) between the observer and the target. When range spotting cannot be determined, the observer makes a request for a lateral correction to place the burst on the observer target line.

3-65. The observer makes range corrections to bracket the target between two successive rounds. The successive bracketing technique is used. After the first definite range spotting is determined, a correction is sent to the FDC to establish a bracket of known distance around the target (such as, one round over the target and one round short of the target). The observer then successively splits this bracket until he is within 50 meters of the target and calls for fire for effect.

3-66. Hasty bracketing is used if the nature of the target dictates that effective fires are needed faster than the above procedures can provide. Hasty bracketing depends on a thorough terrain analysis to give the observer an accurate initial-target location. A bracket is obtained on the first correction in a manner like that used for successive bracketing. Once the initial bracket is established, it is used as a yardstick to find subsequent corrections. The observer sends the FDC the correction to move the rounds to the target and calls for fire for effect. Hasty bracketing improves with observer experience and judgment. Fire for effect consists of one or more rounds from each gun of the unit firing at the target. Dispersion of the guns will cause the rounds to saturate the area with shell fragments. To end a fire mission, the observer states, "End of mission," and reports the results of the fire for effect. The observer may say, "End of mission, three T-62s neutralized, estimate two casualties, over."

Illuminate the Battlefield

3-67. Battlefield illumination can provide MP with enough light to aid in ground operations at night. Illumination can—

- Mark the targets for CAS.
- Increase visibility in areas of suspected enemy activity.
- Furnish the direction to the patrol activity.

3-68. Illumination is called for and adjusted like other indirect fires except the methods of engagement, fire, and control differ. The observer requests illuminating shells. The methods of fire and control differ in that the adjustment is based on how much visibility is needed in the target area. If the observer calls for—

- "Illumination," the observer gets one round from one gun.
- "Illumination, two guns," the observer gets one round each from two guns. The rounds will burst simultaneously.

- "Illumination, range and lateral spread," the observer gets one round each from four guns. The rounds will burst simultaneously in a diamond pattern.

3-69. The initial request for illumination must include the—

- Date when the illumination is needed, if illumination is preplanned.
- Purpose of the illumination.
- Requested time and duration of the illumination (for example, three minutes at 2150 hours or three minutes on call).
- Grid reference and, if needed, the height of the points or areas to be illuminated.
- Method of controls (any restrictions in the time and the place before and during the operation).

OBTAIN ARMY AVIATION FIRE SUPPORT

3-70. Army aviation provides the echelon commander with the ability to move combat resources across the battle space with little regard for the terrain's barriers. These units can provide surveillance or screen over a wide area in adverse weather and at night. Attack helicopter units provide the sustainment area with highly maneuverable antiarmor firepower. They are ideally suited for situations in which rapid reaction time is critical.

Control the Fire

3-71. While en route to a target area, the attack helicopter will contact the caller on the radio. For example, "1L22, this is 1X47, fire team arrives at estimated target area in 4 minutes, over." At this time, a call for fire is transmitted consisting of—

- The target's location and description.

- The proximity of the friendly unit to the target. The words *danger close* must be included when a friendly unit is 600 meters or less from the target. *Danger close* is required because some types of ordnance cannot be used in close proximity to friendly ground forces. When *danger close* is included, MP must mark the unit's location. The method of marking should be one that least reveals the position to the enemy, such as using panels or mirrors.
- The protection of the friendly units (such as good fighting positions, hasty positions, or exposed positions).
- The direction of the friendly unit from the target (cardinal direction).
- Other friendly fire support considerations, including artillery and mortars firing in the area and tactical aircraft (attack direction and altitude).
- Dangers to the flight. Report locations of known or suspected enemy antiaircraft weapons or other dangers to the flight (wires in the target's area, enemy artillery fire impacting in the target's area, or enemy aircraft).

3-72. When the helicopter arrives over the objective, the helicopter's fire team contacts the caller. The caller marks the target and states the method of adjustment. The caller can use the three following methods to mark the target:

- A reference either to a prominent terrain feature that can be identified from the air or to a known point.
- A direction to the target from a reference point, stated in mils or degrees.
- References to friendly fire (such as smoke grenades, tracers, smoke streamers, mortars, artillery, or marking rockets).

Use the Fire Adjustment Method

3-73. The three methods used to adjust the fire of a fire team are impact-observed, impact sound, and observer target. The preferred methods of adjustment for an attack helicopter in support of a ground force are impact observed and impact sound. Once established, the caller does not change the method of adjustment unless the situation dictates. If the method of adjustment is changed, the caller informs the fire team. When any adjustment is 50 meters or less, the observer transmits the adjustment and calls for fire for effect.

3-74. When using the impact-observed method of adjustment, the observer estimates the direction to the target by using a cardinal heading. He estimates the distance from the point of impact to the target in meters. When the observer cannot see the point of impact, he may use the impact sound method of adjustment. For this method of adjustment, the observer transmits, "Adjust fire. Impact sound. Over." The impact sound method differs from the impact-observed method in that the observer senses by sound, rather than sees, the direction of the impact and makes his corrections accordingly.

3-75. Although the impact-observed method is most preferred for adjusting the fire of the attack helicopters, the observer target method, which is less desirable, may be used. When using the observer target method, the observer must mark his location, possibly compromising his location.

3-76. To use the observer target method, the observer senses the direction, left or right, and the distance, in meters, from the impact to a point on the observer target line. Then, he senses the position of the point on the observer target line relative to the target, long or short, and the distance along the observer target line to the target. The sensing of the observer must be converted to corrections, such as right, left, add, and drop, and transmitted to the fire team. The chance of error for this method is greater than for the other methods. Unobserved rounds are handled the same as for the other methods. An example of an exchange of information between an observer and a fire team using the observer target method follows:

- Observer: "Left, five-zero. Add 100. Fire for effect. Over."
- Fire team: "Roger. Out." (The team commits against the target.)

3-77. When the target is suppressed or destroyed, the following transmission would occur:

- Observer: "End of mission. Target suppressed (destroyed). Over."
- Fire team: "End of mission. Out."

3-78. Reference points are used to visually locate the target. The eyes of the pilot are led to the reference point and from the reference point to the target, sometimes through a series of decreasingly obvious reference points. It is much harder for a pilot to find a target than to keep a target in sight. Any reference point must stand out or contrast with its surroundings.

Adjust Fire

3-79. Attack helicopter fire allows the pilot to observe the impact and effect of the ordnance on the target. This simplifies the adjustment procedure. However, the observer must still be prepared to adjust direct aerial fire. When adjusting aerial fire—

- Establish a reference point. The point of impact of the first round is the recommended reference point.
- Adjust for target strike. Do not try to bracket the target. The helicopter crew has direct visual contact with the target and needs only specific directions to fix the location.

- Transmit the corrections.

3-80. The following is a typical exchange of information between the observer and the fire team:

- Observer: "Adjust fire. Impact observed. Over."
- Fire team: "Impact observed. Out." (The team fires at the target.)
- Observer: "Northwest, seven-five. Over."
- Fire team: "Roger. Out." (The team fires at the target.)
- Observer: "North, three-zero. Fire for effect. Over."
- Fire team: "Roger. Out" (The team commits against the target.)
- Observer: "End of mission. Target suppressed (destroyed), over."
- Fire team: "End of mission. Over."

3-81. If the observer does not see the impact, the transmission would be—

- Observer: "Unobserved. Over."
- Fire team: "Unobserved. Over." (The team fires at the target.)

3-82. Adjustments continue until the mission is accomplished. *Table 3-4* shows how to direct a pilot to the target.

Table 3-4. Directing the Pilot to the Target

Ways of Directing the Pilot		How Used
Ammunition	Smoke rounds	Mortars, artillery, or grenade launchers. Phosphorous is usually the best because smoke clouds blossom quickly and are highly visible.
	Ordnance	Ordnance impacting the ground may be an adequate reference point.

Table 3-4. Directing the Pilot to the Target (Continued)

Ways of Directing the Pilot		How Used
	Illumination rounds	Good for guiding the strike aircraft to the target at night, but will not pinpoint small targets if the flares function at the usual height
	Trace fire	Used at night. The intersection of the two streams of traces or the impact point of one stream marks the target.
Fires	Grass or other	Sometimes used near the target as a night reference
Recognizable known points	Terrain features or landmarks	If clearly visible from the air, it can help when used with another location method (streams, roads, bridges, tree lines, cultivated areas, prominent hills).
	Friendly positions	When clearly recognizable from the air, it may be used day or night for locating close-in targets.

OBTAIN AIR FORCE TACTICAL-AIRCRAFT FIRE SUPPORT

3-83. During major enemy incursions in the rear area, fighter aircraft may be available to support ground operations by providing immediate CAS. CAS consists of air attacks against enemy targets that are close to the friendly forces. CAS requires detailed coordination with the maneuver of the ground forces to be effective. The coordination must be responsive, integrated, and controlled. Typical CAS targets are—

- Enemy troop concentrations.
- Fixed positions.
- Armored units of immediate concern to the ground forces.

3-84. CAS missions are flown at the request of the command level. They are planned, directed, and controlled by the Air Force through the Tactical Air Control System.

3-85. Air Force support is directed through a forward air controller who, in turn, talks to the pilots. The controller can be in an aircraft or operating on the ground. In most cases, the controller will come forward to a point where he can see the target. Once the target is in sight, he can adjust the aircraft to the target. If the controller cannot see the target, the observer will have to tell him how it can be identified. The observer must make sure that the controller knows where all the friendly elements close to the target are located.

3-86. If the observer is unable to talk to a forward air controller, he must contact a fire-support team operating in the maneuver area. Fire support teams have the equipment to talk directly to the pilots of the aircraft and are trained observers for CAS.

Mark Friendly Positions

3-87. Friendly positions are marked during close air strikes if there is no danger of compromise to enemy observers. This may require only a message (such as "All friendly positions are south of the target. Nearest are 500 meters."). As a rule, a mark is usually necessary when friendly troops are within 300 meters of the target. Marking of friendly positions may be overt or covert and include the following:

- Weapons fire. Weapons fire is useful as a signal if it is distinguishable from other types of fire. Tracers are especially useful.
- Smoke. Smoke grenades are the most commonly used overt marker, but they have limitations. Wind may move the smoke away from the source. Red and white smoke show up well with most backgrounds while some colors blend with their background.

- Mirrors. Signal mirrors and panels are probably the best covert ground-to-air devices for attracting attention. When the operator is proficient and the sun is shining, pilots can see a mirror flash for many miles away. Mirror signaling requires practiced training.
- Balloons. Balloons make a good covert marking system for use above a thick forest canopy.
- Flares. Rockets or 40-millimeter flares are good for attracting attention at night. If flares are fired in the direction of the aircraft, they can be mistaken as ground fire.
- Lights. Strobe lights produce a bright pulsing flash that is visible at night from 2 to 5 kilometers. Vehicle lights, such as unshielded red taillights, are visible to a pilot for several kilometers at night. Chemical glow lights may also be used.
- Ground commander's pointer. The pointer is a handheld device that is invisible to the naked eye, but its beam is visible through NVDs. It may be used by ground troops to clearly show air elements the location of friendly elements.

Select Attack Headings

3-88. A fighter aircraft is more likely to destroy its target if it attacks along the long axis of the target. Once the fighter aircraft knows where all the friendly units are and where the target is, the forward air controller tells the fighter pilot which attack heading to use. However, if the controller cannot see the target, the observer may have to recommend a direction of approach. The observer must remember that fighters should not attack across friendly positions.

3-89. An attack toward friendly units is undesirable because of ordnance dispersal patterns. An attack from behind and over friendly lines is also undesirable for several reasons. Some fighters dump empty cartridges overboard as they strafe. An empty 20-millimeter case weighs 114 grams and hits the ground at 167 kilometers per hour. An even greater hazard would be an inadvertent bomb release as the pilot repeatedly selects and arms his weapons systems while in the attack pattern.

MOVE

3-90. Movement by MP teams, squads, or platoons in combat is dependent on the mission, the terrain, and the likelihood of enemy contact. MP apply the fundamentals of movement which include—

- Moving on covered and concealed routes.
- Avoiding likely ambush sites.
- Enforcing camouflage, noise, and light discipline.
- Maintaining all-around security, to include air guards.
- Using formations and movement techniques based on METT-TC.

3-91. In addition to applying the fundamentals of movement, MP leaders ensure that they—

- Maximize the capabilities of HMMWVs and ASVs. This includes considering the speed, mobility, and firepower of the vehicles. Fire and move both vehicles as a weapon system.
- Make contact with the enemy using the smallest force possible. MP move with a small force in the lead with the rest of the unit ready to react. A team leads a squad and a squad leads a platoon. One team leads another when two vehicles are moving. This prevents the whole unit from being pinned down by enemy fire and provides the unit with the flexibility to maneuver.

- Use the terrain. The terrain offers natural cover against enemy fire and conceals them from enemy observation. MP leaders must devote constant attention to protect vehicles and prevent them from skylining. MP make use of all natural cover and concealment when moving or stopped. When MP do stop, they stagger their vehicles on the roadway.
- Control subordinate elements. MP leaders issue clear and complete orders to subordinate elements in order to maintain control. They issue OPORDs, which cite MP actions to be taken on contact and the immediate actions the teams should accomplish. The OPORDs also explains how the MP leader will direct subordinate elements through the use of hand and arm signals, pyrotechnics, and other visual signals.

USE MOVEMENT TECHNIQUES

3-92. Movement techniques are designed to minimize the exposure of the platoon to enemy fire and to place the platoon in a good position to react to enemy contact. They provide varying degrees of control, security, and flexibility. The selection of their use is based on METT-TC and the likelihood of enemy contact. Their effective use results in the platoon's contact with the enemy with the smallest platoon element.

3-93. MP employ the following three techniques of movement on the battlefield:

- Traveling.
- Traveling overwatch.
- Bounding overwatch.

3-94. While these techniques provide a standard method of movement, the leader must use common sense and sound judgment when employing them as he performs his missions and encounters different situations. The decision of which technique to use is based on terrain considerations and whether enemy contact is not likely, possible, or expected. The techniques are used in both the mounted and dismounted modes of movement.

Traveling

3-95. An MP leader selects the traveling method of movement when contact with the enemy is not likely and speed is required. This technique allows the lead and trail elements to move together as a unit. It is the fastest but least secure movement technique. Movement is continuous, and interval and dispersion are maintained between the squads as terrain and weather permit. The platoon does not intend to engage in combat, but it is dispersed to prevent destruction in case of unexpected air or ground attack. The distance between the vehicles is based on the factors of METT-TC. This method of movement, with MP mounted, is depicted in *Figure 3-6.*

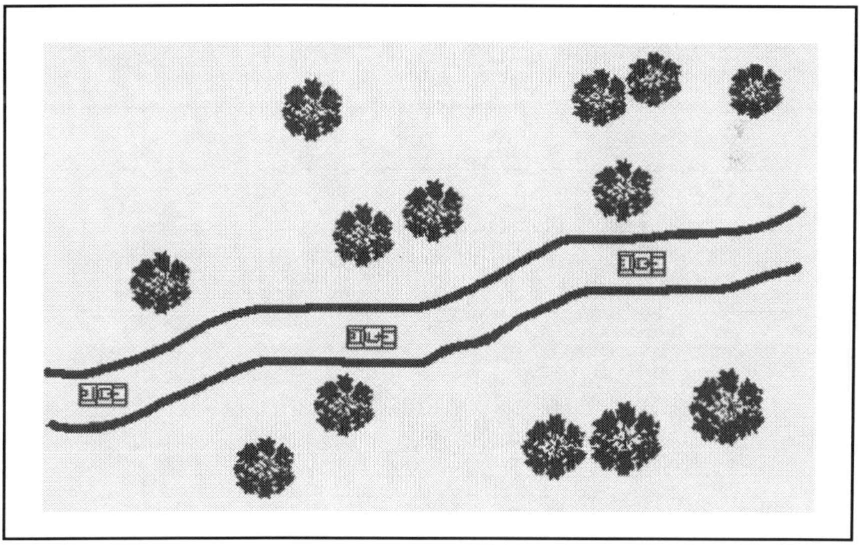

Figure 3-6. Traveling

Traveling Overwatch

3-96. Use the traveling overwatch method of movement when contact with the enemy is possible and speed is desirable. The lead element moves continuously along the best, covered and concealed routes for protection from possible enemy observation and direct fire. The trail element moves at variable speeds, continuously overwatching. It normally maintains contact with the lead element and may stop periodically for better observation. The trail element tries to stay one terrain feature behind the lead element, but close enough to provide immediate suppressive fire and maneuver for support. However, it remains far enough to the rear to avoid contact with the same enemy force that is engaging the lead element. This technique, with MP mounted, is depicted in *Figure 3-7*.

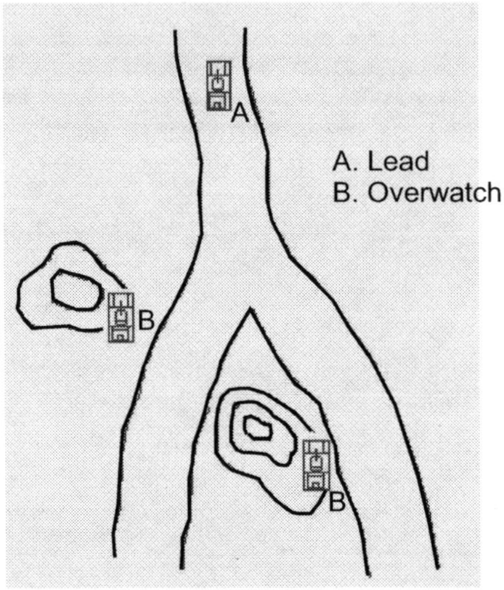

Figure 3-7. Traveling Overwatch

Bounding Overwatch

3-97. When MP expect to make contact with the enemy, they select the bounding overwatch method of movement. It is the slowest, but safest method of movement. In bounding overwatch, the trail element occupies a good, covered and concealed position to overwatch the lead elements. While one element moves, another is always stopped in position to overwatch the bounding element. Although, bounding overwatch is used when enemy contact is expected, use it when time is available regardless of the likelihood of enemy contact. It provides for immediate, direct suppressive fire on an enemy force that engages the bounding element with direct fire. A three-vehicle team uses the V-formation with bounding overwatch. The lead elements advance to a point (first move) where they can support the advance of the overwatch element. On signal, the overwatch element moves forward to a position abreast of the lead elements (second move) and halts. During its move, it is overwatched by both lead elements. The lead elements then move forward, secured by the overwatch. Maximum use is made of folds of the earth and concealment to mask movement from likely enemy positions. MP are mounted in this method of movement as shown in *Figure 3-8.*

PLATOON LEADER RESPONSIBILITIES

3-98. In the conduct of most tactical missions, the MP platoon moves as separate squads under the C^2 of the platoon leader. Regardless of which movement technique is directed, the platoon leader issues the squads an order explaining what each squad will do. This action becomes more critical as the likelihood of enemy contact increases. The platoon leader tells, and if possible, shows the squads—

- The enemy situation as he knows or suspects it to be.

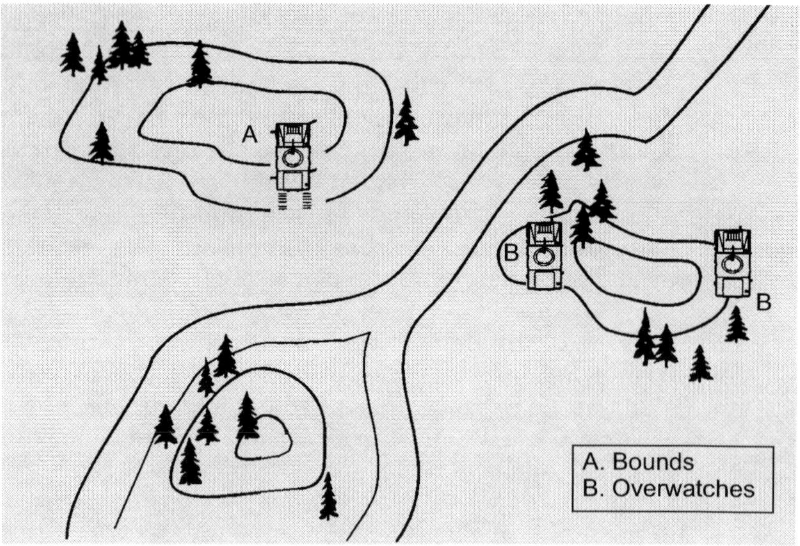

Figure 3-8. Bounding Overwatch

- The next overwatch position (objective for the bounding element).
- The route of the bounding element to that position.
- What he wants the squad to do after the bounding element gets to the next position.

USE EXTENDED AND TEMPORARY HALTS

3-99. When an MP platoon moves as an element, it uses the coil formation for extended halts. This formation provides the platoon with 360-degree observation and fields of fire. The coil is always executed from either the column or staggered column formation. The platoon uses the four-team organization. The lead team assumes the 12 o'clock position (the direction of travel). Teams occupy the 3, 6, and 9 o'clock positions in twos, facing in the appropriate direction. Ensure that there are 50 to 100 meters between the team's vehicles. The interval between the teams will be 100 to 200 meters (terrain dependent). The HQ element occupies the center of the formation. Each platoon must have its own SOP for the formation based on its METL, war plans, and most common organization; it should practice the SOP as a drill to ensure correct execution. *Figure 3-9* depicts an MP platoon in a coil formation.

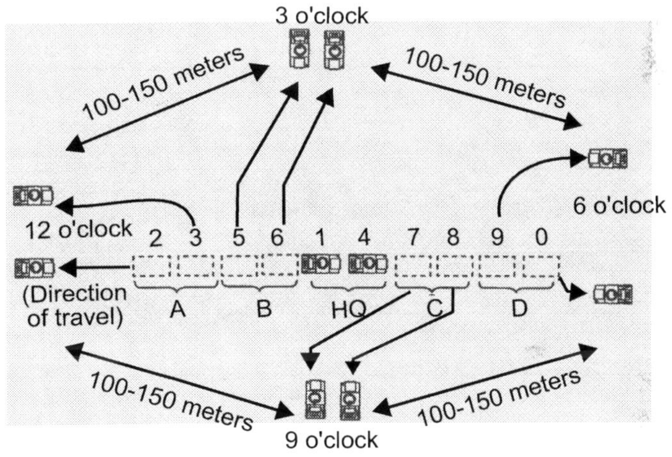

Figure 3-9. Platoon Coil Formation

3-100. MP use the herringbone formation for temporary halts from the march column. It provides them with a 360-degree observation and field of fire *(Figure 3-10).*

USE GRAPHIC CONTROL MEASURES

3-101. MP leaders use graphic control measures to regulate or direct the movement, positions, and fire of the platoon. Control measures—

Figure 3-10. Herringbone Formation

- Are not intended to restrict the exercise of initiative (the function of command). Leaders use control measures to clarify their intent, focus the platoon and squad effort, and ensure synchronization. Each control measure should have a specific purpose that contributes to mission accomplishment. If a control measure fails the purpose test, leaders should not use it.
- Can be drawn on a map, an overlay, a sketch, or a terrain model. Leaders should strive to keep control measures easily identifiable and simple. Graphic control measures may include the AA, the route, the release point (RP) and start point (SP), checkpoints, and so forth. *FM 101-5-1* discusses control measures and provides examples of their use.

CROSS A DANGER AREA

3-102. Danger areas are specific areas where there may be potential danger because of the increased risk of detection. Plan a way in which the patrol crosses danger areas to reduce the chances of a fight. MP make specific plans for crossing each known danger area and general plans for crossing unexpected danger areas. Patrols must be able to quickly modify these plans to fit the tactical situation. Typical danger areas include the following:

- Curves and blind spots on roads and trails.
- Streams.
- Open areas.
- Hill tops.

3-103. MP move cautiously at danger areas, using the bounding overwatch or variations of it to cross them. The MP leader decides how a patrol will cross danger areas based on the—

- Amount of time available.
- Size of the patrol.
- Size of the danger area.
- Fields of fire into the area.
- Amount of security available.

3-104. To cross a danger area, a patrol must designate nearside and farside rally points, secure the nearside and farside, and cross the danger area.

3-105. A small patrol may cross all at once, in pairs, or one element at a time. A large patrol normally crosses its elements one at a time. The leader positions security teams far enough out on both flanks and to the rear of the crossing point to give warning of the approaching enemy and to overwatch the crossing element. Once flank and rear security are positioned, the team crosses the danger area. The team crosses quickly, reconnoiters, and secures the far side of the danger area. The area on the farside must be large enough for a full patrol employment. When the team leader knows the farside is safe, he signals the rest of the patrol to cross. As each element crosses, it moves to an overwatch position or to the farside rally point until told to continue movement. When the patrol has crossed the danger area, the security teams cross and rejoin the patrol.

3-106. Open areas frequently afford the patrol the opportunity to observe the enemy from long ranges. Conversely, they often require that the patrol be exposed to possible enemy observation and fire for long periods of movement. The leader must make maximum use of the terrain and employ effective observation techniques to avoid exposing the patrol to a well-concealed and camouflaged enemy.

3-107. Before moving across a large open area, the patrol takes a thorough visual scan of the area. This should be done both dismounted and mounted, using all available optics. This scan focuses not only on finding potential enemy positions, but also on locating covered and concealed routes for bounding and a covered and concealed position to move to. If time and terrain permit, use dismounted troops to move to the far side of the open area and secure it. In very large open areas, use of dismounts may not be feasible because of the distances between covered and concealed positions.

3-108. Once the area has been cleared using visual means and/or dismounts, the squad or platoon moves across it. The patrol uses bounding overwatch because of the likelihood of enemy contact. If the open area is very large, the overwatch vehicle remains stationary until the bounding vehicle has moved a distance equal to half the effective range of the weapon system of the overwatching vehicle. When that point is reached, the overwatch vehicle must move out, even if the bounding vehicle has not yet reached a position of cover and concealment.

MOVE WHILE IN CONTACT

3-109. Maneuver is the technique used for moving while in contact with the enemy. Maneuver involves two actions that occur at the same time. One element moves to a position where it can engage the enemy while another element supports that movement with a base of fire. A patrol maneuvers to move forward, either to close with the enemy or to gain a better position for firing at the enemy. MP can also maneuver to get more information on the position and strength of the enemy. When the position of the enemy is unknown, it may result in an unexpected encounter known as a *chance contact*. MP use maneuver to move away and withdraw safely.

3-110. When maneuver begins, the MP leader most often goes with the base-of-fire element and controls its fire. The base-of-fire element covers the movement element by shooting at the enemy position. The movement element advances within the supporting range of the base-of-fire element, taking a position from which it can fire on the enemy. The movement element then becomes the base-of-fire element, and the former base-of-fire element begins moving. Depending on the distance to the enemy position and the amount of cover and concealment available, the base-of-fire element and the movement element alternate roles as needed to continue moving.

3-111. MP can maneuver mounted, dismounted, or in a combination of both. A fire element using the MK19 GMG will have difficulty moving dismounted. MP move mounted when the terrain protects them from enemy fire, and look for covered and concealed routes.

3-112. When receiving direct fire, the movement element uses maneuver while the base-of-fire element suppresses enemy fire. If the movement element is not receiving direct fire, it uses bounding overwatch or maneuvers internally.

3-113. MP of a dismounted movement element move based on the intensity of the enemy's fire. When facing intense enemy fire with little or no cover, MP may be forced to crawl. They will use the low or high crawl depending on the situation, the need for speed, and the example set by the leader. Although crawling is slow, it reduces exposure to enemy observation and fire. When MP are not moving forward, they place suppressive fires on the enemy. They may need to advance all the way into and through enemy positions by crawling.

3-114. Dismounted MP can use short rushes from one covered position to another when enemy fire allows brief exposure. To do this they should—

- Advance by short rushes to avoid the enemy's fire.
- Try to stay up no more than 3 to 5 seconds. This keeps the enemy from having time to track and engage them.
- Select the next covered position before beginning the rush.
- Rush from cover to cover.
- Not hit the ground just because 3 to 5 seconds are up.

3-115. MP of a mounted maneuver element move based on enemy fire and the terrain. When they move, they should—

- Use the terrain to mask their movement.
- Move quickly between protected positions so that the enemy cannot engage their vehicles.
- Dismount when the terrain no longer provides protection.

EXECUTE ACTIONS ON CONTACT

3-116. When MP encounter enemy forces, they must quickly execute actions on contact. Whether they remain undetected or are identified by the enemy, MP must first take measures to protect themselves, find out what they are up against, and then decide on a COA. To properly execute actions on contact, MP must take action consistent with the following fundamentals of reconnaissance:

- Remain focused on the reconnaissance objective.
- Report quickly and accurately.
- Maintain contact with the enemy.
- Retain the freedom to maneuver.
- Develop the situation rapidly.

3-117. The platoon leader specifies the actions on contact for the platoon. These specific instructions include the engagement criteria and the desired COA, based on the size and activity of the enemy force encountered. By knowing these details ahead of time, MP can develop the situation more rapidly and arrive at and execute the desired COA. The platoon strives to make contact with the smallest possible element. Visual contact in which the enemy is observed, but MP remain undetected, is the goal. This gives the platoon the greatest possible flexibility to maneuver and develop the situation.

3-118. The steps that make up the actions on contact must be thoroughly trained and rehearsed so that the platoon can react instinctively as a team whenever it encounters enemy forces. The four steps which are executed to allow the platoon to accomplish its mission according to the reconnaissance fundamentals are—

- Deploy and report.
- Evaluate and develop the situation.
- Choose a COA.
- Execute a COA.

Deploy and Report

3-119. The MP patrol that gains initial visual contact with the enemy deploys to covered terrain that has good observation and fields of fire. The MP patrol in visual contact sends a report using the SALUTE format. If the element in contact is unable to report or cannot report quickly, another squad or team must report. The elements not in contact temporarily halt in covered terrain, monitor the report, and plot the situation on their maps. The platoon or patrol leader immediately determines the COA.

Evaluate and Develop the Situation

3-120. The patrol concentrates on defining what they are up against. If they have not sent a SPOTREP at this point, they initially focus on getting enough information to send one. If the enemy has not detected them and time is available, the patrol attempts to confirm or determine in detail the enemy's size, composition, activity, orientation, and the locations of the enemy's weapon systems. They search for any additional information regarding the enemy that can help define the situation and update the SPOTREP.

Choose a Course of Action

3-121. Once the patrol has developed the situation and the platoon or patrol leader has enough information to make a decision, he selects a COA. The COA will be within the capabilities of the patrol and allow the patrol to continue the commander's concept of the operation. The platoon or patrol leader considers various possible COAs, including—

- Breaking contact and bypassing the enemy. This COA may be selected when the enemy sees the patrol before the patrol sees them and comes under—

- Sniper fire. In this situation, the patrol returns fire in the direction of the sniper and conducts the maneuver (fire and movement) to break contact with or destroy the sniper.
- Indirect fire. The patrol quickly gets out of the impact area and does not seek cover, as it may be pinned down by doing so. By continuing to move, the patrol is more difficult to hit. The patrol uses the clock system (described later) to break contact.
- Ambush. In an ambush, a patrol takes immediate action. Personnel in the kill zone return fire immediately and quickly move out of the kill zone. Elements not in the kill zone lay down a base of fire (and smoke if available) to support the withdrawal of the elements in the kill zone. The patrol breaks contact and reorganizes at the last rally point. After or while the elements in the kill zone are being extracted, the leader decides whether to destroy the ambush element or break contact based on the situation and the mission. If no guidance is given, the immediate action of the patrol is geared to breaking contact.

- Maintaining contact to support a hasty attack. This COA is appropriate when the MP patrol discovers enemy elements that the higher HQ wants destroyed. The patrol cannot destroy the enemy because it does not have the combat power (level III threat) or because it has other tasks to perform. In this situation, the patrol maintains contact and continues to develop the situation, focusing on supporting the hasty attack by a TCF. The patrol conducts additional reconnaissance and monitors any changes in the enemy's situation. It focuses on information to support a friendly hasty attack.

- Conducting a hasty attack. If contact is made and the MP patrol and the enemy element become aware of each other at the same time and at such a close range that maneuver is not feasible, the patrol may make an immediate assault. The elements nearest the enemy open fire and shout, "contact, front (right, left, or rear)." The patrol moves swiftly into the assault. It stops the assault if the enemy withdraws and breaks contact. If the enemy fights, the assault is continued until the patrol can break contact, the enemy is destroyed, or the enemy breaks contact. In most cases, the patrol cannot, or should not, mass its combat power to defeat an enemy force. If the patrol concentrates, it risks losing the capability to complete its mission and jeopardizes its ability to conduct subsequent missions. If the patrol attacks an enemy, it should only attack small dismounted formations or lightly-armored or unarmored reconnaissance vehicles. Except in self-defense, patrols should avoid attacking heavily armored vehicles or large formations.
- Establishing a hasty defense. The patrol will establish a hasty defense if it cannot bypass the enemy, all the teams are fixed or suppressed, and the patrol no longer has the ability to break contact by maneuver. Patrols should use a hasty defense when the enemy executes a hasty attack. The patrol maintains contact or fixes the enemy in place until additional combat power arrives or the patrol is ordered to move. If the patrol is required to conduct a hasty defense, the commander then becomes responsible for continuing to develop the situation.

Execute a Course of Action

3-122. The platoon leader updates his SPOTREP to the commander with any new information and then recommends a COA to the commander. The commander approves or disapproves the recommended COA based on how it will affect the parent unit's mission.

3-123. If the commander has anticipated the enemy situation the platoon is reporting, he will already have addressed the contingency in the OPORD and given guidance to his subordinates on what COA the platoon executes. In such a case, the platoon leader can evaluate the situation, choose a COA consistent with the higher commander's intent or concept, and execute it without further guidance. He keeps the commander informed of what he is doing as he executes the COA.

BREAK CONTACT

3-124. To break contact with the enemy without disorder, use the clock system. Use this system when the patrol and a larger enemy element see each other at the same time. The patrol must break contact or be destroyed. The direction the patrol moves is always 12 o'clock. When contact is made, the leader shouts a direction and distance to move (such as "4 o'clock, 300 meters" tells the patrol to move in the direction of 4 o'clock for 300 meters). If contact is broken, the patrol rallies at the designated distance and continues with the mission. If contact is not broken, then another direction and distance is given. This action continues until contact is broken. The leader can also use the clock system to shift or direct fire at a certain location.

CONSOLIDATE AND REORGANIZE

3-125. Once enemy resistance has ceased or the platoon or patrol has broken contact, leaders must quickly take steps to consolidate and prepare to defend against a counterattack. In consolidating on the objective, all-around security is critical because the enemy might counterattack from any direction. The leader must evaluate the terrain thoroughly. Platoons and patrols use the following two techniques when consolidating:

- Clock technique. In using this method, the leader designates either a compass direction or the direction of attack as 12 o'clock. He then uses clock positions to identify the left and right boundaries for squads. The leader positions key weapons along the most likely avenue of approach based on his assessment of the terrain *(Figure 3-11)*.

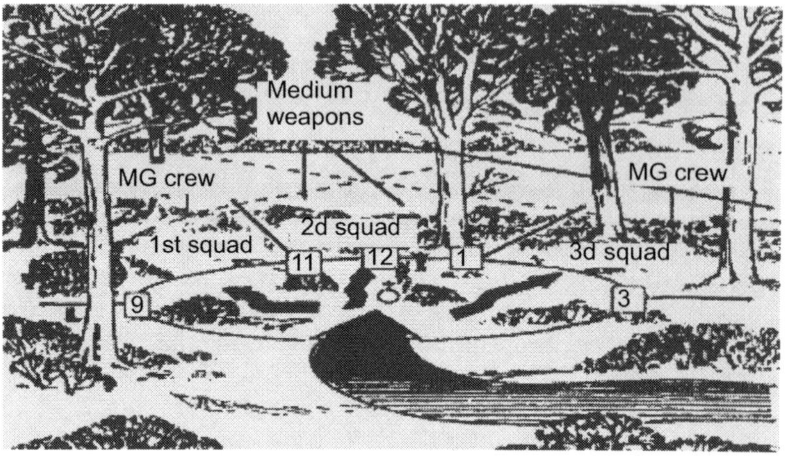

Figure 3-11. Clock Technique

- Terrain feature technique. In a similar manner, the leader identifies obvious terrain features *(Figure 3-12, page 3-58)* as the left and right limits for squads. In both techniques, he ensures that squad sectors of fire overlap each other and provide mutual support for adjacent units.

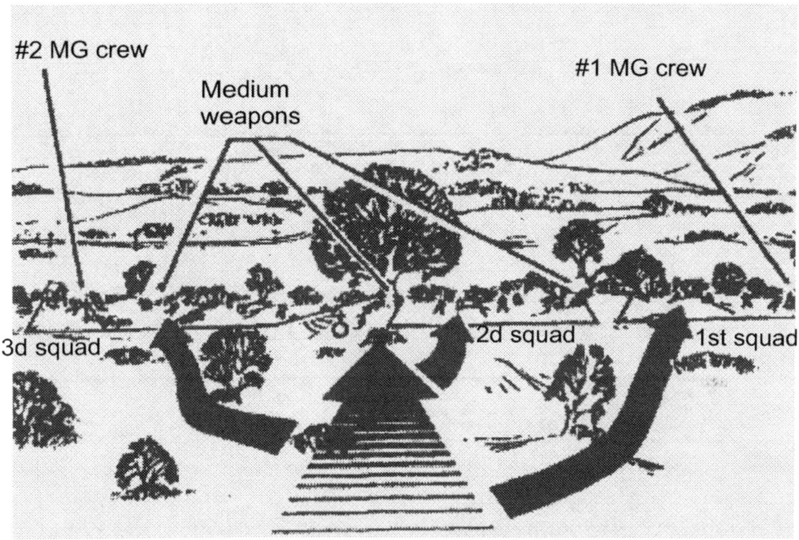

Figure 3-12. Terrain Feature Technique

3-126. Once platoons or patrols have consolidated, they begin to reorganize. Platoons reorganize to continue the mission. Reorganization involves the following:

- Reestablishing C^2.
- Remanning key weapons and redistributing ammunition and equipment.
- Clearing the objective of casualties and EPWs.
- Assessing and reporting the status of platoon personnel, ammunition, supplies, and essential equipment.

COMMUNICATE

3-127. On the battlefield, you must be able to communicate. Communication is the means through which battle command is exercised. MP on the battlefield must be able to communicate to maintain C^2 of their elements, call for fire or request other support, and respond to orders. The chain of command and succession of command must be known throughout the organization. There must be open lines of communication up, down, and laterally. Situations on the battlefield can change rapidly, and losing the ability to communicate for even a short duration can have a major impact on an operation. Commanders must provide for redundancy in means of communications. MP companies have the capability to place backup communication systems at key locations within an area of responsibility. METT-TC and the battlefield situation usually determine communication means. MP use any combination of the following systems:

- Sight and sound.
- Messenger.
- Wire and radio.

USE SIGHT-AND-SOUND SIGNALS

3-128. Visual signals are useful for sending prearranged messages over short distances, during radio silence, or when jamming interferes with radio transmissions. Arm-and-hand signals, flashlights, and pyrotechnics may send quick visual signals. Visual signals have some disadvantages, which include the following:

- They are less effective when visibility is limited.
- They may be seen and intercepted or imitated by the enemy.
- They may be masked by terrain features, reducing the chance of a message being received.
- They are easy to misunderstand.

NOTE: To overcome this last disadvantage, each man in the unit must be able to send, receive, and understand messages using visual signals.

3-129. Although arm-and-hand and light signals are standard throughout the Army, the meaning of pyrotechnic signals must be set in the C^2 portion of the OPORDs and in the SOI. To ensure that a pyrotechnic message was correctly received, confirm the message by some other means as soon as possible. For more information on visual signals, refer to *FM 21-60*.

3-130. Sound signals, like visual signals, work well only for short distances. Simple devices (such as whistles, horns, gongs, and explosives) may be used. Sound signals can be used to—

- Attract attention.
- Transmit prearranged messages.
- Spread alarms.

3-131. A well-known sound signal is the use of metal-on-metal to indicate an NBC hazard or attack. Battlefield noise may blend with or override sound signals, causing confusion and misunderstandings. Sound signals—

- Must be simple to understand.
- May be restricted for security reasons.
- Can be intercepted by the enemy.

USE MESSENGERS

3-132. Using messengers is the most secure way to communicate long messages and documents. However, it is also the slowest and messengers are vulnerable to enemy action. When using a messenger—

- Put the message in writing.
- Make the text clear, concise, and complete.
- Choose the most expedient transportation on hand.
- Encode the message (using the operational code in the SOI) if there is a risk that the messenger might be captured.

- Send a second messenger by a different route if a backup message is needed.

USE WIRE AND RADIO COMMUNICATION

3-133. Often, wire communication is more useful than radio. It is hard to jam, and unlike radio, more than one person can talk at one time. It is used most often for communicating with static posts. Wire communications cannot be secured and may be cut by the enemy. When a wire line has to be checked, MP are sent out in pairs. One MP looks for the cut and the other MP provides overwatch security. The enemy can take prisoners by cutting a line and capturing the soldiers who go to repair it.

3-134. Use a radio to communicate with mobile or distant elements. A secure voice radio is best. The enemy can intercept messages on an unsecured radio net. Regardless of the radio type, if the transmission is heard, the enemy may be able to detect the location of the radio or learn what the unit is doing. Keep the transmissions short. Know and use signal security and electronic counter-countermeasures. MP deny the enemy information from friendly telecommunications by following the SOI. To keep transmissions secure, MP—

- Authenticate all transmissions.
- Use only authorized codes.
- Use secure voice transmissions.
- Use encoded messages.

3-135. The enemy is kept from disrupting radio communications, and friendly emitters are protected from enemy detection, location, and identification by—

- Setting radios at low power.
- Placing antennas where terrain blocks the enemy's interception.
- Using remote radios and antennas.
- Using directional antennas.

- Using wire whenever possible.
- Observing listening silence.
- Using short transmissions.
- Using a random transmission schedule.
- Transmitting only when there is a need to do so.

Chapter 4

Combat Operations

This chapter provides the MP leader with the information needed to successfully complete a combat mission. When MP platoons conduct the tasks addressed in this chapter, they will mirror the actions of the company.

PREPARE FOR COMBAT

4-1. Units that are not directly engaged in combat often find it necessary to travel in order to position themselves for combat operations. During these movements, the battalion supports the company and the company supports and monitors the platoons with the movement plan. The move's success depends on the unit's discipline and ability to execute the plan. (Refer to *Appendix E* for precombat inspection checklists.)

MAINTAIN AND MOVE A COMBAT LOAD

4-2. Combat load is the quantity of supplies, in all classes, that the company must have on hand to sustain operations for a prescribed number of days. The company must be capable of moving the combat load, using organic transportation assets, into combat in a single delivery. To save time, the company combat loads vehicles while the quartering party readies the new site. The combat load ensures that a unit is ready for combat even when it is on the move.

4-3. The principles of combat loading are standard. All equipment, ammunition, and gear are loaded on the vehicles in a logical order and put in predesignated places. Knowing the location of each item allows for quick retrieval during the move. Combat loading also lends to a fast set up at the new site. Like the basic load, the company's combat load is mission-dependent. No single load plan can satisfy all the situations. MP leaders must consider the following:

- METT-TC.
- Vehicle and trailer capacities.
- Weight limits of the unit's vehicles and trailers, being careful not to overload them.
- Whether or not the equipment will fit (cube out). For equipment data, see the applicable technical manual (TM).

4-4. Unit SOPs has load plans tailored for various mission activities. Having a choice of load plans for various deployments reduces the load time. Load plans and diagrams are modified to suit METT-TC and vehicle and trailer capacities. The modifications are shown on the load diagram in the vehicle. HMMWVs may be loaded in many configurations, which include—

- Loading the basic equipment in the mounted standard brackets on the vehicle.
- Modifying and moving the brackets to meet mission or unit requirements.

PLAN A TACTICAL ROAD MARCH

4-5. The basic considerations in planning any road march are the METT-TC factors (especially the enemy situation and the mission); the march order; and the type, number, and characteristics of vehicles available for the movement. A company conducts a tactical road march for relocating and facilitating rapid movement of the vehicles with a prescribed speed and interval between the vehicles. When preparing for a tactical road march, the company uses the following planning sequence when time permits:

- Prepare and issue the WO as early as possible to allow maximum time for preparation.

- Prepare an estimate of the situation, specifying the organization of the march column.
- Organize and dispatch reconnaissance and quartering parties.
- Prepare detailed movement plans based on the organization of the march column and a review of the available reconnaissance information.
- Use the reconnaissance information to—
  - Choose sites for halts and RPs.
  - Spot problem areas along the route.
  - Select bypasses or alternate routes.
- Select fairly secure locations for halts.
- Choose areas that provide cover and concealment.
- Avoid choosing highly populated areas, curves in the road, or other hard-to-secure areas.
- Plan the timing so that the unit arrives at the SP just before it is scheduled to cross it. The time a unit must cross the SP is provided to the unit. As other units may be planning to use the route, each unit must cross the SP on time.
- Prepare and issue the march order.
- Prepare the overlays and issue them to the vehicle commanders and subordinate leaders. The road march overlay includes, at a minimum, the location of the SP, RP, scheduled halts, and checkpoints at critical points along the route.

COORDINATE AND DIRECT THE MARCH

4-6. The chain of command controls the column. The march leader—

- Coordinates the road march through the chain of command with the local movement control unit.
- Determines if the convoy needs a movement credit or a clearance to use the given route. If so, the march leader submits a *Department of*

Defense (DD) Form 1265 through the appropriate movement control center.

- Informs higher HQ and the supported units of the dates and times that the operations will stop at the old site and begin at the new site.
- Tasks subordinate leaders to attend a briefing to discuss unit readiness and load plans and forecast support requirements.
- Submits requests for support based on the forecast developed during the briefing. Requests may include fire support, refueling, vehicle recovery operations, and other support needed to complete the march.
- Issues an OPORD for the movement.
- Requests HQ personnel to prepare a movement table. Refer to *Appendix E* of *FM 55-10* for detailed information on the movement tables.
- Requires unit personnel to analyze the route reconnaissance information looking for likely enemy ambush sites.
- Ensures that a strip map, which may be included as an annex to the OPORD, is prepared. The strip map shows SPs; RPs; route numbers; place names; critical points; directional arrows; distances between the points; scheduled halt locations; and petroleum, oils, and lubricants (POL) refill points. Give copies to the unit drivers.
- Ensures that the drivers and assistant drivers are briefed.
- Ensures that radio communication is kept to a minimum during movement.
- Tasks subordinates to ensure that the road march plan is followed.
- Ensures that safety briefings are conducted and understood.

CONDUCT THE MARCH

4-7. The commander sets the conditions under which military traffic moves at night. The march leader ensures that personnel are aware of and abide by the set lighting conditions when the company moves at night. Conditions that are more restrictive may be imposed contingent on the threat environment (such as air raids). Lighting conditions may include normal lighting, reduced lighting, or blackout. If the situation warrants, travel by total blackout (use of night-vision goggles [NVGs]) may be prescribed. More often, travel is under partial blackout, using only enough light to see the road and to be seen by other road users. Minimal lighting reduces visibility from the air while it permits drivers to—

- Travel as quickly and safely as possible.
- Apply brakes in time.
- See the side of the road.

Movement Techniques

4-8. During a tactical road march, the unit leaders (such as the march leader and the platoon sergeant) travel in separate vehicles. This decreases the chance of a unit's primary leaders being lost in one enemy action. The convoy moves en route by close or open column march or by infiltration. In dusty conditions, space the vehicles so that the dust from one vehicle does not blind the driver of the next.

4-9. **Close Column**. In a close column, the elements are close together. Use a close column for marches during limited visibility, blackout conditions, and radio silence. Under these conditions, space the vehicles so that the driver can see the two lights in the blackout marker of the vehicle ahead. Visibility determines the set distance between the vehicles. A close column—

- Reduces the time it takes for the column to pass points on the route.

- Requires fewer guides, escorts, and markers for control than an open column does.
- Enhances movement through congested areas or over poorly marked routes.

4-10. **Open Column**. In an open column, elements are widely spaced as a passive defense measure. Generally, an open column is used during daylight, but may be used at night with infrared lights, blackout lights, or night-vision equipment. A distance of 50 to 100 meters or more may be designated between vehicles depending on METT-TC. Use an open column—

- When enemy contact is likely.
- To enhance security.
- Over dusty roads. Reducing dust is especially important when moving through areas contaminated by radioactive fallout.

4-11. **Infiltration.** Infiltration is the best passive defense against enemy observation and attack, although it may be difficult to control. To move by infiltration, vehicles are dispatched one at a time or in small groups at irregular intervals to keep traffic density low and to prevent undue massing of vehicles. Use infiltration—

- When time and road space allow.
- When maximum security, deception, and dispersion are needed.
- To maintain security during the march. When the unit approaches likely danger areas (such as bridges and tunnels), have one or more teams dismount. The teams check both sides of the road before the vehicles pass. This is critical when only a map reconnaissance was conducted before the move.

Movement Considerations

4-12. Consider events and control measures that are used during the move in order for the tactical road march to be successful. These measures include the SP and RP, checkpoints, restrictions, speed control, halts, disabled vehicles, and mined areas.

4-13. **Start Point**. A SP gives the vehicles of a march column a common point for starting their movement. When vehicles use more than one route, each route has an SP. The SP is a recognizable place along the route of march (such as a road intersection). It should not be in a defile, on a hill, or at a sharp curve in the road that could cause movement to slow. Ensure that the SP is far enough away from the AAs to allow the vehicles to be organized and move at the prescribed speed when they reach it. Elements of the company reconnoiters the route to the SP to determine the times needed for the serial to arrive at and clear the SP before starting the march.

4-14. **Release Point**. A RP provides all the vehicles of the march column with a common point for reverting to the commander's control. It is a point on the march route that is easy to recognize on the map and on the ground. Guides meet the vehicles as they arrive at the RP and lead them to their new positions, usually in an AA. Multiple routes and cross-country movement from the RP to the assembly areas allow vehicles to disperse rapidly. When leaders select a RP, avoid hills, defiles, and sharp curves that may cause elements to slow or stop on the route. Ensure that vehicles are not required to countermarch or pass through another element to reach their new position.

4-15. **Checkpoints**. Use checkpoints on a route for reference when providing instructions and identifying places where interference with movement might occur or timing may be critical.

4-16. **Restrictions**. Restrictions are points along the march route where the movement may be limited or obstructed during certain time periods (such as bridges, intersections, ferries, or bypasses). The march planner—

- Starts the move early enough to pass such a point before a restriction begins.
- Delays the start of the move to pass a restriction after it has ended.
- Plans to halt the column along the route until the restriction is lifted.

4-17. **Speed Control**. Vehicles in a column of any length may simultaneously encounter different types of routes and obstacles. This causes sections of the column to move at different speeds at the same time, producing an undesirable accordion or whip effect. The movement order specifies the march speed, march rate, and the maximum safe catch-up speed to reduce *column whipping*. The lead vehicle must not exceed the authorized maximum speed of the slowest vehicle in the column. To minimize vehicle congestion on the nearside of an obstacle, vehicle commanders and drivers must be alert and maintain the prescribed minimum following distance. Vehicles should make only gradual speed changes while maintaining their prescribed interval. Vehicle commanders must constantly be aware of the vehicle interval to their front and rear and adjust their speed accordingly.

4-18. **Halts**. Halts are conducted for various reasons. They—

- Allow following traffic to pass.
- Provide time for rest, mess activities, and personal comfort and relief.
- Permit refueling and maintenance of vehicles.
- Provide for maintenance and inspection of equipment.
- Allot time for adjustments in the schedule.

4-19. The time and duration of the halts are usually specified in the movement order or prescribed in the unit SOP. The SOP prescribes actions to take during halts. A short rest halt of 15 minutes is usually taken after the first hour of marching. A 10-minute short halt is taken every 2 hours thereafter. The prescribed march rate includes the time required for short halts. When possible, march elements using the same route should stop at the same time. Route characteristics may make it necessary to halt at a particular point on the route rather than simultaneously at a fixed time.

4-20. Long halts are planned in advance. The length of the halt is added to the total travel time. Locations for long halts are normally selected to allow all the vehicles to clear the road and permit proper dispersion. The unit commander schedules halts for refueling in advance.

4-21. The herringbone formation is used to provide security for the march column during temporary halts. During temporary halts, the MP teams move their vehicles to alternate sides on or off the road in a herringbone pattern that lets vehicles pass down the center of the column. Movement commanders give permission for execution of unscheduled halts. The first priority at any halt is local security. OPs are established and sectors of fire are assigned to each vehicle. These actions should be automatic and part of the unit SOP.

4-22. **Disabled Vehicles**. Disabled vehicles must not obstruct traffic. Their crews must move them off the road and report their status immediately to the PSG. Crews must immediately signal the follow-on vehicles to bypass and continue movement. They then establish security and post guides to direct traffic. If possible, crews repair their vehicles and rejoin the rear of the column just ahead of the trail element. Vehicles that have dropped from the column return to their positions only when the column has halted. The trail party recovers vehicles that cannot be repaired by their crews.

4-23. **Mined Areas**. When a company encounters mined areas, it must remember that the safety of the unit is the most important factor. It bypasses mined areas whenever possible, but considers how the delay will affect the outcome of the mission. Remember to—

- Be cautious. Mines may be used to force an element to take an alternate route into an ambush site.
- Screen the bypass route, if possible, before diverting an element.

NOTE: Refer to *Appendix H* for countermine operations.

4-24. Ensure that all efforts are made to bypass mined areas; however, if the element must cross a mined area when engineer assets are not available to breach the minefield, act quickly and cautiously. Mined areas, like other obstacles, are often covered by enemy fire. Before crossing—

- Detonate mines from a protected position.
- Detonate mine trip wires by rigging an object near the trip wire to fall on the wire.
- Use a hand grenade or direct fire to detonate mines.
- Detonate pressure-sensitive mines by rigging an A-frame over the mine and placing a heavy object, attached to a rope, over the mine. Take cover and allow the object to fall on the mine.
- Devise other methods to detonate the detected mines.
- Send a report to the next higher command when mines have been neutralized. Refer *FM 20-32* and *FM 21-75* for information on mines and countering mines.

ESTABLISH A NEW OPERATIONAL SITE

4-25. MP elements most often will collocate as part of an established base or base cluster. On occasion, MP may need to set up a base of their own. To set up at a new location, whether as part of an established base or base cluster or separately as a company or a platoon base, you must—

- Reconnoiter new sites.
- Pick the most favorable site and its alternate. Choose a site that—
  - Is easily accessible.
  - Can accommodate all the unit's vehicles and equipment.
  - Has a firm, well-drained surface.
  - Has some natural cover and concealment.
  - Is relatively easy to defend.
- Prepare and secure the site.
- Complete the move and establish communication.
- Establish local security to sustain survivability.

USE A QUARTERING PARTY

4-26. A quartering party is needed whenever a unit relocates. The quartering party's mission is to—

- Reconnoiter the area for enemy presence, booby traps, NBC contamination, and other hazards.
- Establish the dismount point and perimeter security.
- Establish communication.
- Identify the company CP.
- Designate vehicle locations.
- Position guides at the RP to meet the main party.
- Prepare the area for occupation and assist the units with occupation.

4-27. The platoon leader or PSG designates vehicles and personnel from the platoon that will be part of a battalion or company quartering party. The entire platoon may serve as the battalion or company quartering party.

4-28. While the unit loads for deployment, the quartering party moves to and readies the new site. The quartering party's job ends when the last vehicle in the main body arrives at the new site. The size of a quartering party is based on the tactical situation and the amount of work required to prepare the site for occupancy. A quartering party for platoon relocation would be smaller than that of a company. A company quartering party is likely to have personnel from the unit's HQ, each platoon, the maintenance and dining sections, and communications.

4-29. The quartering party leader ensures that the equipment and supplies are available to clear, secure, and set up the new site. A quartering party may need—

- NBC detecting and monitoring equipment.
- Mine detectors.
- Saws or axes to clear the wooded areas.
- White engineer tape.
- Portable route signing material.

4-30. The quartering party leader assigns tasks to the teams based on the size of the quartering party, the work to be done, and METT-TC. He ensures that each team has the equipment needed to complete its tasks (refer to *FM 7-10*) and that they are at the proper mission-oriented protection posture (MOPP) level if they are operating in an NBC environment.

March Halts

4-31. At march halts, quartering party teams set up local security. If the vehicles can leave the road, the teams form a 360-degree perimeter around the convoy. If the vehicles cannot leave the road, they are parked at an angle so alternate vehicles face opposite sides of the road. Each team is assigned a sector to observe, with sectors overlapping between the vehicles. Each team member has a specific area of responsibility. The troops remain alert and ready to take action on contact with the enemy. All personnel watch for enemy aircraft.

Site Arrival

4-32. When the quartering party arrives at the site, it clears and then secures the site. One or more teams, after dismounting their vehicles, search the area for mines, booby traps, items of intelligence value, or other signs of enemy presence.

4-33. If nuclear weapons have been used, at least one team using radiation detection, indication, and computation (RADIAC) meters monitors the site for radioactive contaminants. Monitoring for chemical and biological agents must be continuous because it is difficult to detect their first dispersal.

4-34. In urban areas, team members clear buildings that will be used by the unit. Team members clear the structures outside the perimeter if there is a possibility of enemy presence. The priority of buildings to be cleared and the number of teams needed are based on METT-TC. Refer to *FM 90-10-1*.

4-35. When the area is cleared, one or more teams perform the following functions:

- Set up the OPs and the LPs.
- Set up defensive positions on likely enemy avenues of approach. These positions provide early warning and limited protection during the occupation of the new site.
- Prepare the new site for the main body's arrival.

Company Move

4-36. When setting up a company site, the quartering party—

- Chooses a tentative location for the company CP.

- Sets up the company CP where it can best control the company, be well defended, and have LOC to the subelements.
- Uses buildings (in an urban area) to conceal the CP.
- Considers cover and concealment when choosing the CP location.
- Makes use of natural cover and concealment when possible.
- Uses camouflage screens and man-made cover and concealment where needed.
- Sets up the wire communication net. Marks those areas where other unit elements will be positioned, using signs or materials that cannot be easily seen by the enemy.
- Picks roads and trails that permit an easy flow of traffic.
- Chooses alternate exits and marks them for use as emergency exits.
- Designates parking areas for the heaviest, most awkward vehicles, such as 5-ton trucks.
- Selects a troop area and—
    - Marks the areas where latrines, garbage dumps, and tents will go.
    - Uses ground guides for vehicle movement in areas where troops are sleeping. (For safety, unit personnel should sleep only in the troop area).
    - Chooses a structure (in an urban area) that protects the troops from natural elements and has adequate latrine facilities.
- Locates the following:
    - The food service section inside the perimeter, well away from interior roads to keep dust from contaminating the food. Locate the serving line to take advantage of

cover and concealment. In urban areas, select a building for service and meal activities.

- The latrines away from the bivouac area. Place latrines at least 30 meters down slope from wells or other water sources, and at least 100 meters from the dining facility, downwind and down slope, if possible. In urban areas, use existing latrines if they can serve at least 8 percent of the unit at one time.
- The maintenance section where vehicles can arrive easily from the main road through the site. Ensure that vehicles are able to enter the maintenance tent at one end and exit at the other. Use existing garages for maintenance operations in urban areas.
- The supply section to meet space, roadway access, and drainage needs. In urban areas, use warehouse-type buildings for supply operations.
- The tactical communication section where it has space enough to support the whole operation. Usually it collocates with the maintenance section or the operations section.

4-37. When the main body arrives, the quartering party—

- Maintains security as the main body moves into the site.
- Maintains noise and light discipline.
- Ensures that the vehicles rapidly clear the approach route while maintaining vehicle intervals.
- Dismounts all personnel except drivers at the dismount point.

- Briefs the leader of the main body on the situation and the operational status.

4-38. The leader of the main body informs the higher HQ that the move has been completed. He reports the location coordinates for both the CP and the alternate CP by messenger or other secure means. The leader ensures that the entire party prepares fighting positions and other defensive measures.

Platoon Relocation

4-39. A quartering party, in advance of the platoon's relocation, has the same considerations scaled to size and need, as one in advance of a company. An MP platoon may collocate with a company HQ or an existing base. However, more often, a platoon's base must be set up where a platoon HQ can best—

- C^2 its squads.
- Communicate easily with its squads and higher HQ.
- Link squads, the company CP, and the supported unit.

4-40. The platoon HQ can operate from a static base or vehicles. If the platoon elements are going to operate in one location (as they would for an EPW holding area), the platoon leader sets up a static platoon HQ base. If the platoon elements must operate dispersed over a large area, the platoon leader must remain mobile. In such cases, a platoon leader could elect to set up a temporary platoon base as a rally point to report, resupply, and reorganize the platoon's resources.

4-41. The platoon bases are set up the same. The PSG selects a site that offers good cover and concealment. The site must be defendable and allow the HQ vehicle to be located near the tent. A small tent houses the platoon's HQ. Use a radio set control group to relay remote communication into the tent. An antenna increases the transmission distance and is located based on OPSEC principles. Wire communication is limited to the platoons that can hook into an existing wire net.

CONDUCT MILITARY POLICE BASE SELF-DEFENSE

4-42. When collocated with a base or a base cluster, the platoon is integrated into that base's or base cluster's self-defense planning and operations. Although bases and base clusters are more prevalent at corps and echelon above corps (EAC), the same principle applies to MP located at the division or brigade support areas. When an MP base is set up on its own, the base is responsible for its own security and protection.

Collocated

4-43. An MP platoon collocates with a base or a base cluster for logistical support and a means to conduct operations. When collocated, the platoon leader coordinates with the base defense operations center (BDOC) and the base cluster operations center (BCOC) to integrate the platoon's efforts with the base's and base cluster's efforts. The platoon's portion of the base's or base cluster's defense is to help provide early warning of the threat by area security and/or maneuver and mobility support in the area near the base or the base cluster. Because MP resources are austere, the platoon only shares sector efforts on the base's perimeter. MP are used as static posts (such as gates) only under extreme conditions. An MP platoon may be tasked to serve or augment the base cluster's response force. Before the platoon leader accepts this tasking, he consults with the company commander to—

- Ensure that the company commander knows that he may have one less platoon if he is tasked for the TCF or response force for the battalion.
- Allow the company commander to advise the base commander that the MP platoon could

serve as a response force to the base if it is not committed.

4-44. Each base has a BDOC that plans, coordinates, and supervises base defense operations. The BDOC initiates contingency planning that enables the base to—

- Increase the manning posture of the base contingent on the threat.
- Detect and defeat the threat within their capabilities.
- Hold against heavier enemy forces until response forces arrive.
- Maintain control of the fight within the base.
- Support the fire and movement of the response force operating outside the base.

4-45. Each base cluster has a BCOC to monitor base defense plans and establish the base cluster reaction force. The BCOC—

- Provides C^2 of the resources for planning, coordinating, and supervising the base cluster's defense.
- Coordinates base defense operations.
- Maintains communication with bases within the cluster as well as MP, BDOCs, and the sustainment area operations center. A great deal of intelligence is provided to a BDOC and BCOC through the rear operations net, which helps in planning the defense.

4-46. The platoon's plans for the interface of MP support into the base's self-defense plans address—

- Cover and concealment of personnel and equipment.
- Signal security.
- Reliable and redundant communication systems at all guard locations (such as land lines, radio

links to the BDOC, and telephone hookups to the center switch).

- Deception.
- Contingency planning.
- Improvement of base defense positions.
- Assistance of the area MP.
- Coordination with the BCOC or rear-area operation center (RAOC), as required.
- OPs/ LPs.
- Noise and light discipline.
- Immediate reaction to enemy threat or attack.
- Rehearsals of defense measures.

4-47. All plans and overlays depicting MP support are forwarded to the BCOC. There they are consolidated and forwarded to the RAOC. (If a base is not part of a base cluster, the base forwards all the plans and overlays directly to the RAOC.)

Separate Setup

4-48. Set up an MP platoon separately only when there is no other alternative. This is the least desirable means for a platoon to set up. When the platoon sets up as a base separately, it must be able to defend against a wide range of enemy activity. It integrates the defense of its base (including indirect-fire systems, air defense artillery, and tactical aircraft) with the defense efforts of other bases in the sustainment area. Engineers, dismounted troops, armored vehicles, and helicopters contribute to the overall security of the bases. Bases coordinate and synchronize their defense efforts to enhance their strengths and reduce their vulnerabilities. A base's defense priorities include—

- Establishing initial base security.
- Positioning crew-served weapons and troops on assigned sectors of fire.
- Clearing fields of fire and preparing range cards.

- Preparing fighting positions.
- Installing communication.
- Emplacing obstacles and mines.
- Improving primary fighting positions to include overhead cover.
- Preparing alternate and supplementary positions.
- Stockpiling ammunition, food, and water.
- Preparing routes and trenches between positions.
- Developing a counterattack plan.

4-49. Using the IPB process can help the platoon predict threats to base security. Platoons must be aware of the enemy's location, organization, direction of movement, and strength. In the same regard, platoons must have effective OPSEC to deny similar friendly information to the enemy. Platoons can continually improve base defenses by considering what avenues of approach and methods of attack the enemy could use, given the vulnerabilities of the platoon's base. The base defense plan must have overlays depicting weapons positions, sectors of fire, final protective fires, and reaction force contingencies. Plans are updated as often as possible.

4-50. The base's reaction force efforts are coordinated with the designated-area response force. Platoon leaders develop detailed employment plans and exchange as much information as possible with the response force and TCF commander before they are needed. Although the base's reaction force usually would not fight beyond the perimeter of the base, the reaction force must be ready to assist the response force or TCF when it arrives. The following should be considered:

- Command relationships before, during, and after linkup.
- Coordination of fire support before, during, and after linkup.

- Recognition signals and communication procedures to be employed.
- Follow-on operations required.
- Area damage control.

Setup in a Hide Position (Temporary Defensive Position)

4-51. If the platoon steps down from sustained continuous operations and cannot return to its base or base cluster, it may need to operate briefly from a temporary defensive position. When used properly, a temporary defensive position can enable the platoon to rest, recover, repair damaged equipment, and plan for future operations. It offers concealment with little chance of detection by the enemy. Platoons want to get the best security they can while tasking a minimum of soldiers to provide security. When the platoon leader decides to operate from a temporary defensive position, he notifies the company HQ.

4-52. Locate the position in or near the area of normal operations so that sustained operations can be resumed immediately, on order. METT-TC should be a primary concern. Easily defendable positions are preferred over those that are more difficult to defend. Ensure that the position has more than one exit route and provides communication with the next higher HQ that are enhanced or at least not interfered with by terrain. While built-up, urban areas afford suitable concealment for temporary defensive positions, it is essential that the capability to communicate be assessed before the selection of such a site. Other considerations of a temporary defensive position include—

- Keeping vehicles secure and available in a nearby location.
- Positioning vehicles so that key equipment can be moved or removed without displacement of the entire unit.

- Concealing equipment from the sides and overhead. This prevents detection from aerial observers and some side-looking airborne radar.
- Covering and concealing to reduce security and/or defense requirements.
- Providing enough space between vehicles to allow a vehicle to bypass another vehicle that is inoperable.
- Ensuring that the platoon follows signal security and uses noise and light discipline.
- Setting up fighting positions if the situation calls for them.

SET UP LOCAL SECURITY

4-53. Self-defense planning and coordination is done as soon as the base is set up. Prior planning and mission analysis are essential elements of a base defense. MP must be able to defend the site even before occupation is complete. When an MP element is located as part of an established base, it helps defend a portion of the larger unit's perimeter. Elements that set up separately usually defend their sites by deploying in a 360-degree perimeter. The techniques and principles of defense are the same for defending a separate squad, platoon, company, or base. To plan a perimeter defense, evaluate the situation. Analyze the terrain in terms of observation and fields of fire, cover and concealment, obstacles, key terrain, and avenues of approach (OCOKA). Defenses are placed where the threat is greatest.

4-54. The platoon leader establishes the CP and the OP. He locates the CP and the OP where he can best see and control the platoon. If this is not possible, he locates it where it can cover the most likely avenue of enemy approach. An alternate CP and OP, operated by the PSG, is placed where it can control the portion of the perimeter that cannot be seen or controlled by the main

CP and OP. The platoon leader then decides what other security measures and means of communication to use.

4-55. Platoon leaders must plan more than cover and concealment to counteract threat infrared, radar, thermal, and other sensors. The platoon leader uses the principles of camouflage and counters the recognition factors that make an object stand out from its background by—

- Locating soldiers, equipment, or structures where they are least discernible. This alone can reduce or eliminate many recognition factors.
- Using any mix of hiding, blending, disrupting, or disguising that conceals visibility.
- Maintaining camouflage discipline continuously.

4-56. When the number of soldiers that will defend a 360-degree perimeter is small, the platoon leader must—

- Vary the size of the defensive sectors.
- Identify alternate fighting positions.
- Retain flexibility of thinking.
- Decide what equipment—
  - Is needed to set up a perimeter defense.
  - Should stay in the vehicles.
  - Must be requisitioned or picked up later.

4-57. The equipment to improve defensive positions includes concertina wire, sandbags and tape (for cover and concealment), trip flares, pyrotechnic devices, mines, or PEWS, and other lethal and nonlethal technologies.

4-58. Platoons must be able to defend during day or night, in reduced visibility, and in a variety of weather conditions. The platoon leader or sergeant ensures that the platoon has the equipment it needs to defend under these conditions. The platoon must know how and when to use the equipment. During reduced visibility—

- Take steps to keep the enemy from observing or surprising the platoon.
- Require OPs and LPs. There should be at least one OP and one LP per squad. OPs and LPs report the enemy's advance and call for illumination and supporting fire. As in a daylight defense, MP manning OPs and LPs withdraw before they become engaged in close combat.
- Use patrols, illumination, PEWSs, and NVDs to help detect the enemy's advance.
- Use trip flares to provide warning and give some illumination. As a rule, do not fire until the targets are visible.
- Use camouflage, movement control, and light and noise discipline.
- Limit radio traffic to essential information.
- Ensure strict fire control to keep from disclosing the fighting positions.
- Ensure that gunners with crew-served and antiarmor weapons use NVDs.
- Provide illumination by using handheld flares or grenade launchers with illuminating rounds. Added light may be provided by fire support.
- Ensure that platoon leaders plan the use of messengers, visual signals, personal contact, or whistles to communicate with the squad leaders. Squad leaders plan to communicate with their team leaders and teams using personal contact or sound and visual signals.

CONSTRUCT FIGHTING AND SURVIVABILITY POSITIONS

4-59. Fighting positions help protect soldiers and their equipment from the enemy and from the enemy's small-arms fire and fragmentation weapons, while allowing soldiers full weapon system engagement. A fighting position provides cover and concealment from which to engage or defend against the enemy. See *FM 7-8* for individual fighting position construction.

4-60. Fighting positions do not protect against the destructiveness of artillery and other area weapons. Nevertheless, a dug-in fighting position may be the key to survivability. Digging in cannot remove a soldier's vulnerability by itself. It does reduce exposure to the enemy's acquisition, targeting, and engagement systems. Platoons must be able to construct their survivability positions, often without engineer assistance.

4-61. Locate fighting positions for crew-served weapons where gunners can stop dismounted attacks. The sectors of fire must cover infantry avenues of approach and provide the most grazing fire across the platoon or squad front. Overlap the sectors of fire with each other and those of adjacent squads. Prepare fighting positions so that their primary sectors of fire have the guns firing across the front of the unit. Prepare secondary sectors of fire so that the guns fire to the front.

4-62. Usually, one MP team occupies an M249 or MK19 fighting position. One member is the gunner, one is the assistant gunner, and one is the ammunition bearer or rifleman. Each gunner has a primary and a secondary sector of fire. The gunner fires in the secondary sector only on order or when there are no targets in the primary sector. Each gunner uses aiming stakes to set his weapon for a final protective line (FPL) or a principal direction of fire (PDF) within the primary sector. The FPL and PDF are control measures to help defend a position. In an attack, the gunner knows the primary areas. He engages the greatest threat and, on the order of the platoon leader or PSG, fires the FPL.

PREPARE SECTOR SKETCHES

4-63. After the crew-served weapons are in position, the squad leader positions the remaining MP to protect the gunners and to cover areas not covered by the gunner's. Using the range cards, the squad leader makes a squad sector sketch. He includes a rough sketch of the terrain around the weapon *(Figure 4-1)*. The squad sector sketches are used to plan defense and to control fire. Squad sector sketches show the following:

- The main terrain features in each sector of fire and the ranges to the features.
- Each primary fighting position.
- The primary and secondary sectors of fire for each position.
- MK19 and M249 FPL or PDF.
- The type of weapon at each position.
- The OP and LPs and squad leaders' positions.
- Dead space.
- Mines and obstacles.

4-64. The squad leader checks the range cards and the squad sector sketch for gaps or other flaws in the fire plan. He adjusts the weapons or the sectors as necessary. If the squad leader finds dead space, he takes steps to cover it with mines, grenade-launcher fire, or indirect fire. He then prepares two copies of the squad sector sketch. He keeps one copy and forwards the other copy to the platoon leader who makes a platoon sector sketch *(Figure 4-2, page 4-28)*. The platoon sector sketch shows the following:

- Squad sectors of fire.
- The crew-served and antiarmor weapons positions and sectors of fire, including FPL or PDF for the crew-served weapons and target reference points for the antiarmor weapons.
- Positions of the mines and the obstacles.

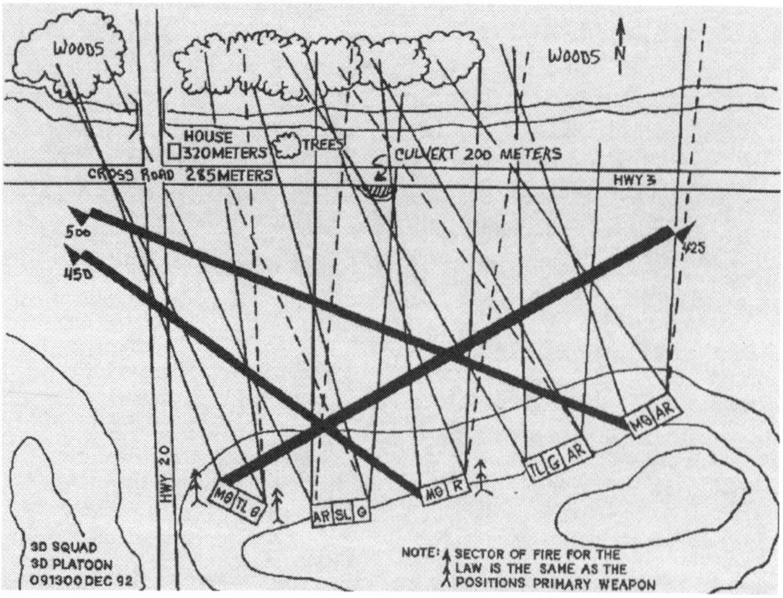

Figure 4-1. Squad Sector Sketch

- Indirect fire planned in the platoon's sector of fire.
- The OP and LPs and patrol routes (if any).
- The platoon CP and OP.

4-65. The platoon leader coordinates with the nearby units. He usually coordinates from left to right and from front to rear. The fires of units within the perimeter must be closely coordinated with the platoon's defensive fire plan. Squad leaders coordinate their fire plans with adjacent squads. All positions and units near the platoon are mutually supporting. The platoon leader makes sure gaps between the units are covered by fire, observation, patrols, OPs and LPs, or sensors. The units exchange information on—

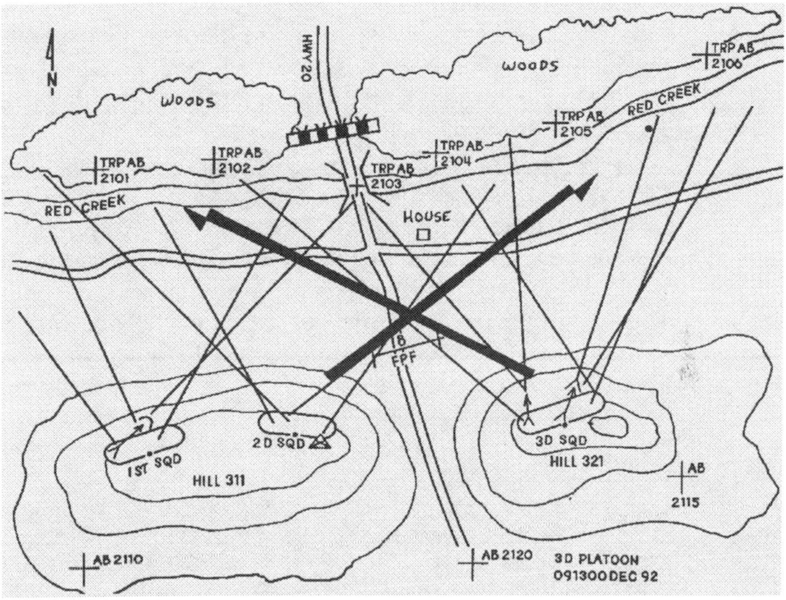

Figure 4-2. Platoon Sector Sketch

- The location of dead space between the elements and how to cover it.
- The locations of primary, alternate, and supplementary positions and sectors of fire for automatic weapons, antiarmor weapons, and subordinate elements.
- The locations of the OPs and LPs.
- The locations and types of obstacles and how they are covered by fire.
- Any patrols to be conducted, giving their size, type, times of departure and return, and routes.

PREPARE RANGE CARDS

4-66. The FPL for the M249 is the line where an enemy assault is to be checked by interlocking fire from all weapons. Use the M249 on the FPL for grazing fire no more than 1 meter above the ground, about hip high, across the front of the element. Use the MK19 or M203 to cover the dead space. To figure the dead space on the FPL, the gunner watches a person walking down the FPL and marks spaces that cannot be grazed. The gunner records all the dead space data on the range card *(Figure 4-3, page 4-30).* He prepares at least two copies of the range card, keeping one card at the position and giving one copy to the squad leader. Fire on a gunner's FPL is its final protective fire (FPF). FPF is usually used as a last resort to stop an enemy assault. All weapons fire on command, continuously, until the call to stop FPF is given.

4-67. When terrain prevents the use of an FPL, the gunner uses a PDF instead. He directs his fire toward the most threatening avenue of approach that leads to his position. His weapon is positioned to fire directly on this approach rather than across the squad's front.

4-68. Construct fighting positions for the MK19 like M249 fighting positions. However, be aware that it takes added effort to keep the M3 tripod from moving because of the recoil of the MK19. If gunners are using the M249 machine gun, they should use the tripod when firing at an angle and the bipod when firing to the front. When gunners change their fires from the oblique to the front, they must move the machine gun, but leave the tripod in place. If gunners are using the MK19, they position the tripod toward the primary sector of fire. However, because there is no bipod for the MK19, gunners must be prepared to adjust both the weapon and the tripod to the secondary sector, if required. After a crew is positioned and is assigned an FPL or a PDF, the team—

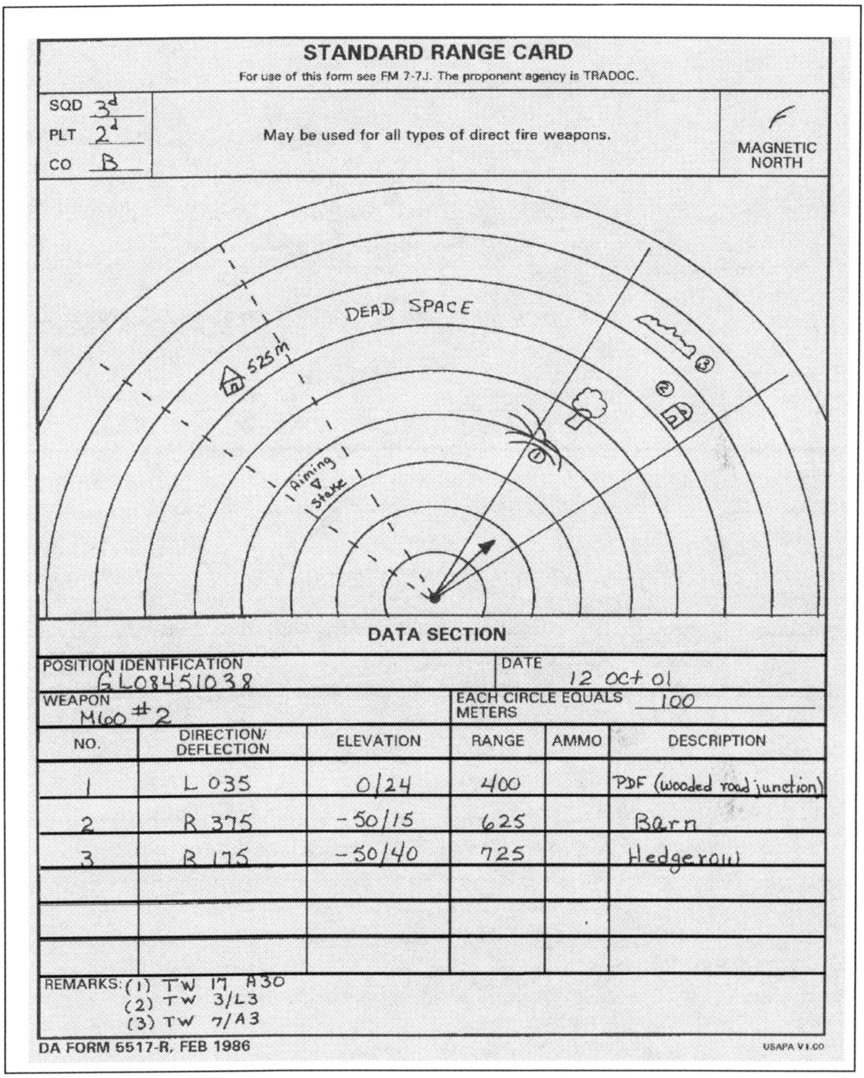

STANDARD RANGE CARD

For use of this form see FM 7-7J. The proponent agency is TRADOC.

SQD 3d
PLT 2d
CO B

May be used for all types of direct fire weapons.

MAGNETIC NORTH

DATA SECTION

POSITION IDENTIFICATION GL08451038

DATE 12 Oct 01

WEAPON M60 #2

EACH CIRCLE EQUALS 100 METERS

NO.	DIRECTION/ DEFLECTION	ELEVATION	RANGE	AMMO	DESCRIPTION
1	L 035	0/24	400		PDF (wooded road junction)
2	R 375	-50/15	625		Barn
3	R 175	-50/40	725		Hedgerow

REMARKS: (1) TW 17 A30
(2) TW 3/L3
(3) TW 7/A3

DA FORM 5517-R, FEB 1986

USAPA V1.00

Figure 4-3. Standard Range Card

- Marks the tripod's position and the limits of their sectors of fire with aiming stakes.
- Outlines the hole.

- Digs the firing platform first. This lessens their exposure if they have to shoot before the position is completely constructed. Dig the firing platform at a level that allows the gun to traverse the sectors of fire.
- Lowers the gun to reduce the gunner's profile. This also reduces the height of the frontal cover needed.
- Digs the hole deep enough to protect itself and still allow the gunner to shoot in comfort (usually about armpit deep).
- Places the dirt where frontal cover is needed.
- Uses the rest of the dirt to build the flank and rear cover when the frontal cover is high enough and thick enough. Sandbags, wire, hatchets, or saws can be useful for building overhead cover or improving the fighting positions.

4-69. The ammunition bearer digs a one-man fighting position to the flank. He positions himself where he can see and shoot to the front and the oblique. Usually the ammunition bearer is on the same side as the FPL or the PDF. From there he can see and shoot into the machine gun's secondary sector. He can also see the gunner and the assistant gunner. The ammunition bearer connects his position to the machine gun position by a crawl trench. This allows him to provide ammunition or replace one of the gunners.

SELECT FIGHTING POSITIONS IN BUILT-UP AREAS

4-70. Planning a defense of a platoon on urban terrain is similar to planning a defense in the countryside. Defensive positions must cover likely enemy avenues of approach, be mutually supporting, and provide cover and concealment. Use AT weapons on mounted avenues of approach. Machine guns cover dismounted approaches. AT4s and M203 grenade launchers work well in built-up areas. They are likely to hit enemy armored vehicles on the top or the side where armor is thin.

4-71. The method of defense (such as in-depth or linear) in the two areas is based on the same considerations. Use obstacles to canalize the enemy into kill zones or to deny key terrain. Orders must be very specific. Due to limited resources, use obstacles to channel, divert, or impede movement.

Position Locations

4-72. Select defensive positions in urban areas based on METT-TC. Often a squad occupies a building, but larger buildings may be defended by a platoon. Select buildings that—

- Are well built. Concrete and steel construction is preferred.
- Have strong floors to keep the structure from collapsing under the weight of debris.
- Have thick walls and floors so that the enemy cannot shoot through roofs and walls to kill defenders.
- Are constructed of nonflammable material. Avoid wood. Strong, fireproof construction provides protection from a nuclear attack as well as conventional firepower.
- Have few glass windows (or break and remove the glass).
- Provide good fields of fire. Buildings located next to vacant lots, alleys, and parks allow better fields of fire than buildings located next to other buildings.
- Allow mutual support between the buildings. No building should be subject to attack without troops in another building being able to provide supporting fire.

4-73. Locate positions so as not to establish a pattern. Avoid obvious firing locations like church steeples (remember the elements of OCOKA) by—

- Placing MK19s in the building where they can cover assigned sectors of fire and FPL.
- Having the squad automatic riflemen and grenadiers cover enemy approach routes to the building.
- Placing most rifle positions at or near ground level to have overhead protection and provide grazing fire on approaches.
- Positioning some MK19 gunners higher to get a longer range. In addition, they can fire into areas that would be dead space for ground-level weapons.
- Positioning AT4s (remember the back blast) so that they can fire down on tracked infantry fighting vehicles and wheeled scout reconnaissance vehicles.

Building Improvement

4-74. Change the outside of the building as little as possible, but inside the building—

- Improve the fighting positions to provide overhead and frontal cover. Use firing ports to avoid enemy observation.
- Cut or blow holes between rooms and floors so the soldiers can move quickly by a covered and concealed route to other firing positions in the building.
- Seal off unused basements to prevent enemy entry.
- Barricade doors, halls, and stairs and take down fire escapes to keep the enemy out of the building.
- Reinforce positions with sandbags, solid debris, beds, furniture, and so forth.

- Screen or block windows and other openings. This keeps the enemy from seeing which windows are manned and throwing hand grenades into the building. When firing from the windows or holes in the walls, be sure the muzzle of your weapon does not protrude beyond the wall. This conceals the muzzle flash.
- Remove combustible materials to limit the danger of fire.
- Turn off electricity and gas.
- Stockpile water and dirt to fight fires.
- Wear armored vests, earplugs, and goggles for protection from dust and debris.

Other Considerations

4-75. Operating in urban terrain can be challenging, so consider the following:

- The employment of weapons is different (shorter ranges).
- The position locations are different, such as LAWs and AT4s go on the upper floors of the buildings (refer to *FM 23-25* for the safety considerations).
- The target acquisition is more difficult (such as, aiming stakes and layered fires are used extensively).
- The CSS will center more on stockpiling materials in positions rather than on traditional resupply methods.
- Controlling indirect fire is more difficult.
- Primary communication must be by messenger, wire, or visual signs rather than radio.
- Avenues of approach are more canalized.
- The three dimensions of the enemy (aboveground, ground level, or below ground)

may be in use. The enemy can easily isolate subordinate units.

- Civilians and fire hazards may be present.

ESTABLISH AND OPERATE AN OBSERVATION POST/LISTENING POST

4-76. OPs/LPs are selected locations from which to look and listen for enemy activity within an assigned area of observation. The OPs/LPs, the primary means of maintaining surveillance of an assigned avenue or a named area of interest (NAI), are positions from which MP observe the enemy and direct and adjust indirect fires against him. From the OPs/LPs, MP send SALUTE reports *(Figure 4-4, page 4-36)* to the commander when observing enemy activity. Use OPs/LPs for the following:

- On key terrain when the surveillance of a specific area is required.
- To prevent the enemy from a surprise attack on other friendly forces.
- As an early warning security measure in a defensive perimeter.
- For the monitoring of likely enemy avenues of approach, drop zones (DZ), and landing zones (LZ).

4-77. OPs/LPs can be performed either mounted or dismounted. A dismounted OP provides maximum stealth and has the greatest likelihood of remaining undetected by the enemy. The disadvantage of the dismounted OP is the time it takes to remount and move if necessary. If rapid movement or displacement is anticipated, the OP mounts or remains mounted.

4-78. A mounted OP/LP offers the advantages of rapid movement and protection because the enemy can easily detect them; however, it is potentially much less effective than a dismounted OP/LP.

SIZE: Give the size of the enemy unit as the number of troops or vehicles seen. Report 10 enemy infantrymen (not an infantry squad). Report three enemy tanks (not an enemy tank platoon).

ACTIVITY: Report what the enemy was doing. "They are emplacing antipersonnel mines in the road."

LOCATION: Report where the enemy was seen. Report the grid coordinates and the direction the enemy was heading. If a map is not available, relate the location to key terrain, such as the enemy's location is "on the Hahn Road, 300 meters south of the Kell River Bridge."

UNIT: An enemy soldier's unit may be hard to determine. Report markings or other distinctive features seen on the vehicles. Some countries have special uniforms and headgear. Some have colored tabs on the uniforms to show the type of unit, or the unit's actions may show its type. The kind of equipment it has may be peculiar to a certain type of unit. For example, a scout reconnaissance vehicle may indicate a reconnaissance unit; an amphibious tracked infantry-fighting vehicle may indicate an airborne unit.

TIME: Report the time the enemy was seen, not the time you are reporting.

EQUIPMENT: Report all the equipment the enemy is wearing or using. If you do not recognize an item of equipment or a type of vehicle, sketch it. Submit the sketch with the report.

Figure 4-4. Salute Report

SELECT OBSERVATION POST/LISTENING POST SITES

4-79. The platoon leader selects the general location for the platoon's OP/LP after analyzing METT-TC factors. From his analysis, he determines how many OPs and LPs to establish. He decides where they must be positioned to allow long-range observation along the avenues of approach assigned by his commander and to provide depth through the sector. Section and squad leaders select the exact positions for each OP/LP on the ground. The OP/LP must have the following characteristics:

- Covered and concealed routes to and from the OP/LP. Ensure that MP can enter and leave their OP/LP without being seen by the enemy.
- Unobstructed observation of the assigned area or sector. Ideally, the fields of observation of adjacent OPs/LPs overlap to ensure full coverage of the sector.
- Covered and concealed positions that are effective. MP select positions with cover and concealment to reduce their vulnerability on the battlefield. MP may need to pass up a position with favorable observation capability, but with no cover and concealment, to select a position that affords better survivability.
- Located where they will not attract attention. Do not locate OPs/LPs in such locations as a water tower, an isolated grove of trees, or a lone building or tree; these positions draw enemy attention and may be used as enemy artillery TRPs.
- Located where they are not silhouetted. Avoid hilltops and position OPs/LPs further down the slope of the hill or on the side, provided there are covered and concealed routes into and out of the position.

MAN THE OBSERVATION POST/LISTENING POST

4-80. Ideally, an MP team should man an OP/LP to maintain team integrity. Position OPs/LPs down the slope or on a flank of a hill, if there are covered withdrawal routes. Each of the OP's/LP's fields of observation overlap those of adjacent OPs/LPs. MP may have to selectively clear fields of observation. Ensure that MP are not seen when entering and leaving an OP/LP. Equip OP/LP teams to observe the area, report information, protect themselves, and call for and adjust indirect fire. OPs/LPs on a defensive perimeter need secure communications. Use of field phones or secured radios are usually best. However, messengers can be used. OPs/LPs may use portable radios to supplement wire communication. One MP observes the area while a second MP provides local security and records and reports information. The third MP rests or provides backup security. The team members switch jobs every 20 to 30 minutes because the observer's effectiveness decreases quickly after that time. The observer needs—

- A map of the area.
- A compass.
- Communication equipment (wire and radio).
- Observation devices, such as binoculars, observation telescope, and NVDs.
- An SOI extract.
- The report formats contained in the SOP.
- A radio (this may be the only means of communication from a remote site like a DZ or an LZ).

POSITION OBSERVATION AND LISTENING POSTS

4-81. Place OPs/LPs either in a linear configuration or in-depth. Linear placement *(Figure 4-5)* allows the platoon to observe the assigned sector from several OP/LP sites, reducing the chance of the enemy entering the sector without being observed. This method works well when the platoon has been assigned a large sector with few avenues of approach or is in desert-type terrain. In-depth OP/LP placement *(Figure 4-6, page 4-40)* allows the platoon to observe the entire sector by placing OP/LP sites where the platoon can observe the most likely avenues of approach in the sector as well as along the sector flanks. This method works well when the platoon is assigned a sector with several avenues of approach or is in heavily wooded terrain. In-depth placement allows for redundancy in observation and better sector coverage.

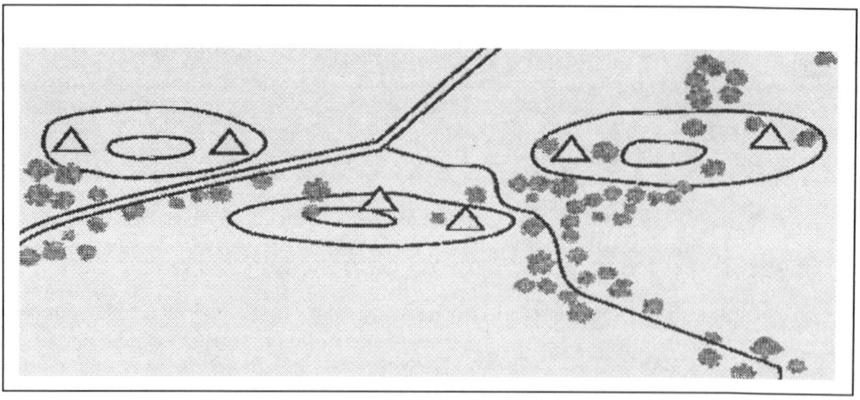

Figure 4-5. Linear OP/LP Placement

4-82. OP/LP team emplacement at night depends on METT-TC factors. The platoon leader places OPs/LPs close to the perimeter and within direct fire range of the defensive perimeter for protection. The team leader designates a specific location and primary direction of fire for the crew-served weapon. The OP/LP team builds a hasty-fighting position or a prepared-fighting position depending on METT-TC. The team leader designates a covered and concealed location behind OPs/LPs for the vehicle. The OP/LP team has a covered and concealed withdrawal route to the vehicle from the fighting position. The team camouflages the OP/LP and their vehicle while the gunner clears a field of fire and prepares a range card. The squad leader establishes communication with higher HQ and tells the team when and how to report. He tells them—

- If and when they should fire at the enemy.

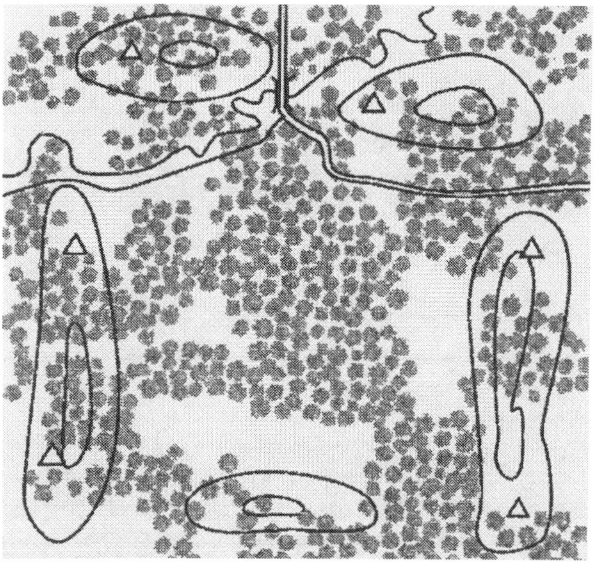

Figure 4-6. In-Depth OP/LP Placement

- How to get back to the squad if they must withdraw.
- What reentry signals to use.
- When they will be replaced, if known.
- To fight or withdraw according to his instructions.
- To be careful not to be drawn away by a small enemy element while the main element attempts to penetrate the perimeter.
- When to pull back or under what conditions they can withdraw without his order.

4-83. The frequency of relief for the OP/LP team depends on the team's physical condition and morale, the weather, the number of troops available, and the next operation. The squad leader carefully plans how each soldier receives rest. When OP/LP team is part of a defensive perimeter, it—

- Ensures that it has rearward cover.
- Builds fighting positions for protection and concealment.
- Uses trip flares, noisemaking devices, and NVDs to detect the enemy.
- Emplaces claymore mines for added protection.
- Coordinates with the perimeter on the reentry procedures to the perimeter from the withdrawal route.

DEFEND A SITE

4-84. Vigilance is the watchword for local security. When the OPs/LPs detect enemy elements, they notify their superior who calls for indirect fire, if it is available. When the enemy's advance threatens the OP/LP, order the OP/LP to withdraw. As the enemy approaches platoon positions, have the platoon increase its volume of fire. The platoon leader determines if the platoon can destroy the enemy from its assigned positions. If the platoon leader determines that the platoon can destroy the enemy, the platoon continues to fight with the following actions:

- The platoon leader or the FO continues to call for indirect fire as the enemy approaches. The platoon normally begins engaging the enemy at the maximum effective range. It attempts to mass fire and initiate them at the same time to achieve surprise. Long-range fire should disrupt enemy formations, channelize the enemy toward engagement areas, prevent or severely limit the enemy's ability to observe the location of friendly positions, and destroy the enemy as it attempts to breach tactical obstacles.

- The leaders control fire using standard commands, pyrotechnics, and other prearranged signals. The platoon increases the intensity of fire as the enemy closes within range of additional weapons. Squad leaders work to achieve a sustained rate of fire from their positions by having buddy teams fire their weapons so that both are not reloading them at the same time.
- The platoon and squad leaders consider the following when controlling and distributing fires:
  - The enemy's range.
  - The priority of the targets (what to fire at, when to fire, and why).
  - The nearest or most dangerous targets.
  - Shifts to concentrate fires on their own or as directed by higher HQ.
  - The ability of the platoon to engage dismounted enemy with grazing fires, and flank shots against enemy vehicles.
- The platoon leader initiates FPF as the enemy closes on the platoon's perimeter. The following actions occur at the same time:
  - The automatic weapons fire along interlocking PDF or FPLs. Other weapons fire at designated PDF. The M203 grenade launchers engage enemy in dead space or against enemy attempts to breach the protective wire.
  - The platoon continues the fight with claymore mines and hand grenades.
  - The platoon leader requests indirect FPF in support of his positions, if applicable.
  - The platoon continues to defend until the enemy is repelled or until the platoon is ordered to disengage.

4-85. If the platoon leader determines that the platoon can not destroy the enemy, he—

- Reports the situation to the company commander.
- Repositions the platoon to—
  - Continue fires into the platoon sector (engagement area).
  - Shift to alternate or supplementary positions.
  - Reinforce other parts of the company.
  - Counterattack locally to retake lost fighting positions.
  - Withdraw from an untenable position using fire and movement to break contact. (The platoon leader should not move his platoon out of position if it will destroy the integrity of the company's defense.)

NOTE: In any movement out of a defensive position, the platoon must employ all direct and indirect fire means available to suppress the enemy long enough for the platoon to move.

4-86. The platoon reorganizes after it has completed the initial battle with the enemy or relocated. The platoon positions key weapons, reestablishes security, provides first aid and prepares wounded soldiers for evacuation, and redistributes ammunition and supplies. The platoon relocates selected weapons to alternate positions if the leaders believe that the enemy may have pinpointed them during the attack and adjusts other positions to maintain mutual support. The platoon also reestablishes communication. It reoccupies and repairs positions and prepares for renewed enemy attack. The platoon repairs damaged obstacles and replaces mines and booby traps. When the platoon reorganizes, it performs the following actions:

- The squad and section leaders provide ammunition, casualty, and equipment (ACE) reports to the PSG. Team leaders provide fuel status. The PSG consolidates the ACE reports, reviews the consolidated ACE report with the platoon leader, and forwards it to the company commander.
- The platoon leader reestablishes the platoon's chain of command.
- The PSG coordinates for resupply and supervises the execution of the casualty and EPW evacuation plan.
- The platoon continues to improve positions. The platoon quickly reestablishes the OP/LP resumes patrolling as directed.

4-87. If the enemy gets through the FPF, repel it by close combat. If the perimeter is penetrated, move teams to block the penetration and cover friendly troops moving to alternate or supplementary positions. Even though the counterattack capability is limited, try to restore the perimeter. When the enemy is repelled—

- Reestablish security.
- Send patrols forward to maintain contact.
- Call for indirect fire on areas where the enemy is likely to regroup.
- Reorganize squads.
- Evacuate seriously wounded MP.
- Redistribute and resupply ammunition.
- Repair positions and continue to improve them.
- Keep the next higher commander informed throughout the conduct of the defense.

LAY HASTY PROTECTIVE MINEFIELDS

4-88. When possible, lay a hasty protective minefield as part of the unit's defensive perimeter. It can stop, delay, or restrict movement. MP often lay mines to restrict enemy movement near a defensive perimeter or at ambush sites. In the defense, platoons and squads lay hasty protective minefields to supplement weapons, prevent surprise, and give early warning of enemy advance. Hasty minefields must be covered by fire. Ensure that adjacent units are informed of the mine locations.

4-89. Platoons and squads must have permission from higher HQ to install hasty protective minefields. Higher HQ may, however, delegate approval authority to the company commander for emplacement of a hasty protective minefield. Requests for permission go through the normal chain of command.

4-90. If the company is not authorized mines in its basic loads, a special request may be needed. The enemy threat to the rear area requires commanders to issue mines as an additional protective measure. The M18A1 antipersonnel mine (claymore) and the M21 AT mine are the two mines most likely to be available to rear-area units for a hasty protective minefield. Refer to *FM 20-32* and *FM 21-75*.

4-91. MP generally will have claymores available to them, which is mainly a defensive weapon. However, the ways in which the claymore is used is limited only by the imagination. Plan the use of claymore mines to suit METT-TC. Emplace the mines—

- On likely dismounted avenues of approach.
- To cover dead space not covered by FPF of crew-served weapons.
- Outside the hand grenade range, but within the range of small-arms weapons.
- Where they are covered by observation and fire.
- Where back blast will not injure friendly forces.
- Beside buildings or other sturdy structures in urban terrain.

- Strapped to boards (for detonation from around corners).

4-92. MP record the exact location of the mines on *DA Form 1355-1-R (Figure 4-7)* when emplacing the minefield. This enables anyone to recover them. If possible, the unit should recover the mines before the unit relocates, and the same persons who emplaced them should recover them.

PATROLS

4-93. MP are organized and equipped to conduct mounted operations. The primary offensive weapon system is the MK19. It is designed as a mounted or static position weapon. However, an MP platoon may conduct limited dismounted operations with its other organic weapons. Refer to *FM 7-8* for more information about patrol operations, including organization, planning, and execution. In general, an MP platoon may be required to conduct reconnaissance and combat patrols.

RECONNAISSANCE PATROLS

4-94. Dismounted reconnaissance patrols are directed by higher HQ and conducted to gather detailed information on the enemy, terrain, specific NAIs, or avenues of approach. When executed as part of a screen or other security mission, a reconnaissance patrol can ensure the security of an OP/LP or the platoon's defensive perimeter. This is also referred to as a security patrol. Refer to Chapter 6 for area and zone reconnaissance.

COMBAT PATROLS

4-95. An MP platoon may conduct a combat patrol to establish an ambush on a dismounted enemy avenue of approach.

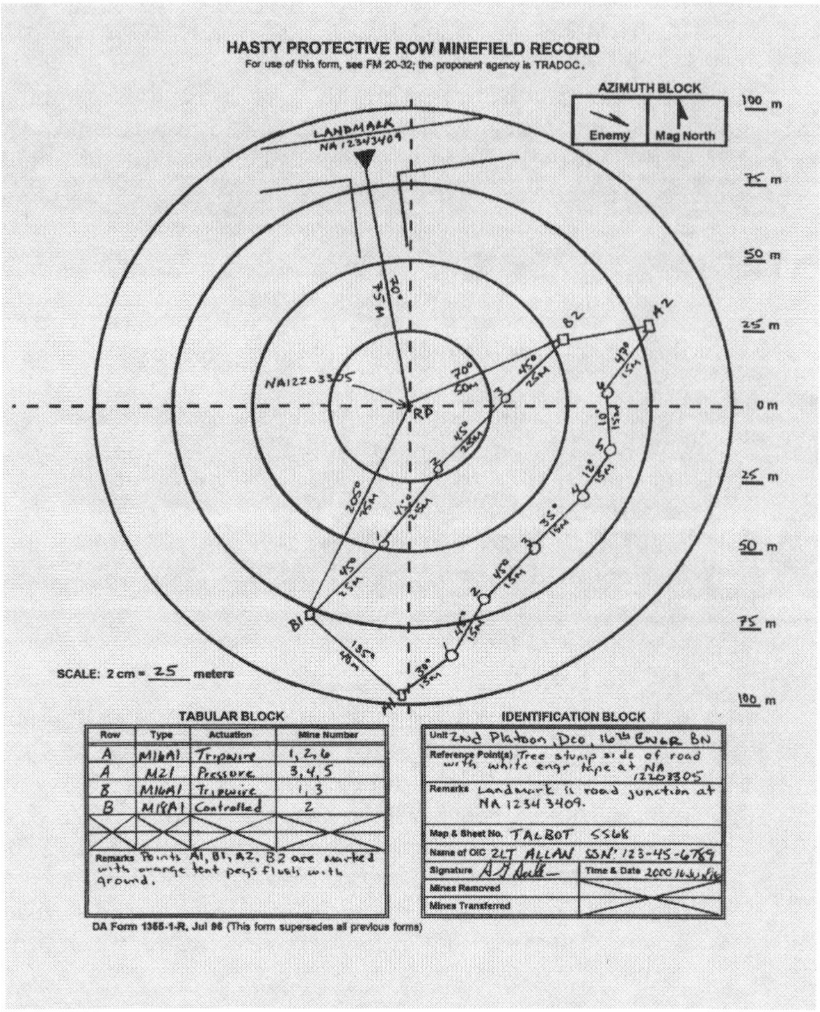

HASTY PROTECTIVE ROW MINEFIELD RECORD

For use of this form, see FM 20-32; the proponent agency is TRADOC.

AZIMUTH BLOCK

Enemy	Mag North

SCALE: 2 cm = 25 meters

TABULAR BLOCK

Row	Type	Actuation	Mine Number
A	M16A1	Tripwire	1, 2, 6
A	M21	Pressure	3, 4, 5
B	M16A1	Tripwire	1, 3
B	M18A1	Controlled	2

Remarks Points A1, B1, A2, B2 are marked with orange tent pegs flush with ground.

IDENTIFICATION BLOCK

Unit 2nd Platoon, D Co, 16th ENGR BN

Reference Point(s) Tree stump side of road with white engr tape at NA 12203305

Remarks Landmark is road junction at NA 1234 3409.

Map & Sheet No. TALBOT 5566

Name of OIC 2LT ALLAN SSN: 123-45-6789

Signature | Time & Date 2000

Mines Removed

Mines Transferred

DA Form 1355-1-R, Jul 96 (This form supersedes all previous forms)

Figure 4-7. Sample DA Form 1355-1-R

Ambush

4-96. MP elements, normally no smaller than a platoon, use an ambush along suspected enemy routes and elsewhere against Level II threats in the rear area. An ambush enables a small unit with light weapons to harass or destroy a larger, better-equipped unit. An ambush may be a surprise attack from a concealed position on a moving or temporarily halted target or an attack by fire only. A successful ambush requires the following:

- Surprise. The platoon must seize and maintain control of the situation.
- Coordinated fire. The platoon must deliver a large volume of fire into the kill zone, using individual and crew-served weapons, mines, demolitions, and indirect fire (if available), and isolate the kill zone to keep the enemy from escaping or being reinforced.
- Control. Before, during, and after the ambush all elements must be able to communicate effectively with the platoon leader, primarily by using hand and arm signals.

4-97. The platoon leader organizes the platoon into assault, support, and security elements. An ambush is laid on an enemy's expected approach route. The platoon leader selects the site and members are positioned to provide—

- Good visibility of the avenues of approach and the kill zone.
- Good fields of fire into the kill zone.
- Cover and concealment.
- Obstacles between the teams and the kill zone.
- Covered and concealed withdrawal routes.

4-98. A good ambush site restricts the enemy's movement to one flank by natural or man-made obstacles. Natural obstacles include cliffs, steep embankments, swamps, steep grades, sharp curves in the road, narrow trails, streams, and heavily wooded areas. Man-made obstacles can include mines, booby traps, and roadblocks. The ambush is configured to suit the—

- Type of ambush.
- Terrain.
- Troops available.
- Weapons.
- Equipment.
- Ease of control.
- Overall combat situation.

4-99. To conduct an ambush, the platoon leader is positioned so he can best control the ambush elements, normally with crew-served weapons or the AT4, especially if the enemy has armor. The platoon leader—

- Positions the flank security elements.
- Emplaces obstacles and mines.
- Improves fighting positions, if time permits.
- Places a crew-served weapon to cover the left and right limits of the kill zone. These weapons must ensure that once an element is in the kill zone, it cannot leave it laterally.
- Reports to higher HQ when the ambush is in place.

4-100. The platoon leader initiates the ambush with a casualty-producing weapon, such as a claymore mine or a crew-served weapon. He ensures that there is a back-up method in case the primary means fails. The remainder of the platoon opens fire once the ambush has begun.

4-101. Most often, platoons will deploy a squad-size element for an attack on a single kill zone (a point ambush). If the company is deploying a platoon-size force to conduct a number of coordinated, related ambushes (an area ambush), the principles are the same. An area ambush works best where close terrain keeps enemy movement largely limited to trails or roads. For an area ambush—

- Choose one central ambush site around which you can control and organize the outlying ambushes.
- Select outlying ambush sites on the enemy's possible avenues of approach and escape from the central site.
- Set up and maintain communication with all the outlying sites.
- Assign the general locations of the outlying sites to the squad leaders. They will each set and conduct a point ambush.
- Direct the squad leaders to let the enemy pass through the kill zones until the central ambush begins.
- Provide specific instructions to the squad leaders in case the enemy detects an outlying site before the central ambush begins.

Ambush Types

4-102. There are three types of ambushes. They are the line and L-shaped ambush formations and the antiarmor ambush (refer to *FM 7-8* for more information).

4-103. **Line Ambush Formation**. A line formation (refer to *FM 7-8* for more information) is easy to control and is useful in all levels of visibility. The assault and support elements parallel the long axis of the kill zone to engage the enemy with flanking fire. The target may be so dispersed that it extends beyond the kill zone. Leaders must—

- Position the assault and support elements parallel to the enemy's movement route (such as on a road or trail or at a stream).

- Limit the kill zone to the size area that the ambush can cover with a great volume of fire.
- Place obstacles (such as claymore mines or existing natural obstacles) between the kill zone and the ambush element to prevent counter-ambush actions.
- Leave access lanes through the obstacles so the kill zone can be assaulted (if directed).

4-104. **L-Shaped Ambush Formation**. An L-shaped formation is useful on a straight stretch of a trail, road, or stream. It also works well at a sharp bend in a trail, road, or stream. The assault element is the long leg of an "L," paralleling the kill zone to provide flanking fire. The support element is the short leg, capping the end of the kill zone at a right angle to the assault element. The support element provides enfilade fire to interlock with fire from the other leg.

4-105. **Antiarmor Ambush**. The principles for an antiarmor ambush are the same as for an area or point ambush. MP are likely to encounter bypassed enemy armor in the rear area. The primary antiarmor weapon for MP is an AT4. This is a light antiarmor weapon with limited capability against medium and heavy armor vehicles. However, MP may be required to set up a hasty antiarmor ambush to destroy one or two light enemy armor vehicles. An antiarmor ambush is best performed by a platoon. The platoon leader positions the antiarmor weapons where they can engage the target from the rear, flank, or top. Multiple AT4s are used to ensure destruction. The platoon provides support and security.

CLEARING TECHNIQUES

HIGH-INTENSITY VERSUS PRECISION CLEARING TECHNIQUES

4-106. Precision clearing techniques do not replace other techniques currently being used to conduct building and room clearing during high-intensity combat. Specifically, they do not replace the clearing technique in which a fragmentation or concussion grenade is thrown into a room before US forces enter. Use precision room clearing techniques when the tactical situation calls for room-by-room clearing of a relatively intact building in which enemy combatants and noncombatants may be intermixed. They involve increased risk in order to clear a building methodically, rather than using overwhelming firepower to eliminate or neutralize all its inhabitants.

4-107. From a conceptual standpoint, standard high-intensity room-clearing drills can be thought of as a deliberate attack. The task is to seize control of the room, with neutralization of the enemy in the room the purpose. The fragmentation or concussion grenade can be thought of as the preparatory fire used before the assault. As in a deliberate attack against any objective, the assaulting element moves into position using covered and concealed routes. The preparatory fire is initiated when the assaulting element is as close to the objective as it can get without being injured by the enemy. The assault element follows the preparatory fire (fragmentation or concussion grenade) onto the objective as closely as possible. A rapid, violent assault overwhelms and destroys the enemy force and seizes the objective.

4-108. Compared to the deliberate attack represented by high-intensity room-clearing techniques, precision room-clearing techniques are more conceptually akin to a reconnaissance in force or perhaps an infiltration attack. During a reconnaissance in force, the friendly unit seeks to determine the enemy's locations, dispositions, strength, and intentions. Once the enemy is located, the friendly force is fully prepared to engage and destroy it, especially if surprise is achieved. The friendly force retains the options of not employing preparatory fire (fragmentation or concussion grenades) if it is not called for (the enemy is not in the room) or if it is inappropriate (there are noncombatants present also). The attacking unit may choose to create a diversion (use a stun-hand grenade) to momentarily distract the defender while it enters and achieves domination of the objective.

4-109. The determination of which techniques to employ is up to the leader on the scene and is based on his analysis of the existing set of METT-TC conditions. The deliberate attack (high-intensity techniques), with its devastating suppressive and preparatory fire neutralizes everyone in the room and is less dangerous to the assaulting troops. The reconnaissance in force (precision techniques) conserves ammunition, speeds up the clearing process, reduces damage, and minimizes the chance of noncombatant casualties. Unfortunately, even when well executed, it is very stressful and hazardous for friendly troops.

4-110. Certain precision room-clearing techniques, such as methods of squad and fire team movement, the various firing stances, weapon positioning, and reflexive shooting, are useful for all combat in confined areas. Other techniques, such as entering a room without first neutralizing the known enemy occupants by fire or explosives, are appropriate in only some tactical situations.

4-111. Generally, if an alerted enemy force that is determined to resist occupies a room or building, and if most or all noncombatants are clear, employ overwhelming firepower to avoid friendly casualties. In such a situation, use supporting fires, demolitions, and fragmentation grenades to neutralize a space before friendly troops enter.

4-112. In some combat situations, the use of heavy supporting fire and demolitions would cause unacceptable collateral damage or unnecessarily slow the unit's movement. In other situations, often during stability and support operations, enemy combatants are so intermixed with noncombatants that US forces can not in good conscience use all available supporting fire. At such times, room-by-room clearing may be necessary and precision room-clearing techniques are most appropriate.

PRINCIPLES OF PRECISION ROOM-CLEARING

4-113. Battles that occur at close quarters, such as within a room or hallway, must be planned and executed with care. Units must train, practice, and rehearse precision room-clearing techniques until each fire team and squad operates smoothly. Each unit member must understand the principles of precision room-clearing, such as surprise, speed, and controlled violence of action.

Surprise

4-114. This is the key to a successful assault at close quarters. The fire team or squad clearing the room must achieve surprise, if only for seconds, by deceiving, distracting, or startling the enemy. Sometimes stun grenades are used to achieve surprise. These are more effective against a nonalert, poorly trained enemy than against alert, well-trained soldiers.

Speed

4-115. This provides a measure of security to the clearing unit. Speed allows soldiers to use the first few vital seconds provided by surprise to their maximum advantage. In precision room clearing, speed does not mean incautious haste. It can best be described as a careful hurry.

Controlled Violence of Action

4-116. This eliminates or neutralizes the enemy while giving him the least chance of inflicting friendly casualties. Controlled violence of action is not limited to the application of firepower only. It involves a soldier's mind-set of complete domination. Each of the principles of precision room-clearing has a synergistic relationship to the others. Controlled violence coupled with speed increases surprise. Hence, successful surprise allows increased speed.

FUNDAMENTALS OF PRECISION ROOM-CLEARING

4-117. The ten fundamentals of precision room-clearing address actions soldiers take while moving along confined corridors to the room to be cleared, while preparing to enter the room, during room entry and target engagement, and after contact. Team members should—

- Move tactically and silently while securing the corridors to the room to be cleared. Carry only the minimum amount of equipment.

NOTE: Rucksacks and loose items carried by soldiers tire them, slow their pace, and cause noise.

- Arrive undetected at the entry to the room in the correct order of entrance and be prepared to enter on a single command.
- Enter quickly and dominate the room. Move immediately to positions that allow complete control of the room and provide unobstructed fields of fire.
- Eliminate the entire enemy in the room by fast, accurate, and discriminating fires.

- Gain and maintain immediate control of the situation and all personnel in the room.
- Confirm whether enemy casualties are wounded or dead. Disarm and segregate the wounded. Search all enemy casualties.
- Perform a cursory search of the room. Determine if a detailed search is required.
- Evacuate all wounded and any friendly dead.
- Mark the room as cleared using a simple, clearly identifiable marking according to the unit SOP.
- Maintain security and be prepared to react to more enemy contact at any moment. Do not neglect rear security.

COMPOSITION OF THE CLEARING TEAM

4-118. Execute precision room-clearing techniques by the standard four-man fire team. Because of the confined spaces typical of building- and room-clearing operations, units larger than squads quickly become unwieldy. When shortages of personnel demand it, conduct room-clearing with two- or three-man teams; four-man teams are preferred. Using fewer personnel greatly increases the combat strain and risks.

BREACHING

4-119. An integral part of precision room-clearing is the ability to gain access quickly to the rooms to be cleared. Breaching techniques vary based on the type of construction encountered and the types of munitions available to the breaching element. Techniques range from simple mechanical breaching to complex, specialized demolitions.

Shotgun Ballistic

4-120. A useful method of breaching is the shotgun ballistic breach for forced entry of standard doors. Use a

12-gauge shotgun loaded with buckshot or slugs to breach most standard doors quickly. When done properly, the shotgun breach requires only a few seconds. The two standard techniques of shotgun breaching are the doorknob breach and the hinge breach. When attempting either technique, the gunner approaches the door from an angle, avoiding standing in the area directly in front of the door. While holding the stock of the shotgun in the pocket of his shoulder, the gunner places the muzzle tightly against the door, and aims down at a 45-degree angle.

4-121. **Doorknob Breach**. For the doorknob breach, *(Figure 4-8, page 4-58)* the aim point is a spot halfway between the doorknob and the frame, not at the doorknob itself. The gunner fires two quick shots in the same location, ensuring that the second shot is aimed as carefully as the first. Weak locks may fly apart with the first shot, but the gunner should always fire twice. Some locks that appear to be blown apart have parts still connected that will delay entry. If the lock is not defeated by the second shot, the gunner repeats the procedure.

4-122. **Hinge Breach**. The hinge breach technique is performed much the same as the doorknob breach, except the gunner aims at the hinges. He fires three shots per hinge—the first at the middle, then at the top and bottom *(Figure 4-8)*. He fires all shots from less than an inch away from the hinge. Because the hinges are often hidden from view, the hinge breach is more difficult. Regardless of which technique the gunner uses, immediately after he fires, he kicks the door in or pulls it out. He then pulls the shotgun barrel sharply upward and quickly turns away from the doorway to signal that the breach point has been cleared. This rapid clearing of the doorway allows the following man in the fire team a

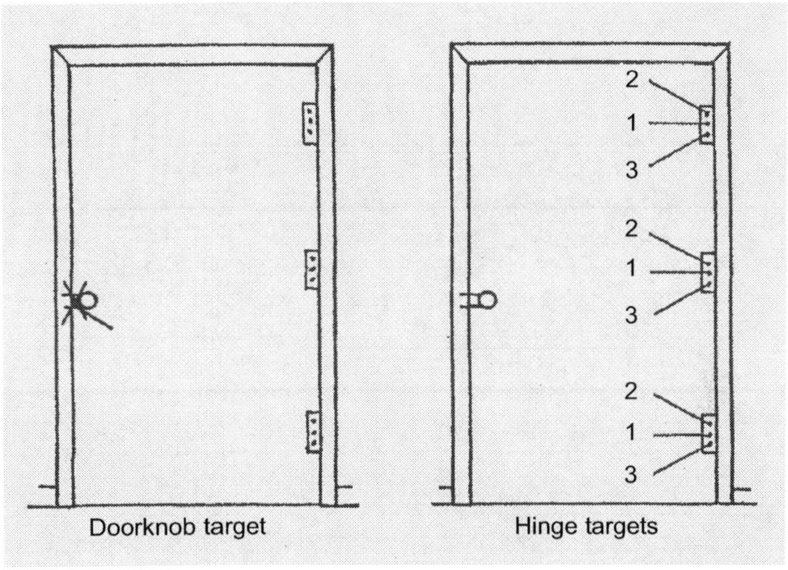

Figure 4-8. Aim Points for a Shotgun Breach of a Standard Door

clear shot at any enemy who may be blocking the immediate breach site.

WARNING
Do not use small arms (5.56 or 7.62 millimeters) as a ballistic breach on doorknobs and hinges except as a last resort. It is unsafe and could result in death.

4-123. Demolitions are often needed to defeat more elaborate barriers or to produce a desired effect to aid the initial entry.

Mechanical Breaching

4-124. Mechanical breaching is planned as a backup to a ballistic or explosive breach. Mechanical breaching is an assumed capability within all units. Taking the time to defeat weak barriers such as doors or windows by means of crowbars, saws, sledgehammers, battering rams, or axes is a decision that must be made based on the conditions of METT-TC.

Order of Movement

4-125. Clearing-team members must approach the breach point quickly, quietly, and in standard order. This approach preserves the element of surprise and allows for quick entry and domination of the room. The order of movement to the breach point is determined by the method of breach and intended actions at the breach point. The members of the fire team are assigned numbers 1 through 4, with the team leader normally designated number 3. If one member of the clearing team is armed with the squad automatic weapon (SAW) rather than an M16 rifle or carbine, he should be designated as number 4.

4-126. **Ballistic (Shot Gun) Breach**. The order of movement for a shotgun breach has the gunner up front, followed by the number 1 man, number 2 man, and then the number 3 man (team leader). After the door is breached, the gunner moves to the rear of the lineup and assumes the position of the number 4 man.

4-127. **Explosive (Demolition) Breach**. The order of movement for an explosive breach without engineer support is number 1, number 3 (team leader), number 2, and then number 4. The number 1 man provides security at the doorway. The number 3 man (team leader) carries the demolition charge and places it. The number 2 man provides security overhead, and the number 4 man provides rear security. After the demolition charge is placed, the team moves to covered positions and prepares to enter in the standard 1, 2, 3, 4 order.

4-128. **Mechanical Breach**. A suggested order of movement for a mechanical breach is the initial assault team in order, followed by the breach man/element. At the breach point the assault team leader will bring the breach team forward while the assault team provides local security. After the breach is conducted, the breach team moves aside and provides local security as the assault team enters the breach.

ENTRY CONSIDERATIONS

4-129. The entire team enters the room as quickly and smoothly as possible and clears the doorway immediately. If possible the team moves from a covered or concealed position already in their entry order. Ideally, the team arrives and passes through the entry point without having to stop.

4-130. The door is the focal point of anyone in the room. It is known as the *fatal funnel*, because it focuses attention at the precise point where the individual team members are the most vulnerable. Moving into the room quickly reduces the chance anyone will be hit by enemy fire directed at the doorway.

4-131. On the signal to go, the clearing team moves from covered or concealed positions through the door quickly and takes up positions inside the room that allow it to completely dominate the room and eliminate the threat. Team members stop movement only after they have cleared the door and reached their designated point of domination. The first man's position is deep into the far corner of the room. The depth of his movement is determined by the size of the room, any obstacles in the room such as furniture, and by the number and location of enemy and noncombatants in the room.

4-132. To make precision room-clearing techniques work, each member of the team must know his sector of fire and how his sector overlaps and links with the sectors of the other team members. Team members do not move to the point of domination and then engage their targets. They engage targets as they move to their designated point. However, engagements must not slow movement to their points of domination. Team members may shoot from as short a range as 1 to 2 inches. They engage the most immediate enemy threats first. Examples of immediate threats are enemy personnel who are—

- Armed and prepared to return fire immediately.
- Blocking movement to the position of domination.
- Within arm's reach of a clearing team member.
- Within 3 to 5 feet of the breach point.

CLEAR A SINGLE ROOM

4-133. Precision military operations on urbanized terrain (MOUT) room clearing is generally accomplished using a four-person team, consisting of three riflemen and one SAW gunner or breacher. Entry does not always require four personnel. If a specific room is to small or there are less than four clearing members due to attrition, the room may be cleared with less. However, never clear a room with less than two soldiers. If the leadership determines that the use of four-soldier teams is impractical for the situation, use the standard three-soldier team filling the numbers 1, 2, and 3 entry positions. The term *point of domination* refers to the two corners of the room assigned to the numbers 1 and 2 soldiers. These points allow the team to gain complete control of the objective. Each team member is assigned a different, but interlocking field of fire. This ensures mutual supporting fires. This type of entry is dynamic in nature. When completed with precision, this technique overwhelms the enemy and allows the team to move on to the next objective very quickly. When a room is cleared, the exiting procedure for leaving the room varies depending on the location of the entry point (either in the middle or at the corner of the room).

> **WARNING**
> **The muzzle of the clearing team's weapons should always be pointing wherever that soldier is looking. The clearing team should remain one foot away from all the walls to avoid injury due to bullets traveling down the wall (the path of least resistance).**

Stack Position

4-134. Once the approach to the room or building is completed, the team stacks *(Figure 4-9)* parallel to the outside wall. The number 1 soldier provides security on the entry point. This point is normally a door in a precision MOUT environment. The number 2 soldier provides security to the front of the team. The number 3 soldier (normally the team leader) is in the high-ready position. The high-ready position consists of the soldier placing the buttstock of the weapon in his shoulder pocket and raising the muzzle of the weapon from a 90-degree angle to a 45-degree angle. The number 4 soldier is in the low-ready position. The low-ready position consists of the soldier placing the buttstock of the weapon in his shoulder pocket and lowering the muzzle of the weapon to a 45-degree angle. The high- and low-ready positions aid in ensuring that the soldiers do not flag their fellow team members with their weapons during entry or clearing. A support element covers the teams from a distance, concentrating on the team's blind spots.

Initial Entry

4-135. When the situation allows (time, concealment, and so forth), it is important to ensure that each team member is prepared to enter the initial-entry point to the building before the assault. This is achieved while in

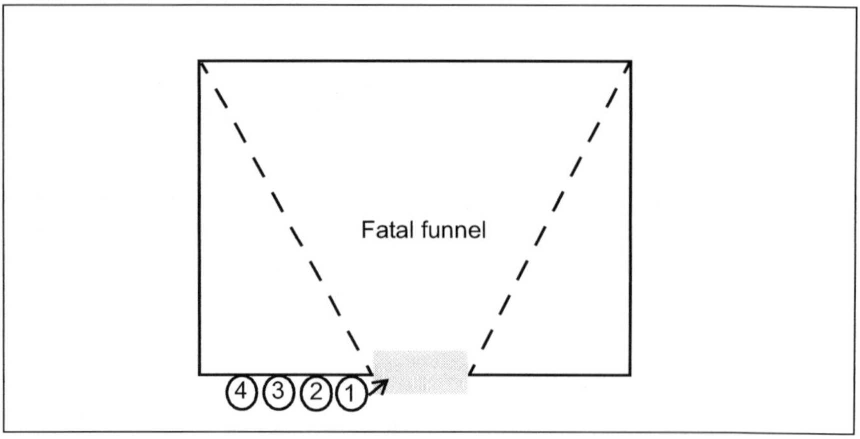

Figure 4-9. Stack

the stack position. When ready, the number 1 soldier nods his head while maintaining security on the door. When soldier number 4 is ready, he squeezes the shoulder of soldier number 3 with the nonfiring hand. This action continues up the line. When the signal reaches soldier number 1, he silently counts to three and then enters the building with the rest of the team following. If someone inside opens the door before this sequence is completed, the team will immediately enter the structure.

Initial-Entry Breach

4-136. *Figure 4-10, page 4-64,* depicts the process of a nonexplosive breach from the standard stack. Soldier number 4 moves up to the side of the door that contains the locking mechanisms. When soldier number 3 gives the signal, soldier number 4 breaches the door (by shotgun, battering-ram, and so forth) and kicks it open. Soldier number 4 then moves back from the door to allow the rest of the team to enter. Soldier number 4 follows them into the room. Refer to *FM 90-10-1* for shotgun breaching techniques.

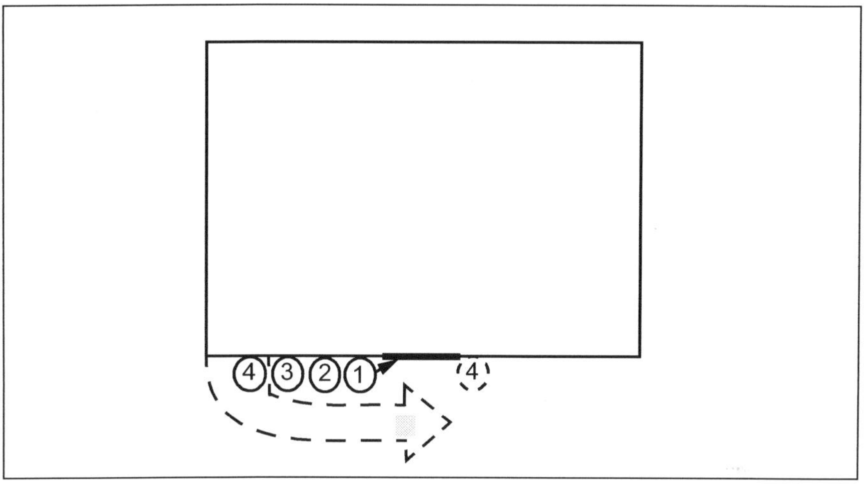

Figure 4-10. Nonexplosive Breach

Room Entry

4-137. Clear a room in the following manner when the entry point is located in the middle of the room:

- Soldier number 1 *(Figure 4-11)* enters the room using the path of least resistance, (moves across the doorway, into the room, and does not make a turn around the doorframe) clearing the fatal funnel as fast as possible. The term fatal funnel is derived from the natural reaction of the enemy soldiers inside a room (the large end of the funnel) to focus on and fire at the doorway (the spout of the funnel). The rounds are funneled into the doorway from almost any position in the room. An attacker should never move slowly through (or stop in) the fatal funnel. Besides providing a silhouetted target to the enemy, these actions will slow the entry of the rest of the team (reducing supportive fires). The number 1 soldier's first responsibility (in this scenario) is to clear the bottom right corner. This soldier will not be concerned with the middle of the room unless a threat is located to his direct front as he

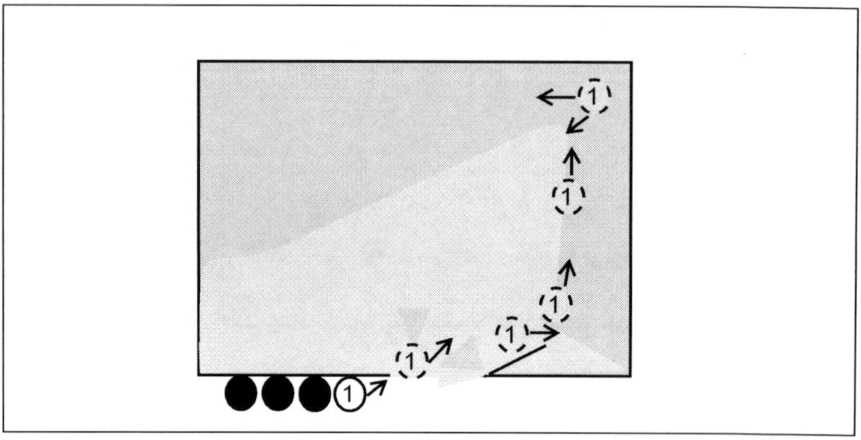

Figure 4-11. Number 1 Man, Middle-Door Entry

enters the room. If the number 1 soldier assesses this threat as being capable of hindering his movement to the top right corner of the room (his point of domination), he may engage that threat in a manner that avoids losing momentum. If the door obstructs or partially obstructs the route, soldier number 1 pushes hard through it. This helps to dissipate any loss of momentum and knocks an enemy soldier off balance if he is hiding behind the door. Once the number 1 soldier visually clears the corner, he moves up the wall to his front, clearing it as he goes. The number 1 soldier stops at his point of domination. He then turns and begins a scan from the top left corner to 1 meter off the number 2 soldier's weapon, located in the bottom left corner in this scenario (number 1 has farther to go, soldier number 2 should be in position once number 1 arrives).

- Soldier number 2 *(Figure 4-12, page 4-66)* enters the room immediately after the number 1 soldier enters. Soldier number 2, in this case, executes a hard left turn and moves in the opposite direction of soldier number 1. Soldier number 2's first concern is to clear the bottom left corner. Once this corner is visually cleared, he continues to clear up the left wall. Once soldier number 2 reaches this corner, he has reached the second point of domination. Soldier number 2 stops and scans from the top left corner to 1 meter off soldier number 1's weapon.

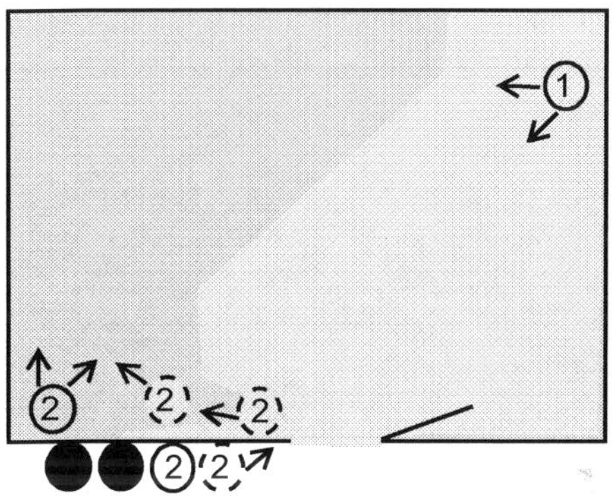

Figure 4-12. Number 2 Man, Middle-Door Entry

- Soldier number 3 enters the room at the same time as the number 1 and number 2 soldiers *(Figure 4-13)*. He moves his weapon down from the high-ready position, between soldiers number 1 and number 2 (using his height advantage, if it exists). Soldier number 3's first and primary concern is the middle of the room. If soldier number 3 enters the room quickly, effective supporting fires will be provided for the number 1 and number 2 soldiers as they clear the corners and move to their points of domination. Soldier number 3 begins with a

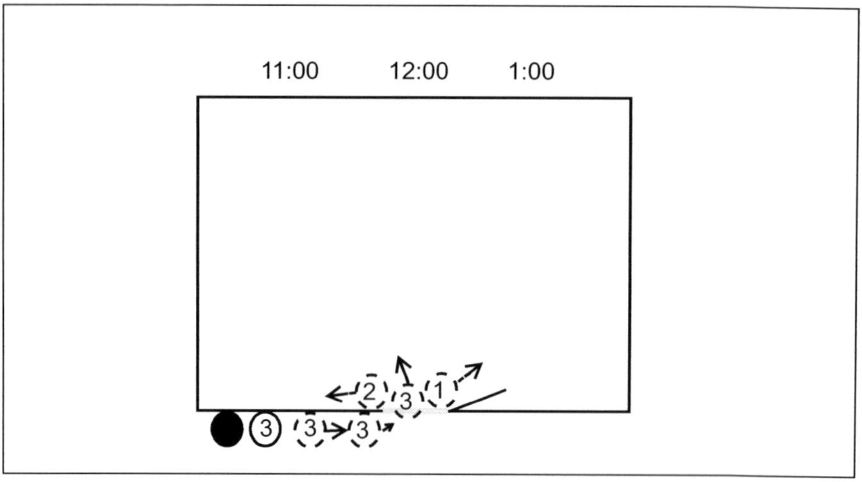

Figure 4-13. Number 3 Man, Middle-Door Entry

point of aim (in this scenario) at the 11 o'clock position. He indexes his weapon down the opposite wall, while moving laterally (cross step) out of the fatal funnel, in the opposite direction of soldier number 2 *(Figure 4-14)*. Soldier

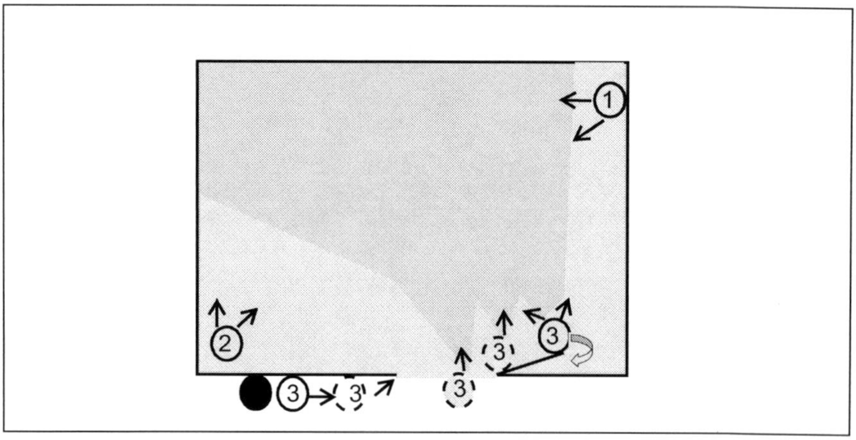

Figure 4-14. Number 3 Man, Middle-Door Entry

number 3 stops when out of the fatal funnel and begins to scan from 1 meter off soldier number 1's weapon to 1 meter off soldier number 2's weapon. Once one full scan is completed, soldier number 3 clears behind the door and then returns to scanning the room.

- Soldier number 4 clears in the manner depicted in *Figure 4-15* if the room is large enough for the SAW gunner or breacher to enter. He enters the room directly behind soldier number 3 and clears the fatal funnel as quickly as possible, moving in the opposite direction of soldier number 3. Soldier number 4's primary concern is the middle of the room. He begins with an overlapping point of aim at the 1 o'clock position and indexes his weapon down the far wall as he moves laterally towards the number 2 soldier. Soldier number 4 stops, once clear of the fatal funnel, and continues to scan from 1 meter off the number 2 soldier to 1 meter off the number 1 soldier.

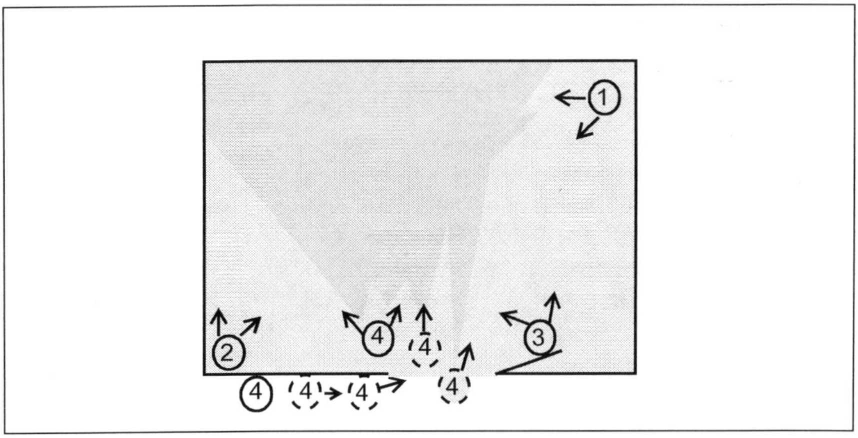

Figure 4-15. Initial Coverage

NOTE: The M249 or shotgun are not precision weapons. The number 4 soldier only engages when a clear shot at a confirmed combatant presents itself. The M249 is also very useful when the situation has degraded and heavier support is required. Leaders should train SAW gunners or breachers to be prudent in the use of their weapons in a precision MOUT environment.

- *Figure 4-16, page 4-70,* shows the coverage area of each team member once they have reached their positions. The room can be visually cleared very quickly because it has been dominated in depth. This is especially important considering the fact that most rooms contain furniture (hiding positions) and many are irregular shaped. A majority of the room is covered at all times, should an undiscovered or wounded enemy soldier attempt to engage the clearing team. The bottom right corner is the only area not effectively covered by fire at this time. Both the number 1 and number 3 soldiers cleared this area during the initial entry. If the number 1 soldier or number 3 soldier determines that a threat might exist in this corner, it is soldier number 3's responsibility to provide security in this area.

Communication

4-138. When the room has been dominated, communication is required between the team members in order to clear the room quickly and move on to the next objective. Whether scanning the room or providing security on possible threats (open doors, unidentified persons, or dead or wounded enemy), the team members cannot shift their gaze around the room to ascertain the current situation. To solve this problem, the team leader calls for a status when he determines that no immediate threat exists. This status or situation report (SITREP)

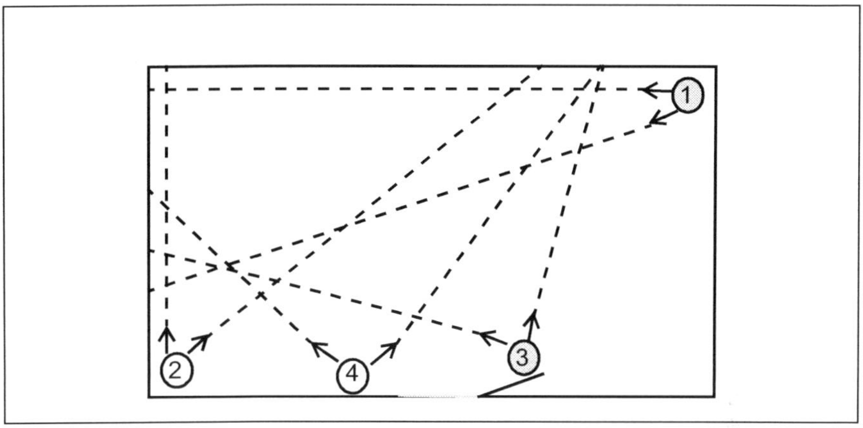

Figure 4-16. Number 1 Man, Center-Door Entry

consists of each clearing team member's current condition and any possible threat in his sector. The number 1 soldier states his situation to the rest of the team. The number 2 soldier follows with his report and so on down the line. The team leader absorbs all this information and then gives directions based on the information. The conversation should be at a level that each team member may hear, but no louder. If soldier number 1 fails to sound off, soldier number 2 states his status. If soldier number 1 is down, the team will know it. If soldier number 1 fails to sound off due to a mental lapse, he may give a SITREP once the other team members have given theirs. An example of this type of communication is as follows:

- Team leader (TL) (soldier number 3 in this scenario): "Status."
- Soldier number 1: "One up!"
- Soldier number 2: "Two, one down to my front!"
- Soldier number 3: "Three up!"
- Soldier number 4: "Four up!"
- TL: "Soldier number 1 secure the subject."

4-139. This scenario is based on a one-room dwelling objective for simplicity. Once the subjects are secured, they are extracted them from the dwelling by a support team and taken to a secure area where their status can be determined and first aid may be rendered, if needed. Meanwhile, the clearing team continues its mission. This would include (dependant on the mission) searching the room for any hidden threats, contraband, or enemy information or equipment. If there are multiple rooms to clear, the TL either directs one of the clearing team members to remain in the room as security (such as "soldier number 1 security") or (if available) calls for a reserve, who is positioned outside of the room, to enter the room and secure the subjects inside it until the building is cleared. The team would then be free to move on to the next room.

Corner-Door Position

4-140. When an entry point is located in the corner of the room, slight modifications are made to the middle-door entry method. The number 1 soldier *(Figure 4-17, page 4-72)* enters through the path of least resistance and clears the path to the corner to his front and stops. This is soldier number 1's point of domination. Soldier number 1 does not continue on to the next corner, as in the middle-door method, because it would take to long for him to reach this location. The number 2 soldier makes a hard turn around the doorframe and clears the path to the corner diagonally opposite of soldier number 1. This is soldier number 2's point of domination. The number 3 and number 4 soldiers use the far corner of the room as the 12 o'clock reference point *(Figure 4-18, page 4-72)*. All other actions in the room are the same as the middle-door method.

Rules

4-141. No matter what the room configuration, there are a few rules that should always be adhered to. These rules include the following:

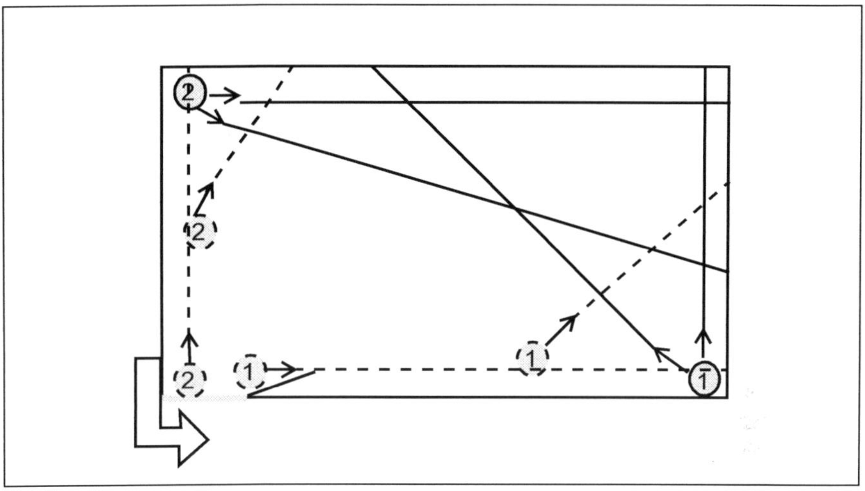

Figure 4-17. Number 1 and 2 Man, Corner-Door Entry

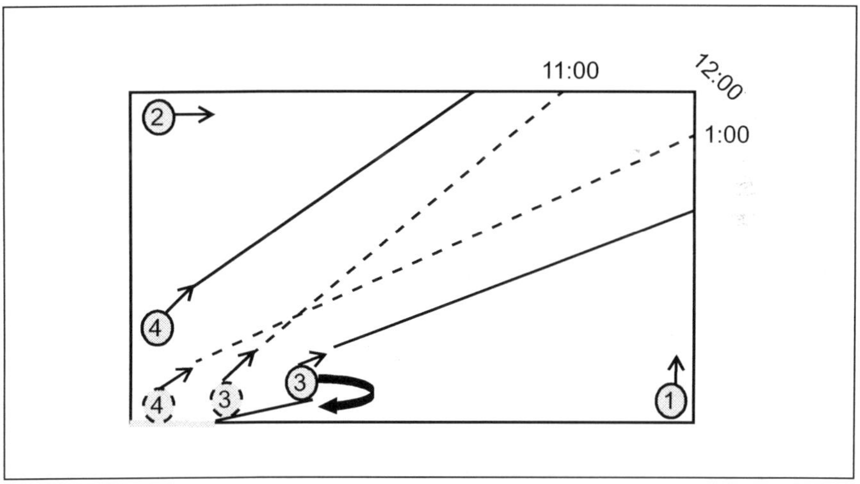

Figure 4-18. Number 3 and 4 Man, Corner-Door Entry

- Know that it does not matter if the person in front of you goes the wrong way when entering the room. Just go the opposite way as the soldier in front of you and it will work out.

- Enter the room as quickly and smoothly as possible and do not waste movements. Remember smooth is fast. The faster each team member picks up its initial point of aim, the more difficult it becomes for the defender. Even a prepared defender can be caught off guard.
- Clear (do not stop) the fatal funnel. The number 3 and number 4 soldiers are especially prone to stopping in the funnel.

NOTE: Ensure that the doorway is completely cleared before assuming a final position in the room.

- Stay focused. Never stop scanning your sector for targets unless—
  - You identify a threat in your sector. This threat could be an open door leading to an uncleared room, a person in the room other than one of the team members, an obstacle that cannot be cleared visually from your position, or anything else that you may determine as a threat. If such a threat exists, provide security for the team by covering it with your weapon.
  - You are ordered by the team leader to perform another task.
- Ensure that if you fall down while entering the room, you stay down and do not move. Do not get up until a team member places his hands on you and lifts you up. If a team member arbitrarily gets up, he may become an impediment to the rest of the team's movements. Another possibility is that he would rise into the path of a team member's bullet.
- Rehearse communication. As stated above, speed and momentum will make or break this type of

action. Poor communication techniques will slow your building clearing to a crawl.

- Ensure that each team member knows the procedures for each position. After the first room is cleared in a multiroom objective, a team member may find that he is in a different position in the stack than when the assault started. The team leader should always attempt to avoid being the number 1 soldier when entering a room. This can normally be accomplished, but in some situations it will be unavoidable.
- Ensure that you never enter a room alone. Two personnel are the minimum room-clearing team. If two soldiers are entering a room (either due to the size of the room or attrition), they should assume the number 1 and number 2 soldiers' positions.
- Ensure that you never flag a fellow team member. Flagging is defined as pointing your weapon at or sweeping your weapon across another team member. Muzzle awareness, at all times, is imperative.

CLEAR A CONNECTED ROOM

4-142. Clearing multiple rooms using connecting doors is preferable to using hallways. This action reduces exposure and compliments speed and momentum.

4-143. *Figure 4-19* shows a team that has just cleared a room using the corner-door technique. The door to the adjacent room is open, providing the number 2 and number 4 soldiers the opportunity to partially clear the second room. If a threat is identified in the adjoining room, it may be engaged from the cleared room. Once this firing has stopped, the SITREP will be restarted if necessary.

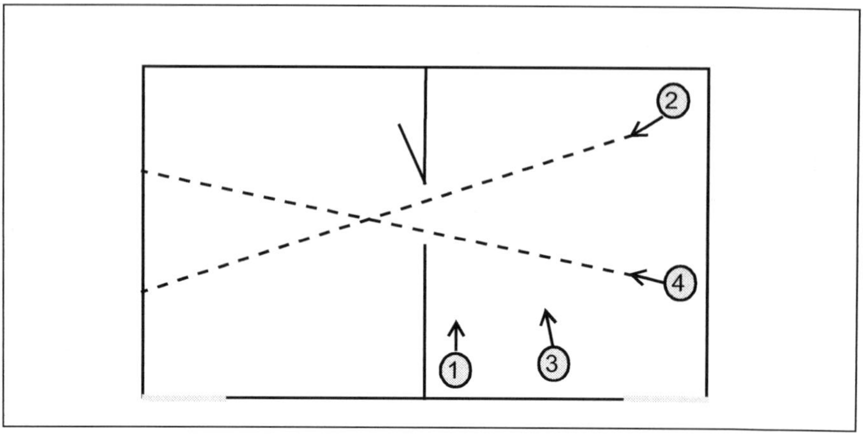

Figure 4-19. Partial Clearing of a Connected Room

4-144. *Figure 4-20, page 4-76,* shows how the team stacks in preparation for entry into the next room. When the team leader declares the first room clear, the team will automatically stack on the entrance to the next room. In this example, crossing the doorway would not only place a soldier in the fatal funnel unnecessarily, it would also mask the cover fire of the number 4 soldier. Therefore, soldiers number 1 and 3 stack to the left of the door and soldier number 2 stacks to the right. Soldier number 4 remains in a position to cover a portion of the adjacent room. In this example, soldier number 1 remains the first soldier to enter the room. He awaits a squeeze from soldier number 3. This squeeze acts as a signal to soldier number 1 that at least one other team member is behind him and is ready to enter the room. This form of communication reduces the time spent discussing the order of entry and eliminates the need for a verbal queue (an early warning for enemy soldiers inside the room) to enter the room. If the door had been positioned elsewhere in the room, it could have just as easily been the number 2 or number 3 soldier entering the adjacent room first.

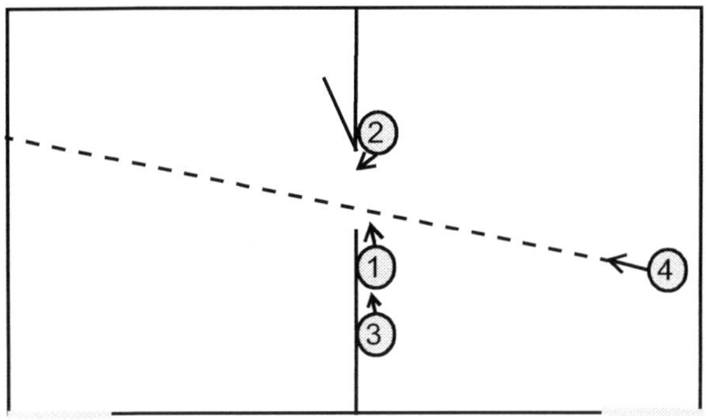

Figure 4-20. Team Stack

4-145. *Figure 4-21* depicts the position of the team members while clearing the second room. The soldiers' original numbers are left in place to demonstrate the difference in the positions from the first room. As the team enters the room, soldier number 4 lowers his weapon and moves quickly to the entry point to prevent flagging any team member. He then moves up to the entrance and follows the rest of the team into the next room.

4-146. *Figure 4-22* depicts a closed door to the adjacent room. A closed door is considered locked in all cases when a breacher is present (this is assuming the breacher has the means available to defeat the lock). Once the other team members are stacked, the breacher moves to the door and breaches it on the hand signal of soldier number 3. Soldier number 4 then steps back in order to clear the path for soldier number 2 (the third soldier to enter the room). If no breaching equipment is available, a quick attempt to open the door, from the side may be made. If the door is locked, kick it open. However, this should be the last choice for defeating a locked door. Kicking generally requires more than one

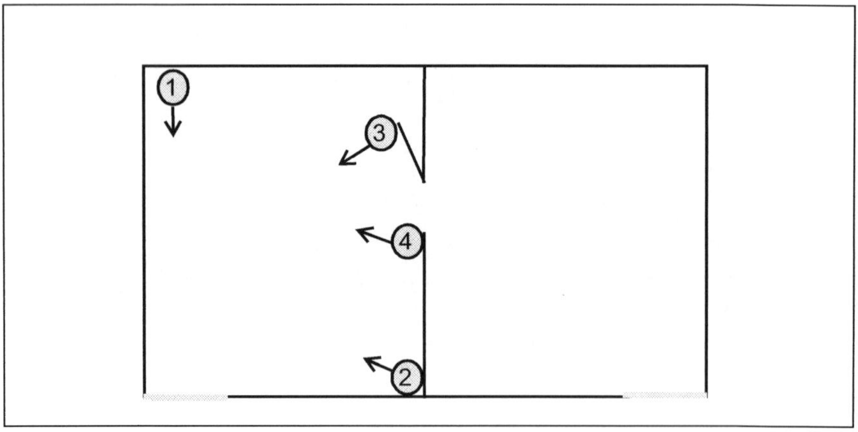

Figure 4-21. Team Members' Positions

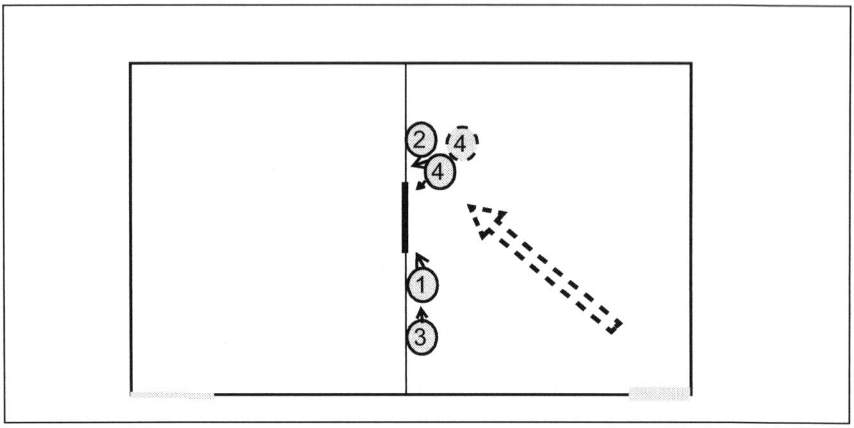

Figure 4-22. Team Positions When the Door to the Adjacent Room is Closed

attempt to defeat the lock. After the first attempt, the breacher's chances of drawing fire through the door are significantly increased.

NOTE: Do not attempt to open the door by hand. If the door is locked, this action will alert any enemy soldiers inside of the room of your intentions and

possibly expose the breacher to fire through the doorway.

4-147. *Figure 4-23* depicts the team preparing to enter the next room or a hallway. Soldier number 1 moves to a position just inside the doorway. He would have had security on the doorway from his original point of domination. The other team members stack to the left, remaining just inside the doorway. Soldier number 1 is now in the position to partially clear the next room or hallway. The team enters the room the same way as described in the initial coverage.

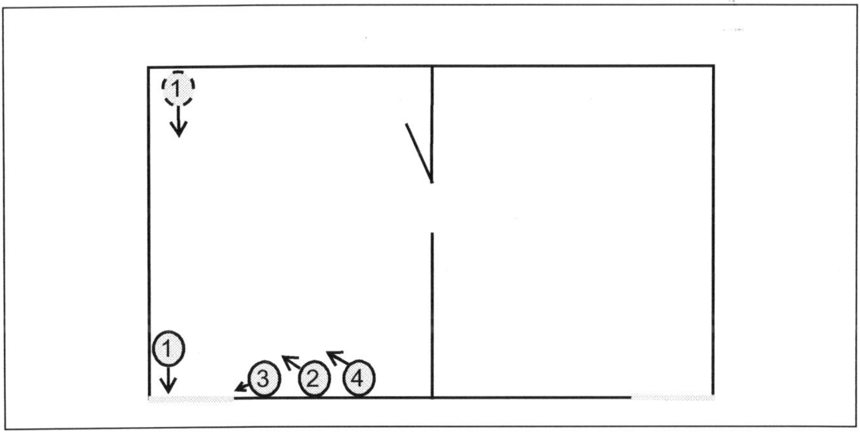

Figure 4-23. Stack Variation

CLEAR A FLOOR

4-148. When clearing a single-floor complex, the principles discussed in single- and connecting-room clearing are used. The only added features are the hallways and an additional squad, when available, to combat attrition. *Figures 4-24* through *4-28* depict the sequence of events in a floor-clearing mission.

4-149. *Figure 4-24* gives an example of a two-squad (four teams, plus HQ element) assault force, preparing

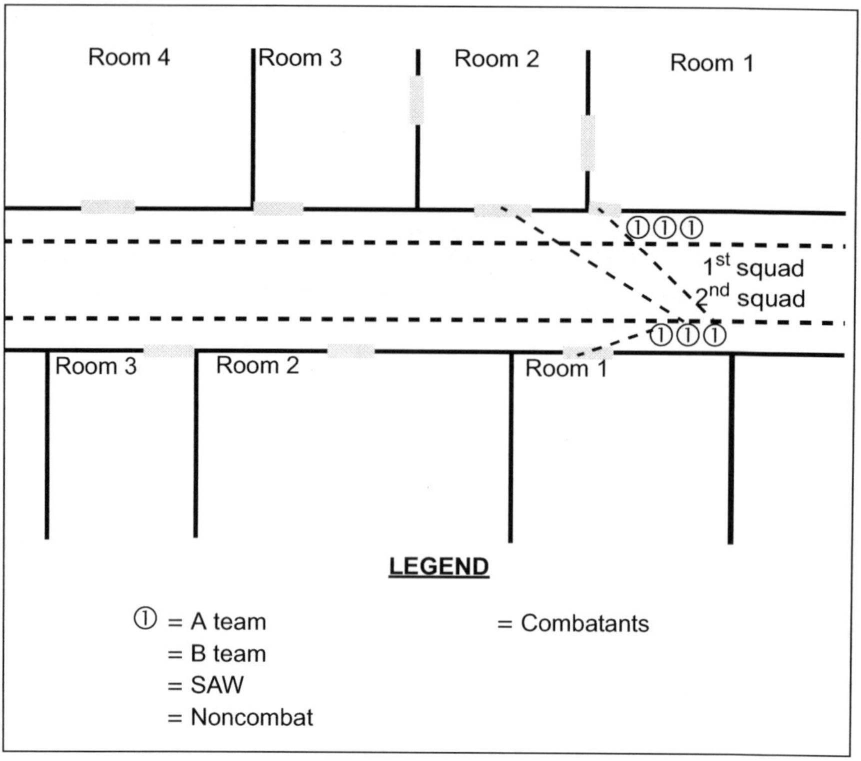

Figure 4-24. Clearing a Floor

to clear a floor. The first and second squads are stacked on adjacent opposite walls in the hallway. The first squad prepares to enter or clear the first room on their side. Soldier number 1 (first squad) is continually providing security on the door to his direct front. The second squad halts before exposing itself to the open door in front of the first squad. The second squad's number 1 soldier provides security on the first doorway to his direct front. If no entry point to his direct front exists, soldier number 1 provides security down the hallway. Soldier number 2 (second squad) provides security on the second doorway to the front of the first squad. Soldier number 3 (second squad) provides security on the door to the direct front of the first squad. He lifts that security when the first squad begins to enter that room. The SAW gunners in both teams provide security down the hallway. The HQ RTO (not depicted) provides rear security.

4-150. In *Figure 4-25,* the first squad clears the first room and moves into the second adjoining room. As the second room is cleared, soldier number 3 calls shot and engages an enemy threat in the next room. The B team of the first squad moves into the first room after the A

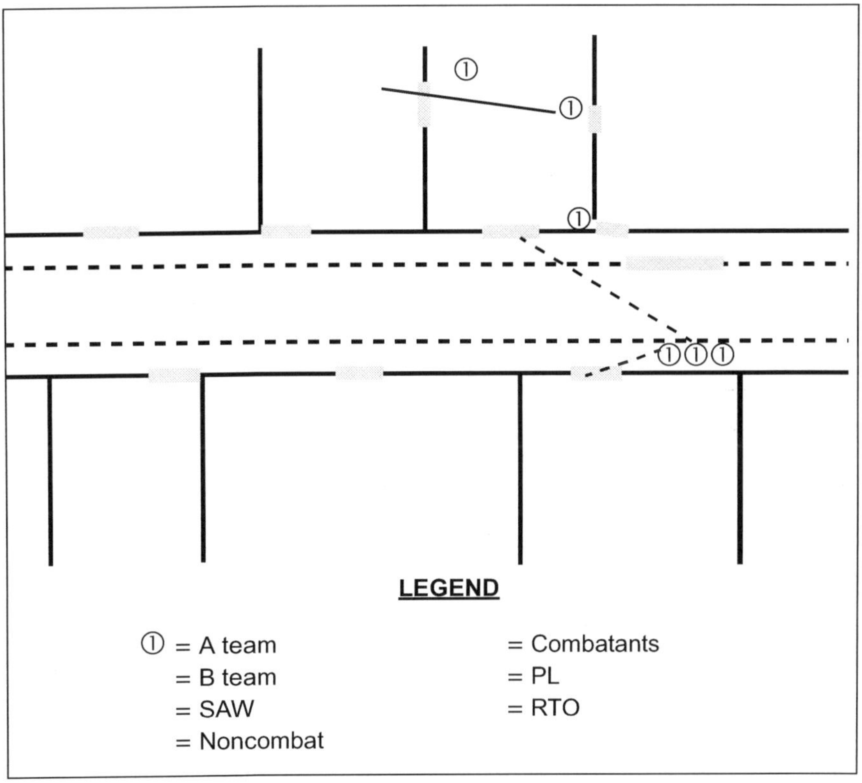

Figure 4-25. Clearing Adjoining Rooms with Hallway Security

team begins to clear the second room. The B team awaits permission from the A team leader to enter the second room. The HQ SAW gunner picks up the hallway security. The second squad has not moved from their original position. A team should never move past an uncleared room. In this case, the second squad will not receive a room clear signal from the first squad until the first squad reaches and clears the third room. This is because entry points interconnect the rooms on this side of the hallway. These entry points should be used in lieu of the hallway, whenever possible.

4-151. In *Figure 4-26, page 4-82,* the first squad has cleared the first three rooms. In this example, one soldier from the B team has been detached to secure the first room, a second has been left in the second room to secure the noncombatant, and a third is securing a noncombatant and a downed enemy soldier in the third room. Soldier number 1 (first squad) remains inside the doorway and provides security on the entry point of the second room to the second squad's front. Soldier number 2 provides security on the third room to second squad's front. The second squad receives the all-clear sign from the first squad (radio, thumbs-up, or hand and arm signals) and moves to their first-room entry point.

NOTE: If there are not enough personnel available to leave security in cleared rooms (especially in multilevel clearing), disarm and secure all combatants and noncombatants using handcuffs or flex cuffs.

4-152. If more than one floor is involved, leave a two-person security team on each floor while the building is cleared. In this instance, each room must be systematically recleared once the building is secured. A marking system (such as chalk or chemical lights) must be implemented in order to identify the rooms that have been cleared.

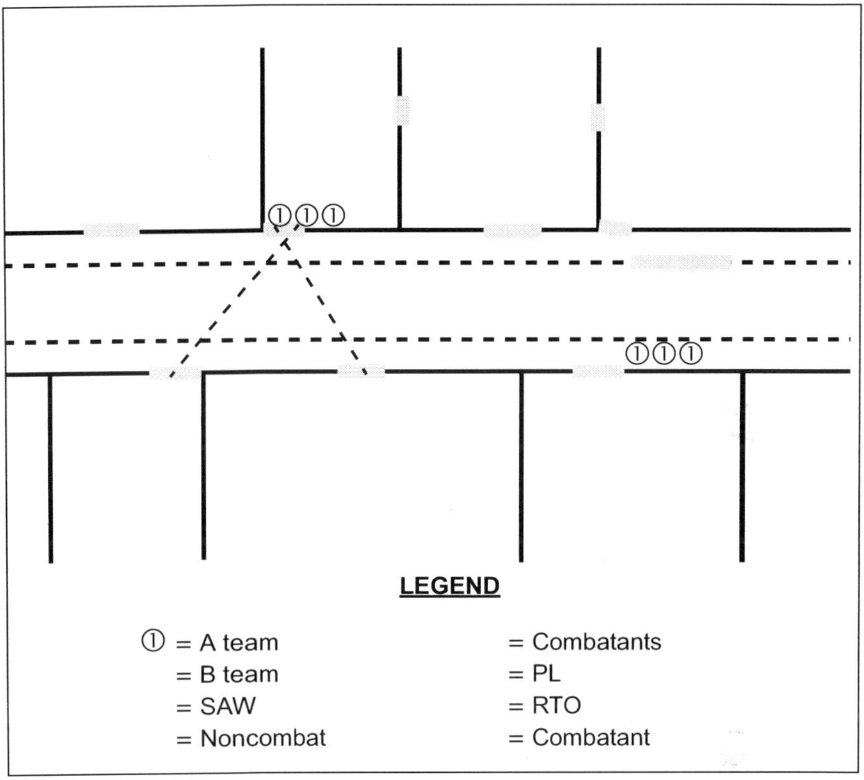

Figure 4-26. First Squad Clearing a Room with Hallway Security

4-153. In *Figure 4-27* the second squad clears all three rooms on their side of the hallway. The second squad was forced to clear each room by using the hallway as an access. In this example, the second squad's B team would not follow A team into the room. When the room is cleared, the first member in the B team stacks and (when instructed) enters in order to relieve the A team of security responsibilities. Once this takes place, A team is free to exit the room and move down the hallway to the next room.

NOTE: Before the team exits the room into the hallway, the lead team member announces that the team is about to exit the room. They should wait for permission to come out into the hallway.

4-154. Once the second squad has finished clearing their side, the number 1 soldier provides security on the next uncleared room on the right (first squad) side. The first squad receives the all clear signal and then moves up to the next room. Both B team SAW gunners continue to provide security down the hallway.

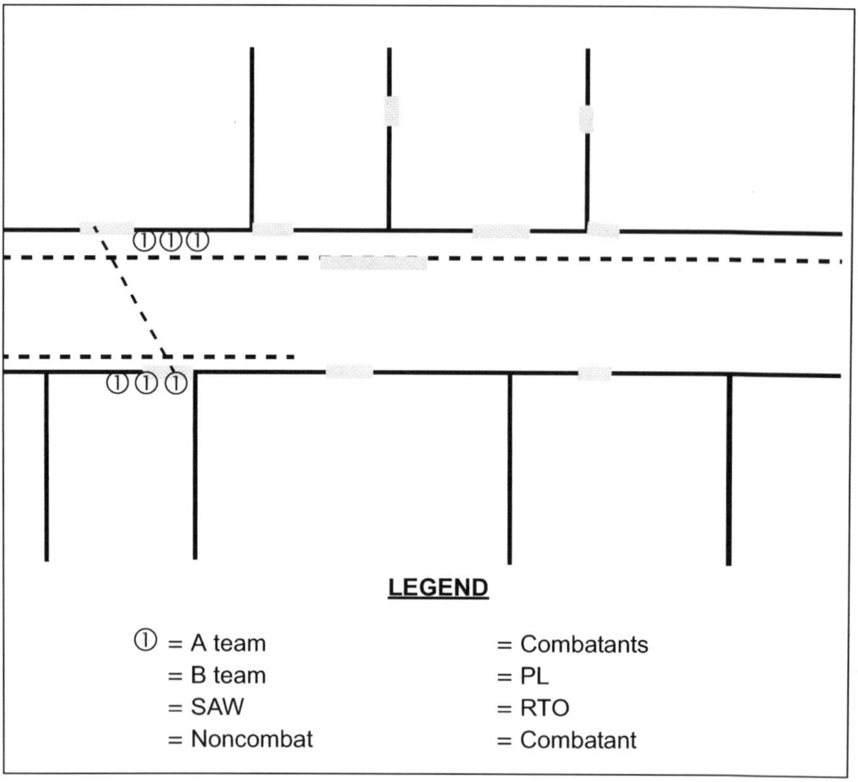

Figure 4-27. Second Squad Clearing Rooms

4-155. *Figure 4-28* depicts the events that take place once the building has been cleared. Two SAW gunners are dispatched to each end of the hallway for security. The rooms containing noncombatants or enemy personnel are searched first (mark them to indicate that they have been searched). Once this is completed, it is usually best to remove all of the combatants and noncombatants encountered during the clearing operation (using a small detachment) along an extraction route that is covered by a support element. The support team escorts the noncombatants to a designated holding area to determine their status, and then searches the remaining rooms.

CLEAR A STAIRWAY

4-156. When multilevel structures are encountered, stairs become an added obstacle that will require maneuver *(Figure 4-29, page 4-86)*. One of the more dangerous stair situations that a team is likely to encounter is a stairway with a turn between floors. Besides the blind spot at the turn, these stairways often have a loft that overlooks the bottom portion of the stairway. If the team is ascending, the number 1 soldier (as always) provides security to his direct front. Soldier number 2 secures the top row of stairs. Soldier number 3 secures the loft area. If a loft area does not exist, soldier number 3 secures the top of the second row of stairs. If the team is descending, each soldier has the same area of responsibility, except that the bottom of the stairway is secured, as opposed to the top.

CLEAR BLIND SPOTS

4-157. When a blind spot is encountered (such as stairs or a sharp corner), it can be cleared with a technique known as cutting the pie *(Figure 4-30, page 4-87)*. The soldier clears the dead space by slicing it up into pieces and then clearing one piece at a time. By doing this, the soldier exposes only a very small portion of his frame to

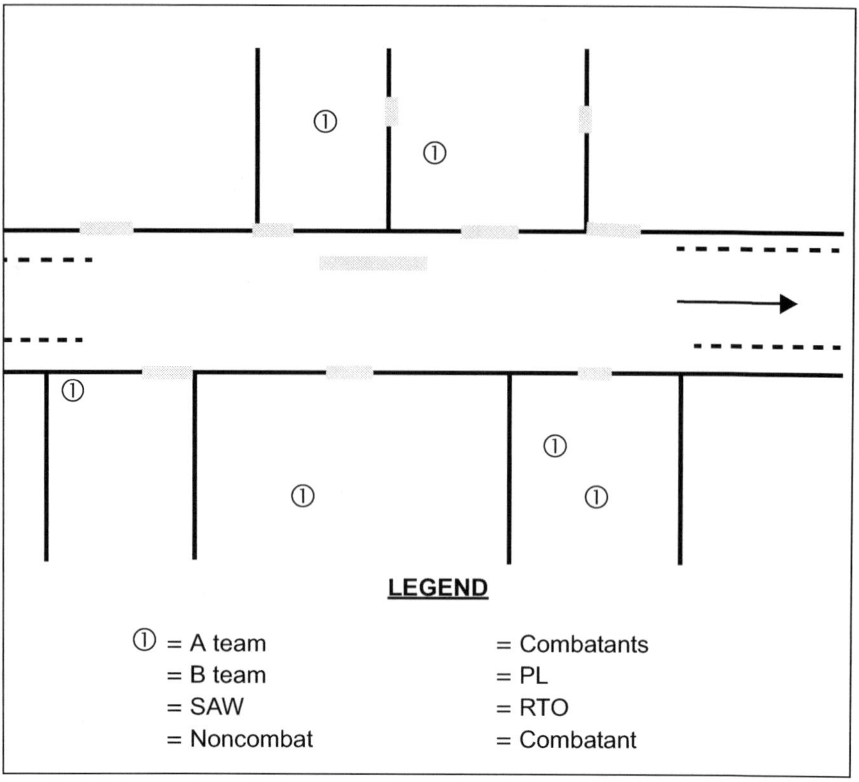

Figure 4-28. Events After Building is Closed

any enemy combatants, while placing himself in a position to methodically clear the dead space.

SECURE PERSONNEL

4-158. The clearing team secures and identifies all personnel found in a room or building that is being cleared or searched. Until identified, (which will not normally occur during the clearance of the building) they approach noncombatants in the same manner as combatants. They do not assume that the frail old man in civilian clothes, cowering in the corner, presents no

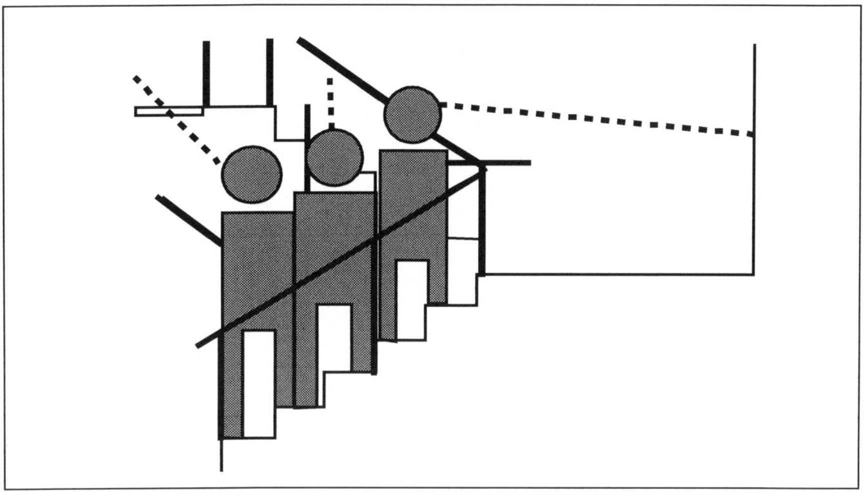

Figure 4-29. Clearing Stairs

threat. On contact, the soldier that first indexes his weapon onto the individual orders the individual to his knees, with his hands on his head, and his fingers interlocked. This soldier stays out of arms reach at all times. Do not allow the unidentified individual near any exits. The individuals will remain in that position, with security present, until the building or room has been cleared. When it is time to secure the individual, the soldier that has been providing security does all the talking. In a slow, loud (but calm) voice, the soldier gives short, easy to understand, instructions to the individual, such as "look at me," "sit up straight," and so forth. This talk has a dual purpose. It—

- Gives directions that helps in securing the individual and calms him down.
- Keeps the individuals attention oriented on him and not on the second securing soldier.

4-159. As the first soldier talks to the individual, a second team member unholsters his pistol for close-in work. The second soldier approaches the individual from

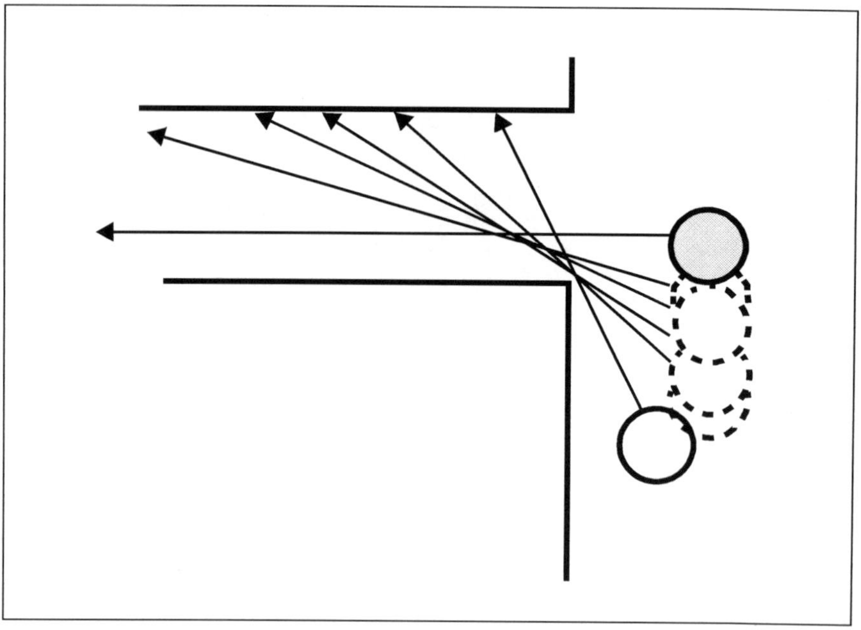

Figure 4-30. Cutting the Pie

a blind spot, grasps the middle three interlocked fingers, and places a knee in the middle of the individual's back. Once the individual is under control, the second soldier holsters the pistol and secures the hands of the individual behind his back. If there are numerous individuals to secure and there will be several minutes before they can be exfiltrated from the building, these individuals can be placed on their knees and tilted forward so that their foreheads are against a wall. Their bodies should be at such an angle that their heads support most of the weight. If the prisoner attempts to get up, he will fall to the floor. This position enables a few soldiers to watch several prisoners. If an enemy soldier becomes engaged and the combatant does not appear to be conscious, it must be determined whether the individual is deceased or not. The body should be approached in the same manner as described above. If a kick does not gain a response, a simple eye gouge or sternal rub should identify whether the combatant is indeed dead. It is always best to secure the hands of the individual before applying an eye gouge or sternal rub. Beware of the enemy soldiers *playing opossum*.

TRAINING FOR MASTERY

4-160. It is obvious that the techniques described above will work best in an environment that has quick-thinking and intelligent soldiers. If the above techniques are practiced using the crawl, walk, and then run method, squads will find themselves catching on very quickly. When the one-room technique is mastered, everything else will fall into place. As teams work together, they will come up with their own systems of signaling, security, and change up in positions. The clearing of a building will be one smooth flow. The platoon leader starts training with single rooms outlined in engineer tape. This system allows for ease in critiquing, and all personnel will be able to observe and learn from the positive and negative performances of each team. The only building layout that cannot be reproduced by tape is a multilevel building (due to the presence of stairwells). Once the platoon is proficient in the one-room technique, the platoon leader moves to connecting rooms outlined in engineer tape. He continues this process until a complete floor in a real building can be cleared smoothly. Once multilevels are covered, he adds simulated breaching, furniture, and combatant or noncombatant scenarios. Once this type of training is completed, an assault team should have the base knowledge required to overcome specific situations that are not covered in this chapter. The platoon leader remembers that this is a perishable skill. As time passes, new platoon members must be integrated into this training and older members must be retrained.

Chapter 5

Maneuver and Mobility Support

MMS, formerly known as battlefield circulation control, consists of those measures necessary to enhance combat movement and the ability to conduct movement of friendly resources in all environments. These measures ensure that commanders receive personnel, equipment, and supplies as needed. MMS is conducted across the full spectrum of military operations. The primary focus of MP during MMS is to ensure swift and uninterrupted movement of combat power and logistical support.

MANEUVER SUPPORT

5-1. Maneuver is the employment of forces on the battlefield in combination with fire (direct or indirect) or fire potential. It is the movement of combat forces to gain a positional advantage, usually to deliver or threaten delivery of direct and indirect fires. MP tasks that support maneuver include—

- MP support to river crossings.
- MP support to breaching operations.
- MP support to a passage of lines.
- Straggler control.
- DC control (refer to *Chapter 7* for more information about DC operations).

SUPPORT FOR RIVER CROSSINGS

5-2. A river is a significant obstacle that may slow, stop, or impede a unit's ability to maneuver. Units are restricted to moving in column formations along limited routes that come together at crossing sites. Friendly forces are vulnerable while crossing water obstacles. The challenge is to minimize the river's impact on the commander's ability to maneuver. The three types of river crossings include—

- Hasty.
- Deliberate.
- Retrograde.

5-3. MP traffic control is essential to help reduce exposure time and speed units across any obstacle. In addition, effective traffic control contributes to the flexibility of the crossing plan by enabling commanders to change the sequence, the timing, or the site of the crossing units. MP can switch units over different routes or hold them in waiting areas as directed by the tactical commander. This support is vital in reducing congestion, speeding the crossing of any obstacle (not just water), and enabling the maneuver forces to maintain momentum.

Hasty River Crossing

5-4. A hasty river crossing is a decentralized operation using organic, existing, or expedient crossing means. It is the preferred river crossing method. Conduct a hasty river crossing as a continuation of an attack to ensure little or no loss of momentum by the attacking force. The MP platoon in direct support of a crossing maneuver brigade, may be required to support the crossing without additional support.

Deliberate River Crossing

5-5. A deliberate river crossing requires planned and augmented MP support. Conduct a deliberate river crossing when a hasty crossing cannot be made successfully, normally when offensive operations must be renewed at the river, and when enemy forces must be cleared from the area. A buildup of firepower and equipment is needed on both entry and exit banks. Normally, MP support from corps is required to augment the division MP company.

Retrograde Crossing

5-6. Closely plan and control a retrograde crossing. Massed crossing forces could slow momentum or exceed bridge classification limits. Forces moving to the rear may retrograde to defensive positions beyond the water obstacle and may be slowed as they set up to defend the exit bank. MP support retrograde crossings the same as they do deliberate crossings.

River Crossing Planning

5-7. The crossing force commander plans the river crossing operation. He prepares an OPORD and specifies what support is required. The PM, based on the OPORD, plans MP support for the river crossing. The plan includes how MP assets will be used and what additional resources are needed. The MP commander supporting the operation plans and supervises the mission based on the OPORD and guidance from the PM. The OPORD normally gives OPCON of all units entering the crossing area to the crossing commander.

5-8. The MP leader supporting the crossing site develops a traffic control plan to support the circulation control plan. He must plan for—

- Traffic control posts (TCPs) and temporary route signs at—
  - Major crossroads on the MSR and near crossing sites and lateral boundaries to control traffic from adjacent unit areas that could interfere with division surface movements.
  - Staging areas and engineer regulating points (ERPs) to provide directions and

information, control movement to and from staging areas according to planned times, and relay messages between traffic HQ and the moving unit.

- Holding areas on the entrance bank to direct traffic to crossing sites; on the exit bank, inside the traffic regulating line (TRL), to control movement; and on the exit bank, outside the TRL, to temporarily hold sections of a convoy or a unit until it can reassemble and continue its movement.

- Mobile patrols to operate along primary routes to control traffic, spot problems, guide and escort vehicles, and reroute traffic when necessary.
- Temporary EPW collecting points. Set up the collecting points outside the TRL. Evacuate EPWs through the crossing areas as quickly as possible so their transit does not impede the movement of friendly forces.

5-9. For brigade crossings, the MP leader may collocate with the brigade staff to form a small, temporary traffic control cell located at the brigade main CP or the brigade TOC. The brigade main CP controls the maneuver support force that consists of corps engineers, bridge companies, MP, and chemical units.

Control Measures

5-10. To ease control of large, fast-moving forces, the river crossing plan usually allots one crossing area for each maneuver brigade. The commander uses control measures to delineate areas of responsibility for subordinates and to ease traffic control. *Figure 5-1* shows the following control measures.

5-11. **Release Line (RL)**. As used in river crossing operations, RLs are used to delineate the crossing area. RLs are located on both the far shore and nearshore and

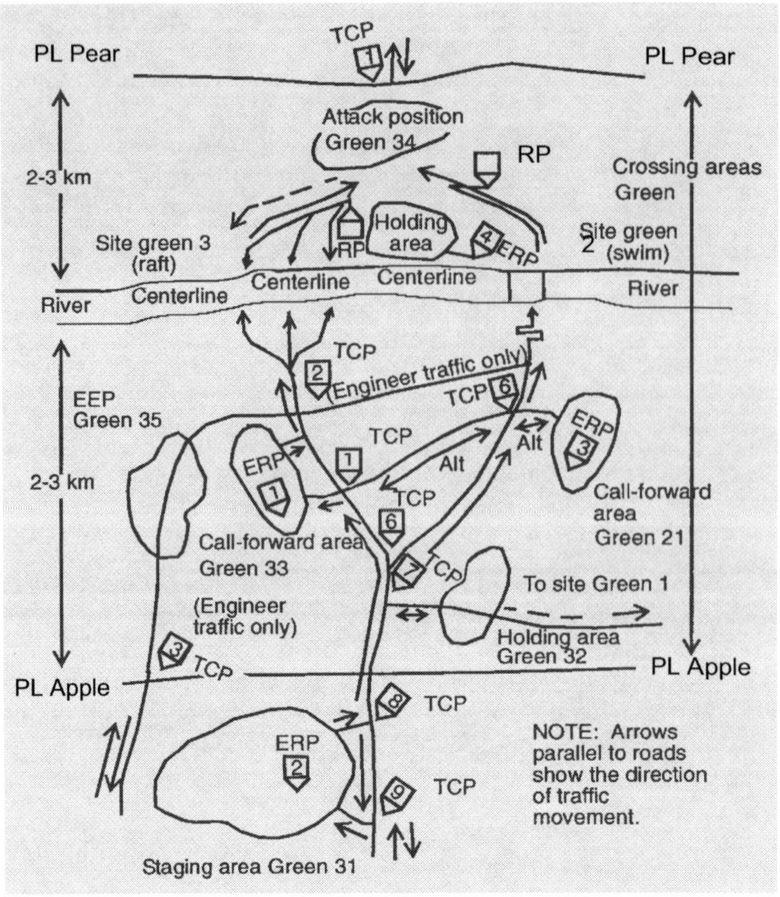

Figure 5-1. River Crossing Control Measures

indicate a change in the HQ that is controlling the movement. RLs are normally located within 3 to 4 kilometers of the river and on easily identifiable terrain features, if possible.

5-12. **Crossing Areas**. Crossing areas are controlled-access areas that decrease congestion at the river. This permits swift movement of the forces. Each lead brigade has a crossing area on both sides of the river that is defined by brigade boundaries and the RL. Crossing areas normally extend 3 to 4 kilometers on each side of the river, depending on the terrain and the anticipated battle.

5-13. **Waiting Areas**. Waiting areas are located adjacent to the routes or axes of advance. Commanders use the following waiting areas to conceal vehicles, troops, and equipment while waiting to resume movement or make final crossing preparations:

- Staging areas. These are battalion-size waiting areas outside the crossing area where forces wait to enter the crossing area. The brigade traffic control cell handles the units' movement into the staging areas. The crossing area commander (CAC) controls movement from the staging areas into the crossing areas. MP operate TCPs at the staging areas according to the crossing and traffic circulation plans. They emplace temporary signs along the route from the staging area through the crossing area to guide the convoys. Units make crossing preparations and receive briefings on vehicle speed and spacing in the staging areas. Staging areas—
  - Are located to support the crossing concept.
  - Are far enough back to permit the rerouting of the battalion along other roads or to alternate crossing sites.
  - Are easily accessible from major routes.
  - Have enough area for dispersing a battalion-size unit.
  - Provide concealment.
- Call-forward areas. These areas are company-size waiting areas located within the crossing

area. Engineers use them to organize units into raft loads; crews use them to make final vehicle crossing preparations. The CAC controls movement from the staging area to the call-forward area. The crossing site commander (CSC) directs movement from the call-forward area to the crossing site and on to the far-shore attack position. As a minimum, each CSC operates his own call-forward area. Call-forward are—

  - Located to support the crossing plan.
  - Company size within the crossing area.
  - Easily accessible from routes.
  - Planned with a minimum of one per crossing site.
  - Collocated with ERPs.
  - Used to organize units into raft loads.
  - The final preparation areas before going to the crossing site.
  - Normally operated by engineers.

- Holding areas. These areas are waiting areas that forces use during traffic interruptions. Units move into these areas when directed by TCP personnel and disperse rather than stay on the roads. Holding areas are battalion size outside of the crossing area and company size within it. Far-shore holding areas are used to organize return traffic. MP operate holding areas according to the crossing and traffic circulation plans and—
  - Are used as call-forward areas for return traffic from the far shore.
  - Are located to support the crossing plan.
  - Are easily accessible from routes.
  - Have enough area for dispersion.
  - Provide cover and concealment.

- Are defensible.
- Maximize traffic flow with minimum control.

- Attack positions. The attack positions are the last positions occupied or passed through by the assault echelon or the attacking force before crossing the line of departure. Within the bridgehead, the attack position is the last position before leaving the crossing area or bridgehead line.
  - Assembly areas. These are the areas where forces prepare or regroup for further action.

5-14. **Engineer Equipment Parks (EEPs)**. These are areas located a convenient distance from bridging and rafting sites for assembling, preparing, and storing bridge equipment and material. They are at least 1 kilometer from the river and hold spare equipment and empty bridge trucks that are not required at the crossing sites. EEPs should be located where they do not interfere with the traffic to the crossing sites and where equipment can be concealed and dispersed. Ideally, routes leading from the EEPs to the crossing sites are not the same routes used by units crossing the river.

5-15. **Traffic Control Posts**. In river crossings, TCP personnel assist the crossing-area HQ in traffic control by reporting and regulating the movement of units and convoys. TCP personnel relay messages between the crossing-area HQ and the moving units. The PM identifies locations that need or require TCPs. MP operate TCPs on both banks of the river to control traffic moving toward or away from it. TCPs are operated at major or critical crossroads and road junctions, staging areas, holding areas, and ERP.

5-16. **Engineer Regulating Point**. ERPs are technical checkpoints used to ensure that vehicles do not exceed the capacity of the crossing means. They help maintain traffic flow. Vehicles not allowed to cross are removed so that they do not cause a traffic backup at the actual crossing site. Engineers man the ERPs and report to the CSC. TCPs are collocated with the ERP to ensure that all vehicles clear the call-forward areas. An additional duty of ERP personnel is to give the drivers final instructions on site-specific procedures and other information, such as speed and vehicle intervals. As a minimum, each crossing site requires an ERP at its own call-forward area. If enough engineer assets are available, an ERP may be established at far-shore holding areas to regulate rearward traffic.

Route Execution

5-17. MP must be prepared to establish holding areas along movement routes on order. If the road network sustains damage, vehicles will need to be routed into the holding areas until traffic can be restored or rerouted. Refer to *paragraph 5-104* for more information about holding areas.

5-18. MP mobile patrols operate along primary routes, monitoring traffic, spotting problems, and rerouting traffic as necessary and conducting AS around the crossing area. They make frequent checks of temporary signs to prevent the enemy from tampering with them.

5-19. MP may be directed to screen the crossing unit's flanks and rear. The size of such an element is determined by METT-TC. In most environments this mission requires at least a squad. MP conduct screening missions to provide early warning of enemy approach and to provide real-time information, reaction time, and maneuver space for the crossing unit. The squad fights only for self-protection and remains within its capabilities. Refer to *Chapter 6* for more information about screening missions.

5-20. Include at each crossing site a temporary EPW collection point. Initially the collection point will be on the entry bank. Once MP cross as part of the support force, a temporary collection point is established on the exit bank. A division central collection point is established outside of the crossing area. Refer to *Chapter 7* for more information about division forward collection points.

5-21. Rigid control of civilian movement is necessary to preclude congestion on movement routes. The PM coordinates for HN police support to ensure that the civilians who live in the crossing area are kept in place or, if necessary, quickly moved to designated areas away from the river. Normally, civilians are not allowed to cross the river or move along the edge of the river during the river crossing operation. Refer to *Chapter 7* for more information about DC resettlement.

MILITARY POLICE SUPPORT TO BREACHING OPERATIONS

5-22. Breaching operations are conducted to allow maneuver despite the presence of obstacles. Obstacle breaching is the employment of a combination of tactics and techniques to advance an attacking force to the far side of an obstacle that is covered by fire. Breaching operations begin when friendly forces detect an obstacle and begin to apply the breaching fundamentals. Breaching operations end when the battle handover has occurred between the follow-on forces and a unit conducting the breaching operation.

Support Planning

5-23. MP support to breaching operations is similar to MP support to river crossing operations. The employment of MP is based on METT-TC, available resources, and the commander's priorities. MP support to breaching operations includes—

- Operating TCPs at the breaching site and along routes leading to or departing from the breaching site.
- Operating holding areas.

- Providing mobile guides to escort the units.

5-24. The platoon leader coordinates with higher HQ and the engineer forces conducting the breach for essential information that includes the—

- Azimuth and distance to the final-approach marker or the 8-digit grid coordinate of the final-approach marker that is entered into the teams Global Positioning System (GPS) receiver.
- Lane marking pattern currently emplaced.
- Type of final-approach marker used.
- Traffic control plan and march order.

5-25. A combined-arms breach is a complex operation and requires precise synchronization. Breaching operations normally require the maximum use of TCPs to assist support, breach, and assault forces to move along various lanes. Refer to *paragraph 5-88* for more information about TCPs. Lanes are marked to safely pass units through the obstacle. The three levels of lane marking are—

- Initial.
- Intermediate.
- Full.

5-26. MP may provide TCPs and guide support to lanes at any level of marking. However, the main effort of MP support may come in later phases of the operation, when larger units (battalion and above) are passed to subsequent objectives, and time permits marking improvements to be made. The increase in traffic and the more diverse forces with different levels of driver experience will increase the need for MP traffic control operations. MP guides are simply mobile MP teams that escort units from one control measure or point to another. Guides and TCPs are essential when there are multiple lanes. *Figure 5-2, page 5-12* shows the flexibility that the combinations of multiple lanes and guides or TCPs provide the commander.

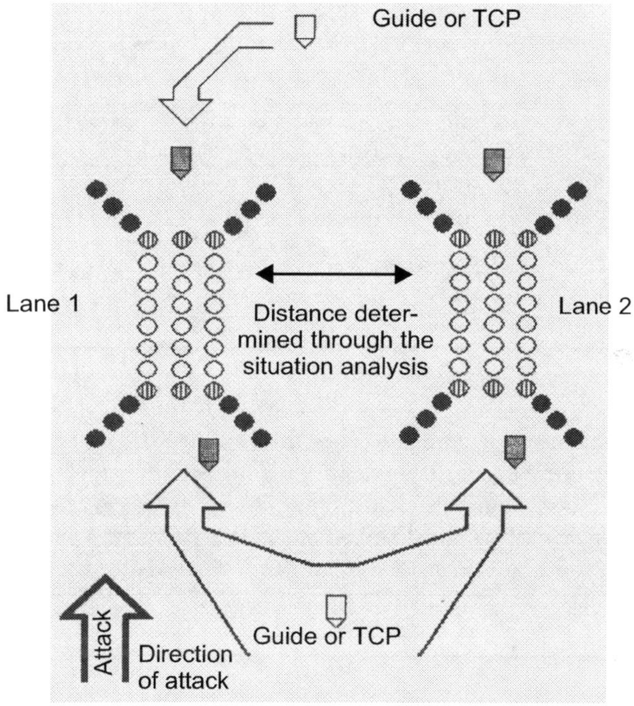

Figure 5-2. Multiple Lanes (Two-Way Traffic)

Movement Execution

5-27. The commander sets the priority of movement based on the situation. MP may concentrate their efforts on assisting the immediate passage of larger combat forces. Or their priority may quickly shift to ground evacuation of casualties or vehicle recovery operations. MP traffic control operations give the commander the ability to make last-minute changes in the traffic flow or lane usage

5-28. MP may be required to establish unit holding areas (battalion and company size) in the event that traffic is disrupted on the lanes due to enemy activity or the need to do maintenance or upgrade a lane. Refer to *paragraph 5-104* for more information about holding areas.

5-29. The commander collocates guides or TCPs at the far recognition marker when he feels the situation requires more positive control.

5-30. Guides and TCPs are briefed on this information and are kept up to date on changes to the traffic control plan and enemy activity in the AO.

5-31. The platoon leader plans for the possible need to establish a forward EPW collection point near the breaching operation. Refer to *Chapter 7* for more information about division forward collection point. He must also plan for an increase in the number of TCPs needed during limited visibility or in restrictive terrain. Refer to *FM 3-34.2* for more information about breaching operations.

PASSAGE OF LINES SUPPORT

5-32. This area describes how an MP leader is to plan and conduct MP support to the passage of lines. The MP elements described in the following paragraphs are supporting the passing and stationary units. MP conducting a battle handover or passage of lines to a TCF is discussed in *Chapter 6*.

5-33. A passage of lines is a tactical event normally associated with a battle handover. A passage may be designated as a forward or rearward passage of lines. Moving a maneuver unit through the positions of an emplaced unit that is in contact with the enemy is a critical action. It requires detailed coordination; planning; and close, continuous supervision of the movement.

5-34. The main focus of MP support to a passage of lines is normally employing special traffic control measures that include—

- TCPs.
- Temporary route signing.
- Checkpoints and roadblocks.
- Defiles.

5-35. MP may also provide guides to escort the passing unit en route to a release point or AA. Similar to MP support to breaching operations, guides provide the commander a means to change the sequence, timing, or lanes of the passing units.

Passage of Lines Planning

5-36. MP support the passage of lines operation to assist a maneuver unit in contact with the enemy to maintain movement. Depending on the scope of the operation, a division MP company may not be enough to support a passage of lines operation. METT-TC may necessitate the need for additional corps MP support.

Control Measures

5-37. When planning control measures for a passage of lines, MP leaders must consider the placement of the following:

- AAs where units prepare for further action.
- The battle handover line (BHL) where the stationary force assumes responsibility for the sector from the covering force.
- The forward edge of the battle area.
- Passage lanes along which the passing units move to avoid stationary units and obstacles.
- Passage points where units will pass through one another. They are located where the commanders want the units to execute the

passage of lines. Designate multiple passage points to help eliminate congestion.

- Contact points (designate an easily identifiable terrain feature) where the units will physically meet.
- SPs where unit elements come under the control of the commander responsible for the movement.
- Phase lines, used in controlling the timing of the operation, are usually recognizable terrain features extending across the zone of action.
- RPs where unit elements revert to their respective commanders and continue moving to their destinations.
- Travel routes from the point of origin to the destination.
- Checkpoints used to coordinate friendly movement. (Checkpoints are not used as reference points for reporting enemy locations.)

Passage of Lines Execution

5-38. MP support for a passage of lines is conducted at the platoon level. The company monitors the platoon and coordinates with higher HQ. The company operations section is required to conduct detailed coordination with the passing and stationary units that includes—

- Communication requirements.
- Recognition signals.
- Route SP.
- Time of passage.
- Passing lanes.
- Control measures to include TCPs, escort and guide vehicles, temporary route signs, or a combination of these.
- EPW and DC evacuation.

5-39. The platoon leader continuously coordinates with the company operations section to confirm the following:

- The size of the passing unit.
- Locations of AAs.
- Recognition signals.
- The actual time that the passage of lines will commence.

5-40. MP support for a passage of lines begins at the route SP, which serves as a contact point. Recognition signals are displayed at the SP. MP monitor the passing unit's command net during the entire passage. Radio silence is maintained during this time.

5-41. Depending on the situation, MP can support the passing unit with the following:

- TCPs.
- Escort and guide vehicles.
- Temporary route signs.
- A mix of these measures.

5-42. The platoon leader selects the method that best supports the moving unit's passage through the stationary unit. Temporary route signing will decrease the number of TCPs needed, but if routes are not well defined or they cross congested areas, expect to provide TCPs or escort vehicles.

5-43. MP must be prepared to initiate vehicle holding areas at designated locations along movement routes. If the road network sustains damage, vehicles will be routed into a holding area until traffic can be restored or rerouted. (Refer to *paragraph 5-104* for more information about holding areas.)

5-44. The platoon leader plans for the need to establish a temporary EPW collection point, and establishes it near the passage area but out of the view of the operation.

5-45. Strict control of the movement of civilians is necessary to preclude congestion on routes used for the operation. When necessary, MP establish evacuation routes to move DCs to a designated area rear of the staging areas. Local nationals who live in the immediate area will remain in place or be evacuated primarily by HN police from the area.

STRAGGLER CONTROL

5-46. MP conduct straggler control operations to assist commanders in maintaining combat strength by locating and returning stragglers to their units. MP identify stragglers at the TCPs, checkpoints, roadblocks, defiles, or while patrolling the MSR. For large numbers of stragglers, special collecting points are set up along the MSR.

Operate Straggler Posts

5-47. When operating a straggler CP, MP teams need to know what units are located or operating in their AO. Most stragglers are soldiers who have become accidentally separated from their command. Stragglers are identified by checking the following:

- Uniforms.
- Unit insignia.
- Bumper markings on the vehicles.
- Identification cards or tags.
- Passes or other authorization documents.

5-48. For each straggler identified, as a minimum, MP record—

- The straggler's name, rank, social security number (SSN), and nationality.
- The straggler's unit.
- The straggler's category ("injured" or "uninjured").
- Whether the straggler is armed or not.
- Where the straggler was coming from and his destination.
- Why and when the straggler left the unit.

- The location where the straggler was sent.

5-49. MP administer first aid to the injured, wounded, or ill stragglers. Seriously ill or injured soldiers are evacuated. Stragglers who have information of immediate tactical value are reported to higher HQ. Soldiers fit for duty who mistakenly became separated are returned to their units or a HQ within their chain of command. The soldiers unit is responsible for any transportation requirements.

5-50. MP treat deliberate stragglers, those who have deserted or are attempting to desert or are absent without leave (AWOL), with caution. These stragglers may resort to violence to avoid military control. MP search, disarm, and detain them. They hold these stragglers until transport and escort can be arranged to take them to their unit, the straggler collecting point, or another facility set by the SOP or the straggler control plan. MP safeguard confiscated property and documents, and dispose of them according to the straggler control plan.

5-51. MP handle stragglers from the HN or other allied forces the same as US stragglers. If the PM has coordinated with other national forces to set up joint straggler posts, allow MP from other national units to handle stragglers from their own forces.

Operate Straggler Collecting Points

5-52. When large numbers of stragglers exist and TCPs, mounted patrols, and straggler control posts are not able to handle the straggler flow, MP may be tasked to operate a straggler collecting point. MP temporarily hold stragglers at collecting points while they process them for return to their units, placement in medical channels, or placement in other military channels.

5-53. The PM operations section plans the location of a straggler collecting point, placed along a key MSR or at an intersection of the MSR. This allows quicker access to the straggler collecting point to aid in moving stragglers to their appropriate destination.

5-54. Often it is collocated where elements of medical, transportation, and MP units can share efforts to ease the disposition of stragglers. At a straggler collecting point, MP may need food, water, clothing, and shelter for stragglers. If a medical facility is not close by, request extra medical supplies and be prepared to administer first aid.

5-55. Guards separate the injured stragglers from the uninjured. They process each soldier at the collecting point. Guards record the key information on each soldier for a straggler report. They search, segregate, and guard stragglers who refuse to return to their unit. Guards assemble and forward the report to wherever the straggler control plan directs. They hold stragglers at the collecting point until transport arrives.

5-56. Detain stragglers who refuse to return to their unit until their unit provides an escort or until they are transported to a detainment facility.

MOBILITY SUPPORT

5-57. Mobility is the capability of military forces to move from place to place while retaining the ability to fulfill their primary mission. It includes those activities that enable a force to move personnel and equipment on the battlefield without delays due to terrain or obstacles. MP activities that support mobility include the following:

- Route reconnaissance and surveillance.
- MSR regulation enforcement.
- Special circulation control measures.

ROUTE RECONNAISSANCE AND SURVEILLANCE

5-58. MP conduct route reconnaissance and surveillance operations to gain detailed information on specific routes to be used as deployment routes, MSRs, or movement corridors. Mobile MP teams record and report the condition of the MSR and other critical roadways, identifying effects of weather on road surfaces, damage to routes, NBC contamination, and the presence of enemy activity. Platoon leaders use this information to develop a road reconnaissance report and a route reconnaissance overlay.

5-59. The platoon leader provides the MP reconnaissance reports and overlays are used to assist the division PM with the development of the division traffic control plan. Movement planners use information gathered from the MP route reconnaissance to update the highway traffic section's (HTS's) traffic circulation control plan and to formulate the highway traffic regulation plan. Refer to *FM 55-10* for more information about traffic circulation control plans and highway traffic regulation plans.

Plan

5-60. MP leaders plan route reconnaissance operations by examining intelligence reports and maps of the area surrounding the route to be reconnoitered. When time is critical, MP conduct a hasty route reconnaissance to obtain specific information only. The MP leader must clearly understand the following critical tasks to be accomplished:

- Find and report all enemy forces that can influence movement along the route.
- Determine the trafficability of the route.
- Reconnoiter any special areas that could influence movement on the route. These areas may consist of highly populated areas or key

terrain features. Additional teams may be needed to cover these areas based on METT-TC.

- Inspect all bridges on the route.
- Locate fords or crossing sites near all bridges on the route.
- Inspect all overpasses, underpasses, and culverts.
- Locate areas suitable for short halts and holding areas.
- Locate mines, obstacles, and barriers along the route.
- Locate a bypass around built-up areas, obstacles, and contaminated areas.
- Report route information.

5-61. A more detailed route reconnaissance would include additional information concerning the terrain, potential hazards, or obstacles and would include key terrain and built-up areas 2 to 3 kilometers on either side of all MSRs. A route reconnaissance this detailed requires considerably more time and personnel.

Execute

5-62. One MP squad can conduct a hasty route reconnaissance of only one route, (about 30 kilometers). MP platoons reconnoiter three routes within the boundaries of the platoon AO, if route reconnaissance is their primary focus.

5-63. The size of the reconnaissance patrol is determined by using METT-TC factors. In most environments route reconnaissance operations are not conducted with less than a squad. The squad leader organizes the squad into security teams and a reconnaissance team. The reconnaissance team records the information, completes a *DA Form 1248*, and prepares a reconnaissance overlay. The other teams provide security.

5-64. MP do not engage the enemy when conducting route reconnaissance except in self-defense or when ordered to do so. They report any visual contact with the enemy and maintain surveillance while gathering as much information as possible. They break visual contact only on order from proper authority.

5-65. Often the main purpose of a reconnaissance is to confirm information already known about a route. Additional information can be obtained from the division PM, the division transportation officer, and from HN police. MP teams talk to convoy commanders, vehicle drivers, local nationals, and highway control regulating teams to gain or verify information about well-traveled roads. Although this information is not as reliable as information gathered from driving a route, its reliability increases when several road users report the same condition.

5-66. Mounted MP patrols continuously collect data at the level of detail required by the PM or commander and report it by the fastest secure means available. The patrols travel the routes within the AO to—

- Identify and locate the recommended route.
- Check the driving time and distances between easily recognized points.
- Look for obstructions and restrictions (bridges, tunnels, steep grades, sharp curves, ferries, snow blockage, defiles, flooding, rock falls, and slides).
- Note the location and type of possible ambush sites on the route.
- Look for terrain where direct fire from the enemy could stop movement on the route.
- Identify natural defense, counterambush, and assembly locations.

- Identify areas where terrain restricts communications.
- Watch for enemy situations that could affect route security or conditions, such as—
  - Enemy elements positioned on key terrain.
  - The enemy emplacing mines and other obstacles.
  - Frequency changes or type of enemy fire in the area.
  - Enemy aerial interdiction.

5-67. To keep from overlooking critical terrain data, the squad leader prepares a checklist of items that may be included on the reconnaissance overlay, such as—

- The route classification formula.
- The identification and location of the reconnoitered route.
- The road distances between the points that are easily recognized both on the ground and on the map.
- The presence and lengths of steep grades (having a slope of 7 percent or greater).
- Curves having radii of less than 45 meters.
- Military load classifications (MLC) and limiting dimensions of bridges. Include suitable bypasses, classifying them as easy, difficult, or impassable.
- Locations and limiting data of fords, ferries, and tunnels. Include suitable classification of bypasses.
- Route restrictions (like underpasses) below minimum standards and, if appropriate, the distances such restrictions extend.
- Areas suitable for short halts, holding areas, or bivouacs that offer easy access to the roadway and adequate dispersion, cover, and

concealment. Include information on the shoulders.

- Rock fall and slide areas that may present a traffic hazard.
- Overhead clearance of less than 4.3 meters.
- Civil or military road numbers or other designations.
- Obstructions to traffic.

5-68. Roads that bisect heavily wooded areas are likely obstacle and ambush sites. Heavily loaded vehicles are vulnerable to ambushes and unable to circumvent obstacles easily. Steep grades and numerous S-turns, where logistical vehicles that are heavily loaded with supplies slow to a crawl, make good ambush points. For more information on route classification, refer to *FM 5-170* and *Appendix I* of this manual.

5-69. If enemy activity is suspected along a route, the squad should—

- Use caution when approaching critical locations.
- Deploy using traveling overwatch or bounding overwatch. Choose movement techniques according to the latest information on suspected enemy activity.
- Avoid danger areas.

5-70. Use caution when approaching a sharp bend or a defile in the road. Such areas are often mined and are ideal sites for an ambush. When necessary, the squad leader has the reconnaissance element conduct a dismounted reconnaissance of these areas while the security element provides overwatch.

5-71. MP check bridges for mines and booby traps. Before crossing a bridge, MP have the security element move to an overwatch position. They have the reconnaissance element dismount and check the bridge and its approaches for mines, booby traps, and demolition charges. If any are found, MP move to a covered and concealed area, report the information, and request engineer support to clear the mines. MP maintain surveillance of the bridge until the mines are cleared. They stop friendly forces and civilians from using the bridge until the engineers have cleared it. When the bridge is cleared, MP have the reconnaissance element gather critical data on the bridge's characteristics and continue the reconnaissance.

5-72. MP reconnoiter key terrain and built-up areas near the route. They move on and off the road to identify enemy activity. The type of terrain dictates whether a reconnaissance is conducted mounted or dismounted. Reconnoitering terrain can be time-consuming. The mission order and the time available for the reconnaissance determine how many and which terrain features are reconnoitered.

MAIN SUPPLY ROUTE REGULATION ENFORCEMENT

5-73. MP traffic control activities support movement control by enforcing highway regulation plans. Traffic enforcement measures, such as speed control and safety inspection checkpoints, help protect the force and ensure that only authorized traffic uses controlled routes. MP employ special circulation control measures, such as temporary route signing, TCPs, holding areas, defiles, and checkpoint and roadblock operations, to support combat and sustainment operations. Refer to *Chapter 6* for more information about checkpoints and roadblocks.

5-74. Highway regulations are set by the agency having jurisdiction over the road network. MSR regulation measures are stated in the command's highway regulation plan. They also may be in the unit SOPs and command directives. The HTS sets the route classification.

5-75. Mounted MP teams patrol MSRs to monitor traffic and road conditions. They gather information on friendly and enemy activity and assist stranded vehicles and crews. Road condition changes and enemy activity are reported immediately through MP channels.

SPECIAL CIRCULATION CONTROL MEASURES

5-76. MP limit, control, block, or direct mounted or dismounted forces traveling on the MSR, by employing special circulation control measures. Special circulation control measures include the following:

- Temporary route signing.
- TCPs.
- Holding areas.
- Defiles.

5-77. Many of these measures are used in MP support to breaching operations, river crossings, and passage of lines.

Temporary Route Signing

5-78. A signed military route system, like the signed US highway system, can enable road users to reach their destinations by following route signs and road markings displayed along the roadside. MP patrols monitor signs on a routine basis, checking specific signs before critical moves. Engineers erect permanent signs, but signs can be damaged, destroyed, or moved by weather, saboteurs, and battle.

NOTE: For more information about temporary route signing refer to *Appendix I*.

Traffic Control Post

5-79. TCPs are used to support MMS only when needed. They are used to preclude the interruption of traffic or unit movement along designated routes. TCPs are communication links to units using the MSR. Show on the traffic control plan and the traffic circulation plan the placement of TCPs. MP activities at a TCP include the following:

- Monitoring and assisting traffic authorized to use the MSR.
- Redirecting unauthorized vehicles to the road network they need.
- Providing route security for the MSR at critical locations or intersections.
- Monitoring for NBC contamination.
- Rerouting traffic as needed.
- Gathering information and reporting it.
- Providing information to passing units.
- Assisting stragglers and DCs.

5-80. **Plan**. An MP squad can operate one TCP for an extended period or three TCPs for a short duration. The platoon leader uses METT-TC to analyze the mission and estimate the situation. He decides the appropriate weapons, materials, and equipment needed and considers such factors as the movement of traffic and the degree of control required. If HN police support is needed, the company operations section or the PM arranges for it. The platoon leader uses overlays and the traffic control plan to determine the location for the TCP. The squad leader plans for emergency destruction of documents and equipment in case the TCP is attacked. The squad leader plans for continuous operations by—

- Developing an adequate sleep plan.
- Arranging for maintenance and refueling.
- Arranging for additional rations.
- Constructing fighting positions.
- Camouflaging all vehicles and equipment.

5-81. **Execute**. TCPs are manned at points where two or more MSRs converge or where confusion could affect vehicle movement. They are used to help protect the force at critical locations where civilian or military traffic can cause an accident. Operations in which TCP will be maximized include the following:

- Deliberate river crossings.
- Deliberate breach operations.
- Defile operations.

5-82. When METT-TC requires a TCP to be manned by one MP squad, the squad leader—

- Analyze the terrain location.
- Positions the teams.
- Directs the squad's vehicles to a covered and concealed position near the squad's fighting position. Use camouflage nets, if needed.
- Selects a fighting position from which the squad can cover and secure the TCP.
- Maintains communication.

5-83. Once the squad has occupied the actual TCP site, the squad leader establishes security and provides a grid coordinate to higher HQ. One team in the overwatch covers the TCP while another team watches the flow of traffic from a covered and concealed position near the road. When needed, a member of this team moves to the center of the road to direct the flow of traffic while the other members provide security. If necessary, the squad leader has the third team resting in a covered and concealed position. The squad leader ensures that all three teams communicate by wire (the preferred method) or radio. If neither is available, they use arm and hand signals.

5-84. At a TCP, the main purpose is to ensure smooth and efficient use of the road network according to the traffic circulation plan. The plan contains—

- Military route numbers and directions of travel.
- Light lines and blackout areas.
- Highway regulation points and MP TCP.
- Route control classification.

5-85. Vehicles too wide or heavy for a road will be denied access. MP reroute them to alternate MSRs. No authorization is needed for travel on an open route, but use of a classified route may be restricted to certain units, operations, or types of vehicles.

5-86. All vehicles on the dispatch route will have a current movement credit issued by the HTS. On a supervised route, normally a column of 10 or more vehicles or an individual vehicle of exceptional size or weight will have movement credit from HTS.

5-87. MP stop vehicles or convoys that are not following MSR regulations. They tell the convoy commander why the vehicles are halted. The convoy commander makes immediate corrections. When immediate corrections cannot be made, MP record the key information about the incident and report it through MP channels.

5-88. TCP is used to disseminate information about the AO. MP provide information to authorized personnel only. They support the commander's force protection program by providing friendly forces with current information on route conditions and enemy activity. At the TCP, MP disseminate information on the locations of contaminated areas, supply points, medical facilities, and other units on a need-to-know basis and should not volunteer more information than is needed to avoid creating security and/or intelligence issues.

5-89. MP actively seek information from road users by asking drivers what they have seen of suspected or actual enemy activity along the MSR. When language is not a barrier, MP talk to local civilians and to the HN civilian and military police to gain information about the road network in an AO. MP pass this information up the chain of command so that it can be verified as reliable.

5-90. MP are constantly on watch for enemy aircraft and suspicious activity by the local populace. When such activity is observed they relay SPOTREPs through MP channels. MP use the SALUTE format to report this information.

5-91. When the movement control agency requests it, MP keep track of military movements that pass through a TCP. This helps movement planners keep track of the progress of convoys. This information is compiled into a passing report that includes the—

- TCP location.
- Date.
- Convoy identification (unit or serial number).
- Time the first vehicle passed the TCP.
- Time the last vehicle passed the TCP.
- Number of vehicles in the convoy.

5-92. Usually, TCP passing reports are picked up at the TCP or transmitted by secure radio. The platoon leader compiles the TCP passing reports into one report. He forwards the report through MP channels or as directed by the commander. The report may be written or transmitted. If transmitted, a report is encoded according to the unit SOP. In some instances, the platoon leader may permit a squad leader to bypass the usual report channels and submit a passing report directly to the movement control agency.

Holding Areas

5-93. MP operate vehicle-holding areas to help regulate the traffic flow. Holding areas can be used as independent measures or with other measures like defiles or checkpoints to support large operations like river crossings or passage of lines. When MP operate holding areas, they direct vehicles, convoys, and troops into and out of the holding areas.

5-94. **Plan**. METT-TC and the size of the holding area determine the number of teams needed to operate it. When one MP squad operates a holding area, the squad leader designates one team to control the entrance to the holding area, another team to control the exit from the holding area, and one team to provide security. He also assigns each squad member a fighting position.

- The general location for a holding area may be designated by the echelon movement control unit, PM, or MP company commander. The exact location is selected by the MP leader with the mission. The holding area's location is noted on the traffic control plan and passed to the echelon movement control unit to keep the traffic circulation plan current. Select a site where—
- Parked vehicles can face the exit so that they can be driven from the area quickly.
- Vehicles can be dispersed.
- There is easy access to and from the roadway.
- The surface of the area is firm enough to hold the weight of the vehicles.
- The area is large enough to allow vehicles to be covered and concealed from air and ground observation.
- The area can be defended.

5-95. **Execute**. MP teams at the entrance and exit to the holding area are positioned in concealed locations. When vehicles approach the holding area, one MP moves to the center of the road and directs the vehicles in. He tells the vehicle driver or convoy commander where to park. He moves back to the concealed location when all vehicles have entered. Vehicles moving in the holding area should be ground-guided by members of the convoy.

5-96. The MP team at the exit operates in a similar manner. MP move from their cover and concealment only when necessary. Depending on the tactical situation, MP teams may use red lens flashlights to direct traffic at night. Colored chemical light sticks are posted to help drivers identify their locations within a holding area if the situation permits.

5-97. Ideally, the MP leader controls the holding area from a position overlooking the entrance and exit. The MP leader receives instructions on when to allow vehicles to pass from higher HQ. When the holding area is to support a river crossing site, the MP leader has a movement schedule to follow. He receives his movement information from the echelon movement control office or the crossing area commander.

5-98. When operating a large holding area, the MP leader may need additional personnel inside the holding area to direct traffic and parking and ensure that the units comply with the flow plan. Large holding areas will have a simple control plan, such as a subdivision system.

5-99. Take the following steps when using a subdivision system:

- Make a map or a sketch of the area showing the road net, trails, and major obstacles.
- Outline the holding area on the map or sketch.
- Divide the area into equal subdivisions and assign a letter or a name to each subdivision. This helps direct units to their section of the holding area.
- Erect signs showing the outline of each area.
- Develop a traffic flow plan and erect directional signs to help users.

5-100. Keep a count of vehicles in the subdivisions by number, size, and unit designation for each vehicle. At night, use chemical light sticks to identify the sections within the holding area and the exit.

5-101. Vehicle holding areas are controlled sites. MP instruct convoy commanders to keep drivers with their vehicles or assign them fighting positions. They do not allow convoy personnel to openly move around within the holding area. They use signs to help control traffic and maintain communications between positions by wire or hand and arm signals.

Defiles

5-102. Defiles keep traffic moving smoothly despite narrowed passageways. Controls at defiles ensure that traffic moves through the passage, one direction at a time, first from one end and then the other. A defile may be considered a critical site and its security a priority.

5-103. **Plan**. METT-TC and the estimated length of the operation are used to determine the actual size of the element needed to operate a defile. In most environments, a defile will not be operated with less than a squad. An MP platoon may be needed to conduct a large defile operation for extended periods. At a defile *(Figure 5-3)*, MP do the following:

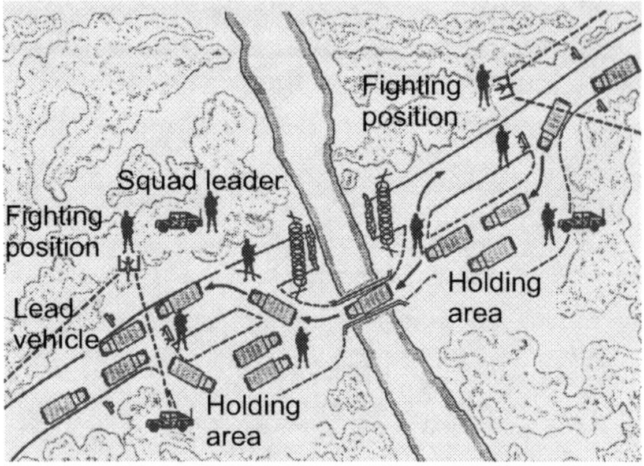

Figure 5-3. Defile

- Secure and defend the site.
- Brief drivers about obstructions.

- Control access so vehicles move through quickly.
- Ensure that vehicles enter one at a time.
- Reroute traffic when necessary.

5-104. Terrain or traffic needs may dictate a need for vehicle holding areas and signs or a TCP. Larger defiles require a holding area at each end. The placement of holding areas depends on the site available and the ease of communicating between the sites and the defile.

5-105. **Execute**. Because defiles involve restricted movement, they are an ideal target for the enemy. Security of a defile is a priority. Before a defile is put into operation, an area reconnaissance is conducted to detect the presence of enemy activity in and around the location of the defile. Once the area has been reconnoitered, the squad leader establishes security and reports the exact location of the defile, if not previously known, to higher HQ. He selects the crew-served weapons fighting position, picking key terrain that overlooks the defile. He ensures that the squad's vehicles are covered and concealed. As in most stationary MP operations, communications within the squad is primarily wire and arm and hand signals.

5-106. MP use control measures to keep traffic flowing smoothly. They use the simplest method of control. Plan for the use of two different control measures. This provides a back-up method, if needed. Control measures can include the following:

- Visual signals to tell traffic when to move, such as hand and arm signals, flashlights, or a handheld flags. Use any technique that shows vehicles when to move. Visual signals work best for small defiles where holding areas are not needed.
- Wireless or wire communications to tell teams in holding areas to hold or start traffic through the defile. Link communications directly between holding areas, or route them through the leader

at the defile site. Use wire communications as the main means of communication. Use wireless communications as a backup or when no other means are available.

- A flag to identify the last vehicle moving through a defile. Give the flag to the last driver or attach it to the last vehicle entering the defile. Another MP removes the flag when the vehicle reaches the end of the defile. This serves as a signal for traffic to start in the opposite direction. This is repeated as often as needed.
- An MP rider to indicate the last vehicle of a column. The rider stays in the last vehicle until the column reaches the opposite side. He dismounts and rides back in the last vehicle returning. This technique ensures that all the vehicles clear the defile.
- MP lead and trail vehicles in the front and rear of a column to guide it through the defile. After the column clears the defile, the vehicles guide a column moving in the opposite direction. MP use this method when movement through a defile is complex and requires an escort. The trail vehicle ensures that all the vehicles clear the defile. A single lead or a single trail vehicle can also be used, depending on the number of vehicles and the complexity of the defile.

5-107. To ensure traffic flow is not interrupted by a disabled vehicle, MP plan for a recovery vehicle to stand by at the defile. If a recovery vehicle is not available, they use field expedient measures.

5-108. Regardless of what special control measure is employed, MP leaders coordinate with the PM for the location, duration and special criteria for the execution of the control measure. Once operational, MP monitor its effectiveness and ensure that force protection and security measures are followed and maintained.

Chapter 6

Area Security

MP conduct AS operations to protect critical functions, facilities, and forces. They synchronize efforts with base and base cluster defense planners within a specific AO to ensure that support and sustainment operations are not interrupted. The HN, when capable, retains responsibility for security of all areas outside US bases. However, US commanders are always responsible for the defense and security of US forces and bases regardless of HN support. AS is conducted by MP across the full spectrum of army operations to protect the force, impose order, and ensure freedom of movement. MP activities that support AS include reconnaissance operations, ADC, base and air base defense (ABD), response force operations, and critical site asset and high-risk personnel security.

RECONNAISSANCE OPERATIONS

6-1. MP plan and conduct area and zone reconnaissance, screening and surveillance missions, and counterreconnaissance.

6-2. MP conduct reconnaissance and screening missions to obtain information about the activities and resources of an enemy or potential enemy or to secure data concerning the characteristics of a particular area. MP reconnaissance, screening, and surveillance efforts include area, zone, and route reconnaissance and counterreconnaissance. These missions may be conducted primarily in the rear area, but may occur anywhere sustaining operations are conducted. Refer to *Chapter 5* for more information about route reconnaissance. MP employ NBC detection equipment to determine the absence or presence and extent of NBC contamination. Refer to *Appendix J* for more information about NBC reconnaissance.

AREA RECONNAISSANCE

6-3. Area reconnaissance is performed to obtain detailed information concerning the terrain or enemy activity within a prescribed area, such as a town, ridgeline, woods, or any terrain critical to the operations. MP conduct area reconnaissance to help guard against unexpected enemy attack in the rear area. Area reconnaissance and surveillance are vital to maintaining AS and contribute to the commander's intelligence collection plan. MP area reconnaissance is a composite of actions. It is initiated from observations and reports gathered over time by MP patrols and information gained through coordination with HN police and other friendly forces. Refer to *FM 7-8*.

6-4. Reconnaissance patrols may differ slightly, depending on the type of reconnaissance to be performed. However, all reconnaissance patrols have a reconnaissance and security team. The size of the patrol is determined by METT-TC. Other considerations to determine the size of the patrol include—

- Size and number of reconnaissance objectives.
- Requirement to secure the objective rally point (ORP) and other points.
- Time allowed for conducting the mission.

6-5. MP leaders plan area reconnaissance based on the IPB and the commander's critical information requirements. Information on enemy activity and likely avenues of approach is coordinated with military intelligence (MI). MP monitor likely enemy avenues of approach and LZ and DZ in critical areas to give early warning of rear-area enemy activity.

6-6. MP area reconnaissance plans include areas near facilities that are designated as critical by the commander, such as—

- NAIs.
- Air bases.
- Bases and base clusters.
- Communications centers.
- Logistic support clusters.
- Key terminals, depots, and bridges.
- Critical terrain features.
- High-value assets.

6-7. When leading an area reconnaissance patrol, in addition to using troop-leading steps and following the general principles for making a reconnaissance, the patrol leader—

- Uses a scheme of maneuver.
- Secures and occupies an ORP.
- Conducts a leader's reconnaissance of the objective area to confirm or change the plan.
- Returns to the ORP, completes the plan, and briefs the soldiers.

6-8. The security elements leave the ORP before the reconnaissance element. The security element leader places security teams at the ORP and on enemy avenues of approach into the objective area. The reconnaissance element conducts the reconnaissance by moving to several vantage points around the objective.

6-9. The reconnaissance element leader may have a small reconnaissance team move to each vantage point instead of having the entire element move as a unit from point to point. This reduces the chances of being spotted.

6-10. After the objective has been reconnoitered for the details outlined in the order, all elements return to the ORP. Teams share their information, consolidate it, and report it, then return to the patrol HQ or continue to the next mission.

ZONE RECONNAISSANCE

6-11. A zone reconnaissance is a directed effort to obtain detailed information concerning all routes, obstacles (to include chemical or radiological contamination), terrain, and enemy forces within a zone defined by boundaries. A zone reconnaissance is normally assigned when the enemy situation is vague or information concerning cross-country trafficability is desired. Zone reconnaissance techniques include the use of moving elements, stationary teams, or a series of area reconnaissance actions. Refer to *FM 7-8* and *FM 17-98*.

6-12. The four methods used to conduct a zone reconnaissance are—

- Box method.
- Fan method.
- Converging routes method.
- Successive sector method.

Box Method

6-13. To use the box method *(Figure 6-1)*, the leader sends his reconnaissance and security teams from the ORP along the routes that form a boxed-in area. He sends other teams along routes through the area within the box. All teams meet at a linkup point at the far side of the box from the ORP.

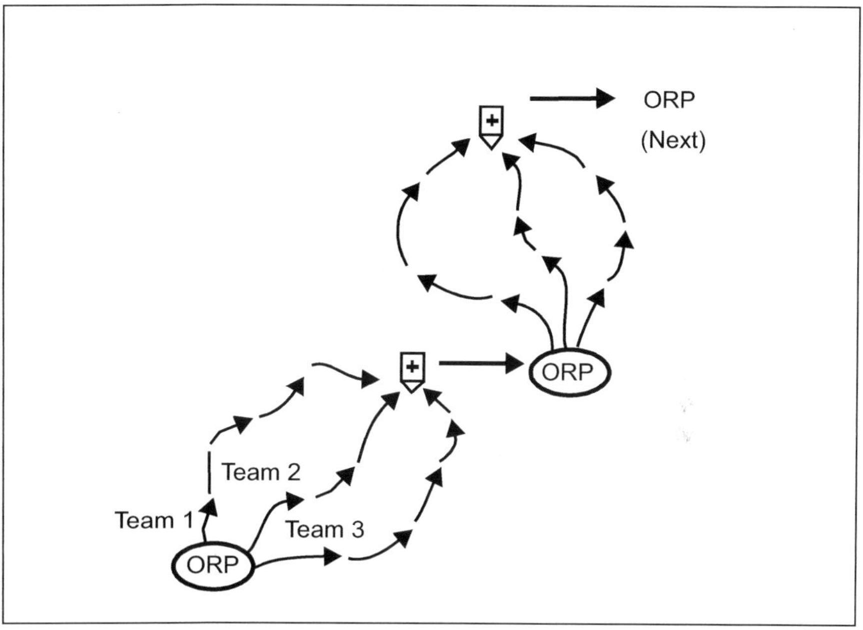

Figure 6-1. Box Method of Zone Reconnaissance

Fan Method

6-14. To use the fan method *(Figure 6-2, page 6-6)*, the platoon leaders selects a series of ORPs throughout the zone. At the first ORP halt and set up security. After confirmation of the patrol's location, the platoon leaders selects reconnaissance routes out from and back to the ORP.

NOTE: These routes form a fan-shaped pattern around the ORP. They must overlap to ensure that the entire area has been rcconnoitered.

6-15. Once the routes have been selected, send out reconnaissance elements along the routes. Do not send out all the elements at once. The platoon leader keeps a reserve at the ORP. He sends elements out on adjacent

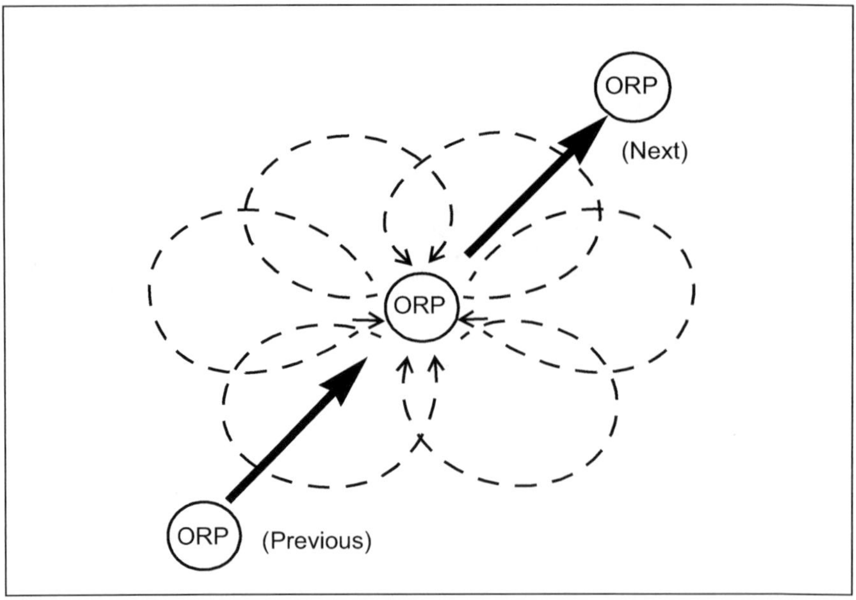

Figure 6-2. Fan Method of Zone Reconnaissance

routes to keep from making contact in two different directions.

6-16. After the entire area (fan) has been reconnoitered, report the information then move the patrol to the next ORP. Repeat this action at each successive ORP.

Converging-Routes Method

6-17. To use the converging-routes method *(Figure 6-3)* (which incorporates the fan method), select an ORP and reconnaissance routes through the zone and the rendezvous point.

NOTE: The rendezvous point is a place where patrol members link up after the reconnaissance.

6-18. Halt the patrol at the ORP and set up security. Confirm the patrol's location. Designate a route for each

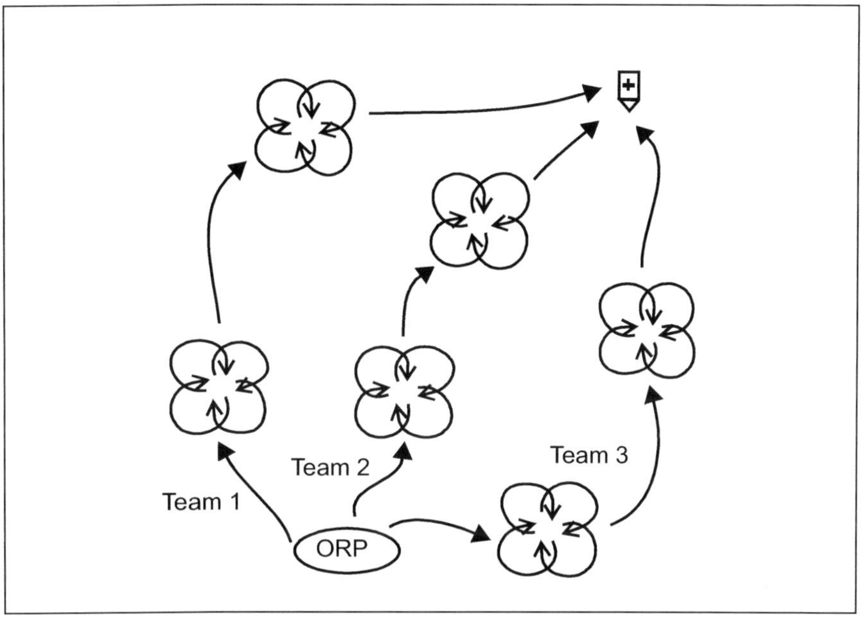

Figure 6-3. Converging Routes Method of Zone Reconnaissance

reconnaissance element, a location for the rendezvous, and a linkup time at the rendezvous point. Send a reconnaissance element to reconnoiter each route (usually using the fan method). The leader moves with the center element.

6-19. At linkup, the patrol secures the rendezvous point as it did the ORP. While at the rendezvous point, information gained by each member is exchanged with all the other members. This provides backup to ensure that all information is passed onto higher HQ. The patrol then returns to friendly lines or continues on to another mission.

Successive-Sector Method

6-20. To use the successive-sector method *(Figure 6-4, page 6-8),* build on the converging-routes method. Select an ORP and a series of reconnaissance routes and rendezvous points. Use the converging-routes method from each ORP to each rendezvous point.

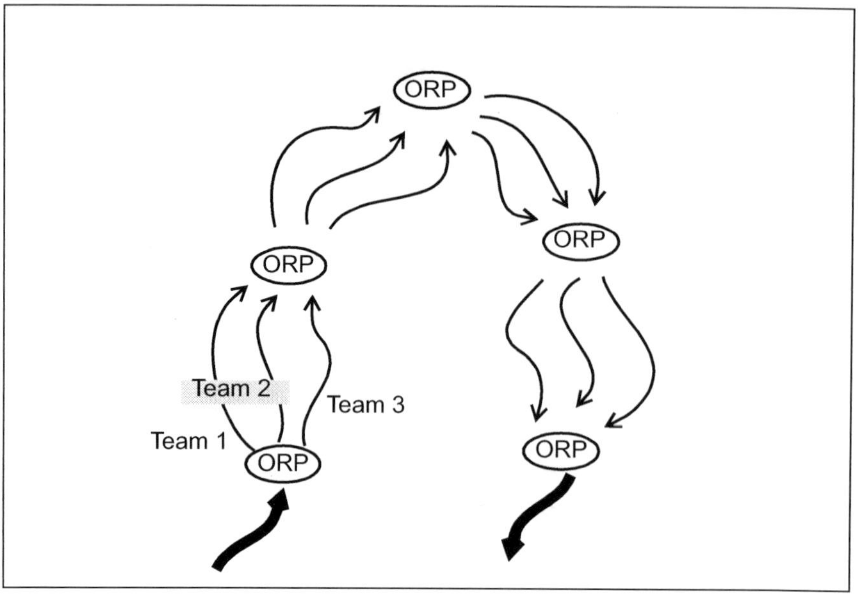

Figure 6-4. Successive Sector Method of Zone Reconnaissance

6-21. Each rendezvous point becomes the ORP for the next phase. Designate reconnaissance routes, a linkup time, and the next rendezvous point when the patrol links up at the rendezvous point. Continue this action until the entire zone has been reconnoitered.

6-22. Regardless of the type of method used, report the information through proper MP channels as soon as possible. Commanders rely on fast, accurate reconnaissance information to plan successful operations.

SCREENING MISSIONS

6-23. Screening missions are defensive in nature and largely accomplished by establishing a series of OPs and conducting patrols to ensure adequate surveillance of the assigned sector. Division cavalry units normally conduct security missions that include a screen for maneuver units during offensive operations.

6-24. MP conduct screening missions for friendly forces in the rear area to provide early warning of enemy approach and to provide real-time information and reaction time for stationary units. In the event of a Level III threat, MP may come under the OPCON of a TCF, which is also referred to as a combined-arms maneuver unit. In this role MP may provide limited security missions, such as a screen to the flank or rear of the main body, with the primary mission of providing early warning and disrupting or destroying enemy reconnaissance vehicles.

6-25. Generally, MP are tasked to observe specific avenues of approach or, more precisely, NAI. The area to observe should be identified in either the reconnaissance and security plan that the platoon leader receives or in the OPORD from higher HQ. If the platoon does not receive an IPB product, the higher OPORD must specifically state where it must focus the screening operation. If the platoon is assigned multiple requirements, the higher HQ must prioritize them.

SURVEILLANCE

6-26. On order, MP maintain continuous surveillance of all the assigned NAIs or enemy reconnaissance avenues of approach into a particular sector. This is accomplished by setting up a series of OPs. MP may conduct active mounted patrols to extend their observation limits or to cover dead space and the area between OPs. Refer to Chapter 4 for more information about setting up OPs/LPs.

6-27. Once the platoon leader understands what his surveillance requirements are, he task organizes the platoon and any assigned assets to achieve the most effective surveillance of the avenue or NAI.

6-28. Unlike a scout platoon, MP focus on providing early warning of enemy reconnaissance elements rather than gaining and maintaining contact with the enemy's main body or destroying it. During screen missions, it is important to understand that an MP platoon by itself does not have enough assets to both acquire and kill an enemy reconnaissance larger than the engaging element. Generally, other assets will be given the specific mission of killing these forces. If MP are ordered to engage enemy forces, they do so by engaging at the maximum effective range of their organic weapons. If available, the platoon leader also plans for and uses CAS and indirect fire.

6-29. During surveillance the platoon's ability to report is critical. Effective early warning requires detailed planning for uninterrupted communications. The platoon leader considers communication distances and significant terrain features to identify potential wireless communication problems. If problems exist, he requests support from the higher HQ.

COUNTERRECONNAISSANCE

6-30. MP contribute to the commander's concept of operations by conducting security and reconnaissance missions designed to detect, disrupt, and impede enemy reconnaissance elements. Counterreconnaissance is not a distinct mission; rather, it is a combination of measures taken by friendly forces to reduce the threat's ability to gather information. It contains both active and passive elements and includes combat action to destroy or repel enemy reconnaissance units.

6-31. Counterreconnaissance prevents enemy reconnaissance forces from observing the main body of friendly forces by defeating or blocking the enemy forces. In the execution of counterreconnaissance, MP operate either offensively or defensively using whatever tactics best accomplish the mission. The principal techniques used are—

- A hasty attack.
- An ambush.
- Indirect fire support.

6-32. MP must task organize to defeat enemy reconnaissance forces. Enemy reconnaissance capabilities in any given situation must be compared to the MP unit's capabilities to determine if additional maneuver or CS assets are required.

6-33. Conventional reconnaissance elements are usually squad-size or smaller. However, special-purpose reconnaissance forces can consist of mechanized forces up to company size. In all counterreconnaissance operations, the goal is to acquire, identify, and kill the enemy reconnaissance force after it has penetrated the initial screen line. Defeating such forces usually requires combined-arms forces, but this is dependent on the type, size, and capabilities of the reconnaissance element.

6-34. MP platoons are not organized or equipped to fight for extended periods or to destroy enemy armor vehicles. MP employ AT weapons, such as AT-4s, for defensive purposes (self-protection and breaking contact). However, MP teams are highly skilled at reconnaissance and surveillance and providing early warning of enemy activity.

6-35. A scout platoon acquires and identifies enemy reconnaissance forces along a screen line, which is a control measure usually named as a phase line, and is an established forward of the main body. MP conduct their counterreconnaissance efforts in a similar manner in the rear area or anywhere sustainment operations are taking place.

6-36. In most cases, the scout platoon cannot be expected to acquire, identify, and defeat enemy reconnaissance elements. As a CS asset, MP can assist a scout platoon by locating the enemy reconnaissance element, freeing the scouts or TCF to perform the killing function of counterreconnaissance on larger mechanized enemy reconnaissance elements. MP activities that contribute to counterreconnaissance include—

- Area reconnaissance.
- Zone reconnaissance.
- Route reconnaissance.
- OP operations.
- Physical security and vulnerability risk assessment.
- Critical asset security.
- OPSEC.
- Deception operations.

6-37. Conventional threat reconnaissance elements push far out in front of their combat unit to gain intelligence on their rear area objective. Unconventional threats such as terrorists, criminals, or gangs may try to observe installations, deep-water ports, or other facilities to obtain information. Conventional threat reconnaissance efforts are concentrated on gaining intelligence on the capability of friendly forces. Unconventional threats try to obtain information to plan sabotage or criminal activity, or to simply disrupt the efforts of friendly forces.

6-38. MP leaders plan measures to counter enemy reconnaissance by coordinating with various staff sections and agencies that include—

- MI for information on enemy capabilities, likely rear area targets and objectives, likely enemy reconnaissance avenues of approach, and the

  commander's critical information requirements (CCIR).
- The PM and criminal investigation division (CID) for HN police information on local gangs, known criminals, and criminal activity.

6-39. To assist commanders with their counterreconnaissance efforts, friendly forces such as MP, CID, engineers, and MI conduct physical security surveys and vulnerability and risk assessments of bases and base clusters, deep-water ports, and air bases. They advise commanders of these facilities on security measures designed to prevent the threat from gaining access to friendly forces and facilities.

6-40. During AS missions, MP conduct security activities around NAI; critical assets, such as communications nodes; and air bases. MP perform area and zone reconnaissance of all terrain that dominates critical facilities. They concentrate their efforts on locating enemy reconnaissance forces. MP deny the enemy the opportunity to observe friendly forces by reporting their location, maintaining surveillance, and assisting in their destruction if required.

6-41. Enemy reconnaissance forces are not likely to use primary reconnaissance avenues of approach to gather information on friendly forces. MP teams are more likely to come in contact with enemy reconnaissance forces operating on trails, rough terrain, and dead space that allows mounted movement. They use the cover of darkness for their operations. MP must make maximum use of NVDs and illumination to help detect their movement. They put the devices on key terrain and along avenues of approach to critical bases, and cover the area with crew-served weapons. Enemy reconnaissance teams are most vulnerable during the day. MP concentrate daytime mounted or dismounted operations on locating their base camp or hide positions.

Once they are discovered, if ordered to do so, MP can lay ambushes on likely routes to destroy them. Refer to *Chapter 4* for more information about ambush patrols.

6-42. Mounted MP patrols use overlapping search techniques to make it difficult for enemy reconnaissance teams to reach their objectives without being exposed. Overlapping searches provide random coverage not easily predictable by simple observation.

6-43. If contact is made, MP may be directed to maintain contact or surveillance until enough force can be assembled to counter the threat. They disrupt their movement and delay them until a large enough force is available to defeat them. If directed to disrupt or delay, they clearly identify the enemy before engaging them; there are generally numerous friendly forces operating in the rear area. MP must consider fratricide preventive measures. Refer to *Appendix F* for more information about fratricide avoidance.

6-44. MP make good use of terrain and maximum use of weapon systems. They use harassing fire from mounted crew-served weapons (MK19/M2), but do not become decisively engaged. They deceive the enemy with heavy harassing fire from crew-served weapons. If available, MP destroy the threat with indirect fire. MP make the threat believe he has encountered a defense. The enemy must not realize these actions are only delaying tactics.

AREA DAMAGE CONTROL

6-45. ADC is basic to successful rear-area operations. ADC measures are taken before, during, and after hostile actions or natural or man-made disasters to minimize effects and reduce damage. All commanders try to limit the impact of enemy actions and reestablish unit operations as quickly as possible. Commanders at each level plan ADC operations.

6-46. ADC operations integrate the functions of many specialized units. Engineers plan and coordinate ADC operations at the site. Medical teams help sort and treat mass casualties and assist in initial evacuations. Army aviators help evacuate casualties and provide emergency resupply, communications relay operations, area damage assessment, and C^2 actions. MP expedite and control battlefield movement into, around, or through damaged or contaminated areas. Signal elements reestablish the signal system. HN civil efforts, like clearing rubble and providing facilities and services, can greatly aid ADC.

6-47. The amount of MP support needed for ADC depends on the extent of the damage, the importance of the affected area, and the effect of the damage on the movement of troops and logistical supplies. Enemy attacks on key military facilities can leave them unprotected. Heavy damage in urban civilian areas can disrupt local government services.

NOTE: Protecting civilian facilities is a HN responsibility. However, MP preserve law and order in such an area if so doing protects friendly forces and resources.

6-48. Route and area reconnaissance are key to determining the trafficability of the routes into, out of, and around affected areas; obtaining a damage assessment, and having early warning of the continued presence of the enemy. MP determine the level and extent of NBC contamination and identify the location of critical points affected by damage.

6-49. The enemy's damage to the terrain determines the degree and kind of MP support needed and where to place the priority of effort. Downed trees, urban rubble, damaged or destroyed bridges, cratered roads, and contaminated road networks affect circulation control. If the roads can be traveled, MP provide circulation control operations locally in the affected area. However, damage to an area may be so great that roads must be closed and MSR traffic rerouted.

6-50. MP give priority of movement to ambulances transporting wounded and engineers clearing debris. They maintain law and order in the affected area by employing measures to stop looting and unlawful behavior, according to the ROE and the use of force requirements. They disseminate key information such as the location of first aid stations, emergency shelters, and other emergency facilities.

6-51. Security patrols may be needed around key facilities. MP may need to set up an OPs/LPs to observe sectors of the affected area. The OP/LP teams can watch for enemy agents trying to exploit the effects of a conventional attack through arson, sniper fire on firefighters, or other disruptive acts. The OPs/LPs also watch for theft, pilferage, or arson against military property.

6-52. MP ADC operations may be either part of an ongoing operation or a separate requirement. During ADC, MP—

- Perform route and area reconnaissance in affected areas.
- Evaluate the serviceability of the road network.
- Note and report the development of critical points caused by damage to bridges, tunnels, and the like.
- Monitor the flow of DCs from the damaged area.
- Report and block off affected areas.
- Provide AS for involved critical facilities.
- Reroute battlefield movement to alternate road networks.
- Check for and report NBC hazards and contamination.
- Prevent sabotage, looting, and pilferage in the damaged area.

- Protect property, contain panic, and enforce emergency restrictions.
- Direct persons to first aid stations, emergency shelters, and other emergency operations.
- Post temporary signs to prevent entry into unsafe buildings or redirect activity to temporary locations.
- Help establish populace control in affected areas.
- Operate mounted and dismounted mobile patrols, checkpoints, and roadblocks to—
  - Enforce emergency restrictions on movement into, within, and out of the affected area.
  - Direct DCs.
  - Collect stragglers.
  - Enforce curfews, stand-fast orders, and movement authorizations.

BASE DEFENSE

6-53. Base defense is the local military measures, both normal and emergency, required to nullify or reduce the effectiveness of enemy attacks on, or sabotage of, a base to ensure that the maximum capacity of its facility is available to friendly forces.

6-54. The base commander organizes and controls the forces assigned to a base to best capitalize on their capabilities. These forces may be joint or single-service forces operating in the joint rear area (JRA). If a base, base cluster, or air base comes under attack, the basc commander responds within his capability. Defense fundamentals, as they pertain to the defense of the bases, include the following:

- Understand the enemy. Defenders must be familiar with the capabilities and limitations of the enemy forces, weapons, equipment, and tactics.

- Use situational awareness. Intelligence operations are key to assembling an accurate picture of the battle space.
- Use the defender's advantages. The defender's advantages may permit a numerically inferior force to defeat a much larger one. These advantages include—
  - The ability to fight from cover.
  - A more detailed knowledge of the local terrain and environment.
  - The ability to prepare positions, routes between them, obstacles, and fields of fire in advance.
  - The ability to plan communications, control measures, indirect fires, and logistic support for contingency operations.
  - The ability to deceive enemy forces about friendly defensive capabilities, dispositions, and execution of operations.
- Concentrate at critical times and places. Conduct the defense of a base along interior lines, permitting the timely and secure movement of forces to engage the most critical threats. The commander must mass combat power at points of decision by economizing in some areas, retaining a reserve, and maneuvering to gain local superiority at critical points.
- Conduct counterreconnaissance and counterattacks. Fixed bases have well-established perimeters that usually have limited depth. Counterreconnaissance and counterattack add depth to the battle, outside the perimeter, allowing the base to continue its primary mission with minimal interference.

- Extend counterreconnaissance patrols 3 to 5 kilometers outside the perimeter, based on terrain. This reduces the risk of fratricide, especially at night.
- Coordinate critical defense assets. Synchronization of indirect fires, air defense resources, tactical aircraft, engineers, dismounted troops, armored vehicles, naval surface fire support, and helicopters can produce a combined-arms effect.
- Balance base security with political and legal constraints. This fundamental is especially critical in a low-intensity conflict environment.
- Know the law of war and ROE. Base commanders and their subordinates must comply with ROE. In joint-service operations, reconcile inconsistencies with ROE.

6-55. A base or base cluster has certain inherent capabilities which allows it a degree of defense. Each base has a BDOC that plans, coordinates, and supervises base defense operations. The BDOC initiates contingency planning that enables the base to—

- Increase the manning posture of the base as the threat level increases.
- Detect and defeat the threat within its capabilities.
- Defend against heavier enemy forces until the response force is committed.
- Maintain control of the fight within the base.
- Support the fire and movement of the response force operating outside the base.

6-56. Each base cluster has a BCOC to monitor base defense plans and establish the base cluster response force. The BCOC—

- Provides C^2 of resources for planning, coordinating, and supervising the defense of the base cluster.

- Coordinates base defense operations.
- Maintains communications with the BDOC within the cluster, MP, and the supporting RAOC.

6-57. MP leaders plan for the interface of their support into the base plan. The plan requires—

- Cover and concealment of personnel and equipment.
- Secure and redundant communications systems at all guard locations (a landline, radio links to the BDOC, and a telephone hookup to the center switch).
- Deception.
- Contingency planning.
- Improvement of base defense positions.
- Coordination with BCOC or RAOC as required.
- OPs/LPs.
- Noise and light discipline.
- Immediate reaction to enemy threat or attack.
- Rehearsals of defense measures.

6-58. All plans and overlays depicting MP support are forwarded through normal MP channels to the BCOC.

6-59. Any threat to the bases must be dealt with swiftly. Few sustainment units can continue their mission while under attack by even a low-level threat.

6-60. Sustaining operations rely on MP to stay apprised of enemy activity near bases. MP forward early warnings of enemy activity to the operations cell at the rear CP. The rear-operations cell immediately notifies the subordinate RAOC and the base and base clusters. It may also alert the response force.

6-61. If a base or base cluster comes under attack, the base commander responds within his capability. However, interruptions of base sustainment operations must be kept to an absolute minimum. If a threat exceeds a base's ability to defeat it, the base commander requests response force support.

DANGER
Base defense forces must be aware of the disposition of other friendly forces in front of them. Fire discipline between the response force, mobile reserve, and TCF is imperative. Positive control must be maintained at all times.

AIR BASE DEFENSE

6-62. The Army is responsible for defending air assets from ground threats outside the boundary of the air base and its area of responsibility (AOR). ABD forces are tasked to defend particularly critical air bases. They may be forces operating in the area specifically on call for ABD or they may be the rear-area response force.

6-63. Air Force security forces (AFSF) are responsible for the internal security of US air bases. HN and US Army forces are primarily responsible for external defense of US air bases. Given the size of an air base and the high priority as a threat target, providing ABD as a dedicated asset requires a force larger than a company. MP do not have the force structure to be dedicated solely to ABD operations. However, an echelon commander could decide to redirect MP mission priorities to ABD and deploy his limited MP assets to serve as part of a larger composite force, such as a TCF. Or an MP response force may be required to respond to threats beyond the base's capabilities.

6-64. MP operating as part of a larger ABD force may provide support that includes —

- Setting up roadblocks and checkpoints to limit access to the base.

- Occupying OPs/LPs and fighting positions on key terrain and likely avenues of approach to the air base.
- Conducting screening missions.
- Conducting combat patrols (security and ambush).
- Conducting DC operations.

6-65. A typical ABD layout consists of a base perimeter that surrounds the runway and the facilities that support air base operations. The tactical perimeter of a base is the outer most ABD tactical AOR. It is where defense forces cover, patrol, or occupy likely avenues of approach to the base and other key terrain. *Figure 6-5* shows a typical ABD layout.

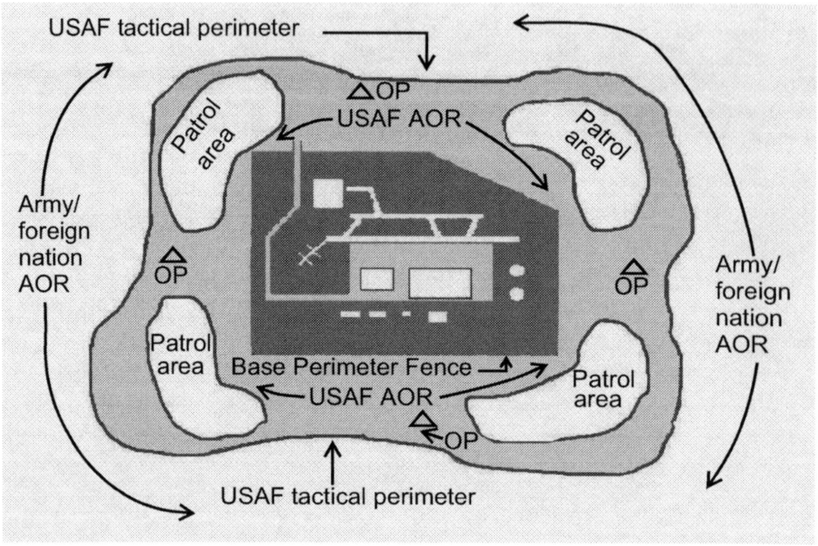

Figure 6-5. Typical ABD Layout

6-66. Commanders develop unit SOPs and contingency plans for ABD. SOPs should address C^2 and fire support issues that are different for ABD. However, AS plans for ABD provide for increased security patrols and static security measures around an air base the same as they would for other critical facilities.

6-67. The depth of the defense area surrounding the air base depends on METT-TC and other defensive planning factors. MP activities that support ABD are concentrated in the tactical perimeter and on key terrain adjacent to the tactical perimeter.

FIRE SUPPORT

6-68. Organic AFSF fire assets typically available are the M29 81-millimeter mortar, the MK19, and the M2 .50 caliber machine gun (MG). These fire systems are incorporated into the base defense plan for both the security force (SF) and MP. This ensures compatibility with all ABD measures in the surrounding area and mutual support to other bases when practical.

6-69. The use of mortars and other indirect-fire weapons located in the defense area is coordinated through a 5-man FDC located at the air base BDOC. The AFSF primarily uses its mortars for illumination, suppression of enemy indirect-fire weapons, and suppression of enemy surface-to-air missile (SAM) capability. AFSF weapons controlled by the FDC vary. The number of weapons available depends on the size of the air base.

6-70. CAS is obtained from aircraft at the base. Aircraft may be sent aloft to provide CAS or targets may be assigned to aircraft already airborne. CAS provides suppressive fire or increased detection capabilities. Among the CAS aircraft that may be available, the AC-130 Spectre gunship and the A-10 Thunderbolt II both have night-flying capability. Both have compatible frequency modulated/modulation communications equipment. Both are especially effective for use against targets considered danger close to ABD forces. However, they are very susceptible to SAM.

DANGER
ABD forces must be aware of the disposition of other friendly forces in front of them. Fire discipline between AFSFs response forces, mobile reserves, and the TCF is imperative. Positive control must be maintained at all times.

COMMUNICATIONS

6-71. Joint communications for all forces supporting ABD must be compatible. Communications must be secure, have redundancy, and possess the ability to integrate all security forces.

6-72. Effective coordination must be an ongoing process. Call signs, frequencies, and SOI are exchanged between all joint, combined, and coalition forces supporting ABD.

MILITARY POLICE RESPONSE TO AIR BASE DEFENSE

6-73. Air bases are normally located in the rear area. Threats that are beyond the capability of the base may require the deployment of an MP response force. In general, the response force responds to an air base the same as any other base or base cluster. However, as a joint service operation there are C^2 and fire support differences. Company and platoon tactical SOPs for ABD should be developed to help speed the response time.

6-74. The response force leader coordinates with the Air Force defense force commander (DFC) for the current tactical situation and threat. When committed, the MP response force may transition to the operational control of the DFC. This ensures that MP actions do not interfere with air sorties being generated by the air base commander.

6-75. The DFC is in charge of all aspects of ABD operations. He provides guidance on where the response force is needed and informs the response force leader of the tactical situation. The response force leader directs the tactical operations of MP elements, keeping the DCF informed. If a TCF must be called to defeat a Level III threat, all units, Air Force and Army, are under the OPCON of the TCF commander until the threat is defeated.

6-76. For the local ground defense of their base, the Air Force provides a BDOC which operates similar to an Army BDOC. It is responsible for combat intelligence, combat operations, logistics, and personnel administration. It collects the ground combat intelligence within the ABD area.

6-77. The AFSF maintains a mobile reserve force. The mobile reserve's purpose is to mass enough firepower to destroy threat forces within the air base boundary or, at least, delay the threat until a larger force can be assembled. The mobile reserve is usually a mounted force. It is under the direct control of the BDOC. The mobile reserve tries to contain any direct landing by threat forces on the airfield. It responds to penetrations of the defense area by forces that have managed to elude external defense forces.

6-78. If the enemy penetrates deeply into the defense area, the response force leader can request that the BDOC commit the mobile reserve to battle. This may prevent the premature commitment of the TCF. Contact points and AAs are established around the base to assist in integrating the mobile reserve during combat operations outside the ground defense area.

6-79. The response force must keep the enemy from destroying resources on the ground and interrupting or stopping air operations.

Night Operations

6-80. Air bases are most vulnerable when visibility is limited. In response to the threat, the bulk of both internal and external ABD operations must be dedicated to detecting threat forces during periods of limited visibility. The large gaps between defensive positions facilitate enemy infiltration through external defenses. To narrow these gaps during limited visibility, external defensive efforts may have to be set up closer to the perimeter. External defensive efforts should integrate the extensive use of—

- Night vision equipment.
- Reconnaissance and surveillance.
- Electronic detection devices.
- Tactical deception.
- OPs/LPs.

Water Obstacles

6-81. Many air bases border water obstacles. Air bases located next to rivers, lakes, or oceans provide unique ABD problems. Threat efforts directed against an air base from the water include infiltration, reconnaissance, and standoff-weapons attacks. As a result, water obstacles that penetrate the defense area may require ABD forces to set up continuous battle positions between the water and any critical air base facility near the water. ABD obstacle plans should include fences, mines, and sensor employment along the portion of the perimeter that borders water. Planned fires are coordinated to defeat watercraft. Floating expedient barricades may have to be assembled. Warning buoys should also be anchored offshore. Underwater approaches provide excellent avenues to air bases. They must be considered during ABD planning. Joint support from the Navy and the Coast Guard may also be needed.

Urban Terrain

6-82. Air bases bordered by urban areas pose a special challenge for ABD operations. Increased security is a necessity. MP must be aware of underground approaches, such as sewers, that could bypass defensive positions. Threat forces, saboteurs, and special-purpose forces will use these avenues of approach to gain access to the air base.

6-83. Infiltrating groups will be very difficult to detect because they avoid direct confrontation and usually move at night. These teams have the greatest potential for disrupting air operations throughout the rear area.

CHECKPOINTS AND ROADBLOCKS

6-84. Checkpoints and roadblocks are set up to control the movement of vehicles, personnel, and material and prevent illegal actions or actions that aid the enemy. They are used to prevent trafficking in contraband and stop the movement of known or suspected belligerents. Checkpoint and roadblock operations contribute to the commander's information and intelligence collection process. Additionally, they assist friendly forces in detecting and establishing the behavioral patterns of the local populace.

6-85. When conducted improperly, checkpoints and roadblocks can foster resentment toward US forces. MP minimize the negative impact that checkpoints and roadblocks may have by treating local nationals with dignity and respect. Whenever possible, checkpoints and roadblocks should be conducted with interpreters, HN police, or other HN authorities.

6-86. In most stability and support operations there exists the potential for threat forces to attempt to exercise some level of force against US forces. Checkpoints and roadblocks may become prime targets for threat forces (both traditional and nontraditional).

METT-TC determines the level of self-protection needed at a checkpoint or roadblock. However, MP leaders must always plan force protection measures that allow their forces to quickly implement adequate self-defense when threatened.

6-87. Checkpoints and roadblocks must be designed so that potential threat forces are discouraged from attempting to disrupt its operation or try to inflict casualties. MP use armored vehicles whenever possible to support the dismounted troops operating the checkpoint or roadblock. The MP security element assigned to protect the site should have an antiarmor capability. The vehicle search area should be located in an area that provides blast protection.

Checkpoints

6-88. Checkpoints are part of a police action. As such, the ROE and the use of deadly force must be clearly understood by all soldiers participating in the operation. MP use the minimum intrusion and imposition necessary to accomplish the mission and protect the force. They allow for a vehicle escape route and plan to destroy a hostile element who uses it. If the checkpoint is completely sealed off, an enemy will have to penetrate it by attempting to run over the barricades. This puts the search team in a position to have to defend itself and fight back.

6-89. MP establish and operate two types of checkpoints—deliberate and hasty.

6-90. **Deliberate Checkpoint.** A deliberate checkpoint is a fixed position set up on a main road in a rural or built-up area. It can be classified as either a *heavy-* or *light-traffic checkpoint*, depending on how much traffic is expected to pass through it. A heavy-traffic deliberate checkpoint normally requires a platoon. An MP squad can only operate a light-traffic checkpoint for a short duration (12 hours or less).

6-91. To operate a heavy-traffic checkpoint, task organize the platoon into—

- A HQ element. The HQ element is responsible for C^2 and maintaining communications with subordinate elements and higher HQ.
- A security element. The security element is an MP squad that provides early warning to the search and assault element, watches for and reports suspicious activity, and monitors traffic flow up to and through the checkpoint. The security element should have an antiarmor capability to protect the site from an enemy armor threat.
- A search element. The search element is an MP squad that halts vehicles at the checkpoint, guides them to the designated search point, conducts personnel and vehicle searches, and directs cleared vehicles on through the checkpoint.
- An assault element. The assault element is an MP squad responsible for destroying (consistent with the ROE) any hostile element that forces its way past the search team. The squad leader places his soldiers beyond the search point and emplaced zigzag obstacles and barriers. The soldiers prepare and occupy fortified fighting positions. When confronted by a threatening vehicle, the search element gets out of the way and allows the vehicle to pass. The vehicle passes through the escape lane, and the assault element makes the decision whether to engage or not. If the assault element has to engage, the battle will occur away from the checkpoint. This reduces the possibility of fratricide to friendly forces or injury to innocent civilians.

6-92. In smaller checkpoint operations, a squad can be organized in a similar fashion. MP elements use handheld portable radios, if available, or wire communications. However, much of the needed signals at a checkpoint or roadblock can be easily accomplished using arm and hand signals.

6-93. A deliberate checkpoint is organized into sections *(Figure 6-6)*. The physical layout and detail of preparation depend on the amount of traffic that will pass through it and the duration of its operation. Normally, a deliberate checkpoint will require engineer support to construct obstacles, barriers, escape lanes, and possibly fighting positions. All checkpoints consists of—

- Obstacles or barriers.
- Search areas (personnel and vehicle).
- Security overwatch and fighting positions.
- Holding areas.

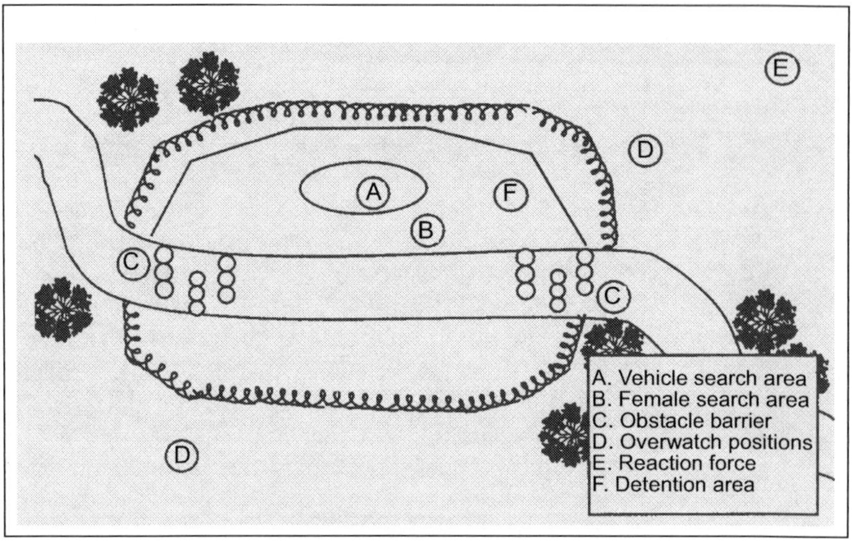

Figure 6-6. Deliberate Checkpoint

6-94. **Hasty Checkpoint.** MP set up hasty checkpoints to achieve surprise. They are temporary and should be moved often. The materials used to construct these checkpoints are carried by the platoon. The platoon or squad uses its vehicles, reinforced with concertina wire, as the obstacle. MP may employ tire deflation devices or road spike stripes. These devices are more effective than concertina wire and may be less intrusive in peace operations. They are commercially available or can be locally fabricated.

6-95. MP position the vehicles to partially block the route *(Figure 6-7)*. MP conduct the search in the area

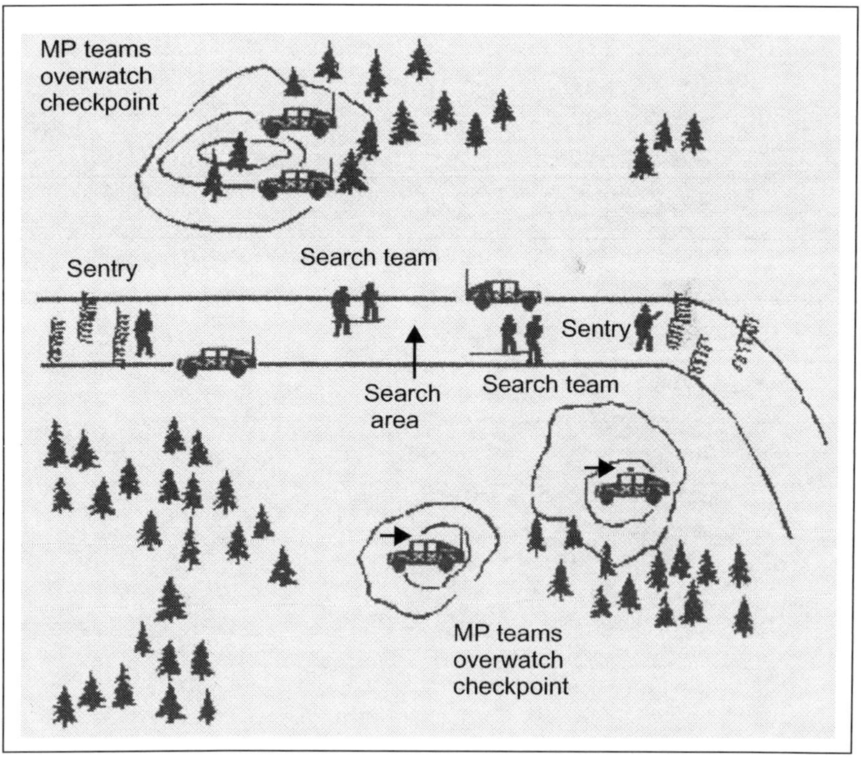

Figure 6-7. Hasty Checkpoint

between the vehicles. MP are positioned at each end of the checkpoint. They cover them by mounted or dismounted automatic weapon positions. MP conceal a reaction force (at least one team) nearby to react in case the site is attacked.

6-96. MP establish hasty checkpoints where they cannot be seen by approaching traffic until it is too late to withdraw. Good locations to set up hasty checkpoints include—

- Bridges.
- Defiles.
- Highway intersections.
- The reverse slope of a hill.
- Just beyond a sharp curve

6-97. **Vehicle Searches**. Two members of the search team position themselves at both rear flanks of the vehicle undergoing a search, putting the occupants at a disadvantage. These soldiers maintain eye contact with the occupants once they exit the vehicle and react to any threat attempts by the occupants during the vehicle search.

6-98. The actual search is conducted by two MP armed with pistols only. One MP conducts interior searches and the other performs exterior searches. They instruct the occupants to exit the vehicle during the interior search and instruct the driver to watch the vehicle search. Once the interior search is complete, they escort the driver to the hood of the vehicle and instruct him to open it. Once the engine compartment has been examined, they instruct the driver to open the other outside compartments (tool boxes, gas caps, trunks, and so forth). The driver removes any loose items that are not attached to the vehicle for inspection. Members of the search team rotate positions to allow for mental breaks.

6-99. MP use MWD teams, mirrors, and metal detectors to thoroughly search each vehicle for weapons, explosives, ammunition, and other contraband. Depending on the threat level, the vehicle search area provides blast protection for the surrounding area.

6-100. **Personnel Searches**. MP may be required to conduct personnel searches at the checkpoints. Every attempt should be made for HN authorities to conduct, or at least observe, searches of local nationals. Additionally, MP leaders must plan for same-gender searches. Personnel searches are conducted only when proper authorization has been obtained, usually from higher HQ, according to the ROE, Status of Forces Agreement (SOFA), or HN agreements. This does not preclude MP from searching individuals that pose a threat to US or other friendly forces.

6-101. MP may have to detain local nationals who become belligerent or uncooperative at the checkpoints. The OPORD and the ROE must address the handling of such personnel. In any case, self-protection measures should be planned and implemented according to the orders from higher HQ.

6-102. Searches of local nationals should be performed in a manner that preserves the respect and dignity of the individual. Special consideration must be given to local customs and national cultural differences. In many cultures it is offensive for men to touch or even talk to women in public. Searchers must be polite, considerate, patient, and tactful. MP leaders must make every effort not to unnecessarily offend the local population. Such situations can have a very negative impact on peace operations and can quickly change popular opinion toward US and other friendly forces.

6-103. Each captive is searched for weapons and ammunition, items of intelligence value, and other inappropriate items.

NOTE: When possible, conduct same gender searches; however, this may not always be possible due to speed and security considerations. Therefore, perform mixed gender searches in a respectful manner using all possible measures to prevent any action that could be interpreted as sexual molestation or assault. The on-site supervisor carefully controls soldiers doing mixed-gender searches to prevent allegations of sexual misconduct.

6-104. MP conduct individual searches in search teams that consist of the following:

- A searcher. A searcher is the MP that actually conducts the search. He is in the highest-risk position.
- Security. Security includes at least one MP to provide security. He maintains eye contact with the individual being searched.
- An observer. The observer is a MP leader that has supervisory control of the search operation. He also provides early warning for the other members of the team.

6-105. The two most common methods that are used to conduct individual searches are the frisk search and the wall search.

- Frisk search. This method is quick and adequate to detect weapons, evidence, or contraband. However, it is more dangerous because the searcher has less control of the individual being searched.
- Wall search. This method affords more safety for the searcher because the individual is searched in a strained, awkward position. Any upright surface, such as a wall, vehicle, tree, or fence may be used.

6-106. If more control is needed to search an uncooperative individual, the search team places the subject in the kneeling or prone position.

6-107. Strip searches should only be considered when the individual is suspected of carrying documents or other contraband on his person. This extreme search method should be conducted in an enclosed area and by qualified medical personnel when available.

Additional Checkpoint Considerations

6-108. The effective use of all task organization elements is vital to the success of checkpoint operations is. Roles and responsibilities must be well defined and rehearsed. Additional considerations when conducting checkpoint operations include—

- Preparing and emplacing signs in the local language instructing drivers what to expect and do at the checkpoint (for example, "You are entering a military checkpoint. Prepare to stop your vehicle, and have your identification papers ready for inspection.").
- Determining if it is necessary to apprehend or detain those who see the checkpoint ahead and turn around to avoid it. If it is, HN police are responsible for this mission, if they are available. If they are not available, it may be necessary to position a respond force close to the approach route to block or detain vehicles that try to avoid the operation.
- Clearing and maintaining control of all buildings and terrain that dominate the checkpoint.
- Staying alert for any change of scenery around the checkpoint. A parked car that was not there before, crowds gathering for no apparent reason, or the media waiting for an event are all indicators that something may happen.

- Using artificial illumination for night operations, arrange the lighting to keep those passing through the checkpoint in the light and our forces in the shadows as much as possible.

6-109. If HN personnel are used to assist, commanders ensure that they do not represent a national, ethnic, or religious group of faction feared and hated by the majority of the local population.

6-110. Leaders avoid setting patterns by moving the checkpoint location and changing the method of operation at random.

6-111. MP can gain valuable police, criminal, and combat information while operating checkpoints. They use a checklist to standardize the information collection effort. The following information is included in the checklist:

- The number and type of vehicles stopped. Report identifying markings, license plate number, and any signs displayed on the vehicle.
- The number of passengers in the vehicle. Report the nationality, the ages, and the sex mixes of passengers.
- The type and quantity of cargo.
- The point of origination and destination of the vehicle.
- The stated reason for travel by passengers.
- A description of arms, ammunition, explosives, and sensitive items found and confiscated from the vehicle.
- Possible or actual sightings of weapons, explosives, or threat forces by the passengers.
- The condition of passengers (general health, dress, attitude).
- Anything unusual reported by the passengers.

6-112. When conducting checkpoint operations, MP will need the support of the following:

- Engineers to build obstacles and barriers to channel traffic.
- Linguists familiar with the local language.
- A civil affairs officer or HN police.
- Trained interrogators.

6-113. Sometimes MP are tasked to operate a checkpoint to assist convoys of friendly forces. These checkpoints are set up at the entrance to the controlled route and will have a vehicle holding area to accommodate large convoys. Refer to *Chapter 5* for more information about holding areas.

6-114. At the convoy checkpoints, MP check convoy vehicle movement credits issued by the local movement control unit to ensure that the convoy is moving on the correct route at the correct time. When convoys are ahead of schedule, MP hold them near the checkpoint in a vehicle holding area until it is their scheduled time to pass. They allow convoys that are behind schedule to proceed if route traffic permits and assist drivers who have taken the wrong route.

Roadblocks

6-115. Roadblocks are set up to stop, slow, or limit movement of vehicles along a route. They are used to limit access to certain areas or roads. A roadblock can be established separate from a checkpoint or used to channel vehicles and personnel to a checkpoint.

6-116. MP use their vehicles with concertina wire, barbed wire, and warning signs as a roadblock *(Figure 6-8, page 6-38)*. If available, armored vehicles make excellent roadblocks. Place a roadblock where unauthorized vehicles or enemy personnel cannot bypass it. If possible, position it so that obstacles like cliffs, swamps, or rivers restrict movement to its flanks or around it. Build man-made obstacles to tie in with and reinforce natural obstacles.

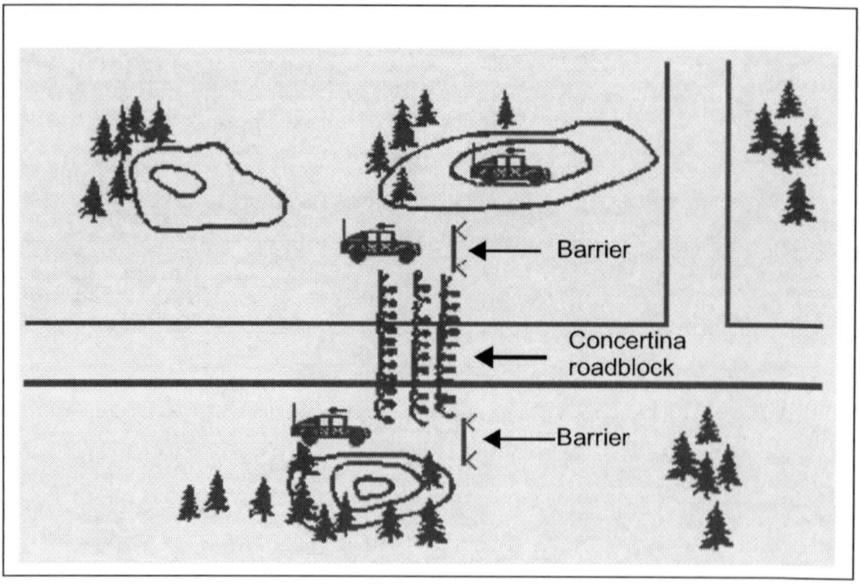

Figure 6-8. Roadblock

6-117. When using a roadblock to channel traffic to a checkpoint, place it where drivers cannot see the roadblock until after they have passed all possible turnoffs. When using a roadblock to close off a road, place it at an intersection to let drivers change to another route with little delay.

6-118. Select a defendable site for the roadblock. Cover the roadblock with automatic weapons. Ensure that defensive positions—

- Include a fighting position for the crew-served weapon to provide overwatch for the roadblock.
- Have fields of fire that cover the approaches to the roadblock to keep it from being breached.
- Are not accessible to an attacker and provide cover and concealment.
- Are placed at an intersection to let drivers change to another route with little delay.

- Are placed near an area where drivers can turn their vehicles around easily.

RESPONSE FORCE OPERATIONS

6-119. MP response forces may respond to threats to a base, a base cluster, or an air base and possibly respond to downed aircraft. MP are normally the commander's designated force to respond to attacks beyond the capabilities of a base or base cluster. The objective of response force operations is to eliminate the threat or make the enemy disengage without requiring the premature commitment of the TCF. MP are organized and equipped to defeat Level II threats. These threats exceed base and base cluster self-defense capabilities but do not require the commitment of a TCF.

6-120. With prior coordination and appropriate supporting fires, MP can defeat the following:

- Small combat units conducting raids, ambushes, and reconnaissance operations.
- Unconventional forces conducting diversionary and sabotage operations.
- Special warfare intelligence collection and operational missions.

6-121. MP response forces are prepared to delay and disrupt Level III threats, protect the flank of a TCF, or allow a base time to establish defenses in greater depth.

Plan

6-122. METT-TC, the number of MP available, and the commander's intent determine the size of a response force. Normally, an MP response force requires at least a full-strength platoon. However, response force operations conducted in built-up areas or in support of ABD may require a much larger force. MP platoons are not organized and equipped to fight for extended periods. The commander task organizes a platoon for response force operations. A response force should have increased antiarmor capability, integrate available fire support, and have CAS available. The result is a force capable of causing an enemy to break off its attack. The response force can then fix and destroy the threat using close combat techniques and the application of artillery, Army aviation, and CAS.

6-123. MP must know the terrain and be able to use it against the enemy. They mass combat power quickly to destroy the enemy or delay them until a TCF can arrive. The rear CP fire support element, along with the main CP fire support cell, establishes procedures by which MP can call for fire support.

6-124. The response force identifies the location of the bases within their AO, and must know which bases are the most critical and which are the most vulnerable. They include this information in the local IPB. In coordination with the rear CP operations cell and the affected RAOC, they position the response force where they can best—

- Detect enemy incursions (near the DZ, the LZ, and the like).
- Interdict enemy forces enroute to their targets.
- Consolidate quickly in response to threatened key assets in the rear area.

6-125. The response force leader must have the following readily available:

- The base defense status of each base.
- Locations of any obstacles or mines near the base.
- Locations and direction of fire of the crew-served weapons.
- The signal for FPF.
- Locations of TRPs and preplanned fires.
- Methods for contacting the BDOC or the BCOC, to include call signs and frequencies.

- Locations of the OPs/LPs and friendly patrols if employed.
- Call signs and frequencies for supporting artillery and Army aviation units tasked to respond.
- Call signs and frequencies for the TCF, RAOC, and the rear CP.
- Fire support targets that are on the approved list.
- Locations of the nearest medical treatment facility (MTF), NBC decontamination site, and ammunition supply point (ASP).

6-126. Continuous communication with the base and base clusters is the key to knowing how and when an MP response force is needed. Bases and base clusters establish 24-hour communications with the RAOC, if they are located in the corps rear area, or the rear CP, if they are located in the division. This liaison and interface allows timely response and information dissemination.

6-127. The response force will be effective only if it can react swiftly and is thoroughly familiar with the base defense plan. They coordinate all response force actions for a base through its BDOC. (Go through the BCOC for base clusters.) They coordinate the following with BDOC:

- Call signs and frequencies.
- Base defense plans and layouts.
- Positions of critical internal assets, external coordination points, and no-fire areas.
- Indirect-fire support.
- Engineer support, if needed, to help prepare defensive positions or for ADC.

6-128. In coordination with the main CP fire support cell, the operations cell sets procedures by which MP can call for fire support.

Execute

6-129. Base response force operations rely heavily on MP tactical skills. The choice of action is based on METT-TC and the base commander's tactical assessment. Once a situation develops that requires a response force, there are several options available to disrupt or eliminate the threat. Response force options include the following:

- Call for fire support to make the enemy break contact.
- Attack into the enemy's flank using the base for fire support.
- Use ambushes along likely escape avenues if it is obvious that the attacking force is not strong enough to overrun the base.
- Augment the base's defense forces and provide support from within a base.

6-130. Once the MP company commander receives the mission to direct response force operations, the company's main efforts shift to supporting the response force. The commander conducts a mission analysis and begins the TLPs, culminating in issuance of an OPORD and a FRAGO. Refer to *Chapter 2* for more information about TLPs.

6-131. The company HQ coordinates for support that includes communications, air evacuation, and logistical support (Class I, III, V, and IX). The company operations section coordinates with follow-on forces and other support that includes the following:

- Fire support to include CAS.
- The location of adjacent units.
- The location of threatened friendly forces.
- The location of noncombatants.
- Coordination with the TCF.

6-132. Once the platoon leader receives the mission to conduct response force operations he initiates TLPs. Normally, the PSG assembles the response force at a prearranged AA near the base under attack.

6-133. The base commander gives the response force an objective and a tactical AO. He may designate a small force to assist the response force. This force would be under the OPCON of the response force leader.

6-134. Normally, the response force acts as a maneuver element against the enemy. Once the platoon leader develops a COA he designates the ORP and selects a scheme of maneuver. The response force moves to the ORP, establishes security, and reports its status to higher HQ. Once the response force has established communications with the friendly forces under attack, the platoon leader reconnoiters the area and assesses the situation. Based on the situation and guidance from higher HQ the platoon leader may decide to—

- Request fire support to destroy the enemy.
- Continue to observe for weaknesses in the enemy's flanks.
- Engage and further develop the situation.
- Attack and destroy the enemy.

6-135. If the decision is made to call for fire to destroy or make the enemy break off the attack, the platoon leader follows the procedures set by the main CP fire support cell. When available, MP can call for and adjust indirect fire provided by artillery. Another option is CAS, if there is a tactical air control party or army aviator trained in joint air attack team operations. Refer to *Chapter 3* for more information about call for fire and fire support. Also refer to *FM 6-20*.

6-136. If the response force is ordered to attack and destroy the enemy, the platoon leader immediately prepares the response force for a hasty attack.

6-137. **Plan a Hasty Attack**. A hasty attack is an immediate action taken without extensive preparations, with the resources at hand and, if by so doing, the attacking element can maintain momentum or take advantage of the enemy situation.

6-138. The platoon leader uses fire support to complement the scheme of maneuver. The goal is to kill as many enemies as possible and suppress the rest to keep them from seeing or shooting the maneuver element as it closes on the objective. The attack plan includes the use of all available direct and indirect fire. The response force leader mainly plans and directs the fire of his organic weapons (MG, M203, AT-4, and MK19). The fire element's makeup depends on how much direct-fire support is needed and what support is to be provided. The initial employment of the fire element includes the following:

- Positioning crew-served and M249 weapons on or near the line of departure (LD), when the objective and route to it can be seen and covered by fire.
- Positioning some crew-served and M249 weapons on or near the LD and some with the moving element, when the objective can be seen and covered by fire from the LD, but only a part of the route is visible, or when the entire route is visible from the LD, but the objective is not visible or is out of range from the LD.

6-139. Plan control measures at the objective and initial employment of the fire element. If time is available before moving to the assault position, make a second reconnaissance to see if the plan needs changing.

6-140. **Conduct the Attack.** The platoon leader moves to where he can lead the force and influence the fight. He moves a fire element into the over watch position and has the attacking element move from the LD to the assault position, using covered and concealed positions. If en route to the objective, the patrol should—

- Receive indirect fire. Move quickly out of the impact area to a preset point.
- Meet obstacles, breach, or bypass them. Report their location if the obstacles might affect follow-on units. Take special care when crossing a danger area like a field, roadway, or creek. Refer to *Chapter 3* for more information about moving in contact.
- Meet direct enemy resistance before reaching the objective and return fire. Call for and adjust fire on the enemy.

6-141. The platoon leader coordinates action so that the response force hits the enemy with its full combat power. He maneuvers the force to strike the enemy's flank or rear. The response force engages the hostile element with the appropriate level of force. He uses overpowering force on armed combat forces. While in contact—

- Immediately assume the best available covered positions, when moving, and at the same time, return fire in the direction of the enemy.
- Locate actual or suspected enemy positions and engage them with well-aimed fire.
- Distribute fires evenly over the objective when no individual positions have been identified.
- Report and monitor the situation and recommend COAs.
- Make visual or verbal contact with soldiers on the left or the right. (Platoon members and team leaders make frequent visual contact with the squad leaders. Squad leaders make frequent visual contact with the platoon leader. Relay all arm and hand signals from the platoon chain of command.)

6-142. When resistance is destroyed, the attacking force continues quickly onward to the assault position. They deploy on line just before reaching the assault position. To maintain the momentum, the attacking force passes through the assault position and attack the objective. They halt at the assault position only to allow all the soldiers to come on line to attack at once. The assault position should be as close to the objective as possible to prevent needless exposure to enemy fire.

6-143. As the moving element attacks, the fire element in its overwatch position will cover the attacking element by shooting at the enemy. Before the attacking element moves beyond support range, the fire element moves up to a position from which it can continue its support and shoot at the enemy.

6-144. With the fire element in overwatch, the platoon leader (commander) gives the command or signal to open fire. The fire leader will control the method and the rate of fire. He and the gunners watch the progress of the attacking element and engage targets that threaten it. When crew-served weapons are close together, the fire leader anticipates the masking of their fire and moves the weapons one at a time. When those weapons are separated, each gunner displaces when its fire is masked or when it can no longer support the attacking element. The platoon leader (commander) times the displacement of weapons in a fire element so that the attacking element has continuous fire support.

6-145. Maneuvering begins when the attacking element makes contact with the enemy. As the attacking element maneuvers near the objective, the fire element in overwatch suppresses the enemy with a high rate of fire.

6-146. When the attacking element reaches the objective and closes with the enemy, the fire element walks its fire across the objective in front of the maneuvering element. Then it shifts its fire to supplementary targets (rear of the objective, escape routes) or lifts its fire to keep from endangering the attack element. Soldiers in the element closing with the enemy move singly, by pairs, by teams, by squads, or by a combination of these, using as much cover as possible. As they close, they fire on selected targets to suppress the enemy. Automatic rifles are fired in short bursts across the forces' front. Rifles, M203s, hand grenades, bayonets, and MG are used to overcome pockets of resistance. The advance continues past the objective far enough to shoot at any withdrawing enemy. Firing continues until the enemy is killed or captured or withdraws.

6-147. Once the threat is eliminated, the platoon leader reconsolidates the response force and assembles it at a secure rally point. Refer to *Chapter 3* for more information about consolidating and reorganizing.

6-148. **Conduct an Attack on Urban Terrain.** In military operations in built-up areas, a response force must be able to find, isolate, and evict the enemy concealed in hardened structures. Normally, responding to a threat in a built-up area will call for a larger response force and should be supported by armored fighting vehicles. When attacking and clearing a building in an urban area, try to—

- Determine and isolate the area around the objective. In urban terrain it is often hard to pinpoint the enemy's position.
- Number the area's buildings and use those numbers as reference points to control movement and for clearing operations.
- Clear buildings from the top down if possible.
- Retain a reserve force to have the flexibility essential for urban combat.

6-149. Refer to *FM 90-10-1* for information about attacking on urban terrain.

Respond to Downed Aircraft

6-150. MP are often required to respond to downed aircraft. This specialized mission requires close coordination with aviation assets. In the event of a downed aircraft, the primary mission is to safeguard the aircraft, render first aid to the crew, evacuate the wounded, and secure the crash site. Normally, this requires MP to be airlifted to the crash site.

6-151. MP leaders plan for downed-aircraft rescue operations as a contingency mission. The primary concern is the safety of personnel onboard the aircraft. The MP leader establishes preplanned lift packages for a mounted or dismounted platoon-size downed-aircraft response force. When organized for downed-aircraft operations, the force must meet the following requirements:

- Possess cross-country capability.
- Be prepared to defend the site against Levels I and II threats.
- Have internal medical support to treat and evacuate injured personnel.

6-152. Engineer support may be necessary to clear lanes to the downed aircraft and medical evacuation (MEDEVAC) LZs.

6-153. Once the response force arrives at the crash site, the platoon leader immediately establishes security and notifies higher HQ of the exact location, the type of aircraft, and the number of wounded or killed personnel. Medical personnel treat the injured and prepare them for evacuation. The platoon leader selects and secures a MEDEVAC LZ. Once all the injured personnel have been evacuated, the response force continues to secure the crash site until relieved.

6-154. Security at the crash site involves MP preventing the handling or disturbance of the wreckage, theft of any parts, or compromising evidence such as gouges or marks on the ground made by the aircraft. Other security measures include—

- Preventing all unauthorized access to the site. This includes media and local authorities.
- Protecting all civil and military property at the site.
- Prohibiting removal of deceased persons until properly identified by military authorities.
- Prohibiting the removal of wreckage.
- Keeping spectators and the media at a reasonable distance from the site for safety.

6-155. All requests for information will be referred to the Public Affairs Office (PAO). MP at the scene will not release information about the crash to the media.

ENEMY DELAY

6-156. A delay is an operation in which a force under pressure trades space for time by slowing down the enemy's momentum and inflicting maximum damage on the enemy without, in principle, becoming decisively engaged. Normally a delay operation is unlikely to be conducted in a sustainment area. However, bypassed enemy forces could create a situation where a MP response force may be required to delay a threat to buy time for a base to establish defenses in greater depth or to delay an enemy until the arrival of the TCF. MP deliberately conduct a delay only when directed to do so by higher HQ. In that case, higher HQ determines the size of the MP force. In the delay, the destruction of the enemy force is secondary to slowing his advance to gain time.

PLAN

6-157. The platoon delays either from successive positions or from alternate positions. When delaying from successive positions, the delaying force will fight rearward from one position to another, holding each position for a given time or until it becomes decisively engaged.

6-158. When delaying from alternate positions, the delaying force will fight rearward with the force deployed so elements can provide overwatch and subsequent maneuver. While the forward element is fighting, the rear element prepares to assume the action. The forward element disengages. It passes through or around the rear element to prepare to assume the action from a position in greater depth. Use the following methods to delay the enemy *(Figure 6-9):*

- Delay forward of a specified line or terrain feature if you have—
  - A narrow sector.
  - Cross-compartment terrain.
  - An area restrictive to armor and the enemy can be canalized into a selected area.
  - Terrain dominating mounted avenues of approach.
- Delay from successive positions when you—
  - Have wide sectors or limited observation, fields of fire, and visibility.
  - Have a mobility advantage simultaneously moving one or two elements at a time.
  - Are in dangerous sectors.
  - Lack good defensible terrain in depth.
- Delay in a sector if you have—
  - An area of responsibility that cannot be adequately covered from one battle position.
  - Multiple enemy avenues of approach.

  - Limited fields of fire and observation.
- Delay from alternate positions when —
  - The sector is narrow or there is long-range observation and fields of fire.
  - The enemy is superior in combat power.
  - There is good defensible terrain in depth, and the terrain facilitates movement by bounds.

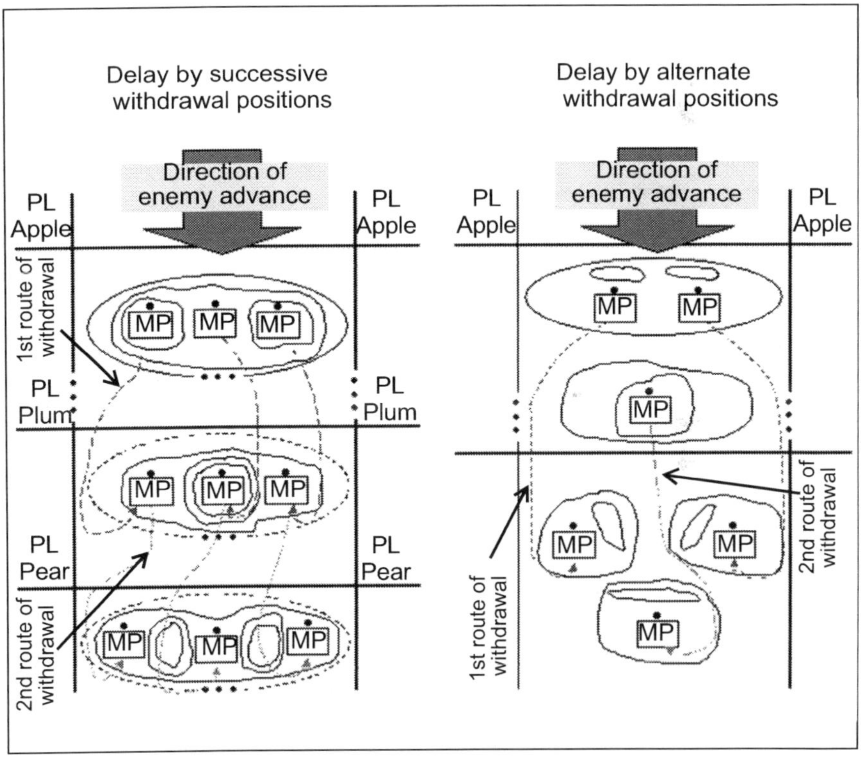

Figure 6-9. Methods of Delaying the Enemy

6-159. The platoon leader develops a plan of action using the factors of METT-TC and the commander's concept and intent. He gives priority to—

- Obtaining a detailed knowledge of the terrain.
- Gearing the terrain analysis to the enemy's intent, avenues of approach, and likely positions and targets.
- Gaining information from aggressive area patrol and reconnaissance.
- Having alternate communications, both sound and sight, ready to be used in case radio communications are disrupted.
- Forecasting time. The response force must consider how fast the enemy will close, how long will it take to move off a position, and how long to get to and occupy the next position. In delaying actions, timing can be critical.
- Requesting fire support along the route.

6-160. In the OPORD to subordinate leaders, includes the—

- Concept of conducting the action.
- Initial delay positions.
- Trigger points identifiable on the ground. ("Squad leader, I plan to initiate fires when the enemy reaches the railroad track.")
- PLs and delay lines. Include the length of time to delay on or forward of the given position or PL.
- Coordination and contact points.
- Indirect-fire plan (smoke, high explosives [HEs] munitions, artillery-delivered scatterable mines).
- Responsibility for adjacent-unit coordination (passage and the like).
- Fire control measures, engagement areas, targets, and rate of fire.
- Disengagement criteria. ("Squad leaders, I plan to start disengagement when the enemy has three armor vehicles across the railroad tracks.")

- Plan for moving after disengagement (signals, breakpoint, use of smoke, and the like).
- Subsequent delay positions and routes and the sequence of disengagement.

6-161. Most of the firepower will be oriented toward the enemy. However, flank and rear security must be provided. The platoon leader plans fire support in front of and on both flanks of the positions. He plans fire to cover the disengagement and movement to subsequent positions. He has fire concentrate on breaking up the enemy's advance at long range.

6-162. Higher HQ will resupply, rearm, and refuel the delaying element. On a platoon action, the platoon sergeant is crucial in ensuring that the platoon has the required support to continue a delaying action. The OPORD provides for a central point where the teams or squads can recover.

EXECUTE

6-163. If directed to delay the enemy, the platoon leader moves to a position where he can best control the action. The platoon leader makes good use of the terrain to mask all movements and provide vantage points for observation and harassing fire.

6-164. The enemy will try to bypass a delaying force. The platoon leader tries to keep the enemy in front of or on an oblique angle to the delaying force's position. He prepares positions as thoroughly as time permits, and reconnoiter routes to subsequent positions.

6-165. When selecting delay positions, look for the following:

- Clear observation and fields of fire onto the avenues of approach.
- Positions offering mutual support and interlocking and flanking fires where possible.
- Restrictions to canalize the enemy.

- Short, covered, concealed, and trafficable routes to rearward positions.
- Positions hidden from enemy observation and fires.
- Defilade positions for HMMWV and ASV.

6-166. If the delaying force has time they—

- Emplace obstacles to slow the enemy while they move to rearward positions.
- Block obvious routes rearward and direct the force to use more hidden routes.
- Continue preparing positions.
- Consider preparing ambush sites along the routes to slow enemy pursuit.
- Select primary and alternate firing positions.
- Have soldiers familiarize themselves with the routes to these positions.
- Place the OPs/LPs, and other security means far enough forward to warn the platoon, particularly if the enemy is using a secondary avenue of approach.

Control Fire

6-167. Fire control is extremely important. The platoon engages the enemy at maximum range with all weapons systems. They fire on a prearranged signal or event (trigger point). They aim for leaders and C^2 vehicles and avoid premature firing. The platoon leader considers not firing until the enemy reaches a preset point on the ground. Then he uses ambush-type fires to quickly destroy lead enemy elements. This also keeps the enemy from learning the friendly force's size and location.

6-168. If the enemy force cannot evade, it will make every effort to envelop and destroy the delaying force. Avoid presenting an exposed flank to the enemy. Use contact points to help ensure flank coordination.

6-169. Each time the delaying force moves back, it will have less time to plan and prepare. Speed, firepower, and maneuver become more important. The platoon leader calls for added fire support or assistance from higher HQ if decisive engagement or envelopment seems likely.

6-170. If additional elements arrive, the platoon leader uses them to augment the element in contact, employing them on the left and right of the element in contact. The platoon leader ensures that as the element in contact moves back, the left and right elements know to also move, but not at the same time. As the element in contact moves back, the flank screens can open harassing fire. Drawing enemy attention to flank elements will allow the element in contact to disengage more easily.

6-171. The platoon leader keeps higher HQ informed of the situation. Forward elements must not become so heavily engaged that they cannot be withdrawn effectively. If contact with higher HQ is lost, the platoon leader uses initiative. He bases each move on METT-TC, the events on the battlefield, and the tactical commander's intent.

Disengage

6-172. Deciding when to disengage is difficult. If the delaying forces remain too long, they risk decisive engagement or envelopment. If they move too soon, they may give up ground unnecessarily and risk pursuit. The decision of which element to move is based on METT-TC. Move the least heavily engaged element first. To disengage—

- Direct supporting elements to engage enemy forces forward of the disengaging element.
- Concentrate all available fire and mutual support.

- Give priority fire, if available, to the disengaging element to rapidly increase its firepower.
- Begin an orderly movement to successive or alternate positions.
- Use all available fire.
- Place the fire far enough forward to avoid impeding personnel leaving the protected positions.
- Place smoke between the delaying force and the enemy.
- Move into subsequent positions from the rear and send a SITREP to higher HQ.
- Report the delaying force's arrival to higher HQ.
- Tell the elements left in contact to disengage and take their next positions.

6-173. If it becomes necessary, replace key leaders, give crew-served weapons a priority of manning, evacuate casualties, and redistribute ammunition. Refer to *FM 17-95* for more information on defensive tactics used for delays.

BATTLE HANDOVER TO A TACTICAL COMBAT FORCE

6-174. When the combined efforts of the base and base cluster defense force and the MP response force are unable to jointly defeat an enemy threat, the operation becomes the responsibility of the TCF. Coordination is essential to determine when, where, and how the response force gives up responsibility for the fight and the TCF takes over. The MP response force and the base-defense force will then hand over the fight to the TCF by conducting a battle handover.

6-175. A battle handover *(Figure 6-10)* is a coordinated operation between two units that transfers responsibility for fighting an enemy force from one unit

ORGANIZATION

Maintain unit integrity for C2.

ORDER OF MOVEMENT

Base the order of movement on the—

- Number of passage points.
- Degree of security required.
- Enemy situation.
- Terrain.
- Prevention of confusion and congestion. Give priorities indicating which units move and when they move.

COMMUNICATIONS

Ensure that the following actions occur:

- Both units have SOI compatibility.
- Both units monitor the rear operations net (critical for recognition signals and fire support).
- Both outgoing and incoming unit elements operate on the outgoing unit's command net.
- The outgoing unit maintains routine radio traffic volume while the incoming monitors it.
- When the responsibility of the battle has changed over, the incoming unit operates on its own net only.

COMMAND AND CONTROL

Ensure that the—

- Number of passage points can determine the method of C2 used.
- Multiple passage points are set up, dictating decentralized control.
- TCF commander decides how he can best influence the action and positions himself accordingly.
- Response force commander collocates with the command group of the TCF during the passage of lines.
- Time or circumstances when responsibility for the battle is transferred are mutually agreed on before hand by the response force and the TCF commanders.
- Responsibility can be passed by radio communications if need be.

Figure 6-10. Planning Considerations for a Battle Handover

COMMAND AND CONTROL POINTS

Ensure that the following occurs:

- All units follow the control measures.
- For a response to force passing through a stationary TCF, control measures include the following:
  - AAs. Used by a unit to prepare for an upcoming mission.
  - BHLs. The single, most important control measure in delineating response force and TCF responsibilities. A PL is placed where the TCF assumes responsibility for the battle from the response force.
  - Attack position. The last position occupied or passed through by the TCF before crossing the LD.
  - Passage lanes. Areas on which the response force moves to avoid the stationary TCF.
  - Passage point. A place where the response force will pass through TCF units. It is located where the commander desires subordinate units to physically execute the passage of lines.
  - Time of passage. The time at which the passage is executed by the leading element of the response force to facilitate control through the passage point.
  - Recognition signals. Established signals to prevent the response force from directing fire on the TCF and vice versa.
  - Contact points. Designated, easily identifiable points on the terrain where two friendly forces are required to meet.
  - RPs. Points on the ground where subordinate units are under the control of their respective commanders while en route to their appropriate destinations.
  - Routes. Routes are the prescribed course to be traveled from a specified point of origin to a specific destination.
- For a TCF passing through a delaying response force, control measures include the following:
  - Battle position. A defensive location established by the response force oriented on the most likely enemy avenues of approach.
  - LD and LC. This control measures places the BHL in that control of the battle is assumed by the TCF once its units cross the PL.

Figure 6-10. Planning Considerations for a Battle Handover (Continued)

to another in close-in battle. It is designed to sustain continuity of the combined-arms fight and protect the combat potential of both forces involved. Battle handover is usually associated with conducting a passage of lines. MP leaders must establish clear tactical SOPs that allow their forces to quickly establish the necessary coordination to preclude a loss of momentum in the attack. In a battle handover, the MP response force will pass rearward through the TCF lines while handing over the fight or the TCF will move forward through the MP response force lines to engage the enemy.

6-176. However the fight is handed off, coordination between the MP response force commander and TCF commander is critical. Use a BHL (refer to *FM 17-95* and *FM 17-98*) and contact points as control measures to set both forces' responsibilities. Both the TCF commander and the response force leader coordinate the BHL location. The BHL is shown on the operations overlay and in the OPORD as a PL or a boundary. It identifies ground as "owned" by the TCF or by the response force. It sets the location where control of the battle passes from one commander to the other. It is where the TCF can use direct fire and observe indirect fire.

6-177. The handover occurs at the time or event coordinated by the TCF commander. The TCF commander decides the method of handover based on METT-TC. Generally, one of the following methods will be used:

- The response force passes rearward through a stationary TCF.
- The TCF passes forward through a delaying response force.
- The response force sets up a final blocking position, with elements of the TCF conducting an attack to destroy the enemy.

6-178. Coordination for battle handover normally flows from the TCF commander to the response force leader. Coordination normally includes—

- Establishing communications.
- Providing updates on both friendly and enemy situations.
- Coordinating passage.
- Collocating C^2.
- Dispatching representatives to contact points.
- Ensuring that there are visual recognition signals for vehicles.
- Ensuring the status of obstacles and routes.
- Coordination of fire support and CSS requirements.

6-179. The most important task for the response force is to maintain contact and continue to engage the enemy. The enemy must continue to see the level of activity that has been established. If the enemy perceives that the handover is occurring, he will attempt to seize the opportunity to destroy the response force.

PASS A RESPONSE FORCE THROUGH A STATIONARY TACTICAL COMBAT FORCE

6-180. The TCF commander coordinates with the response force leader at a designated contact point. The rear CP operations cell dictates the contact point location and notifies both units.

6-181. The contact point should be located forward (50 to 200 meters) of the BHL. It should be near or on easily identifiable terrain features. At a prescribed time, liaison parties from both forces will meet. Normally, a response force representative is sent to the contact point. At the contact point, the TCF commander or his representative briefs the response force representative. Response force teams may remain in the vicinity of the BHL if passage is imminent. TCF security forces screen along the BHL and monitor the response force command net.

6-182. The actions of the elements on the BHL are critical. At the BHL, the TCF scouts, or others assigned the BHL action, pick up the fight from the passing response force elements. If the response force is in contact, its maneuver elements are bound behind the BHL, preferably to a covered and concealed location. At the same time, the TCF elements on the line relieve the pressure on the response force elements.

6-183. At the passage point, where the response force passes through TCF units, response force teams identify passing response force units. They ensure that the passing vehicles are displaying the correct visual recognition signal. TCF scouts on the BHL notify the forward TCF teams by a prearranged signal that contact has been made and friendly forces are at the BHL. Response force elements continue to delay forward of the BHL. Once behind the BHL and covered from enemy direct fire, the response force should—

- Confirm that the proper visual recognition signal is still displayed on all vehicles.
- Orient weapons systems toward the enemy.
- Move quickly, in the directed formation, through the passage points and along designated routes to the RP.

6-184. Stationary TCF elements overwatch this movement. The only time the TCF fires is when positive enemy identification is made. At the completion of the handover, response forces may either be released by the operations cell to resume other priority missions or tasked to remain in OPCON to the TCF and continue the fight.

PASS A TACTICAL COMBAT FORCE THROUGH A DELAYING FORCE

6-185. This technique is similar to the response force passing through the stationary TCF. The response force will have more information on the terrain and enemy situation than the TCF.

6-186. The response force leader sets up contact points, passage points, and routes. At the least, he provides guides. The guides meet the TCF at the contact points and lead them along routes to the passage points and/or RPs near the LD and LC.

6-187. The response force maintains normal radio traffic. The TCF and response force operations elements temporarily collocate. The TCF commander assumes control of the battle handover at this time. TCF company teams may maintain listening silence on their TCF command net. The response force guides notify their commander when the TCF begins moving forward from the contact points. As the TCF company teams deploy across the BHL, the response force leader and the TCF commander transfer responsibility for the battle.

6-188. The TCF fire support officer (FSO) may collocate with the response force forward observer (or the individual appointed to act as forward observer for the artillery unit in the AO). Fire missions are approved by the RAOC supporting the response force until the battle handover has occurred. After that, any fire missions for the response force are cleared through the TCF FSO. Response force units provide direct-fire overwatch according to the TCF commander's scheme of maneuver. It is coordinated directly by the TCF commander and the response force leader at collocated command groups.

6-189. Response force direct-fire support depends on the TCF commander's attack plan and the battle handover.

However, the stationary unit lifts and shifts its direct fire as coordinated by the collocated command groups.

BLOCK FOR A TACTICAL COMBAT FORCE

6-190. When defending from a battle position, the response force positions its elements to achieve the TCF commander's intent. The response force sets up blocking positions to keep the enemy from escaping. It selects battle positions where it can control, maneuver, and concentrate direct fire on the enemy. As blocking positions are set up, the TCF maneuvers to attack the enemy's flank. The TCF commander and the response force commander set up a contact point for coordination and conduct a reconnaissance of this point, time permitting. The TCF commander and response force leader coordinate C^2 and exchange the following information:

- When to lift or shift fire.
- What targets must be destroyed to allow the TCF to close on the enemy.
- The time, location, or activity to pass control of the battle.
- The location of the engagement areas.

6-191. The TCF and blocking force coordinate so that the TCF flanks the enemy. The response force suppresses the enemy from the front allowing time for the TCF to maneuver into position.

6-192. The TCF commander specifies tasks for the response force to achieve the TCF scheme of maneuver. The response force normally will have to orient its weapon systems on the enemy avenue of approach using TRPs or engagement areas (EAs). The response force may be tasked to—

- Destroy a certain enemy force from the battle position.

- Control the terrain or block an avenue of approach by holding the battle position against the enemy assault.
- Reorient weapons systems on a secondary avenue of approach from supplemental positions or avenues of escape for the threat.
- Assist in any other task needed to achieve the TCF mission.

6-193. Refer to *FM 17-95* for more information about battle handover.

CRITICAL SITE, ASSET, AND HIGH-RISK PERSONNEL SECURITY

6-194. MP are often tasked to perform security operations that require integration and prioritization from other MP CS operations. During security operations, MP focus their efforts on measures designed to safeguard resources by conducting special security activities that prevent unauthorized access to installations, equipment, materiel, and documents; and safeguard against espionage, sabotage, damage, and theft. An MP security operation includes protection of CPs, convoys, ASPs, deep-water ports, rail, pipelines, and high-risk personnel.

COMMAND POST SECURITY

6-195. CPs are priority targets for enemy attack, and their security can be crucial to the success of military operations. CPs at division, corps, and EAC are *designated critical sites and assets*. When CPs relocate, MP provide in-transit security. The amount and type of security provided to an established CP depends on whether the CP is dispersed or massed, large or small. In all cases, the focus of MP security efforts is on providing early warning of threat activity. MP possess mobility and communications assets that make them most valuable in operating AS patrols around the CP. Elements do not share a common defensive perimeter in a dispersed CP. Each cell provides its own local security and access control. MP conduct screening operations to provide early warning of enemy activity. In a massed CP where elements share a common perimeter, MP may contribute to perimeter defense with other elements that are collocated at a base.

Division Main CP Security

6-196. All Army of Excellence (AOE) divisions have a main, rear, and tactical CP. Division XXI currently has a division main and tactical CP and a security operations center (SOC). The best use of MP assets to secure any division CP is to employ MP teams to conduct AS around the location of the CP. MP provide early warning of enemy activity by conducting area reconnaissance and surveillance (R&S) of likely enemy avenues of approach to the CP. This security measure also provides the CP with standoff protection from direct enemy fire.

Division Band

6-197. When assigned to augment security operations, the division band should be employed intact, facilitating the commander's ability to quickly return it to its primary mission. The MP commander ensures that the bandmaster and his soldiers know the tactical situation and the defensive plan for the CP when they are assisting in CP security. The band may—

- Assist in perimeter defense of the CP.
- Provide access control on the road to the CP.
- Operate the dismount point for the CP.
- Provide access control to the main CP.

Corps Command Post Security

6-198. Normally, the corps CPs are echeloned into a tactical CP, a main CP, and a rear CP. The corps main

CP is generally located in the general vicinity of the division rear boundary. Traditionally, the corps main CP is massed and is a large organization with attendant mobility and signature problems. The corps tactical CP is small, mobile, survivable, and normally located well forward in the vicinity of a division main CP.

6-199. MP security is provided to both the corps main and tactical CPs. How MP elements deploy depends on the CP dispersion, the particular situation, and the METT-TC. The MP company provides corps CP security. If the CP is massed, one platoon secures the tactical CP and the remainder of the company secures the main CP. Or, depending on the METT-TC,—

- One squad provides access control within the corps tactical operations center (CTOC).
- Two squads serve as a response force.
- Two squads secure the tactical CP.
- Two squads man dismount points or augment other squads.
- The remaining squads operate the OPs/LPs and the security patrols.

6-200. The main security efforts and emphasis of MP are on early warning and screening. Both internal access and screening force measures provide CP security. These measures are enhanced by—

- Internal guard posts.
- Perimeter access control points.
- LPs/OPs and a security post outside the perimeter.

6-201. When a corps main CP is dispersed in cells, focus MP efforts on screening force measures. Allocate MP based on the location of the function and the METT-TC.

CONVOY SECURITY

6-202. MP leaders plan and conduct convoy security (specifically convoy escort) using several different methods to secure them as they move through their AO. Primarily, MP secure vehicle movement by maintaining security in the area that a convoy will pass through. MP also conduct route security on a specific route for a designated period during which multiple convoys move unimpeded.

6-203. The main threats to convoys are mines and ambushes. Convoys are vulnerable to sniper fire, long-range fire from artillery, mortar, threat aircraft, and AT-guided missiles. They are also vulnerable to remotely controlled mines and demolitions. Armored vehicles are better suited to provide convoy escort protection. Leaders carefully evaluate the threat before assigning a convoy escort mission to HMMWV-equipped units.

6-204. When MP conduct convoy escort security missions, platoons normally escort the convoy through their AO and pass responsibility for escort to adjacent platoons as the convoy moves from one platoon's AO to another. However, some situations may require a platoon to escort a convoy from its SP to its final destination.

6-205. Whatever method is used to secure the convoy, MP platoons executing the escort mission operate under the control of the convoy commander. The relationship between the MP platoon and the convoy commander provides for unity of command and effort, if combat operations are required during the course of the mission.

Plan

6-206. The platoon leader plans for security of the convoy in all directions and throughout the length of the convoy. This requires MP teams to be dispersed throughout the convoy formation. When available, locate engineer assets toward the front to respond to obstacles. If a fire-support team (FIST) is attached, locate it near the platoon leader.

6-207. The platoon uses the column formation *(Figure 6-11)* due to its inherent speed and ease of movement.

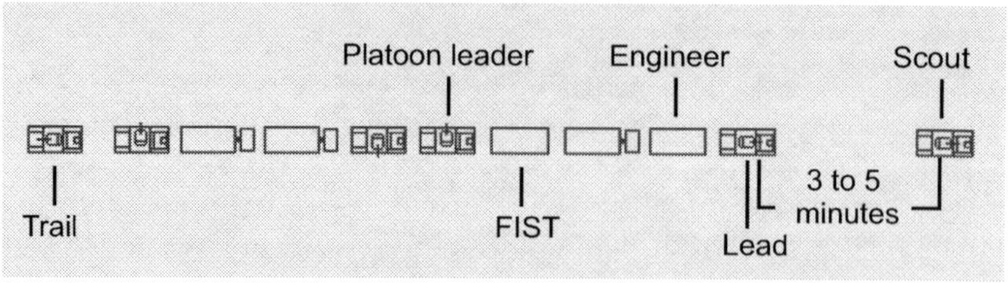

Figure 6-11. MP Platoon Escorting a Convoy

6-208. The platoon leader develops a convoy security plan that includes the following:

- HN support.
- The escorted unit's frequencies and call signs.
- The vehicle number and type and the order of the march.
- Checkpoint identification along the route.
- Security for halts and rest stops.
- Enhanced security at critical points such as, bridges, tunnels, key intersections, and other danger areas.
- An alternate route determination.
- Emergency actions to be taken during ambushes, a sniper attack, an air attack, or an artillery attack.
- Ambush reactions.
- Coordination with the response force responsible for the AO.

6-209. The platoon leader assigns selected teams to remain with the convoy to provide close-in security, if attacked, and other teams to maneuver toward the enemy to deliver a high volume of fire. Use a scout vehicle to travel ahead of the convoy (3 to 5 minutes) to alert the convoy commander of enemy activity, obstacles, or anything that might disrupt the convoys movement.

6-210. If a tracked armored vehicle is not available to lead the escort, the platoon leader assigns one team as the lead security element. Hardening of all vehicles used for convoy escort missions must be SOPs.

6-211. The platoon leader coordinates with the convoy commander to ensure that all convoy personnel are briefed and fully understand the convoy security plan. If time permits, the platoon leader conducts a joint rehearsal with the convoy commander and all security elements. At a minimum, the rehearsal should cover reactions to an ambush and air and artillery attacks.

React to Ambush

6-212. A well-planned ambush can have a devastating effect on a convoy. The reaction to an ambush must be quick, overwhelming, and decisive. The platoon executes its reaction to an ambush as a drill, taking special care to prevent fratricide. Actions to be taken include the following:

- When the ambush is initiated, security vehicles direct their action toward the enemy *(Figure 6-12, page 6-70)*. Selected teams move to covered positions between the convoy and the enemy to suppress the enemy with the highest possible volume of fire and obscure the enemy's view with smoke. At least one team should remain with the convoy and continue close-in security. The platoon leader sends the appropriate reports to higher HQ.

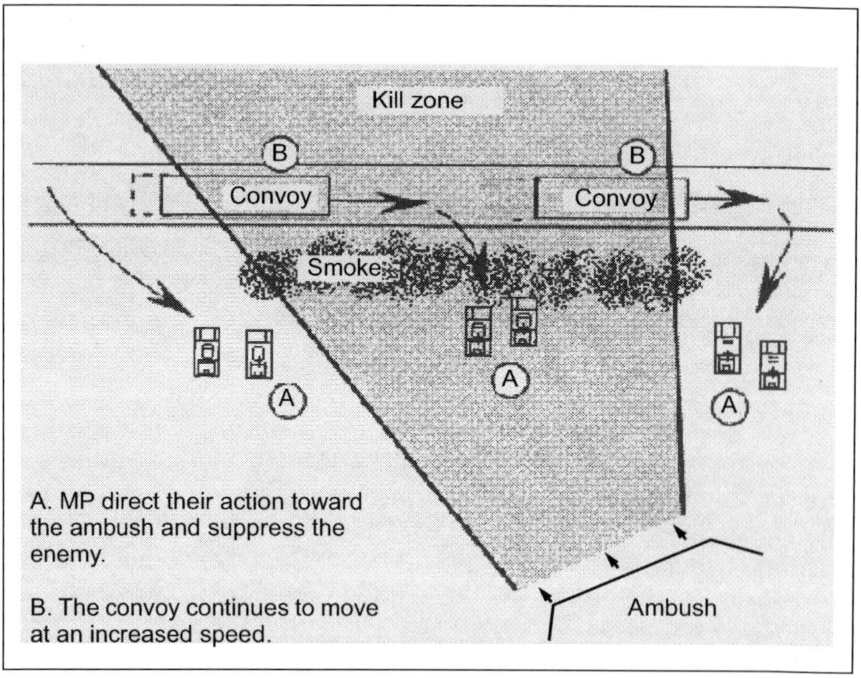

Figure 6-12. Security Vehicles' Actions During an Ambush

- The convoy commander continues to control the convoy and maintains radio contact with the security force (MP platoon leader). Vehicles not yet in the kill zone are stopped before they enter. Vehicles in the kill zone are moved out of the zone at the highest possible speed.
- The scout vehicle immediately selects a safe area for the convoy commander to move vehicles. The vehicles use a Light Vehicle Obscuration Smoke System (LVOSS) or other smoke to mask movement if possible. This safe area will be the linkup location when the convoy resumes.
- The convoy vehicles, if armed, return fire only until the security force positions itself between the convoy and the enemy.

- The damaged or disabled vehicles are abandoned and pushed off the route if they are not mission essential.
- The MP platoon leader submits SPOTREPs. Based on his assessment of the situation he can call for fire or CAS, or request the response force.
- Once the convoy is clear of the kill zone, the MP platoon leader chooses one of the following COAs based on the composition of the escort and the strength of the enemy:
  - Continue to suppress the enemy, while response forces move to support.
  - Assault the enemy.
  - Break contact, move out of the kill zone, and calls for indirect fire.

6-213. Normally, MP equipped with HMMWVs move out of the kill zone as soon as the convoy is clear. ASV-equipped security forces may continue to suppress the enemy or execute an assault. The platoon leader decides when to break contact.

6-214. **Mines**. Command-denoted mines are usually to signal the start of an ambush. When ambush fire comes from one side of the road, the enemy will most likely deploy a second, smaller force on the opposite side. They will deploy mines and obstacles between themselves and the convoy to contain US forces. The ambush force often uses mines to protect its flanks. To avoid damage from mines, drivers must—

- Drive on the track of the vehicle in front.
- Avoid driving on the shoulder of the road.
- Avoid running over foreign objects, brush, or grass in the road.
- Watch local traffic and the reactions of people on foot, which may give away locations of mines and booby traps.

6-215. **Artillery Attack**. If a convoy comes under artillery fire, vehicles must not stop. The convoy continues forward as quickly as possible to clear the area.

6-216. **Air Attack**. If the convoy comes under air attack, vehicles are dispersed, and soldiers attempt to find cover for the vehicles. Personnel begin firing all available weapons in an antiaircraft effort.

6-217. **Obstacles**. Obstacles are a major threat to convoys. Obstacles can be used to harass the convoy by delaying it. If the terrain is favorable, it may stop the convoy altogether. In addition, an obstacle or series of obstacles can be used to channel or stop a convoy to set up an ambush. The purpose of a route reconnaissance ahead of a convoy, is to identify obstacles and either breach or find bypasses. The convoy takes actions to reduce or bypass the obstacle. A convoy becomes more vulnerable when it is stopped and its escort force is occupied with tasks required to overcome or bypass the obstacle. For these reasons, security becomes critical, and actions at the obstacle must be accomplished quickly. Security personnel take the following actions at an obstacle:

- Direct (convoy commander) a halt when the lead security vehicle identifies an obstacle.
- Establish dismounted security and provide security throughout all activities directed by the convoy commander (such as conducting a reconnaissance and breaching the obstacle).
- Establish an obstacle overwatch.
- Ensure that convoy vehicles remain on the road.
- Move escort elements to the flanks to provide security.
- Ensure that all convoy vehicles are aware that the enemy may have buried mines in the area, especially on the road flanks.

6-218. **Halts**. When a convoy has to halt for short breaks it assumes a herringbone formation *(Figure 6-13)*. Security forces take up protective positions 360-degrees around the convoy and orient their weapons outward. They remain alert and establish dismounted local security. The convoy pulls into the protected area in the center of the perimeter. Stagger the convoy vehicles off the road, if there is no threat of enemy mines, but position them so they can easily pull back onto the road.

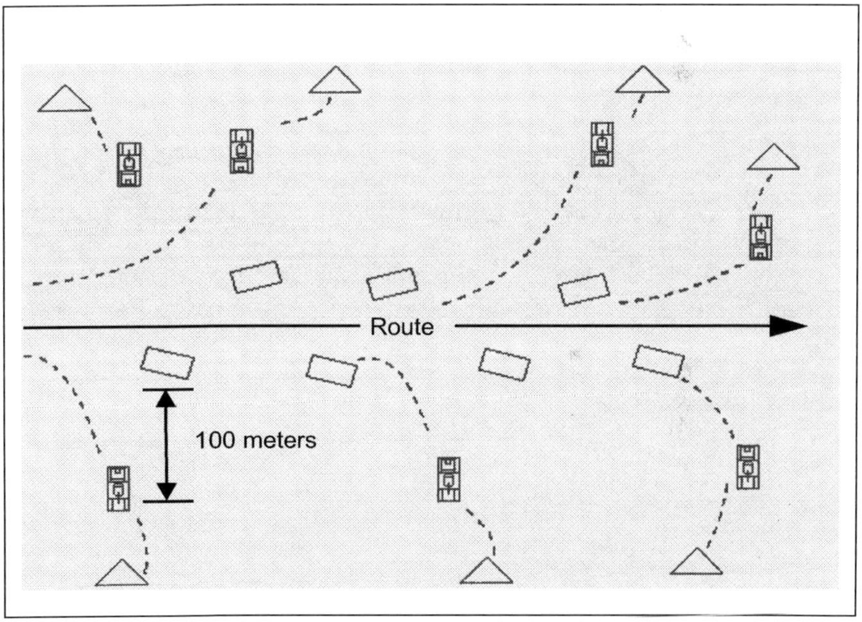

Figure 6-13. Herringbone Formation

6-219. When the convoy is ready to resume, the convoy vehicles reestablish the column formation, leaving space for the security vehicles. Once the convoy is in column, the security vehicles join the column, leaving local security dismounted.

6-220. Once all elements are in column, local security personnel mount and the convoy continues.

CONVENTIONAL AMMUNITION SUPPLY POINT SECURITY

6-221. When an MP platoon has been tasked to provide security for an ASP, the platoon establishes internal and external security. The platoon secures the ASP against enemy or criminal activity so that the ASP is able to maintain normal operations. MP leaders coordinate with ASP personnel to verify communications, distress and duress signals, defense plans, and fratricide prevention.

Internal Security

6-222. When MP conduct internal security, they establish a dismount point and allow only authorized personnel access to the ASP. Access is granted according to the local SOP and access and entry control rosters. MP conduct perimeter patrols ensuring that they do not set a pattern and maintain communications with platoon HQ.

External Security

6-223. When tasked to provide external security for an ASP, MP establish an OP/LP for early warning. They establish a TCP at the intersection of the MSR and the ASP access road. The TCP maintains free traffic flow, screens the traffic entering the access road, and controls stragglers and refugees. To further enhance external security, MP employ movement sensors, establish defensive positions, maintain communications with the platoon HQ, and emplace obstacles and physical security measures.

DEEP-WATER PORT SECURITY

6-224. The very nature of a terminal operation generates some different, and frequently more demanding, aspects of physical security. Ports and harbors are prime targets for enemy and criminal activities. The perimeter areas of these facilities are more vulnerable because of the extensive distance and the exposed beach or pier areas. Terminal areas may include fully developed piers and warehouses or may be an unimproved beach where logistics over the shore (LOTS) or roll-on/roll-off (RORO) operations are conducted.

Security Measures

6-225. When MP provide security for cargo in a port, the main effort is to provide security from the perimeter of the port outward. Security measures focus on aggressive patrolling and reconnaissance designed to detect, report, or destroy enemy threats before they disrupt port operations. Measures may also include—

- Developing police information and intelligence designed to identify criminal threats operating near the port.
- Controlling traffic in the area surrounding the port.
- Conducting mounted or dismounted patrols, with MWD if available, around the port perimeter.
- Establishing an access control and identification section.
- Providing a response force to react to incidents inside the port's perimeter.
- Providing observation and early warning of threat ground and air attacks.

6-226. When providing security for cargo, MP concentrate their efforts on providing AS in the area used to move cargo from the port to its final destination.

6-227. Inside a port's perimeter, MP security forces limit access to cargoes by—

- Operating random mounted or dismounted patrols (with MWD if available).
- Using the combined patrols as a response force for incidents inside the perimeter.
- Controlling access to the most restricted areas.

6-228. The priority is safeguarding the most critical cargoes waiting to be transferred to land transport. To safeguard stored cargo MP—

- Establish access control procedures.
- Search bundles and packages being taken from the area.
- Examine trip tickets and documentation of cargo vehicles.
- Inspect seals on locked containers.

Waterside Security

6-229. If the restricted area is a pier or other maritime environment, access from the water must be controlled as well as from the land. Entry on the landward side of a pier can be limited with fencing, pass control, and aggressive patrolling, but the part of the pier that protrudes over the water is accessible from the sides and below. Methods for securing the pier along its water boundaries include—

- Patrols (both walking on the pier and in small boats).
- Protective lighting.
- Log booms.
- Nets.
- Buoys or floats.
- Anchored or pile-mounted navigational aids and signaling devices.

- Barges.

DANGER
Ports pose special dangers for MP foot patrols. MP leaders must ensure that soldiers are thoroughly briefed on the hazards of patrolling too close to the water's edge.

6-230. While most of the barriers described above will stop or impede access to facilities from boats or swimmers, nets are among the most effective. Well-marked, partially submerged objects are also effective; however, there may be legal prohibitions against the emplacement of barriers that may constitute a hazard to navigation. The barriers should be emplaced only after coordination with and approval by the appropriate legal and HN authorities. Sometimes it is best to just close off the waterside of a pier. A floating boom will keep small boats out. Suspending a cable or chain link net from the bottom of the boom will deny access underwater.

6-231. MP establish at least two security zones on a facility's waterside: the reaction zone and the keep-out or exclusion zone. MP in these zones notify vessels, craft, and swimmers that they are entering restricted waters and should alter their course. MP may stop and search intruders if necessary. Security zones should be extended at least 1,000 meters from the nearest protected asset; however, in some port areas, a security zone this large is not possible. In such cases, other measures, such as boat patrols, must be increased to mitigate the possibility of attack. The two zones are defined as the—

- Reaction zone. The reaction zone extends from the high water mark to a distance beyond the maximum range of anticipated waterborne threats is the reaction zone. Inside the reaction zone, MP will stop and challenge intruders.

- Keep-out zone. This is the zone closest to the protected assets. The keep-out zone extends from the cargo asset to the maximum range of anticipated threat weapons. MP prevent the entry of all unauthorized craft or vessels into this zone. The tactical response force boat, in this case, may be employed if necessary.

Cargo Security During the Transfer

6-232. To keep cargo secure during transfer from one transport method to another, MP control the traffic moving in and out of cargo handling areas by—

- Setting up a single access control point.
- Erecting field-expedient barriers and using truck trailers or other large vehicles to constrict the traffic flow if permanent barriers are not in place.
- Limiting entry to mission-essential personnel, vehicles, and equipment, as designated by the port authority.

6-233. If gates are used by other than cargo vehicles, a *turnout* should be available. Cargo vehicles can pull into it while they are being checked. The turnout must be large enough to handle the volume and size of traffic being inspected. A wooden deck or platform at, or slightly higher than, the level of the truck bed hastens checking; it makes it easier to see the cargo. The platform should be as long as the vehicles being inspected. An empty flatbed trailer could be used.

6-234. Cargo is less likely to be diverted when close observation of cargo documentation and container safety is enforced. Containerized cargo is less likely to be stolen or sabotaged if closely watched as they are filled and sealed. Cargo can be pilfered before the seal is applied. An unsealed container can be moved to a stacking area or someone may apply a false seal, break the seal later, remove cargo, and then apply a legitimate seal. At access control points security personnel—

- Inspect inbound and outbound containers. Look for signs of damage or unserviceability.
- Inspect containers for the presence of seals and/or locks and hinges. Check their serviceability.
- Verify that the document's transport number, container number, and seal number match those numbers on the transportation control and movement document (TCMD).

NOTE: Check seals by handling them, not simply by visually checking them.

- Allow only containers with valid documents to pass inbound or outbound through the control point.

TRAIN SECURITY FOR CARGO MOVED BY RAIL

6-235. Military resources moving by rail are often vulnerable to theft, pilferage, and sabotage. Depending on the threat, the cargo's sensitivity, vulnerability, and MP availability, the echelon commander may require MP to protect cargo moving by rail.

6-236. Most train operating crews consist of four or five people–the engineer, a conductor, a fireman, a senior brakeman, and a brakeman or flagman. This crew controls the train. The conductor is the train commander unless a transportation railway service officer is assigned to the train. The train commander is responsible for the train's operation and security. He makes all decisions affecting the train. MP integrate their security efforts with the security efforts of the train's crew. MP watch for and report any discrepancies or interruption to normal procedures at any time during the movement. Information about the movement is usually sent along the movement route by the chief dispatcher through a telephone circuit.

Safety Considerations

6-237. Many regions use electric trains. These trains use aboveground, high-voltage lines and travel at greater speeds. When these high-speed trains pass parallel to a stopped cargo train and MP security forces are dismounted, the train can create a vacuum strong enough to pull a security force member under the wheels of the passing train. Some of the slower-moving trains are referred to as floaters. They travel at much slower speeds than the express trains, cause less vibration, and are much quieter. These trains may not be heard until they are too close to react. All trains pose a risk to MP working in and around them.

6-238. Track switches are another danger to MP while they are dismounted. Track switches are used to alter the course of a moving train or car onto another rail line. Track switches are remotely controlled and operated. A member of the rail crew activates some switches manually at some sites. The inherent danger to dismounted MP is getting a foot caught or crushed as the track is being switched.

Security Force

6-239. Normally, an MP squad is an adequate force to secure railway shipments of sensitive freight, but additional MP may be needed for movement of critical cargo. When required, MP conduct railway security to prevent pilferage, detect enemy activity directed against the train and, within their capabilities, defend the train against enemy attack. In addition to MP, the shipper may send specially trained personnel with highly sensitive cargo. The number of MP in a train security force depends on the—

- Sensitivity of the freight.
- Priority of need for the freight.
- Terrain over which the train will pass.
- Length of the train.

- Duration of the trip.
- Degree of enemy threat.

6-240. When planning rail cargo security, MP—

- Review the unit SOP to determine guard force security clearance and weapons qualifications requirements and the use of force and rail security operations training.
- Obtain the time schedule for the rail movement.
- Make a map reconnaissance of the route. If time permits, they conduct an aerial reconnaissance. Bridges and tunnels are especially vulnerable.
- Coordinate with the HN and other agencies, such as the military traffic management command (MTMC) and movement control team (MCT).
- Plan actions at scheduled stops or relief points, deploying MP according to these plans.
- Plot the locations of MP units and other friendly forces along the route and note their radio frequencies and call signs.
- Obtain an intelligence report covering the route to indicate sites where sabotage may occur, or attacks may be expected, and if thefts and pilferage are likely.
- Determine weapons and special-equipment requirements.
- Assess border crossing requirements.
- Develop a security cargo checklist.
- Make a guard force serial-numbered weapons list.

6-241. MP prepare and maintain a record, by car number, of guarded cars in the train. MP may be required to ride in—

- A specific car that requires protection.
- The caboose.

- A special train car provided for the MP to travel in.

NOTE: If only one security car is used, it should be near the center of the train; if more than one is used, space the cars to provide the best protection for the train.

Security Measures

6-242. The shipper is responsible for the security of all carload freight until it is turned over to the transportation railway service (TRS).

6-243. Railway cars are sealed after loading. A seal shows that a car has been inventoried and inspected. The standard method of sealing a railway boxcar door (in addition to padlocks or wires) is by a soft metal strap or cable seal that contains a serial number.

6-244. Maintaining rigid accountability of all the seals is necessary to prevent the undetected replacement of an original seal with another. While sealing does not prevent pilferage, a broken seal is a good indicator that the car and its contents have been tampered with.

6-245. When military vehicles are shipped by railcar, sensitive and high-value items must not be secured in the vehicles. Container express (CONEX) and military-owned demountable containers (MILVANs) are ideal for shipping these and other small items on flatcars since they greatly reduce the chance of pilferage. These containers must be locked and sealed and, if possible, placed door-to-door for additional security.

6-246. Cars containing freight that is likely to be pilfered, high-priority cargo, or special shipments are grouped in the train to permit the most economical use of MP security forces. When flatcars or gondolas are used to transport sensitive or easily pilfered freight, the MP leader positions a security element where it can continuously observe and protect these cars.

6-247. The security force commander conducts a joint inventory and inspection of the seals and locks with the MCT or transportation officer on *DD Form 1911*. The security force commander maintains possession of this document and uses it when transferring custody.

6-248. The MP leader ensures that each seal is checked at every scheduled stop (time permitting) and—

- Reports broken seals immediately to help pinpoint the time and place of possible thefts.
- Dismounts and conducts 360-degree security when the train is stopped.
- Checks for damage to the cars.
- Notes and reports irregularities in procedures, the presence and/or actions of unauthorized personnel, and deficiencies and incidents that occur.
- Conducts a joint inspection of the cars when relieved by another security force while en route. The relief force signs for the train (*DD Form 1911*). The record is kept on the guarded cars.

6-249. Security personnel obtain a receipt for the secured cars when they arrive at their destination. The receipt is attached to the trip report that includes the following:

- The date and time of the SP and the RP.
- Any additional information required by local SOP or command directive.
- Any recommendations for correcting deficiencies or for improving future security on the trains.

6-250. Dock workers unload the cars as soon as possible when arriving at unloading points to reduce the opportunity for loss through pilferage and sabotage. MP assets are normally not available for security of freight in railway yards; however, MP must be prepared to conduct stabling and staging operations.

6-251. Stabling operations are conducted when numerous train shipments are held in a rail yard or other locations, such as a marshalling area, for extended periods pending arrival at the final destination. Stabling operations require continuous security that may include establishing perimeter patrols, access control, and a quick-reaction force to respond to internal and external threats.

6-252. Staging operations are not extended like stabling operations. They occur when parts of a shipment enter a secured or unsecured off-load site. The site may be large enough to accommodate the entire shipment or only several rail cars at a time. Security forces may be split and required to maintain security at two locations. Additional security forces may be required based on the METT-TC. Refer to *FM 55-20* for more information about security of rail cargo.

PIPELINE SECURITY

6-253. Pipeline systems are widely used in a theater of operations to transport bulk petroleum products or other liquids. Such systems are open to a number of security threats from the point of entry to the point of final delivery. Pipeline systems are composed of storage and dispersing facilities, pump stations, terminals, and extended pipelines. They also include discharging facilities for tankers at ports or other water terminals.

6-254. The type and extent of risk to a pipeline varies with the level of conflict in the AO. In a communications zone (COMMZ), the chief hazard is likely to be pilferage. Pipelines can be tapped by loosening the flange bolts that join sections of pipe or cutting holes in the hose line. The risk rises if gasoline is scarce and expensive on the civilian market. Sabotage is a security hazard during all levels of conflict. It is committed by any method such as simply opening pipe flanges, cutting hose line, or setting fires and causing explosions to destroy portions of the line.

6-255. In areas of greater conflict, the likelihood of sabotage and interdiction increase. Pipeline systems are vulnerable to air attacks, especially at—

- Aboveground sections of the pipeline.
- Pump stations.
- Storage facilities.

6-256. Deploy MP in the best manner to provide coverage to the most vulnerable portions of the pipeline which are at the greatest risk to enemy, terrorist, partisan, and ground attack. MP are best able to provide pipeline security by conducting mounted and dismounted patrols; establishing a LP/OP for early warning; and establishing defensive positions around terminals, pump stations, and aboveground pipeline sections. They patrol isolated areas and remote pumping stations and—

- Detect, report, and respond to attacks on or sabotage of the pipeline.
- Monitor critical parts of the pipeline on a routine but random basis.
- Monitor ground sensors and other intrusion detection devices. These are often used at pump stations and elsewhere along the pipeline to detect and identify threats to the system.
- Check line pressure devices in the pipeline and pumping facilities. These devices monitor the flow and detect breaks in the line, which may indicate pilferage of gasoline or other petroleum products.

6-257. Dedicated MP are rarely enough in number for surveillance of an entire pipeline system. Available supporting forces, in the course of their normal duties, assist in observing and reporting items of intelligence for further investigation. Examples of suspicious activities in the pipeline area might include the unusual presence of commercial tank trucks, the appearance of gasoline drums or cans, or the increased use of motor vehicles in fuel-scarce areas.

HIGH-RISK PERSONNEL SECURITY

6-258. Security of high-risk personnel is the responsibility of the US Army Criminal Investigation Division Command (USACIDC). However, due to a limited number of CID special agents, in any theater of operation, MP are often tasked to provide close-in security of protected persons.

Security Considerations

6-259. Normally, the MP company commander designates a platoon leader as the officer in charge (OIC) of the security detail. The platoon leader is given full responsibility for all phases of the security mission, to include the coordination. This allows for a single point of contact for the conduct of the mission. The platoon leader prepares a comprehensive plan to ensure the safety of the designated person. Once the platoon leader receives the designated person's itinerary, he selects an advance team and conducts a reconnaissance of the route and locations the designated person is scheduled to visit.

6-260. Normally, the platoon leader uses organic vehicles and equipment to conduct the mission. If special equipment is required, he makes the necessary coordination through his higher HQ. Personnel selected for the mission are thoroughly briefed on the security plan and their conduct and demeanor during the mission. The security plan should include the following:

- Restrictions on individuals approaching the protected person.

- Evacuation procedures if the protected person is attacked and the threat cannot be immediately reduced.
- Control measures such as designated checkpoints.
- Crowd control measures.
- Alternate routes.
- Emergency MEDEVAC procedures.
- Emergency actions to be taken in the event of an ambush or indirect fire.

6-261. Depending on the threat level, selected members of the security team are armed with a M4, a M16, or a M203. When dismounted with the protected person, they provide standoff security out to the maximum effective range of their weapons. Shotguns should be considered shotguns if there is a threat of attack in a congested area where there is danger of injuring innocent persons if long-range weapons were used. Members of the team that are selected to physically escort the protected person should be armed with sidearms only. This allows them to have both hands free to react to close-in threats.

6-262. The security plan must be flexible to allow for itinerary and weather changes and mechanical failures. Once the platoon leader has finalized the security plan, he assembles the security team and conducts detailed rehearsals.

Security During the Visit

6-263. Protection may be accomplished by establishing a series of protective cordons around the protected person. The protective measures used must allow the person freedom of movement and must not interfere with the conduct of his duties. Irritating conduct must be avoided at all times.

6-264. MP must remain alert at all times and provide constant security to the protected person. MP provide security by—

- Restricting the circulation of unauthorized personnel around the protected person.
- Remaining a short distance from the protected person.
- Taking positions where they can observe everything and everyone near the protected person.
- Securing the facility of the protected person, but not entering if the protected person does.
- Maintaining constant communications with higher HQ.

6-265. When required to secure a designated person while moving, the mode of transportation must be guarded at all times. Vehicle security normally requires a protected vehicle to transport the designated person, security vehicles to react to threats, and an advance team to travel ahead of the motorcade to provide early warning of threats or hazards before the main body is affected.

6-266. Regardless of the method of transportation, each situation must be evaluated to determine the degree of security that is practical and necessary. Good judgment on the part of the OIC will be necessary in solving the various situations that may exist. Refer to *FM 3-19.30* for more information about the security of designated personnel.

Chapter 7

Internment and Resettlement

I/R consist of those measures necessary to guard, protect, and account for people that are captured, detained, confined or evacuated by US forces. In any military operation involving US forces, accountability and the safe and humane treatment of detainees are essential. US policy demands that all persons who are captured, interned, evacuated, or held by US forces are treated humanely. This policy applies from the moment detainees become the responsibility of US forces and continues until the time they are released or repatriated. (Refer to the *Geneva Conventions* and *AR 190-8, AR 190-14, AR 190-47, FM 3-19.40,* and *FM 27-10.*)

OVERVIEW

7-1. The task areas that support the I/R function are EPW and CI handling, US military prisoner handling, and populace and resource control.

7-2. Captured, detained, and protected persons fall into several different categories that include the following:

- Enemy prisoners of war. EPWs are members of an enemy armed force or militia who must be guarded to prevent escape.
- Civilian internees. CIs are persons who have committed an offense against or poses a threat to friendly forces and must be guarded to prevent escape, but are kept separate from the EPWs.

- Dislocated civilian. DCs are persons that have been removed from their home because of war, disaster, or other reasons. They may be refugees, evacuee, stateless persons, or war victims. DCs are provided sustenance, safety, and humanitarian assistance. They are kept separate from EPWs and CIs. DCs are controlled to prevent interference with military operations and to protect them from combat or to relocate them to safety. DC operations are discussed later in this chapter.
- US military prisoner. US military prisoners are members of the US armed forces being confined, awaiting trial, or waiting transportation to a confinement facility outside the AO. They must be guarded to prevent escape and cannot be confined in immediate association with EPWs and CIs, detainees, or other foreign nationals who are not members of the US armed forces. Refer to *FM 3-19.40* for more information about field confinement of US military prisoners.

7-3. EPWs are more specifically defined in *FM 3-19-40* and the *Geneva Convention Relative to the Treatment of Prisoners of War, August 1949.*

ENEMY PRISONERS OF WAR AND CIVILIAN INTERNEE

7-4. MP receive EPWs and CIs as far forward as possible to prevent maneuvering units from being burdened with large numbers of prisoners. Prisoners are evacuated from the battle area as quickly as possible. The capturing unit is responsible for guarding prisoners until relieved. They field process captives using the *Five Ss-and-T method (Table 7-1).*

7-5. The capturing unit is usually responsible for delivering the detainees to the collecting point and the

Table 7-1. Five Ss-and-T Methods

Procedure	Description
Search	Search each captive for weapons and ammunition, items of intelligence value, and other inappropriate items. **NOTE: When possible, conduct same gender searches; however, this may not always be possible due to speed and security considerations. Therefore, perform mixed gender searches in a respectful manner using all possible measures to prevent any action that could be interpreted as sexual molestation or assault. The on-site supervisor must carefully control soldiers doing mixed gender searches to prevent allegations of sexual misconduct.**
Silence	Do not allow the captives to speak or let anyone speak to them. Speak only to captives to give orders.
Segregate	Segregate captives by rank, gender, nationally, and status.
Speed	Remove the captives from the battlefield as quickly as possible.
Safeguard	Safeguard the captives according to the Geneva Convention and US policy. Provide medical care as needed.
Tag	Use *DD Form 2745* and include at a minimum the following information: • Date of the capture. • Location of the capture (grid coordinates). • Capturing unit. • Special circumstances of capture (how the EPW was captured, for example, did he resist, did he give up, and so forth).

nonwalking sick or wounded detainees to the nearest medical-aid station for evacuation through medical channels. Medically evacuated EPWs and CIs must be physically segregated from friendly forces. Detainees are normally turned over to MP at the nearest EPW collecting point or holding area. However, MP must be prepared to go forward to accept EPW from capturing units.

7-6. Traditionally, MP operate collecting points in a division AO and holding areas in a corps or EAC AO. However, collecting points and holding areas should be established wherever they are needed. The evacuation chain normally moves from the division forward or the central collecting point to corps holding area, then to internment facilities. When circumstances permit, such as taking advantage of available transportation, EPW evacuation may bypass one or more stations and deliver the detainees directly to a corps holding area or an internment facility.

7-7. At collecting points and holding areas, MP work closely with MI determining if captives, their equipment, or their weapons have intelligence value. MI interrogation teams conduct interrogations during field processing. Other MI interrogations teams conduct interrogations once EPW have been evacuated to more permanent facilities.

OPERATE A DIVISION FORWARD COLLECTING POINT

7-8. The number of MP needed to operate a division forward collecting point is based on the number and rate of captives expected and the METT-TC. A division forward collecting point must be mobile and modular and able to set up, expand, and move quickly with little or no notice. The general location of a forward collecting point is given in the brigade OPLAN or OPORD. It often is located near or in the brigade support area (BSA), but should not allow detainees to observe activities in the BSA. The collecting point should be situated close to an MSR. This makes it easier to get supplies, such as water, food, and barrier material from the BSA. Even a moderate number of detainees will put a strain on the equipment and supplies of an MP company. Minor medical treatment may come from the MP company's combat medical section. However, the company's medical resources are very limited and are primarily used to support medical needs within the company.

Units needed to support the division forward collecting point should be specifically tasked in the brigade OPORD. MP leaders operating the division forward collecting point will—

- Conduct a reconnaissance before selecting an exact location for the collecting point.
- Locate the collecting point far enough from the fighting to avoid minor shifts in the main battle area (MBA) (normally 5 to 10 kilometers from the MBA).
- Notify the BSA TOC and the PM operations section of the selected location. The BSA TOC reports the exact location of the collecting point to the brigade TOC. The brigade TOC notifies subordinate units where the collecting point is located so capturing units with detainees can take them there.
- Coordinate with the MI interrogation team if they are to colocate their interrogation site with the division forward collecting point.
- Request transportation, additional medical supplies, and other support through the forward support battalion.
- Ensure that captives do not remain at the division forward collecting point more than 12 hours before being escorted to the division central collecting point.

7-9. A forward collecting point *(Figure 7-1, page 7-6)* should not be set up near local inhabitants. Existing structures like vacant schools, apartments, or warehouses should be used when possible. This reduces construction requirements and minimizes logistical requirements. If existing structures are not used, detainees, except officers, can be tasked to help construct the collecting point. Prisoners may dig or build cover to protect themselves from artillery, mortar, or air attack. There is no set design for a forward collecting

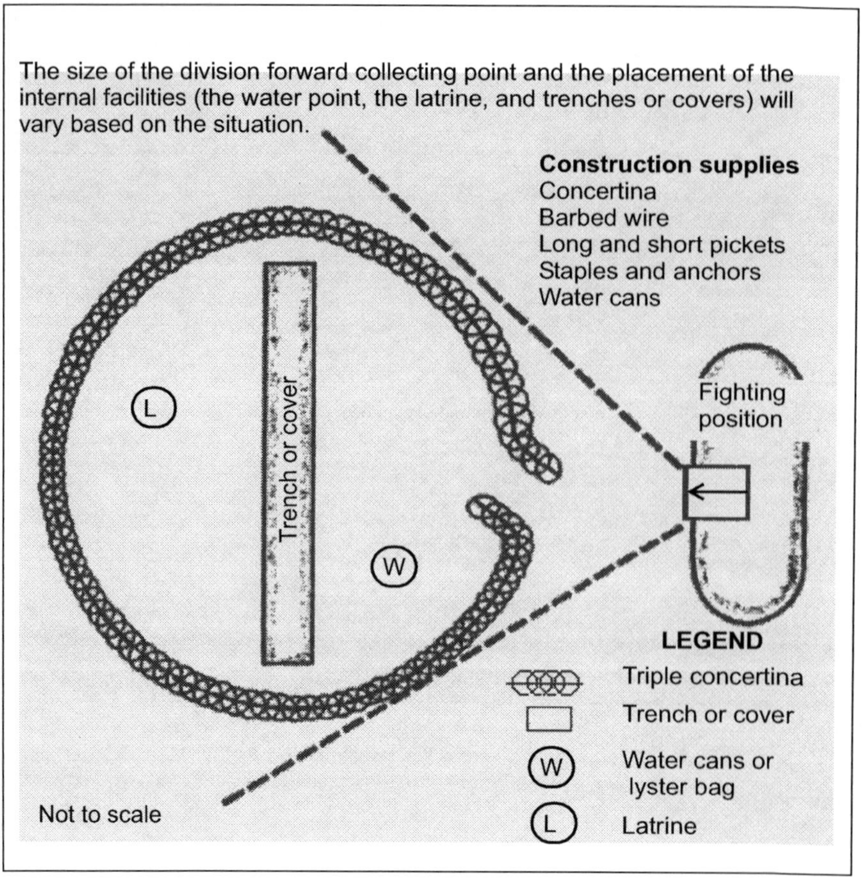

Figure 7-1. Division Forward Collecting Point

point. It can be anything from a guarded, roped off area to a secured, existing structure. The collecting point is built to suit the climate, the weather, and the situation. When selecting a collecting point, consider the following:

- The security of the detainees. The perimeters of the enclosure must be clearly defined and understood by the detainees.

- First aid. Injured or ill detainees require the same treatment that would be given to US casualties.
- Food and water. Detainees may have been without food or water for a long time before capture.
- Latrine facilities.
- Field sanitation. If possible, have detainees wash with soap and water to reduce the likelihood of disease.
- Shelter and cover.
- Language barriers. Provide interpreters and/or instructional graphic training aids (GTAs) in the EPW native language to compensate for the language differences.

7-10. MP at collecting points normally receive detainees directly from the capturing troops. MP then process the detainees using the *stress method*. The six principles of stress are search, tag, report, evacuate, segregate, and safeguard.

7-11. **Search**. Search and inspect every EPW and CI and their possessions. Use males to search male prisoners and females to search female prisoners wherever possible unless, in exceptional situations, an individual of the opposite gender must conduct the search. If this is the case, the search of the opposite sex must be performed in a respectful manner using all possible measures to prevent any action that could be interpreted as sexual molestation or assault. Captives may keep the following items found in a search:

- Protective clothing and equipment (such as helmets, protective masks and clothing) for use during evacuation from the combat zone.
- Retained property, such as identification cards or tags, personal property having no intelligence value, clothing, mess equipment (except knives

and forks), badges of rank and nationality, decorations, religious literature, jewelry, and articles that have sentimental value.

- Private rations of the EPW or the CI (in the early stages of captivity).

7-12. Certain items are confiscated from the EPW or the CI and never returned even if the EPW or the CI is released or repatriated. MP confiscate the following items when searching a captive:

- Weapons and ammunition.
- Items of intelligence value (maps, orders, and so forth).
- Other inappropriate items.

7-13. MP will coordinate with the MI interrogation teams to determine which items that have been confiscated are of intelligence value. Personal items, such as diaries, letters from home, and family pictures may be taken by the MI teams for review, but are later returned to the MP for return to the proper owner.

7-14. Currency will only be confiscated on the order of a commissioned officer (*AR 190-8*) and will be receipted for using *DA Form 4137*.

7-15. Impounded articles are items taken from the EPW or the CI during his internment because the articles make escape easier or compromise US security interests. Items normally impounded are cameras, radios, and all currency and negotiable instruments found on the captives. Refer to *AR 190-8* and *Defense Finance and Accounting Service-Indianapolis (DFAS IN) 37-1* for more information about confiscated and impounded property.

7-16. MP prepare a receipt when taking property from a detainee. The MP leader ensures that both the EPW or the CI and the receiving MP sign the receipt (such as *DA Form 4137*). MP consider bundling a detainee's property or placing it in bags to keep each detainee's property intact and separate. They turn in cleared, confiscated property as far forward as possible. MP maintain a strict chain of custody for all items taken from the EPW or the CI. They ensure that a receipt is obtained for any items you release to any other MP or agency. The escorting MP signs for and transports any remaining property that was taken from the EPW or the CI.

7-17. **Tag**. Each EPW or CI is tagged by the capturing troops using *DA Form 2745* as a way of accounting for them. MP check each tag at collecting points and holding areas for—

- The date and time of the capture.
- The capturing unit.
- The place of the capture (grid coordinates).
- The circumstances of the capture (how the EPW was captured).

7-18. The remaining information on the tag will be included as it becomes available. *DD Form 2745* is a perforated, three-part form which has an individual serial number. It is constructed of durable waterproof, tear-resistant material with reinforced eyeholes at the top of Parts A and C. The capturing unit attaches Part A to the captive with wire, string, or another type of durable material. They maintain Part B in their records and attach Part C to the confiscated property so that the owner may be identified later.

7-19. MP at division collecting points will ensure that *DD Form 2745* has been placed on any captive arriving at the collecting point without it. MP may have to direct the capturing units to complete the capture tag before accepting prisoners into the CP. They ensure that the following is done:

- The tag is filled out with the minimum information listed above (also listed on the back of Part C of the form).

- A statement is on the tag if the captive arrived without a tag.
- The captive is instructed not to remove or alter the tag.
- The capture tag's serial number and the captive's name are annotated on a locally developed manifest.

7-20. MP receive detainees from capturing troops using *DD Form 2708* or a similar document. They ensure that the receipt includes the following:

- The capturing unit.
- The time and date the detainee was received.
- The identification of the detainee. (Use the number on the capture tag when the detainee's name, service number, grade, or date of birth is unknown.)
- The name, service number, grade, unit, and signature of the MP who accepts custody of the detainee.
- A statement in the remarks section about the general physical condition of the detainee. For example, *received without wounds, illness, or injury* or *wounded in upper left arm*.

7-21. **Report**. The number of captives at each collecting point is immediately reported through MP channels. This aids in the transportation and security planning process.

7-22. **Evacuate**. Captives are humanely evacuated from the combat zone through appropriate channels as quickly as possible. MP do not delay evacuation to obtain name, rank, service number, or date of birth. When MP evacuate captives, they give them clear, brief instructions in their own language when possible. Military necessity may require a delay in evacuation beyond a reasonable period. When this occurs, MP leaders ensure that there is an adequate supply of food;

potable water; and appropriate clothing, shelter, and medical attention available.

7-23. MP ensure that EPWs or CIs are not be exposed to unnecessary danger and are protected while awaiting evacuation.

7-24. Medical personnel determine if captives with serious wounds or sickness should be kept in the combat zone. Sometimes prompt evacuation would be more dangerous to their survival than retention in the combat zone.

7-25. **Segregate**. The senior officer or noncommissioned officer in charge (NCOIC) having responsibility for custody of the EPWs or CIs will designate how and at what level to segregate them to ensure their security, health, and welfare. EPWs and CIs are segregated into the following categories:

- Officers, noncommissioned officers (NCOs), enlisted, male, and female.
  - Deserters and those that gave up without a fight may be further segregated for their protection.
  - Nationality, ideologies, and recognized ethnic groups are used for further segregation.
- CIs and/or refugees are physically separated from the EPWs and CIs.
- US military prisoners are physically separated from EPWs, CIs, retained persons (RP), other detainees (OD), and refugees.

7-26. MP do not use coercion of any kind to obtain any information from the captives. This includes basic information, such as name, rank, service number, and date of birth, which they are required to provide under the Geneva Conventions. Coercion or inhumane treatment of any EPW, CI, RP is prohibited and is not justified by the stress of combat or with deep provocation. Inhumane treatment is a serious violation of international law and the Uniform Code of Military Justice (UCMJ).

7-27. MP must not speak to captives except to give orders or directions. Captives must not be allowed to talk to or signal each other. This prevents them from plotting ways to counter security and plan escapes. Uncooperative captives may require a gag in certain tactical situations. However, gags should be used for only as long as needed and should not harm the individual.

7-28. **Safeguard**. In order to safeguard captives according to the Geneva Convention and US policy, MP must—

- Provide first aid and medical treatment for any wounded or sick captive. The wounded and sick will be evacuated separately through medical channels using the same assets as those used to medically evacuate US and allied forces.
- Ensure that the detaining power provides their captives with food and water. These supplies must be the same as to that of US and allied forces.
- Provide firm and humane treatment.

7-29. Protecting detainees from attack, preventing their escape, and quickly removing them from the battle area further safeguards them. Detainees should not remain at the division forward collecting point more than 12 hours, if possible. MP from the division central collecting point move forward to escort detainees back to the central collecting points. When detainees are field processed and ready for evacuation, the MP at the division forward collecting point will—

- Report detainee status to the BSA TOC and through MP channels to the PM.

- Request transport, rations, and water for the detainees from the forward support battalion supply officer (US Army) (S4).
- Ensure that the receipts for the detainees are ready for signing by the escort guards.
- Ensure that items taken from detainees for security or intelligence reasons are signed over to the guards taking the detainees to the rear. Ensure that each item is tagged to identify the owner.

ESTABLISH AND MAINTAIN A DIVISION CENTRAL COLLECTING POINT

7-30. MP in GS are responsible for establishing and maintaining the division central collecting point. They collect detainees from the forward collecting points, then process and secure them until corps MP come forward to evacuate them to the rear. Detainees should be transferred to the corps holding area or directly to an internment facility within 24 hours, if possible. One or more GS MP platoons operate the division central collecting point. The MP platoons are augmented by the division band and/or by the corps MP. Augmentation is based on the number and rate of captives expected.

Band Augmentation

7-31. When necessary, members of the division, corps, or EAC band augment MP for EPW operations. They guard detainees, operate dismount points, and provide perimeter security. When band members are tasked to augment MP for EPW operations they are OPCON to the MP company for the duration of the mission and released at the earliest opportunity to return to their primary mission.

Division Central Collecting Point

7-32. A central collecting point *(Figure 7-2)* is larger than a forward collecting point, but the considerations for setting up and operating the collecting points are

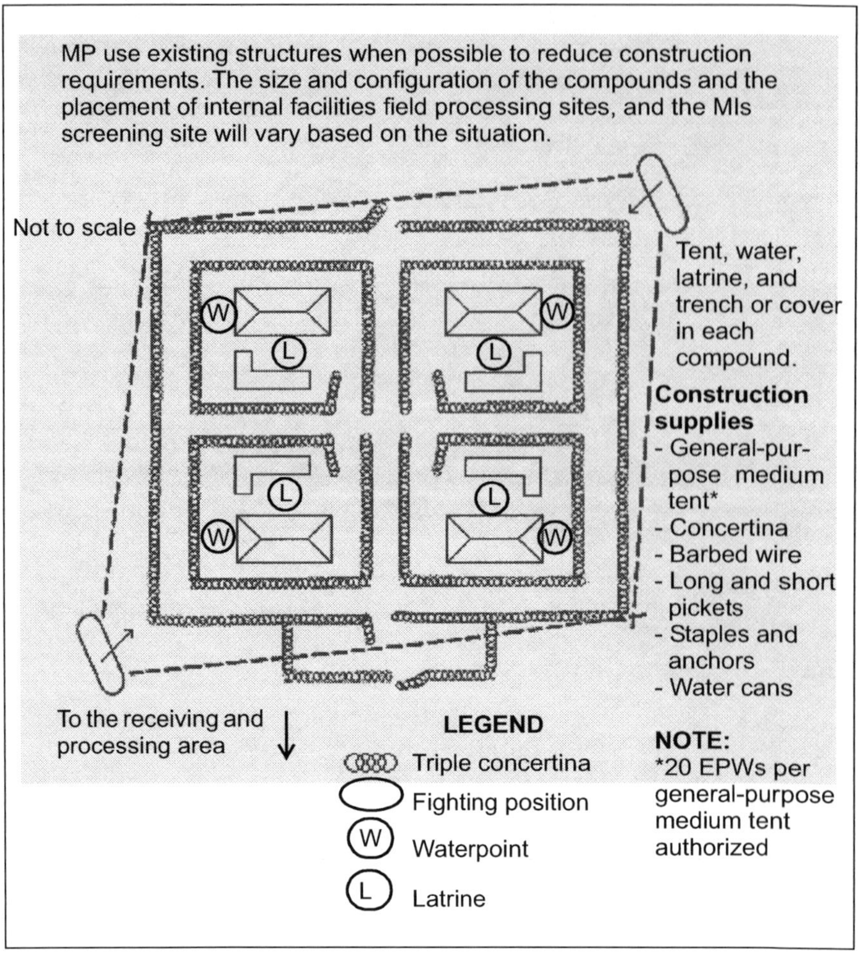

Figure 7-2. Division Central Collecting Point

generally the same. The general location of the central collecting point is given in the division OPORD or OPLAN. It is located near the division support area (DSA), preferably close to an MSR. This makes it easier to obtain supplies, transportation, and additional medical support from the DSA. Non-MP units should be specifically tasked in the coordinating instructions of the division OPORD to provide the support needed for the division central collecting point. MP establishing the collecting point should—

- Coordinate with the unit responsible for the area.
- Conduct a reconnaissance before picking the exact location for the collecting point.
- Notify the PM and the operations cell of the division rear CP (through MP channels) of the collecting point location.
- Coordinate with MI for the location of their screening site.
- Use existing structures when possible.
- Request supplies through the division MP company.

EVACUATE DETAINEES FROM A DIVISION FORWARD COLLECTING POINT TO A DIVISION CENTRAL COLLECTING POINT

7-33. The MP platoon charged with operating the division central collecting point sends MP forward to the division forward collecting point to escort detainees back to the central collecting point. EPWs or CIs must be evacuated from the division forward collecting point as soon as possible, preferably within 12 hours. Before evacuating the detainees, MP checks with MI interrogation teams for any property to be returned to, or evacuated with, the detainees before they are moved.

PROCESS ENEMY PRISONERS OF WAR AND CIVILIAN INTERNEES FOR EVACUATION

7-34. MP consider the physical status of detainees before evacuating them. Categories for consideration are the sick and wounded EPWs and CIs and the able-bodied EPWs and CIs.

Sick and Wounded Enemy Prisoners of War and Civilian Internees

7-35. The MP unit's combat medical section will screen detainees and decide if they will be escorted within MP channels or medically evacuated. Generally, the walking wounded go with the other detainees. Litter patients go through medical channels. US forces provide the same medical care for sick or wounded detainees as that given to US and allied soldiers. Sick and wounded EPWs in the combat zone are either treated and returned to the MP for evacuation or stabilized and moved through medical channels to the rear as far and as quickly as possible. If medically evacuated, MP release the detainees to the medical authorities using *DD Form 2708* or other receipt. The corps medical regulating officer (MRO) and the receiving hospital commander coordinate with the Internment Resettlement Information Center (IRIC) to account for detainees in medical channels.

7-36. MP determine when security is required for sick or wounded detainees. Normally, sick or wounded detainees requiring MEDEVAC are not likely to be a security risk. Detainees well enough to present a security risk can be treated by the combat medical section and evacuated through MP channels as soon as possible.

Able-Bodied Enemy Prisoners of War and Civilian Internees

7-37. Able-bodied detainees are escorted during movement to keep them from escaping. MP planning the movement of detainees consider the following:

- The factors of METT-TC.
- The number of detainees being escorted.
- The condition and morale of the detainees. Fatigued and cooperative detainees may not require as many guards as those who are fresher and more motivated.
- The type of transport to be used. The type of transport may influence the number of guards. A planning consideration is one guard per 5 to 10 detainees. Aircraft is loaded according to the airplane crew's instructions.
- The terrain conditions along the route. Detainees are more likely to attempt escape in close terrain, like dense woods or jungle, and may require more guards than open terrain.
- The level of enemy activity along the route. The more enemy activity in the area, the greater the need for increased security precautions.
- The likelihood or presence of suspected sympathizers and hostile local nationals along the route.
- The scheduled arrival of the transport. Use backhaul transport whenever possible.
- Transportation considerations. Transportation depends on the availability of vehicles delivering cargo in the nearby area.
- The location of MP units or bases and base clusters along the route that could provide assistance during the movement.
- The number and locations of rest stops (based on the type of transportation, distance, and the type of terrain).

7-38. Detainees are evacuated on foot only as a last resort when transport is not available. Transportation for detainees is arranged through the company HQ. At division, the company HQ contacts the local movement control officer.

7-39. Before leaving for the collecting point, the MP in charge of the escort must—

- Conduct a route reconnaissance of the evacuation route.
- Verify the location of the collecting point shortly before departing, since BSAs move often.
- Plan to stop only during daylight and outside towns or installations if possible.
- Designate guards to dismount at halts and supervise the loading of the detainees.
- Segregate detainees by category, if possible.
- Secure the rations and the water. Use captured enemy rations for the detainees, if available. Do not allow the use of utensils or can openers.
- Search detainees and baggage before loading in any transport.
- Use hand irons, leg irons, or special restraining jackets on detainees, if necessary. If hand irons are used, restrain the detainees with their arms in front.
- If prescription drugs are needed, disperse according to the medical officer's instructions.

EVACUATE ENEMY PRISONERS OF WAR AND CIVILIAN INTERNEES FROM A DIVISION CENTRAL COLLECTING POINT

7-40. In order for MP to conduct successful evacuation of EPWs and CIs, MP brief the escorts and the detainees.

Brief the Escorts

7-41. MP or other military personnel may perform as guards in evacuating EPWs and CIs. Escort personnel are briefed on evacuation considerations and escape attempts. Considerations include the following:

- Procedures to ensure that the detainees follow instructions and orders. Escorts must be firm, but will not punish detainees who fail to obey.
- Requirements to inspect passenger areas, latrines, and other places that might be accessible to detainees during transport. Escorts should look for the means of escape or items that could be used as weapons and remove the latches from the latrine doors on transports, if possible.
- The necessity to talk to detainees only to give orders and maintain control.
- Emergency actions to secure and safeguard the detainees in case of enemy contact. Members of the escort element must know in advance which of them will control the detainees and which ones will react to the enemy.

Receive and Brief the Detainees

7-42. The senior MP in the escort element accepts custody of the detainees. Each detainee is tagged and field-processed before being accepted for evacuation. Each detainee is accounted for using *DD Form 2708* or a similar receipt. The senior MP escort ensures that all the detainees (and any equipment) are listed on the custody receipt when custody is accepted. The escort retains a copy of the custody receipt.

7-43. Before moving, the senior MP ensures that the detainees have been briefed on movement discipline in a language understood by them. MP use locally produced GTAs to conduct the briefing. If available, an interpreter should give the instructions to the detainees. They are told—

- The meaning of the word *halt*.
- That the "silence rule" applies at all times (no talking to the guards, no talking to each other).
- The actions to take during an emergency.

COLLECT ENEMY PRISONERS OF WAR IN OTHER OPERATIONS

7-44. Some offensive operations are executed so rapidly that combat forces completely overwhelm the enemy forces. These operations create special considerations when planning EPW operations.

River Crossing Operations

7-45. During river crossing operations, an EPW collecting point is established on the nearside of the river far enough to the rear to prevent interference with tactical operations and to afford reasonable protection against hostile fire. The EPWs are evacuated from the bridgehead area as soon as possible to prevent congestion. The movement of EPWs from the farside of the river must be coordinated with traffic control personnel at the crossing sites. Secondary crossing sites must be used if available.

Amphibious Operations

7-46. During amphibious operations, initially, the shore party or helicopter support teams operate EPW collecting points in the beach support areas or LZs. EPWs are evacuated from the collecting points to designated ships by landing craft, a helicopter, or amphibious vehicles. MP must coordinate with the support force for the handling of EPWs once they have been evacuated from the beachhead. EPWs are retained in the objective area when facilities, supplies, and personnel permit, consistent with the reasonable safety of EPWs from enemy action.

Airborne Operations

7-47. During an airborne operation, the METT-TC considerations for collecting EPWs include the geographical location of the airhead, the tactical plan, the availability of transportation, and plans for linkup with ground forces. EPWs are evacuated primarily by air, especially during the early stages of the operation. The EPW collecting point should be located near a LZ. Plans should provide for the attachment of MP escort guard units from the area EAC to the airborne force to guard EPW during evacuation.

Armored Operations

7-48. Armored units are able to quickly penetrate deep into hostile territory, possibly bypassing enemy strong points. They may leave isolated enemy groups, which would hinder the normal evacuation of EPWs. In this circumstance, it may be necessary to hold EPWs in the area of capture until they can be safely evacuated.

Air Assault Operations

7-49. During an air assault operation, organic military police elements accompany assault elements to the objective areas so that they can collect and evacuate captured EPWs. Collecting points are established as required near heliports or airfields. Arrangements must be made for nondivisional MP to accompany designated incoming or resupply aircraft to guard EPWs during their evacuation from the division.

UNITED STATES MILITARY PRISONER HANDLING (FIELD DETENTION FACILITIES)

7-50. Use field detention facilities *(Figure 7-3, page 7-22)* to hold US soldiers in custody until they can be tried. Use pretrial confinement only to ensure that an accused appears at trial or when the seriousness of the offense or the threat of violence makes confinement essential. Whenever possible, soldiers awaiting trial remain in their units. Only when they are a hazard to themselves or others are they detained in pretrial confinement under MP control.

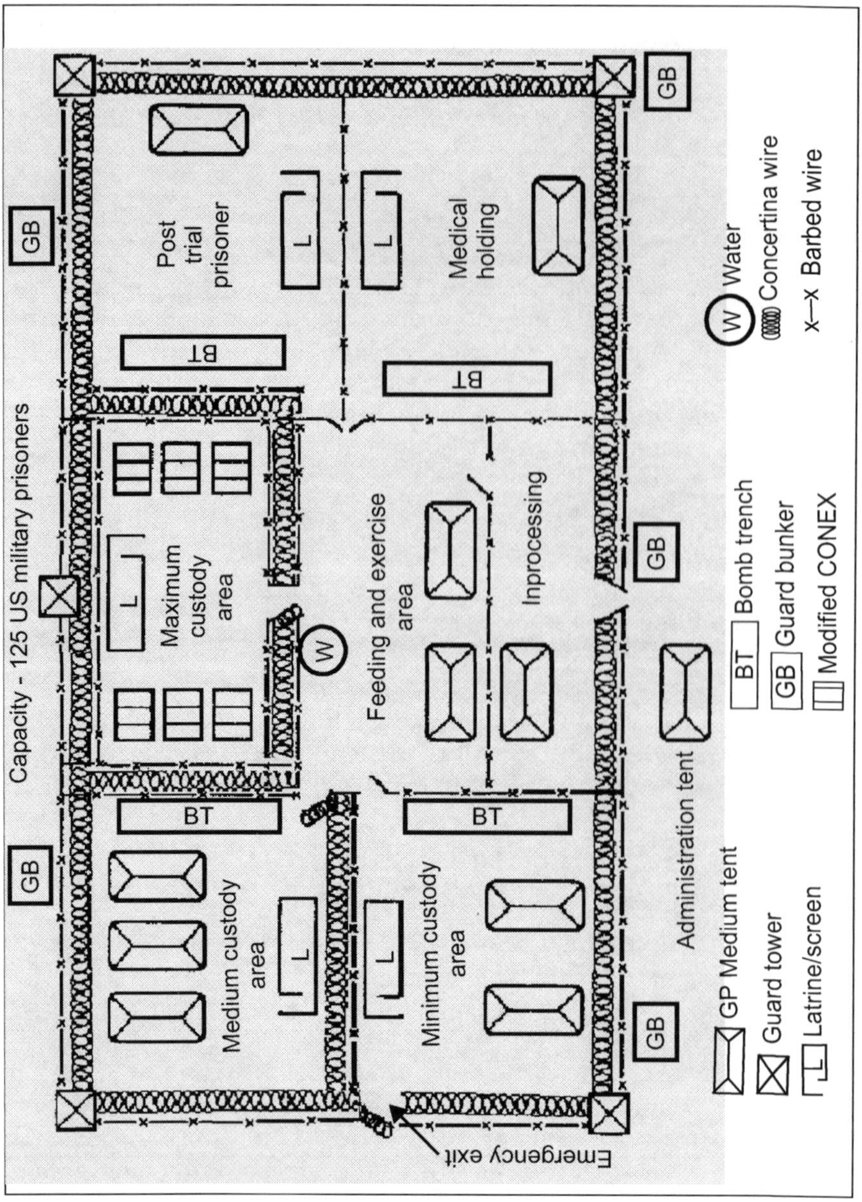

Figure 7-3. Field Detention Facility

7-51. Use field detention facilities to hold sentenced prisoners waiting for transfer to a theater's field confinement facility (FCF) or the continental US (CONUS). After trial, move convicted military prisoners, whenever possible, to confinement facilities outside the combat zone.

7-52. Each echelon commander sets procedures and policies for detaining and confining soldiers. Often US military prisoners in a combat zone are placed under the control of an MP unit operating an EPW collecting point. When small numbers of US prisoners are on hand, a squad operating an EPW collecting point can best take responsibility for the security of US prisoners. US military prisoners must be kept physically apart from EPWs. The policy and procedures for the care and treatment of prisoners and the safeguarding of their personal effects remain the same as that set for other Army confinement facilities.

7-53. When prisoners are retained in-theater, separate temporary detention facilities maybe set up in the corps or division areas. US military prisoners should be held in the division rear area for the shortest possible time. At a division facility MP must—

- Safeguard US prisoners.
- Coordinate for their food and medical care.
- Sustain them until they can be evacuated to a corps facility.
- Transfer them to the corps facility as quickly as possible.

7-54. When the situation permits, MP from a detention facility at the corps come forward to pick up the prisoners at the request of the division's detention facility commander. From the corps, the prisoners are evacuated to the theater confinement facility.

7-55. If a temporary detention facility is set up in the corps, it usually is operated by confinement teams from the confinement battalion in a personnel command (PERSCOM). These teams are organized and trained to perform confinement operations. But when corps detention operations are limited to prisoners being evacuated to a confinement battalion in the PERSCOM, elements from a combat support company can operate a temporary facility.

7-56. A field detention facility usually is located near the MP company CP for food, transport, and supply support. MP request construction materials from the engineers to set up and run a facility. Equipment and supplies must include the following:

- Barbed wire (roll and concertina).
- Fence posts.
- Gates and doors.
- Floodlights and spotlights, complete with wiring.
- An emergency generator.
- Mess equipment and equipment for cleaning mess gear.
- Water cans or lyster bags.
- Computers.
- First aid equipment and supplies. Spare clothing and bedding.
- Hand restrains or leg irons.
- Heating equipment (cold climate).
- Field sanitation supplies.

7-57. MP leaders ensure that the facility is large enough to separate prisoners by prisoner status, custody grade, sex, and rank. They locate the facility away from a base's perimeter or any other area of increased risk.

7-58. The size of the facility is based on the number of prisoners being detained. It may be a room or a tent, as long as it provides shelter equal to that offered to other soldiers in the combat zone. The physical criteria for permanent and temporary structures are the same. MP use existing structures if you can. Otherwise, they use tents. Field-expedient facilities must be approved and periodically inspected by a medical corps officer or a designated representative.

7-59. A team or a squad can operate a field detention facility. A team operating a facility may organize so that the team leader controls the operation and relieves the guards. The other team members perform guard duty in alternating 12-hour shifts. When operating a field detention facility, the team—

- Accepts sentenced soldiers on the written order (currently a *DD Form 2707*) of a court martial convening authority (or the authorized representative).
- Ensures that the order states the name, grade, social security number, organization of the prisoner, offense for which convicted, and the sentence.
- Accepts the accused soldiers on the written order (currently a *DD Form 2707*) of the accused's commanding officer.
- Signs a receipt for each prisoner and his property on the correct and current form.

POPULACE AND RESOURCE CONTROL (PRC)

7-60. PRC operations are the responsibility of the Assistant Chief of Staff, (Civil Affairs) (G5), the Civil Affairs Officer (US Army) (S5), and/or the HN authorities. (Refer to *FM 41-10* for more information about PRC.) PRC is often conducted in stability and support operations where national authority has broken down and the government cannot control the population. MP support PRC by conducting L&O operations designed to restore order and protect the people and property.

7-61. Insurgent organizations often emerge in unstable regions. The aim of such groups is normally to overthrow the established government. The less control the government has, the greater the chance for insurgents to succeed. These organizations try to exploit the population, often through threat and intimidation. When insurgent organizations pose a threat to the population, US forces employ PRC operations that are designed to deny support and assistance to insurgents by controlling the movement of people and goods and restricting access to key facilities. Police activities, such as roadblocks, cordons, curfews, access control, and checkpoints are an important measure in counterinsurgency, but have a high potential for harm if used excessively or incorrectly. MP are specially trained to conduct these operations, as a force focused on security, protection, and assistance. MP are continually trained on the prudent use of force, crisis management, and operations requiring restrictive ROE.

7-62. PRC measures deprive insurgent organizations support and aid in identifying their supporters. Appropriate psychological operations (PSYOP) help make these measures more acceptable to the population by explaining their necessity. The government informs the population that, although its actions may cause inconvenience, the threat posed by the insurgents makes them necessary.

7-63. PRC is often conducted in urban areas. The best use of PRC comes before an organized insurgent movement has the capability for armed conflict. MP intelligence operations support PRC programs. Criminal acts, such as robberies, kidnappings, terrorism, and extortion, may accompany insurgent propaganda or money-raising activities.

7-64. MP employ special control measures to aid populace control that include the following:

- Enforcing curfews.

- Enforcing movement restrictions.
- Verifying travel permits and registration cards.
- Assisting with rescue and evacuation operations.
- Assisting with crowd control.

7-65. MP also employ control measures to aid in resource control. They include—

- Operating roadblocks and checkpoints.
- Inspecting facilities.
- Enforcing local regulations and guidelines.
- Controlling rations.
- Assisting with amnesty programs.

7-66. PRC operations play a vital role in winning support away from the insurgent threat and encouraging support for the goals of the legitimate government.

7-67. MP leaders responsible for supporting PRC must plan for the protection of their forces as these types of operations have a history of turning violent very quickly.

DISLOCATED CIVILIAN RESETTLEMENT

7-68. International law recognizes the humanitarian practice of providing temporary refuge to anyone, regardless of nationality, who may be in imminent physical danger for the duration of the threat. It is US policy to grant temporary refuge in a foreign country to nationals of that country, or of a third nation, solely for humanitarian reasons when extreme or exceptional circumstances exist.

7-69. Civil affairs (CAs) units are trained to plan, coordinate resources for, and monitor the handling of DCs. (Refer to *FM 41-10* for more information about DC operations.) Whenever possible, resources and control should be arranged with the HN, other governmental agencies, and nongovernmental and private organizations.

7-70. To assist in properly identifying DCs, the definition of the five subdivided categories is provided. These subcategories are defined by legal and political considerations as the following:

- Displaced person. A civilian who is involuntarily outside the national boundary of his country in time of conflict.
- Refugee. A civilian who, because of real or imagined danger, has left home to seek safety.
- Evacuee. A civilian removed from his place or residence by military order.
- Stateless person. A civilian who has been denationalized, whose country of origin cannot be determined, or who cannot establish his right to the nationality claimed.
- War victim. A classification that describes civilians suffering injuries, loss of a family member, or damage to or destruction of his home resulting from war.

7-71. DC operations are a special category of PRC. The goal of DC operations is to minimize civilian interference with military operations, relieve suffering, and protect civilians from combat operations or other threats. When the HN cannot or is unwilling to control DCs, MP may be required to collect, evacuate, and resettle them.

7-72. When directed to conduct DC operations, MP leaders coordinate with CA and the PM to establish a traffic control plan. The plan includes the primary and alternate routes used to move DCs and the location for—

- The TCP.
- Holding areas.
- Roadblocks and checkpoints.

7-73. Temporary route signing is not normally an effective control measure during DC operations. Even when the signs are posted in the local language, DCs usually ignore them.

7-74. MP collect and process DCs in the same manner as EPWs with regard to the difference in their status–they are detained personnel, not prisoners. As such, security of the I/R facility should not give the impression that it is a prison.

7-75. MP collect DCs at assembly points located away from the MSR and areas where combat operations are heaviest. They move DCs along selected routes that have the least military traffic. DCs will normally require frequent rest stops. When possible, holding areas for rest stops should be near a source of water and provide protection from the elements or hostile fire.

7-76. MP provide close-in and standoff security for DCs while en route to a civilian camp or collection point. When required, MP provide external and internal security at a DC camp. Within limits, DCs are allowed freedom of movement once they have been resettled. MP and CA must continue to closely monitor DCs at collection points and camps. Recent operations that involve large numbers of DCs have been known to turn violent very quickly. Such a situation can be avoided when DCs are treated with respect and dignity.

EVACUATION OPERATIONS

7-77. In addition to EPWs, CIs, and DCs, MP are often required to evacuate other selected individuals or groups from collecting points, holding areas, or areas of hostility. The safe evacuation of noncombatants, US diplomats, or US military prisoners requires close coordination and extensive planning. This type of operation is referred to as noncombatant evacuation operations (NEO). Such missions require contingency plans and unit SOPs that are specifically designed for these special operations. Unit SOPs should cover the evacuation of designated personnel by ground, air, rail, and water. When directed to conduct evacuation operations, MP leaders task organize based on METT-TC and the availability of MP. They perform extensive rehearsals to successfully conduct evacuation operations.

7-78. When a HN can no longer ensure the safety of US civilians in a foreign nation, US military forces may evacuate them. MP are often required to provide security and escort for evacuees from their point of origin to their destination. A NEO is usually a joint operation conducted with other US and HN government agencies and CA.

7-79. An MP platoon is most likely employed for these operations. The platoon leader conducts direct coordination with CA and the higher HQ. The platoon leader begins TLP, issues a WO to the platoon, and requests information that includes the—

- Screening and identification system being used.
- Number of evacuees, their point of origin, and their anticipated direction of movement and destination.
- Location of marshalling areas, collection points, and AAs.

7-80. If time permits, the platoon and squad leaders conduct a reconnaissance of the evacuation route. The purpose of the reconnaissance is to prepare an overlay which is then used to plan control and special security measures at critical areas along the route. If helicopter support is available, platoon and squad leaders conduct an aerial reconnaissance of the evacuation route. An aerial reconnaissance will normally take less time and the observation of the terrain around the route is much better by air than from the ground. At a minimum, a map reconnaissance must be conducted.

7-81. During the reconnaissance, the platoon leader selects possible rest areas and identifies friendly units along the route. He identifies critical areas, such as defiles, bridges, and areas where enemy activity is likely. He marks these areas on the overlay and plans for special security measures at these locations.

7-82. The platoon leader coordinates with the movement control center to ensure that the evacuation route does not interfere with the movement of friendly forces. If available, he coordinates for fire support along the route, normally at critical sites that were identified during the reconnaissance.

7-83. The platoon leader coordinates for MEDEVAC and vehicle recovery. His plan includes emergency reaction to the following:

- An air attack.
- An artillery attack.
- An ambush.
- A riot.

7-84. NEO are normally carried out according to the guidelines established by CA and US policy. MP may be required to screen for authorized personnel to determine who may actually be evacuated. There are generally three groups of personnel. They are—

- Group I. Group I includes US citizens, officials, dependents, tourist, business persons, and non-US family members when the father, mother or wife is a US citizen.
- Group II. Group II includes foreign nationals holding diplomatic papers, visas, or passports who receive Department of state approval.
- Group III. Group III includes all others, to include HN citizens, who do not fit into the first two categories.

7-85. Persons who fall into any of these groups and require immediate medical attention are always evacuated first. MP should give special consideration to the elderly and children.

7-86. MP brief the evacuees on discipline and the actions to take during emergencies. The briefing must be in a language that all the evacuees understand. The CA unit or HN agency should provide an interpreter. The briefing should cover all aspects of the evacuation. This will help calm the evacuees and instill confidence and cooperation.

7-87. The platoon provides security of the evacuees at the marshalling, evacuation, and holding areas and the reception station. The level of protection depends on the level of the threat. Methods of security depend on the type and location of the facilities used. At a minimum, MP must be prepared to provide interior guards for group areas, establish perimeter security, and operate a dismount point to restrict access to the evacuees.

7-88. During movement, the platoon escorts the evacuees by providing close-in security at the lead, middle, and end of the convoy or in front of and behind a single transportation vehicle. MP use a scout vehicle that travels 3 to 5 minutes ahead of the convoy to alert the main body to danger or delays. If a threat tries to disrupt the evacuation operation or destroy the evacuation vehicle, selected MP teams protect the evacuees as other teams counter the threat within the ROE. Throughout the evacuation operation, MP maintain all-around security, protection, and evacuee accountability.

Chapter 8

Law and Order

This chapter addresses the L&O function across the full spectrum of military operations. Refer to *Appendix K* for more information about civil disturbance control measures.

OVERVIEW

8-1. L&O operations consist of those measures necessary to enforce laws, directives, and punitive regulations; conduct criminal investigations; and control populations and resources that ensure commanders the existence of a lawful and orderly environment. MP enforce laws and appropriate commander directives. They maintain liaison and coordinate joint L&O operations with other DOD police organizations; HN military and civilian authorities; multinational police organizations; and US federal, state, and local police agencies. A coordinated law enforcement effort removes the conditions and opportunities that promote crime, thereby preventing diversion of military resources and maintaining military discipline.

8-2. The evolving criminal threat affects military operations and requires commanders to minimize that threat to preclude negative impacts on forces, resources, and operations. The importance of the criminal threat to military operations is a current, as well as future reality. Nationalist ideologies, the instability of a government to effectively govern and control its population, and the breakdown of government infrastructures will foster the linkage between criminal organizations and the government and its armed forces.

8-3. MP provide the capability to train foreign MP or assist in the reorganize of indigenous constabulary forces as part of stability and support post conflict operations. Under the provisions of and exceptions to *Section 660 of the Foreign Assistance Act*, MP provide initial assistance and training to foreign military and civilian police forces or assist in the creation of these forces where national authority has broken down. Additionally, MP forces provides short-term emergency L&O capabilities until the foreign military and civilian forces are functional.

8-4. MP and the USACIDC are the primary collectors of police information and criminal intelligence. They gather information through contacts made with the local populace and from conducting combined and joint patrols with HN military and civilian police agencies. MP and CID conduct field interviews and gather police information from surveillance operations. They investigate serious offenses and maintain a close liaison with the HN or allied civilian and military police agencies. This police information and criminal intelligence is collected, analyzed, and shared with the intelligence community and contributes to the police information assessment process (PIAP).

8-5. The police activities that support L&O operations include the following:

- Criminal investigations.
- Police information collection and dissemination of the information.
- Traffic accident investigations.
- Antiterrorism force protection support.
- Crowd control.
- US customs operations.
- Use of MWDs.

8-6. The enforcement of military laws, orders, and regulations is a command responsibility, as well as an MP responsibility. Each commander is responsible for maintaining order and discipline in the unit. To support commanders, the PM plans the use of MP assets to help enforce military laws, orders, and regulations.

8-7. MP performing L&O operations enhance and extend the tactical commander's C^2 by—

- Aiding commanders in maintaining combat strength.
- Helping prevent diversion of military resources.
- Suppressing opportunities for criminal behavior by US and non-US personnel or elements.
- Assisting and protecting military forces.
- Helping ensure the discipline of US forces.
- Assisting intelligence organizations in obtaining a true tactical intelligence picture by providing criminal and operational data and intelligence.

LAW AND ORDER AUGMENTATION DETACHMENT

8-8. The technical and supervisory expertise to support L&O operations in a mature theater is provided by L&O teams organized under an L&O augmentation detachment. The L&O teams are designed to give the commander the additional flexibility and capability in any environment to conduct a wide range of force protection mission requirements, to include split-based operations. This enables the supported commander to perform needed L&O missions while other MP assets are conducting MMS, AS, or I/R.

8-9. When these teams are not available or have not arrived in theater, L&O missions are prioritized with other MP combat support missions and are usually not full-scale dedicated operations. The theater commander determines when he can afford to dedicate MP assets to L&O operations.

8-10. A complete 45-person L&O augmentation detachment includes a detachment C^2 team, an operations team, a desk team, five MP investigation (MPI) teams, five traffic accident investigation (TAI) teams, and two force protection teams. Refer to *Figure 8-1*.

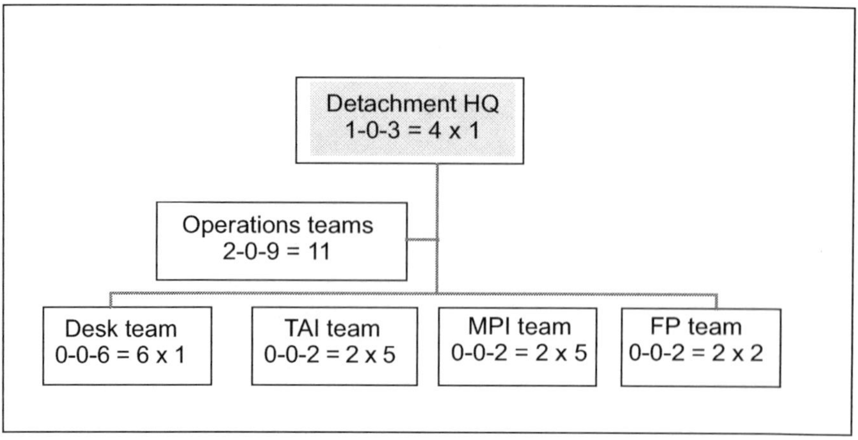

Figure 8-1. L&O Augmentation Detachment

8-11. When deployed, detachment and teams are under the overall operational C^2 of the PM deployed for the operation. Those teams may be attached to MP battalions and companies, as the situation requires. In war, the L&O augmentation detachment is assigned to an MP brigade and further attached to MP battalions and division MP companies as required. The L&O augmentation detachment consists of the following teams:

- Operations. The operations team provides C^2; supervises traffic, force protection, and investigative operations; and provides an evidence custodian to ensure that the chain of custody for evidence is maintained.

- Desk. A desk team establishes and operates the MP desk and coordinates law enforcement patrol activity.
- MPI. MPI teams provide the technical personnel to investigate crimes that do not fall within the investigative purview of the CID and conduct surveillance operations. Each MPI team consists of two MP with additional skill identifier (ASI) V5. These teams have no organic vehicles or communications capabilities.
- TAI. TAI teams provide the technical personnel to conduct TAIs. Based on operational requirements, TAI teams augment PM and MP organizations to ensure unimpeded movement and regulation enforcement on the MSRs. Each TAI team consists of two MP with ASI Q9.
- Force protection. Force protection teams provide technical personnel to assist in the safeguarding of personnel and equipment according to the command policies and field operating procedures. The role of the force protection team is to assist, inspect, and educate. Force protection teams provide expertise for safeguarding personnel and PRC for refugees, displaced persons, and civil disturbance operations. They enhance the unit's ability to ensure that soldiers and units are aware of appropriate personal security safeguards. Additionally, force protection teams assist customs operations teams in the establishment of sterile areas. Each force protection team consists of MP NCOs.

8-12. MP brigade commanders employ L&O augmentation teams based on the METT-TC to support the US forces operating within the brigade AO. While the number of teams is based on the troop population supported, whenever possible the team AO should coincide with the MP battalion and the division AO. The detachment HQ, operations, desk, and force protection teams should be collocated with the supported battalion HQ or division PM. If transportation requirements for these teams exceed the capabilities of the operations team, the supported unit must provide transport.

COMPANY AND PLATOON LEVEL LAW AND ORDER OPERATIONS

8-13. As previously stated, L&O augmentation teams may not arrive until a theater matures and L&O operations become the priority. Corps and division MP units must be prepared to conduct full-scale L&O missions. When ordered to establish L&O operations, the company commander determines mission requirements that include the following:

- Special equipment (vehicles, communications, and barricades or traffic cones).
- Specialized personnel support (a linguist, HN police, and PSYOPS or civil affairs personnel).
- CCIR.
- Threat assessments from MI and the CID.

8-14. The commander and platoon leaders conduct a map reconnaissance of the AO and determine the platoon areas of responsibility. When the tactical situation allows and adequate structures are available, the commander selects a building that is centrally located within the AO to establish the MP operations center. The facility should, at a minimum, offer areas for the following:

- An MP station desk.
- Offenders processing.
- Good radio transmissions.
- Arms, evidence, and property storage.
- Suspect detainment.

- Latrine facilities.

8-15. The company operations section operates the MP operations center and organizes it into MP sections that include—

- MP operations.
- Administration.
- Force protection.
- An MP desk.
- MP investigations.
- TAI.
- MWDs (if attached or assigned).
- Customs (if attached or assigned).

8-16. MP that are school-trained and have been awarded an ASI should be in charge of the appropriate section.

MILITARY POLICE OPERATIONS AND ADMINISTRATION OPERATIONS

8-17. The company operations sergeant is in charge of the day-to-day activities of the MP operations center. He instructs platoons regarding patrol patterns and distribution and coordinates for specialized equipment and personnel. He reviews all MP reports and the MP blotter. The center maintains close liaison with HN and allied military and civilian police agencies. The operations sergeant directs and supervises the administration section and provides guidance to other sections.

FORCE PROTECTION

8-18. The company physical security NCO (ASI H3) coordinates and directs the efforts of force protection for the commander. The force protection section is responsible for reviewing the security measures of all critical facilities and assets within the AO. The force protection NCO reviews points designated as critical by the senior commander and identifies criminal and operational threats. He develops a draft mission-essential or vulnerable area (MEVA) list and submits the list to higher HQ for approval. Once the list is approved, he conducts vulnerability assessments of points on the MEVA list. He applies risk analysis procedures according to *AR 190-51, AR 190-11*, and *DA Pamphlet 190-51* and ensures that appropriate army antiterrorism force protection standards are met according to *AR 525-13*. He reassesses assets for addition to or deletion from the MEVA list or changes in priority. The role of the physical security NCO is to assist, inspect, and advise commanders on force protection measures.

MILITARY POLICE DESK

8-19. The MP desk coordinates MP patrol activity, documents those activities, and reviews incident reports to ensure accuracy and completeness. The MP desk is also responsible for referring incidents to the appropriate investigative agency for further action.

8-20. The MP desk operates continuously during the conduct of L&O operations. It should be located as near as possible to the center of the L&O operational area, and have communications throughout the AO. When possible, existing facilities should be used for the MP station, but requirements for communications, transportation, and proximity to troops and facilities may rule out existing structures. The MP desk should be located within a reasonable distance of the HN police desk during stability, support, or other operations involving interaction with HN authorities. It may be necessary to establish substations if the area covered or the volume of activity is large. The MP station should have, at a minimum, the following:

- Areas and lighting for preparing reports and other paperwork.

- Communications with the MP elements performing L&O and the command and populace that the L&O operations are supporting.
- Detention areas where persons can be secured by a guard and/or physical constraints, such as a detention cell.
- Latrine facilities.

8-21. MP establish joint service, multiagency, or multinational operations when appropriate for the populace being provided L&O support or as directed. The PM establishes agreements before initiating joint operations to prevent conflicts in jurisdiction, documentation, and communication. Those agreements may be informal at first but should be formalized as soon as the situation permits. Refer to *FM 19-10* for more information about operating a MP desk.

MILITARY POLICE INVESTIGATIONS

8-22. Crimes occurring within an AO may require an in-depth investigation, depending on the seriousness of the offense and the availability of investigative personnel. Commanders report minor offenses within the unit area to the MP for statistical purposes, but may still investigate the offense themselves.

8-23. MPI investigators handle the most criminal offenses not investigated by CID or the unit commander. MPI normally investigate those offenses cited in *AR 190-30*. The commander ensures that only school trained (ASI V5) personnel are assigned to the MPI section to investigate crimes. Outside continental US (OCONUS) areas, MPI investigate off-post incidents according to SOFA and/or the US and HN agreements.

8-24. CID investigates serious offenses. It has elements in support of all echelons down to division level. CID operates across the full spectrum of army operations, with an emphasis on logistics security (LOGSEC) during wartime operations.

8-25. When MPI has purview to investigate a crime, they use the following investigation measures:

- Take control of the crime scene, ensuring that the crime scene is protected and secure.
- Identify the personnel involved, both suspects and witnesses.
- Identify the type of offense or offenses that have been committed.
- Process the crime scene using notes, sketches, and photographs.
- Collect and secure the physical evidence.

8-26. Once the investigator has processed the crime scene, he pursues leads and conducts investigation activities that include—

- Interviewing victims, suspects, subjects, and witnesses, and obtaining written statements from them.
- Preparing case documents and the required reports.
- Collecting related documents necessary to support the investigation.
- Coordinating with the Staff Judge Advocate (SJA).
- Submitting the final draft report for review by the MPI supervisor.

8-27. The commander appoints an evidence custodian to maintain evidence obtained during criminal investigations. The evidence custodian coordinates the needed laboratory examinations and ensures the proper disposition of evidence. It may be necessary to coordinate with the CID for this support.

8-28. Joint investigative activities, such as drug or black market suppression teams, may involve investigators from CID, MPI, or other US services or agencies. Such operations may require close liaison and cooperation with HN or allied civil or military agencies.

8-29. Refer to *FM 19-20* for more information about criminal investigation techniques and procedures.

TRAFFIC ACCIDENT AND INVESTIGATION

8-30. MP help reduce nonbattle casualties and the loss of equipment by ensuring that vehicles are operated according to regulations and determining the cause of accidents. Traffic enforcement supports the commander's intent in HNs by reducing the likelihood that military traffic will be seen as a threat to the native population. Thorough investigations of traffic accidents can—

- Identify the personal, environmental, and equipment factors that caused or contributed to the accident.
- Document the facts of the incident for future criminal or civil actions.

8-31. Traffic enforcement measures vary according to the operational and political environment. Those measures may include the following:

- Operator license and dispatch checks.
- Safety inspection checkpoints.
- Vehicle load and route restrictions enforcement.
- Speed control measures.

8-32. Traffic accidents are investigated on the scene whenever the tactical situation permits. MP investigate traffic accidents involving military personnel or vehicles within their AO. When the HN or other authorities have primary responsibility for an accident investigation involving US forces or equipment, MP conduct a concurrent investigation.

8-33. The first MP that responds to a traffic accident is responsible for securing and protecting the accident scene. When necessary, they administer medical attention to the injured persons and implement traffic control measures. MP safeguard classified materials and take the appropriate action to identify and contain hazardous materials at the scene.

8-34. The TAI team takes control of the accident scene when it arrives. Whenever possible a school-trained (ASI Q-9) traffic accident investigator processes the accident scene.

8-35. Once the TAI team receives a briefing from the senior person present, it takes the following steps to investigate the accident:

- Collect physical evidence to include evidence of drug or alcohol involvement.
- Photograph the scene.
- Interview victims and witnesses.

8-36. Once the TAI team has the necessary information, it clears the scene. To do this it ensures that the vehicles and debris are removed and reestablishes traffic flow. The TAI team then prepares a final report and submits it to the proper authority.

8-37. Accidents involving fatalities or life-threatening injuries require a serious incident report (SIR) according to *AR 190-40*.

8-38. Refer to *FM 19-25* for more information about traffic enforcement techniques and conducting TAIs.

MILITARY WORKING DOGS

8-39. MWD teams further enhance L&O and customs operations with capabilities to detect explosive devices and residue and controlled substances, and detect, track, control, and apprehend personnel.

8-40. Mission support requirements determine the number of functional MWD teams that make up a wartime, mission-oriented MWD team. These mission-oriented MWD teams are assigned to the PM. MWD teams conducting explosive or narcotic detection are unable to provide security for themselves and require security by the supporting unit.

8-41. Explosive, narcotics, and patrol teams each consist of three handlers and three working dogs. This allows each team to provide 24-hour support for a mission that requires one MWD or up to three short-duration missions. There is also a kennel master team, which provides technical supervision and is responsible for establishing kennel operations.

8-42. The explosive or patrol team provides the capability to detect explosive devices or residue in support of personal protection, MOUT, health and welfare, crime scene, and customs operations. When not required for explosive detection, this team functions as a patrol team.

8-43. The narcotic and patrol team provides the capability to detect controlled substances in support of the crime scene, health and welfare, and customs operations. When not required for narcotics detection, the team functions as a patrol team. The patrol team provides the capability to search for, track, detect, and control personnel and augment security in support of personal protection, MOUT, criminal apprehension, force protection, EPW, refugee or displaced person, and crowd control operations.

8-44. Refer to *AR 190-12* for more information about the MWD program, to include standards for selecting and retenting handlers and dogs, training teams, constructing kennels and training areas, maintaining and disposing of dogs, using teams, and using force. Refer to *DA Pamphlet 190-12* for more information about the MWD program.

UNITED STATES CUSTOMS SUPPORT

8-45. Units (and individuals) returning to the customs territory of the US (CTUS) have the responsibility to ensure that their vehicles and equipment meet US customs, other governmental agencies, and US Department of Agriculture (USDA) standards. Units request and receive customs support to ensure that they comply with all the established requirements. Trained military personnel, normally from the unit deploying to CTUS, conduct military customs inspections in the absence of US customs personnel. Senior military customs inspectors (SMCIs) train and supervise unit personnel on custom's responsibilities and requirements. SMCIs supervise the customs operations. The SMCIs supporting the theater deploy as part of the customs operations team.

CUSTOMS OPERATIONS TEAM

8-46. The customs operations team supports the commander's requirement to ensure that personnel, equipment, and materiel meet US customs, immigration, USDA, and other federal agency requirements for redeployment back to the CONUS. A complete customs operations team includes an SMCI supervisor team and two 5-member SMCI inspection teams.

8-47. The customs operations team, when deployed, is assigned to the Army service component command (ASCC) PM. It may be under the OPCON of or attached to the theater support command (TSC) PM.

8-48. The SMCI supervisor provides technical supervision, staff planning, and coordination for mission requirements with the Army staffs and agencies, redeploying units, and US federal agencies.

8-49. Two SMCI teams provide the capability to support two ports of embarkation (POEs). Before arrival at the

POE, these teams advise deploying units on requirements and procedures to preclude delays or rescheduling of unit and strategic lift transportation. Team members train and supervise selected redeploying unit personnel as inspectors to augment the SMCI team.

REDEPLOYMENT OPERATIONS

8-50. *AR 700-93* provides guidance on processing and shipping DOD-sponsor retrograde cargo destined for the CTUS. In order for a unit to redeploy successfully, it must—

- Coordinate for site requirements and standards.
- Establish wash and holding areas.
- Determine the existing USDA preclearance requirements.
- Use the expertise of assigned SMCI.

Site Requirements

8-51. A site includes those areas necessary to getting the unit's vehicles and equipment ready for the customs inspection and shipment. A suitable site includes the following areas:

- Staging.
- Download.
- Vehicle or equipment wash area.
- Inspection.
- Holding (sterile).

8-52. **Staging Area.** A staging area is an open area large enough to handle the number of vehicles and equipment being processed. Units often underestimate the amount of space needed for the staging area. An accurate count of vehicles and personnel redeploying to the CTUS must be received to accurately determine the staging area requirement.

8-53. **Download Area.** The download area is an area that must have a base of cement, asphalt, or clean or coarse gravel at least 6 inches deep. Vehicles arriving from the staging area download equipment here. The download area breaks out equipment and vehicles for wash area activities. Vehicles are taken to the vehicle wash area and equipment is taken to the equipment wash area.

8-54. **Vehicle and Equipment Wash Area.** A vehicle and equipment wash area or wash site consists of a hardstand area with adequate drainage, water outlets, high-pressure steam cleaners, and ramps. Most wash sites resemble a birdbath using long tubes with holes in them. A vehicle passes by the tubing while it pressure sprays the sides and undercarriage of the vehicle. Unit personnel clean vehicles to free them of dirt, vegetation, insects, and so forth, and check for and clean up fluid spillage. Equipment is cleaned meticulously, ensuring that there is no debris or soil present. When vehicles and equipment are assumed to be clean, they are reloaded and moved to the inspection area.

8-55. **Inspection Area.** The inspection area is where the vehicles and equipment are inspected for cleanliness. If vehicles and equipment are clean, they are moved to sterile holding areas. If they are dirty, they are moved to a spot-wash area and spot-cleaned. Vehicles and equipment that were spot-washed are reinspected, and if clean, go to the sterile holding area.

8-56. **Holding Area.** Cleaned vehicles and equipment are placed in the holding area. Unit personnel guard the holding area to ensure that unit members or other personnel do not contaminate or place restricted or contraband items in vehicles or equipment. The holding area must be free of USDA concerns. Some of these concerns include the following:

- Weeds. The download area must be vegetation free.

- Soil. The download area must be completely free of soil and soil particles.
- Pests. Insects and their debris or excrement are a major concern in the download area. Ensure that insects are kept out of the download area.

NOTE: SMCIs assist units in identifying and eliminating USDA concerns.

8-57. Vehicles and equipment in the holding area will be loaded onto rail, ship, or other means of transport for redeployment.

MULTINATIONAL LAW AND ORDER OPERATIONS

8-58. Multinational police operations may involve joint patrols with allied forces, HN military or civilian police, or a combination of these options. Commanders may also choose to operate certain patrols or stations with MP forces of one nationality, with all police operations answering to a combined, multinational operations cell. Because each nationality has its own statutes, regulations, policies, and traditions, police forces should enforce laws only on persons normally under their own jurisdiction. This reduces confusion about authority and the potential ill feelings that can result when citizens are controlled by other foreign forces. If a situation occurs involving HN or third-nation citizens, the MP should contain or monitor the situation until the appropriate police agency can respond to control the situation. MP may be required to control the situation (consistent with ROE or ROI) if the appropriate agency is not reasonably available or the situation is so severe that a delay would endanger lives.

8-59. *The Foreign Assistance Act of 1961* prohibits the military from training HN civilian police. However, MP may be directed to provide initial assistance and training to foreign military or civilian police forces or assist in the creation of these forces, where the local national authority has broken down. This is an exception to *Section 660 of the Foreign Assistance Act*. MP, when directed, conduct training for foreign military police and assist in reorganizing local constabulary forces during stability and support operations. MP provides short-term emergency L&O capabilities until the foreign military or civilian police forces are functional.

8-60. US military training assistance of foreign military or civilian police normally includes the following:

- Special forces. These detachments teach basic skills needed to perform security duties such as small arms training, unit security tactics, radio procedures, and human rights observation and safeguard.
- Military police. MP teach investigative techniques, law enforcement skills, and confinement operations.
- Civil affairs. CAs coordinate with the HN to effectively integrate training with the established political system. They also provide language interpreter support.
- Psychological operations. PSYOPS personnel work to gain and maintain acceptance of the training effort by the civilian population and encourage their acceptance of the new or reconstituted civilian police force.

8-61. The US ambassador and country team provide guidance for implementing all US Department of Justice (USDOJ) training, including police training. US military forces conduct training under the direction of the civilian authorities, and only until the International Criminal Investigative Training Assistance Program (ICITAP) staff or another agency can assume training. Contractors working for ICITAP, USDOJ, or another international agency may also provide instruction and assistance, and frequently assume those duties from the military as the situation matures. US military involvement in training foreign military or civilian police should not normally exceed 180 days following the end of a conflict.

8-62. MP train HN military or police forces to maintain L&O, using current operational and investigative techniques. They may also act as the agents to equip those forces with the material needed in police operations and train HN personnel on the use and maintenance of that equipment. The training provided may be formal or informal instruction, or consist of on-the-job training with MP working with, and monitoring the performance of, HN police authorities. Classroom instruction is based on training packages developed locally or by higher HQ, or exportable training packages from a service school.

8-63. Once the basics of law enforcement have been learned or reinforced, new or reconstituted police may benefit most from working together with MP in joint operations. ROE and ROI may vary by situation, but some basic guidelines for working with newly constituted or reconstituted police agencies are—

- Allowing the HN authorities to handle situations involving local nationals. US forces assists HN police and monitor the situation. US forces should not appear to be giving orders to the HN agents.
- Ensuring that US forces take care that they do not embarrass or demean HN agents, especially in view of the populace. Soldiers must understand that negative behavior toward their HN counterparts could seriously damage the relationship between US forces and the HN authorities.
- Ensuring that US forces are aware of and sensitive to the cultural traditions and

standards of the HN. What may be acceptable in one culture may be, at best, ill-mannered and possibly, a serious insult or even criminal behavior in another culture.

- Ensuring that US forces appear publicly to be respectful of its agents and leadership and be confident in their abilities. This helps the HN force be effective and respected by the local populace.

8-64. As HN authorities develop professionally and gain the confidence of the HN population, the role of US forces changes. MP shift from being the sole providers of L&O, to being role models and advisors for local authorities, to being monitors of HN police activity.

8-65. Combined police operations reinforce training and provide HN police with the mobility, security, and communications to operate more effectively while conserving critical personnel resources. MP provide support through training and the following, usually combined, operations:

- Law enforcement, security, and criminal information support to the indigenous police force to include L&O operations and administration in a democratic society.
- Patrol and desk.
- Traffic control.
- PRC and civil disturbance.
- MWD.
- Physical security.
- Personnel security.
- Area and route security.
- Counterdrug.
- Antiterrorism.

8-66. In most cases, US forces will continue to exercise jurisdiction over their own forces, especially for offenses of a purely military nature. Local government, once reestablished, may choose to assume jurisdiction in serious or high interest offenses involving US personnel or according to the applicable SOFA or other agreement. MP, in the absence of competent civil authority, establish control over the movement of personnel and supplies and guard critical food supplies and material during production and storage. As the local police and paramilitary forces become effective and assume more security responsibilities, MP elements reduce their activity. They withdraw their support gradually, ensuring that the HN has adequate resources to carry out ongoing programs.

Chapter 9

Police Intelligence Operations

The PIO support, enhance, and contribute to the commander's protection program, situational awareness, and battlefield visualization by portraying relevant threat information that may affect the operational and tactical environment. MP and the CID gather police, criminal, or combat threat information during the performance of their functions.

OVERVIEW

9-1. MP collect and share information during the conduct of their day-to-day operations. This information, which may be police, criminal, or combat, is provided as input to the intelligence collection effort and turned into action or reports. The PIO function ensures that information collected during the conduct of MMS, AS, I/R, and L&O is reported up through the proper channels so that it can be analyzed. MP perform PIO while conducting combat support tasks, such as—

- Checkpoints and roadblocks.
- TCPs.
- Field interviews.
- Criminal investigations.
- Reconnaissance (zone, area, and route).

9-2. Collecting police information during MP activities under the functions of MMS, AS, and I/R may result primarily in combat information, but may lend to police or criminal information. Likewise, the police collection efforts during full-scale L&O operations may result in combat information. MP apply lessons learned from peacetime, conflict, and war environments to the PIO function to gain the most advantage for performing the function. The MP platoon is capable of operating in all functions, day or night, and in various terrain, weather, and visibility conditions. An MP platoon operates independently over large, dispersed areas. MP, along with MI, engineers, and NBC reconnaissance, are key to the IPB (refer to *FM 34-130* for more information about IPB). They are collectors of information in the rear area during sustaining operations or anywhere throughout the battlefield. MP gather police information from contacts that are often very valuable in substantiating or verifying other sources of information. These sources include the following:

- Daily contact with the local populace.
- Combined police patrols with HN military and civilian police agencies.
- Close liaison with local, HN, and multinational police agencies.
- Field interviews.
- Nongovernmental organizations (NGOs).
- Private volunteer organizations (PVOs).

9-3. Refer to *FM 3-19.1* for more information about the processes discussed in this chapter.

POLICE INFORMATION ASSESSMENT PROCESS

9-4. The PIAP is a tool used to contribute to the PIO function. Information gained through the PIAP may contribute independently or simultaneously to the all-source analysis product (ASAP) and the IPB process. The PIAP, the ASAP, and the IPB enhance and support the commander's force protection program, situational awareness, and battlefield visualization. However, the PIAP independently or collectively—

- Provides the commander with information necessary to improve measures to protect the forces.
- Provides information that clarifies the threat and operational situation.
- Reduces opportunities for threat forces to disrupt military operations and inflict US or friendly casualties.

9-5. The commander and his staff continually monitor the environment at the tactical level consistent with the METT-TC. They apply the military aspects of terrain (OCOKA) as a means of protecting the force. PIO clarify the evolving criminal threat picture for commanders through the PIAP. This helps planners predict threat courses of action against our forces or protected populations.

9-6. MP use PIAP to continuously collect, organize, interpret, and report police and criminal information in support of the IPB. The PIAP consists of 6 steps. Refer to *Table 9-1, pages 9-4 and 9-5.*

POLICE INFORMATION

9-7. MP gather police information actively or passively. Active collecting efforts result from a direct tasking, and passive collecting efforts result from normal, daily MP operations.

ACTIVE MODE

9-8. MP perform the PIO function in the active mode when directed by the higher HQ. In this mode, the MP platoon conducts specific missions with the *intent* to actively collect information. Specific MP activities (such as setting up a checkpoint or roadblock) are performed to specifically fulfill a requirement (such as looking for individuals who are in possession of or have knowledge of others in possession of weapons, US military property,

Table 9-1. Police Information Assessment Process

Step	Action
1.	Determine the scope of the PIAP by— • Conducting a detailed mission analysis. • Reviewing the mission of the higher HQ and the commander's intent. • Reviewing the CCIR. • Reviewing mission priorities. • Determining the AO and area of interest (AI). • Reviewing the IPB estimates of the higher HQ. • Determining the required information products.
2.	Assemble the working aids and— • Post the applicable maps. • Acquire crime statistics and other related data. • Obtain language aids, such as cultural references and interpreters. • Acquire the necessary automation equipment.
3.	Determine information requirements and coordinate access to data by— • Coordinating with the CID for access to available criminal information developed by CID programs. • Determining police and criminal gaps in the designated geographical area. • Determining the requirement for information collectors to fill police and criminal gaps. • Coordinating with the Intelligence Officer, US Army (S2), the Assistant Chief of Staff, G2 (Intelligence), PSYOP, and other agencies to determine if the information is already available.

Table 9-1. Police Information Assessment Process (Continued)

4.	Recommend and supervise police and criminal collection efforts by— • Coordinating with the S2, the G2, the SJA, the CID, and other applicable police agencies before initiating a collection effort. • Determining which police and criminal information tasks will be assigned to the MP and the CID. • Determining the criteria to satisfy information requirements. • Providing collectors with reporting instructions (such as the reporting frequency and the report format). • Monitoring information collection efforts to prevent duplication of efforts.
5.	Process police and criminal raw data by— • Assembling and assessing the reliability of the data (according to the assessment criteria established above). • Integrating information from the collectors (the MP and CID agents in the field). • Evaluating the data to determine if it meets the requirements. • Developing criminal trends and indicators.
6.	Report and disseminate police and criminal information assessment by— • Recommending MP and CID actions to improve and focus future collection efforts. • Reporting police and criminal information assessments to the S2, the G2, and other MP units. • Reviewing and ensuring that the release of police and criminal information assessments do not violate established guidelines and constraints. **NOTE: Refer to *Appendix E of FM 3-19-1* for a further discussion of PIAP.**

and so forth). MP gain valuable police, criminal, and combat information while conducting these operations. A standardized checklist enhances the information collection effort and aids in the analysis of the information collected. The checklist indicates a pattern in the behavior of the local nationals. It shows what the local nationals are transporting, to where they are transporting the items, and so forth. The checklist may include—

- The number and types of vehicles stopped. Identifying marks, license plate numbers, and any signs displayed on the vehicles are recorded and reported.
- The number of passengers in the vehicle. The nationality, age, and sex mix of the passengers are recorded and reported.
- The type and quantity of cargo.
- The vehicle's point of origin and destination.
- The stated reason for travel by the passengers.
- The description of arms, ammunition, explosives, and sensitive items found and confiscated from the vehicle.
- The possible or actual sightings of weapons, explosives, or threat forces by the passengers.
- The condition of the passengers.
- The reporting of anything unusual by the passengers.

PASSIVE MODE

9-9. Every MP conducts the PIO function in the passive mode during their normal day-to-day operations. In the passive mode, PIO are not a stand-alone function and, as such, it cannot be separated from the other MP functions. If while performing MMS, AS, I/R, and L&O, MP receive, observe, or encounter police, criminal, or combat information, they immediately submit a SALUTE, SPOTREP, or other appropriate report to relay information up the chain of command. The information is then integrated into the on-going PIAP and forwarded to the higher echelon S2 and G2 for IPB applications.

RESPONSIBILITIES

9-10. The responsibilities for the PIO function start with the MP company. The company has the overall responsibility of supervising the collecting and reporting of information to the higher HQ. (Refer to *FM 34-2* for more information about the collection management process). Training scenarios can be developed to enhance the PIO information collecting process of both the active and passive mode. Additionally, the scenarios stress collective training (such as processing police information) and individual training (such as improving interpersonal communication and interview skills).

PLATOON LEADER

9-11. When the platoon leader is tasked to conduct PIO collecting and reporting, he initiates TLPs for the mission. The platoon leader—

- Coordinates with the higher HQ for the CCIR, police and criminal information requirements (PCIR), and threat estimates. The PCIR allows additional focus on local threats.
- Establishes liaison with civil and MP forces and law enforcement agencies in the AO.
- Coordinates with the local and HN police to determine the existence of organized crime in the AO and the identification of current and emerging criminal leaders and associates.
- Coordinates with the local and HN police and the populace to identify the types of criminal activity (such as smuggling, counterfeiting, narcotics, extortion, and so forth) in the AO.
- Reports information of potential intelligence value by—

  - Gathering collected information from squads and teams.
  - Consolidating original reports and sketches of potential intelligence value and forwarding them through intelligence channels for analysis and a determination of their intelligence value.

PLATOON MEMBERS

9-12. The platoon sergeant, squad, and team leaders, down to the junior member of the platoon share in seeking out sources of information by—

- Collecting R&S information that supports the commander's overall PIOR and PCIR collection efforts while on a routine patrol. They accomplish this by—
  - Noting terrain information pertaining to streets, roads, canals, subterranean systems, built-up areas, cities and villages, and the impacts of weather on the terrain.
  - Collecting information on progovernment and antigovernment individuals and groups who might disrupt L&O during protests, strikes, riots, and other spontaneous or organized efforts.
  - Identifying private establishments that may be a target or whose presence or operations contribute to the disruption of L&O (such as gun shops, pawnshops, and liquor stores).
  - Identifying critical infrastructures such as power stations, water works, radio and television stations, telephone and communication facilities, public transportation, and other establishments that may be critical to the sustenance of the community.

- Identifying EPWs, stragglers, and DCs who may have information of potential intelligence value and reporting it to the chain of command. Use a SPOTREP, SITREP, or a SALUTE report or a format directed by the chain of command to report information. SOPs may also be used. Refer to *Appendix D* for report formats

Appendix A

Metric Conversion Chart

A-1. Use *Table A-1* to convert from US to metric measurement and metric to US measurement.

Table A-1. Metric Conversion Chart

US Units	Multiplied By	Equals Metric Units
Length		
Feet	0.30480	Meters
Inches	2.54000	Centimeters
Inches	0.02540	Meters
Inches	25.40010	Millimeters
Miles (statute)	1.60930	Kilometers
Miles (nautical)	1.85320	Kilometers
Yards	0.91400	Meters
Area		
Square inches	6.45160	Square centimeters
Square feet	0.09290	Square meters
Square yards	0.83610	Square meters
Volume		
Cubic inches	16.38720	Cubic centimeters
Cubic feet	0.02830	Cubic meters
Cubic yards	0.76460	Cubic meters
Gallons	3.78540	Liters
Fluid ounces	29.57300	Milliliters
Quarts	0.94600	Liters

Table A-1. Metric Conversion Chart (Continued)

US Units	Multiplied By	Equals Metric Units
Weight		
Ounces	28.34900	Grams
Pounds	453.59000	Grams
Pounds	0.45359	Kilograms
Short tons	0.90700	Metric tons
Long tons	1.01600	Metric tons
Metric Units	**Multiplied By**	**Equals US Units**
Centimeters	0.39370	Inches
Meters per second	2.23700	Miles per hour
Millimeters	0.03937	Inches
Kilometers	0.62137	Miles (statute)
Kilometers	0.53960	Miles (nautical)
Meters	3.28080	Feet
Meters	39.37000	Inches
Meters	1.09360	Yards
Area		
Square centimeters	0.15500	Square inches
Square meters	10.76400	Square feet
Square meters	1.19600	Square yards
Volume		
Cubic centimeters	0.06100	Cubic inches
Cubic meters	35.31440	Cubic feet
Cubic meters	1.30790	Cubic yards
Milliliters	0.03380	Fluid ounces
Liters	1.05700	Quarts
Liters	0.26420	Gallons
Weight		
Grams	0.03527	Ounces
Kilograms	2.20460	Pounds
Metric tons	1.10200	Short tons

Appendix B

Media Relations

Military operations are carried out, especially in the initial phases, under the full glare of public scrutiny. The press can distribute reports and pictures faster than the news can be released by the command by using satellites and modern communications technology. Incidents, sometimes embroidered or slanted towards a partisan viewpoint, are screened on the television the same day and in the countries that are parties to the dispute and their allies.

OVERVIEW

B-1. Journalists fall back on speculation when information is withheld. Such speculation, although usually inaccurate, is often near enough to the truth to be accepted as such by large sections of the public and even by the governments. Belligerents find it advantageous to leak part of a story to the press to build public support for their position. On occasion, such activities can grow into a fully orchestrated press campaign.

MEDIA INTERACTION

B-2. Establish procedures for media interaction and incorporate them into SOPs. Know what (such as a simple theme which they can tie their responses back to) to discuss with the media and what to refer to the PAO. Ensure that deploying units receive a predeployment briefing from the installation PAO to clarify what may or may not be discussed with the media. Ensure that this action occurs before any exposure to the media. Do not make off-the-record statements in briefings or discussions with the media. Public or media knowledge of any classified activity associated with an operation does not imply or mean that the information is unclassified or may be released or confirmed.

B-3. Ensure that MP are courteous, respond candidly, speak effectively, and continue the mission when interacting with broadcast and print reporters and photographers. This relationship with the media prevents any misgivings. The media will report more accurately and not feel that information is being withheld. The media provides its audiences with real-time information of varying accuracy and completeness.

B-4. The media can be a valuable tool or a lost asset during military operations. It can potentially have a quick and pervasive impact on the plans and operations of commanders, and can change the public's opinions and perceptions about military operations. It can target audiences whose support is crucial to the desired end state of an operation. MP may be the centerpiece for humanitarian assistance or during operations to relieve suffering. A reporter or a photographer may show up at a checkpoint or a TCP where local nationals (belligerents) are refusing to cooperate. Ensure that the media is not only documenting the activities and behavior of the belligerents, but is also recording the military's response to the incident.

Appendix C

Training Execution Model

The TEM allows numerous rehearsals and identification of the TTP required for mission accomplishment. It causes each soldier, regardless of position, to understand the unit leader's plan with contingencies and the TTP required. It allows junior leaders and soldiers to see the big picture and understand the tactical environment. Place the company as the OC for the subordinate platoons, and place the platoon leader as the OC for the subordinate squads during internal evaluations.

OVERVIEW

C-1. The TEM ensures that soldiers are trained to standard, not to time. It is derived from train-up or certification exercises (CERTEXs) before deploying to the National Training Center (NTC).

TASK IDENTIFICATION

C-2. The TEM takes place after the METL is approved. The approved METL, coupled with an assessment process (such as external or internal evaluations), determines the collective and individual tasks to be covered during training exercises. Once these tasks are identified, implement the following eight-step TEM:

Step 1. Back brief number 1.

Step 2. Back brief number 2.

Step 3. Rock drill.

Step 4. Subunit leaders walk-through.

Step 5. Subunit leaders walk-through with troops.

Step 6. Dry run on the actual lane.

Step 7. Execution.

Step 8. After-action review (AAR).

C-3. The TEM steps complement but do not replace the TLPs. The following outline explains the eight-step model applicable to an MP platoon. The TEM assumes that the platoon's collective task lists are approved at the appropriate level, the OCs are certified and trained, and all resources are available.

STEP 1: BACK BRIEF NUMBER 1

C-4. During step 1 *(Figure C-1),* the OC issues the mission order to the platoon leader. He briefs the platoon's mission using a sand table, which the platoon built. The platoon leader provides a mission back brief to the OC. The back brief eliminates any confusion or misunderstanding at the very beginning of the mission or plan cycle. The platoon leader's back brief is only with the OC. This allows the interaction between the OC and the platoon leader to take place while leaving the platoon leader's credibility intact. With no subordinate present, the environment to learn and ask questions is established.

C-5. In a nearby AA, leaders train soldiers on critical individual soldier tasks that support the platoon's collective tasks. Here, the PSG checks on the focus and progress of those individual tasks that have been assessed as weak. Additionally, AA actions in step 1 focus on PCIs.

STEP 2: BACK BRIEF NUMBER 2

C-6. During step 2 *(Figure C-2, page C-4),* the platoon leader returns to the sand table area where the mission order is received from the OC. He briefs the tentative

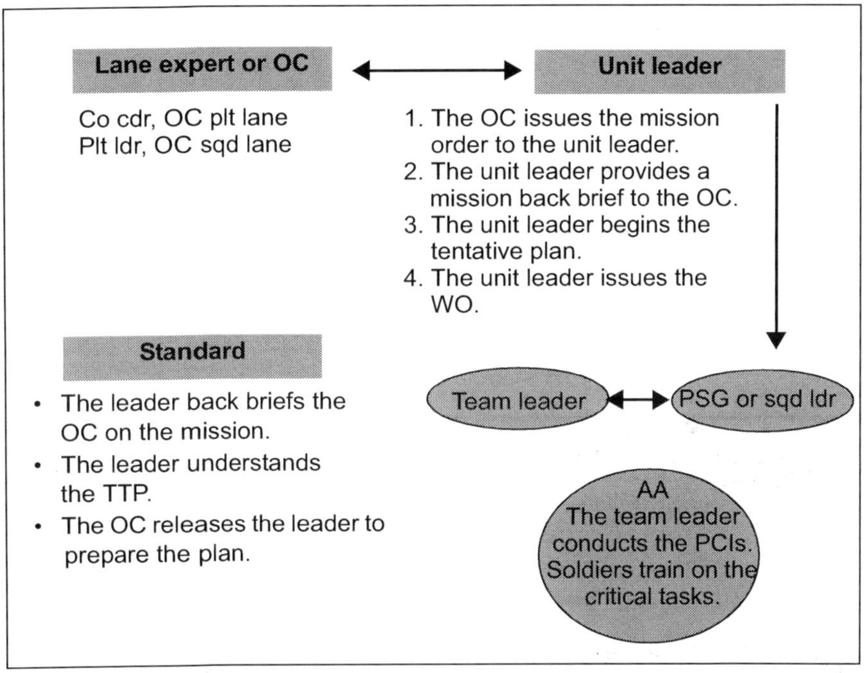

Figure C-1. Back Brief Number 1

plan to the OC. Once the OC verifies the MTP standards and pertinent TTPs, the platoon leader briefs squad leaders on the same sand table. The OC remains present to hear the interaction among the platoon's leadership. After the squad leaders are briefed, the platoon leader conducts the leader's reconnaissance. If any modification to the tentative plan is required based on the reconnaissance, the platoon leader briefs the OC. He briefs the squad leaders again if changes are approved. The interaction and planning get more focused after the leader's reconnaissance.

STEP 3: ROCK DRILL

C-7. During step 3 *(Figure C-3, page C-5)*, the platoon leader and squad leaders conduct a rock drill on a

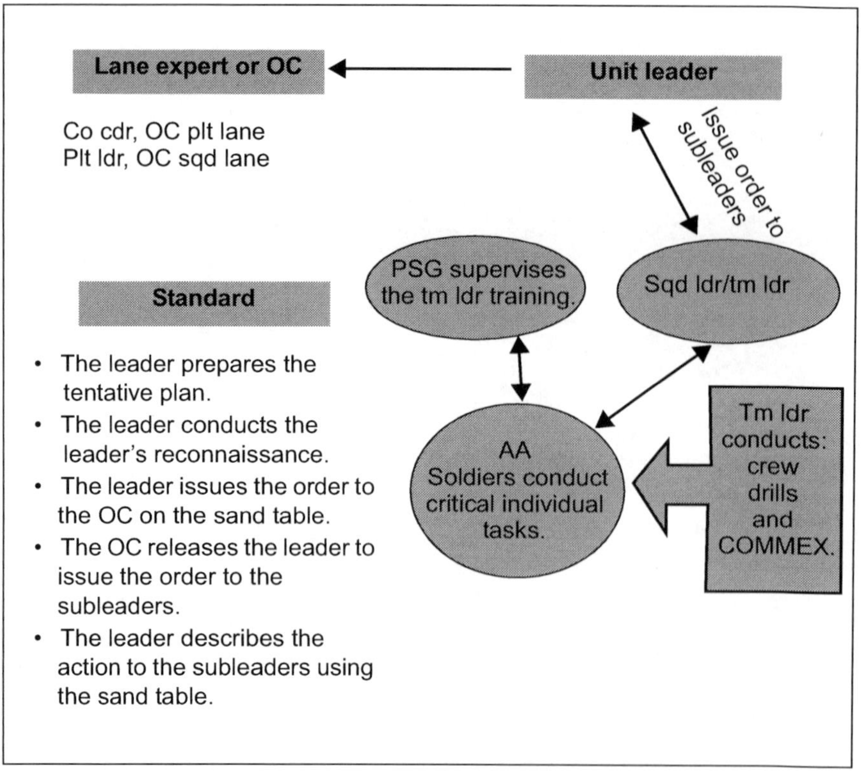

Figure C-2. Back Brief Number 2

terrain model that the platoon constructed of the actual training lane. The OC is in an overwatch position. He always assesses the interaction of the leadership and serves as the OPFOR for the rock drill. The platoon leader talks or walks through the plan on the terrain model with all the squad leaders.

C-8. In reaction to the OPFOR, leaders solidify contingencies based on the interaction on the terrain model. The OC's "what if" drills will usually trigger detailed planning. The credibility of the platoon leadership is still intact because the interaction and pertinent questions are not being asked in front of the soldiers. The soldiers are in the AA continuing their individual training and preparing equipment based on the WO received. Locate the PSG in the AA to coach team leaders in their AA activities. Once the OC is sure that the platoon leaders understand the plan, he releases them to conduct a walk-through.

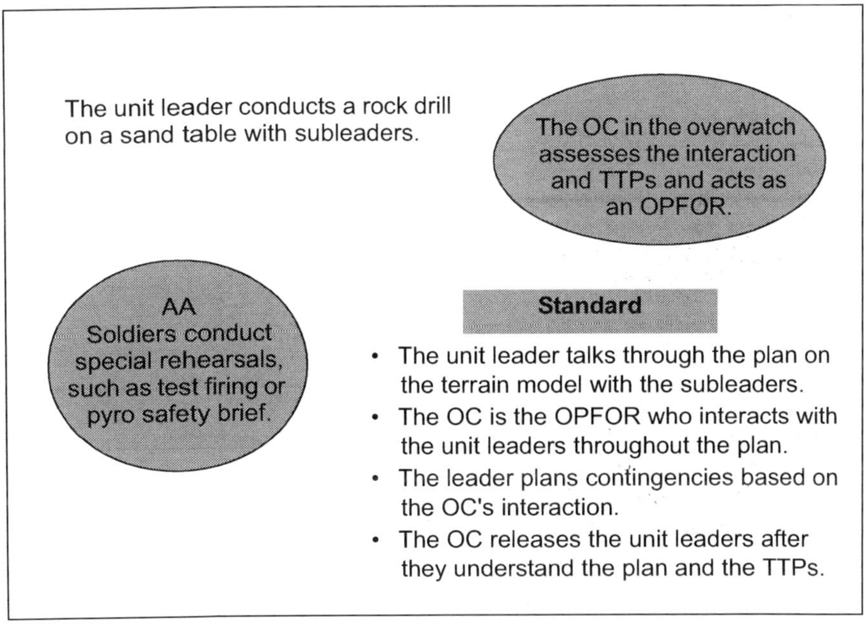

Figure C-3. Rock Drill

STEP 4: SUBUNIT LEADER'S WALK-THROUGH

C-9. During step 4 *(Figure C-4, page C-6)*, the platoon leader displays all terrain features and control measures. During the walk-through, the platoon leadership imitates the movement selected, such as traveling, traveling overwatch, and bounding overwatch. The platoon leadership uses actual radio procedures and call signs according to the platoon's plan. The platoon's leadership demonstrates its grasp of the plan and the mission without troops, showing that credibility is still sound.

C-10. Once the OC is satisfied that the platoon's leadership knows the tasks, he releases them to brief the soldiers on the sand tables previously mentioned. Squad leaders ensure that their team leaders and soldiers understand the mission by showing them on the sand table what their part of the mission is and how it fits into the overall scheme of maneuver.

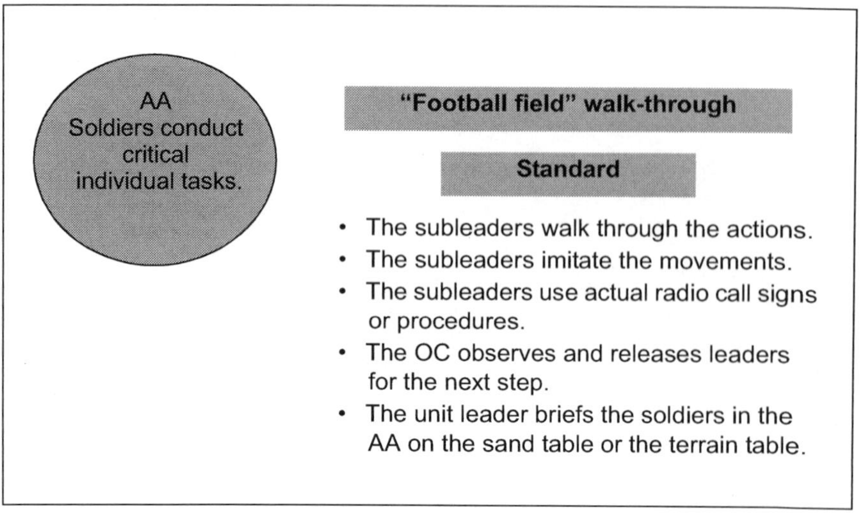

Figure C-4. Subunit Leader's Walk-Through

STEP 5: SUBUNIT LEADER'S WALK-THROUGH WITH THE TROOPS

C-11. During step 5 *(Figure C-5),* the soldiers are assembled and briefed on the mission using sand tables. The MP teams walk and talk through the mission. The platoon walks through its action of the plan; and the soldiers react and dress on their team, squad, and platoon leaders. Shout commands out so that all can hear. Imitate movement as well as radio procedures, call signs, fire commands, calls for suppression, and so forth. The OC who plays the OPFOR and interacts with the platoon through its leaders observes the platoon. Often, while conducting the walk-through with the troops it will be necessary to stop and inform soldiers of any risk hazards and individual responsibilities. The platoon's interaction on the field is a critical component of mission accomplishment. The OC only releases the unit for the next step if he is satisfied that the unit understands the leader's plan and the required TTP.

Issue warning OPORD

Subleaders walk-through

All now in final rehearsal

- The subleader walks through the actions with troops on the sand table.
- The troops act on the leader's commands.
- The subleader imitates movements.
- The subleader uses actual radio call signs or procedures.
- The OC observes and releases the leaders for the next step.

Figure C-5. Subunit Leader's Walk-Through with the Troops

STEP 6: DRY RUN ON THE ACTUAL LANE

C-12. During step 6 *(Figure C-6, page C-8),* the unit follows the leader's plan and conducts a dry run of the mission on the actual training lane. The OC observes to ensure that the platoon follows its plan and conducts the mission to MTP standards. The dry run identifies weak subtasks and allows the platoon to retrain or repeat the dry run. This is a full dress rehearsal and ensures that the platoon is meeting the standard. If the platoon performs satisfactorily on the dry run, the OC releases it to execute at full combat speed.

The unit conducts a dry run of the mission following the leader's plan on the actual lane.

Actual lane dry run.	**Standard**
	• The unit conducts the mission to the MTP standard. • The unit identifies the weak subtasks. • The unit identifies the terrain-induced weakness to the plan. • The unit retrains the weaknesses. • The unit does the lane again. • The OC releases the unit to execute at the full combat speed if satisfied with its performance.

Figure C-6. Dry Run on the Actual Lane

STEP 7: EXECUTION

C-13. During step 7 *(Figure C-7),* the unit executes the mission at full combat speed against an unrestricted OPFOR that is controlled by the OC. This phase involves force on force, controlled by the OC, with a complete simulation system (such as multiple-integrated laser engagement system [MILES]). The OC ensures that soldiers adhere to MTP standards and TTPs. The OC calls for an end of exercise (ENDEX) when the training lane has run to culmination (such as the mission is accomplished and all the training effects took place). The OC has enough information to assess the platoon's performance and initiate the next step—the AAR.

The unit executes the mission.

Actual lane

Standard

- The unit conducts the mission to the MTP standards.
- The unit has full use of OPFOR and MILES.
- The unit uses force on force or is unrestrained.
- The OC controls the OPFOR.
- The OC calls for an ENDEX.

Figure C-7. Execution

STEP 8: AFTER-ACTION REVIEW

C-14. During step 8 *(Figure C-8, page C-10),* the unit conducts an AAR of the lane. If possible, it conducts the AAR on the objective site that overlooks the lane. The OC facilitates the AAR. He or the platoon leader states the tasks, condition, and standards for the training objective. The OC ensures that the AAR becomes a review that is driven from the soldier level up toward the top. He elicits the soldiers to identify their actions, both right and wrong, and gets the leaders to do the same. The OC states what was right and wrong with the mission according to the MTP and the appropriate TTPs, always striving to leave the AAR on a positive note to improve soldier and unit morale and performance.

The unit conducts an AAR of the training event.

Standard

- The OC facilitates the AAR.
- The unit members state the training objectives (task, condition, standards).
- The unit members have a bottom-up discussion.
- The unit members have the soldiers identify their mistakes.
- The unit members have the leaders identify their mistakes.
- The OC assesses the task according to the MTP.

NOTE: The assessment dictates whether to go to the next training or redo the training.

Figure C-8. AAR

Appendix D

Orders and Reports

Orders are written or oral communications directing action. They are based on plans or the receipt of a new mission. Use reports extensively to provide information to higher, lower, and adjacent commands. Although there are many, this appendix is limited to combat orders and reports. Refer to *FM 101-5* for more information about orders.

ORDERS

D-1. Combat orders pertain to strategic, operational, or tactical operations and their service support. The combat orders used by MP are WOs, OPORDs, and FRAGOs.

WARNING ORDERS

D-2. Refer to *FM 101-5* for an example of a WO. WOs provide advance notice of an action so that MP can use available time for preparations, and they—

- Are issued at each level down to the squad.
- Are issued to subordinates in as much detail as possible.
- Are issued as brief, oral, or written messages.
- Are a part of planning the use of available time. WOs should be kept simple.
- Describe the operation and the preparations to be made before the OPORD is issued.

- Include the situation, the mission, special instructions, and the time and place for issuing the OPORD.

OPERATION ORDERS

D-3. OPORDs coordinate actions to carry out the commander's plan for an operation, and they—

- Explain how leaders at different levels want the operation conducted. To ensure that maneuver and other non-MP units carry out actions or provide a form of support that MP plans and operations depend on, state the required actions or support in the coordinating instructions paragraph of the division and brigade OPORDs.
- Have a great impact on how subordinate leaders employ units and perform missions. For example, the MP platoon leader's latitude to employ the unit could be restricted, based on how the company commander wants the operation conducted.
- May be written, oral, graphic (such as traces and overlays), or a combination of these forms.
- Are usually written when prepared at company level and above.
- Have a standardized system of designating days and hours in relation to an operation or an exercise.
- Follow a prescribed written format, that contains a classification, heading, body, and ending.
- Have a prescribed five-paragraph format for the body of the order, which includes the following:
    - Situation.
    - Mission.
    - Execution.
    - Service support.
    - Command and signal.

D-4. The risk management process is greatly enhanced by incorporating the risk assessment into the OPORD. The commander integrates risk management into the order, rather than treating it as an afterthought. Leaders at all echelons assess the effectiveness of their units by reviewing how well hazards are identified and risk controls are specified in oral and written orders. Refer to *Figure D-1, pages D-5 through D-10,* for an example of an OPORD format. (Refer to *FM 101-5, Appendix H*, for further discussion of an OPORD.)

D-5. FRAGOs issue supplemental instructions to a current OPORD or OPLAN while the operation is in progress, and they–

- Contain missions of immediate concern to subordinate units.
- May be either written or oral.
- Provide brief, specific, and timely information without loss of clarity.
- Have no prescribed format. Prevent confusion OPORD.
- May be issued to change an OPORD that has already been issued. As such, only those items from the original OPORD that have changed are included in the FRAGO, as long as clarity is not sacrificed.

STANDING OPERATING PROCEDURES

D-6. SOPs prescribe routine methods to be followed in operations, and they–

- Supplement other combat orders.
- Reduce the number, length, and frequency of other orders. Because the SOP is a standing order, do not repeat the information contained therein in other orders unless emphasis is desired.

- Have no prescribed format. However, ensure that the subordinate unit SOPs follows the format of the next higher HQ SOP if possible.
- Prescribe actions of a recurring nature that lend themselves to definite or standardized procedures. Examples include the following items:
  - Troop safety matters
  - Methods of reporting unit locations.
  - Measures for handling captured personnel or equipment
  - Distribution of supplies.
  - Standard communication procedures for exercising C^2.
  - Other items that lend themselves to standardization.

NOTE: These items are generally the constants in what is otherwise a frequently changing set of circumstances.

REPORTS

MP contribute to the commander's situational awareness by providing timely information to higher HQ. They collect and report information in the form of administrative, operational, or intelligence reports. These reports ensure that the commander receives continuous current information. *Tables D-1* through *D-3, pages D-11 through D-18,* are examples of reports used by MP when conducting CS operations. Refer to *FM 101-5-2* for a complete listing of the standardized Army report and message formats.

(Classification)

Place the classification at the top and bottom of every page of the OPORD.

(Change from oral orders, if any.)

This statement is applicable only if an oral order is issued by the commander. The phrases "No change from oral orders" or "No change from oral orders except paragraph ___ " are necessary here.

Copy ___ of ___ copies
Issuing HQ
Place of issue (coordinates)
DTG of signature

Show the place of issue (location of the issuing HQ) on each copy. Show the name of the town or the place in capital letters, the coordinates in parentheses, and the country in capital letters. Both may encode.

The effective time for implementing the plan or order is the same as the DTG, unless coordinating instructions state otherwise. When orders apply to units in different time zones, use time zone Zulu (Z). In OPORDs and service support orders, list the time zone applicable to the operation in the heading of the order following the references. When an order does not specify the actual date and hour for beginning an operation, apply the proper reference designations.

Message Reference Number

Message reference numbers are internal control numbers that the unit signal officer issues and assigns to all orders. The unit's SOP normally describes the number's allocation and use. Using the number allows an addressee to acknowledge receiving the message in the clear.

OPORD ________ (code name)
(Number)

Orders normally contain a code name and are numbered consecutively within a calendar year.

Reference(s): The heading of the order includes a list of maps, charts, datum, or other related documents the unit will need to understand the plan or order. The user does not need to reference the SOP but may refer to it in the body of the order. The user references a map using the map series

Figure D-1. OPORD Format

number (and country or geographic area, if required), sheet number, name, edition, and scale, if required. Datum is the mathematical model of the earth used to calculate the coordinates on any map. Different nations use different datums for printing coordinates on their maps. Reference the datum in the marginal information of each map.

Time Zone Used Throughout the Order: The time zone used throughout the order (including annexes and appendixes) is the time zone applicable to the operation. Operations across several time zones use Z time.

Task Organization: Describe the allocation of forces to support the commander's concept. Show task organization in one of two places: preceeding paragraph one or in an annex, if the task organization is long and complicated.

OPORD ________ - _________________________
(Number) (Issuing HQ)

(Place this information at the top of the second and any subsequent pages of the OPORD.)

1. **SITUATION.**

a. **Enemy Forces.** Express this information down to two enemy echelons below yours (for example, battalions address platoons or companies address squads). Describe the enemy's most likely and most dangerous COA. When possible, provide a sketch of the enemy COA in lieu of verbiage (Appendix ___ [sketch] to Annex ___ [title]). Include an assessment of terrorist activities directed against US government interests in the AOs. Reference more sources using the final subparagraph to refer the reader to the documentation.

b. **Friendly Forces.** Include the mission, the commander's intent, and the concept of operations for HQ one and two levels up. Subparagraphs state the missions of the flank units (left, right, front, and rear) and other units whose actions would have a significant bearing on the issuing HQ.

c. **Attachments and Detachments.** State when, and for how long, units are attached or detached to the operation. Do not repeat information already listed under task organization or in a task organization annex. State "See Annex ___ (title)" here if a task organization annex is necessary.

Figure D-1. OPORD Format (Continued)

2. **MISSION**. State the mission derived during the planning process. This statement describes the task and purpose of the operation and clearly indicates the action to be taken and its reason. There are no subparagraphs in a mission statement. The mission statement covers on-order missions.

3. **EXECUTION**.

Intent: State the commander's intent derived during the planning process. It is a clear, concise statement of what the force must do to succeed with respect to the enemy and the terrain. It provides the link between the mission and the concept of the operation, which provides the basis for subordinate initiative and decentralized mission execution. Always include the desired end state.

a. **Concept of Operations**. The concept of operations may be a single paragraph or divided into two or more subparagraphs. It is based on the COA statement from the decision-making process and, at a minimum, should address close, deep, rear, and security operations as well as designating the main effort. The commander uses this subparagraph when he feels he must supply enough detail to ensure appropriate action by subordinates in the absence of additional communications or further instructions. When an operation involves two or more clearly distinct and separate phases, the concept of operations may be prepared in subparagraphs describing each phase. Ensure that the concept statement is concise and understandable. The concept describes the following:

- The employment of maneuver elements in a scheme of maneuver.
- A plan of fire support or *scheme of fires* supporting the maneuver with fires.
- The integration of other major elements or systems within the operation, including reconnaissance and security elements, intelligence assets, engineer assets, and air defense.

NOTE: Depending on what the commander considers appropriate, the level of command, and the complexity of any given operation, the following subparagraphs are examples of what may be included within the concept of operations:

(1) **Maneuver**. State the scheme of maneuver derived during the planning process. Ensure that this paragraph is consistent with the operation overlay. Ensure that this paragraph and the operation overlay is

Figure D-1. OPORD Format (Continued)

complementary and adds to the clarity of, rather than duplicating, each other. Do not duplicate information to be incorporated into the unit subparagraphs and coordinating instructions.

(2) **Fires**. State the scheme of fires to support the concept, and include priorities of and restrictions for fire support.

(3) **Reconnaissance and Surveillance**. Specify the reconnaissance and surveillance (R&S) plan and explain how it ties in with the basic concept of operations.

(4) **Engineer**. Clarify the scheme of engineer support to the plan, paying particular attention to the integration of engineer assets and obstacles, including the priorities of effort.

(5) **Air Defense**. State the overall concept of air defense in support of the scheme of maneuver if necessary.

(6) **Information Operations**. State the overall concept of information operations in support of the scheme of maneuver. Refer to the appropriate annexes if necessary.

NOTE: Units required to accomplish specific tasks for information operations and R&S are specified in the appropriate subparagraphs of 3b (tasks to maneuver [subordinate] units).

b. **Tasks to Maneuver (Subordinate) Units.** Clearly state the missions or tasks for each maneuver (or subordinate element) unit that reports directly to the HQ issuing the order. List the units in the same sequence as in the task organization, including attachments. Use a separate subparagraph for each subordinate element. State the tasks that are necessary for comprehension, clarity, and emphasis. Place tactical tasks that commonly affect two or more elements in subparagraph 3c (coordinating instructions).

c. **Coordinating Instructions.** List only those instructions applicable to two or more units (subordinate elements) and not routinely covered in unit SOPs. This is always the last subparagraph in paragraph 3.

NOTE: The following are examples of subparagraphs that are generally included as coordinating instructions. Subparagraphs 1 through 5 below are mandatory.

Figure D-1. OPORD Format (Continued)

(1) Time or condition when an order becomes effective.

(2) CCIR.

(a) Priority intelligence requirements (PIR).

(b) Essential elements of friendly information (EEFI).

(c) Friendly force information requirements (FFIR).

(3) Risk-reduction control measures may include such items as MOPP, operational exposure guidance, vehicle recognition signals, and fratricide prevention.

(4) ROE. When lengthy and complicated, include ROE as a separate annex.

(5) Environmental considerations.

(6) Force protection.

(7) Any additional coordinating instructions.

4. **SERVICE SUPPORT**. Address service support in the areas shown below as needed to clarify the service support concept. Refer to any annexes, where necessary. Subparagraphs can include the following:

a. **Support Concept**. State the concept of logistics support necessary to support the operation. Include the concept for—

- Support HQ or support area locations, including locations of logistical bases if not clearly conveyed in the overlay.
- Commander's support priorities.
- HN support.
- Any significant sustainment risks.
- Unique support requirements in the functional areas of manning, arming, fueling, fixing, moving, and sustaining the soldier and the systems.

b. **Materiel and Services**.

c. **Medical Evacuation and Hospitalization**. At a minimum, include frequencies, call signs, and locations of medical support facilities and units, as well as the plan for casualty evacuation (CASEVAC).

d. **Personnel Support**.

e. **EPW**.

f. **Transportation**.

Figure D-1. OPORD Format (Continued)

5. **COMMAND AND SIGNAL**.

a. **Command**. State the map coordinates for the CP locations and potential future locations for each echelon CP applicable to the operation. Identify the chain of command if not clearly addressed in unit SOPs.

b. **Signal**. List signal instructions not specified in unit SOPs; and identify the specific SOI addition in effect, required reports and formats, and the times the reports are submitted. List primary and alternate means of communications, as well as pyrotechnic signals and their meanings.

Acknowledge: Include instructions for the acknowledgement of the order by addressees. The word *acknowledge* may suffice or a message reference number. Acknowledgement of an order means that it has been received and understood.

Commander's last name
Rank

The commander or his authorized representative signs the original copy. If the representative signs the original, add the phrase "For the Commander." The signed copy is the historical copy and remains in the HQ files. It is always designated "Copy 1 of ______ copies."

OFFICIAL:
(Authentication) Use only when applicable. If the commander signs the original, no further authentication is required. If the commander does not sign, authorization is required by the signature of the preparing officer or individual and only the last name and rank of the commander appear in the signature block.

Annexes: List annexes by letter and title and in the sequence by which they were referenced in the order.

Distribution: Furnish distribution copies either for action or for information. List in detail those who are to receive the order. If extremely lengthy, refer to an annex containing the distribution list or to a standard distribution list or SOP.

(Classification)

Place the classification at the top and bottom of every page of the OPORD.

Figure D-1. OPORD Format (Continued)

Table D-1. Administrative Reports

Types	Who	What	When	Where	How	Content
Casualty	The unit or the section with casualties	The number of dead, wounded, or sick	Upon experiencing casualties or as required by the SOP or HQ	Higher HQ	By the most secure means and encoded if by unsecure radio	*DA Form 1156* or per SOP
Personnel	Submitted at company level and higher (input is needed from the squad or section)	Personnel strength accounting and status	Daily	Higher HQ	*DA Form 5367-R*	*DA Form 5367-R*
Periodic logistic	Submitted at company level and higher (input is needed from the squad or the section)	Supported strength and status of critical supplies	As supplies become depleted or as required by the SOP	Higher HQ	By the most secure means and encoded if by unsecure radio	• Logistical situation • Supply • Maintenance • Transportation • Service

Table D-1. Administrative Reports (Continued)

Type	Who	What	When	Where	How	Content
Journal	Units or sections operating independent of the parent organization	Events about a unit or section during a given or specified period	As events occur over the given or specified period	Maintained locally (provide to higher HQ on request)	*DA Form 1594*	*DA Form 1594* • Item • Time • Incident, message, or order • Action taken • Initials
Closing	The unit leader	Notice of change of the location	Upon arrival at the new site	Higher HQ	Secure the radio	• Unit • Date and time • CP location • Vehicles and radios • Personnel and weapons

Table D-2. Operational Reports

Type	Who	What	When	Where	How	Content
Situation and status	Commander or leader closest to the situation	Tactical situation or status	Immediately after a significant event or as specified	Higher HQ.	By the most secure means and encoded if by unsecured radio	• Enemy • Own situation • CS • General • Commander's evaluation
Force tracking (diversion, holding, or passing)	Designated TCP and mounted patrols	Unit movement along the MSR	As requested or designated in the SOP	Local movement control unit	Secure voice or encoded	Division transportation office (DTO) or the Provost Marshal Office (PMO) adopted SOP format
Offense	TCP and mounted patrols	Violation of MSR regulations	When tasked to conduct MMS and MSR regulations are in force	Through military channels to the driver's commander	Written report. (refer to *FMs 55-10* and *101-5)*	• Date • Driver's name • Name of vehicle commander • Particulars of the offense observed

Table D-2. Operational Reports (Continued)

Type	Who	What	When	Where	How	Content
EPW	Platoon, squad, and team operating the collecting points	The number of EPWs collected and evacuated	As required	Higher HQ	By the most secure means and encoded if by unsecured radio	Per SOP
Intention of laying a minefield	Platoon, squad, or team leader preparing to lay the minefield	Tactical objectives and characteristics of the minefield	Before emplacing the mines	Higher HQ	Written or by a secure radio (refer to *FM 20-32*)	• Purpose of the minefield • Estimated number and type • Location • Proposed start and completion times
Initiation of laying a minefield	Platoon, squad, or team leader of the force laying the minefield	Emplacement of the mine	When emplacement begin	Higher HQ (mandatory)	Written or by a secure radio (refer to *FM 20-32*)	• Time begun • Location and target number

Table D-2. Operational Reports (Continued)

Type	Who	What	When	Where	How	Content
Completion of laying a minefield	Platoon, squad, or team leader of the force laying the minefield	Completion of the minefield	Upon completion of the minefield	Higher HQ	Written or by a secure radio (refer to *FM 20-32*)	• Field is complete and functional (expedite the report and follow up with a hasty protective minefield report)

Table D-3. Intelligence Reports

Type	Who	What	When	Where	How	Content
Spot	Unit, section, or individual observing the enemy (all the echelons)	Enemy activity and area information of immediate value	Upon contact or as requested	Higher HQ	By the quickest means and encoded if by unsecured radio	• Reporting unit • Date and time of the event • Location and grid coordinates (encoded) • Event (SALUTE) • Original source • Remarks

Table D-3. Intelligence Reports (Continued)

Type	Who	What	When	Where	How	Content
Meaconing, intrusion, jamming, interference (MIJI)	Soldier experiencing electronic warfare	Correct report according to the type of interference	As soon as possible after the incident	Through signal channels to the higher HQ.	SOI defines the report and how to prepare it	• Type of report • Frequency or channel affected • Victim designation and call sign • Type of emission or audio characteristics • Coordinates of the affected station
NBC 1	Observer	Observation concerning suspected NBC attacks and resulting hazardous areas	Upon observing a suspected enemy NBC attack	Higher HQ.	Secure radio	• Observer's location • Direction from the observer • Date-time group (DTG) of the attack • Type of burst

Table D-3. Intelligence Reports (Continued)

Type	Who	What	When	Where	How	Content
NBC 4	Element in contact with radiation	Radiation dose rates	Upon contact with radiological contamination or as directed	To higher HQ	Secure radio	• Location • Dose rate • DTG of the measurement
Patrol	Patrol leader	Pertinent information pertaining to the patrol's activity	Upon completion of the patrol	To higher HQ	Written	• Designation of the patrol • Maps • Terrain • Enemy • Results of encounters with the enemy • Condition of the patrol
Enemy minefield location	Soldiers encountering an enemy minefield	Characteristics of the minefield	Upon locating a minefield	To higher HQ	Secure radio (refer to *FM 20-32*)	• Type of minefield • Location and depth • Enemy weapons or surveillance • Routes for bypassing • Coordinates of lane entry and exit • Width of lanes, in meters

Table D-3. Intelligence Reports (Continued)

Type	Who	What	When	Where	How	Content
						• Map sheet designation • Date and time information collected • Coordinates of minefield boundaries • Estimated time to clear the minefield • Estimated material and equipment required to clear the minefield • Other, such as the type of the mine, new mines, or booby traps
SALUTE	Observer	Enemy activity; for example, convoy, patrol vehicles, and aircraft	Upon encountering the enemy's activity	Higher HQ	By the most secure means and encoded if by unsecure radio (refer to *FM 101-5*)	S - Size A - Activity L - Location U - Unit T - Time E - Equipment

Appendix E

Precombat Inspections

Combat missions start with PCIs. These PCIs are an essential part of every mission. The objective of PCIs is to confirm the combat readiness of the unit. Subordinate leaders conduct detailed checks, allowing the unit leader to conduct PCIs in an atmosphere of total preparedness. After the PCI, the unit is prepared to move on the receipt of orders.

MODIFIED PRECOMBAT INSPECTION CHECKLIST

E-1. A modified PCI checklist *(Figure E-1, page E-2)* ensures that the team has the required equipment to conduct its mission when the unit is short on time. The modified PCI allows the team leader to focus on preparing the team for the mission, while the platoon leader issues the mission to the squad leader. Use the checklist as a recovery checklist when the team returns from the mission. Immediate, proper recovery ensures that the team is ready for the next mission. This checklist is only a guide and can be modified based on METT-TC and the unit SOP.

Steps	Actions
1	The squad leader and the team leader receive the WO.
2	The team leader prepares the squad for movement, while the squad leader is receiving the mission from the platoon leader.
3	The team leader follows the same checklist, while the squad leader debriefs the platoon leader if recovering from a mission.
4	The team leader inspects the communications equipment and ammunition in HMMWVs. He— • Conducts a communications check with the platoon CP. • Ensures that proper frequencies are set and that call signs are available. • Ensures that there is a Single-Channel, Ground-to-Air Radio System (SINCGARS), a manpack, accessories, and a battery case. • Ensures that there are batteries for the radios (automated net control device [ANCD], precision lightweight global positioning system receiver [PLGR], OE-254). • Ensures that the Class V basic load is accounted for.
5	The driver conducts preoperation checks and services on the vehicle. He ensures that— • The correct paperwork is completed, such as the dispatch, folder, and signature. Ensures that the correct TM is on hand. • He has a valid driver's license. • Preventive-maintenance checks and services (PMCS) are performed to a -10 standard. • A current *DA Form 2404* is on hand. • The fire extinguisher is sealed, tagged, and dated. • The engine's coolant, oil, transmission fluid, and washer fluid are at the correct level. • The battery level is correct and that cables are secured. • Lights are operational. • Fuel and water cans are filled and that there are meals, ready-to-eat (MREs) available. • There are no exhaust leaks. • Basic-issue item (BII), on-vehicle material (OVM), pioneer tools, locks, and chains are present on the vehicle. • The load plan is followed and that the vehicle's interior is clean.

Figure E-1. Modified PCI Checklist

Steps	Actions
6	The gunner conducts preoperation checks and services on the crew-served weapon and NBC equipment. He ensures that— • A TM for the equipment (-10) is available. • The weapon is oiled, cleaned, and functional. • There is a spare barrel and BII. • There is a T&E mechanism. • There is a wrench. • There is a tripod mount with an adapter. • There are weapon cleaning kits present. • There are night vision devices (NVDs) present. • All the NBC equipment is present and serviceable.

Figure E-1. Modified PCI Checklist (Continued)

PRECOMBAT INSPECTION CHECKLIST

E-2. This PCI checklist is extensive and time-consuming. Use it when the team has a lot of downtime and has a secure environment in which to work. Once combat operations begin, the secure environment and the downtime will be lost. *Figure E-2, page E-4,* is a draft checklist, which may be modified to meet the needs of the unit.

Steps	Actions
1	Soldiers. Soldiers should ensure that they— • Have a seasonal uniform. The uniform should include the following: ▪ Load-carrying equipment (LCE) that is complete. ▪ First aid packet. ▪ Canteens that are full and have a M17 drinking cap. ▪ M40 with a hood, M258A1 kit, M13 kit, and optical inserts (if required). ▪ Kevlar with a camouflage cover and band. ▪ Individual weapon with 6 magazines. ▪ ID tags around the neck and ID cards. ▪ Ear plugs. • Have been briefed on the current situation and the mission. • Possess a current driver's license. • Have been briefed on hot- and cold- weather injuries. Leaders. Leaders should conduct a safety briefing, to include— • Checking for hazards, such as ticks, chiggers, and spiders, and taking preventive first aid measures. • Checking for snake hazards and ensuring that there is first aid. • Checking for poisonous plants, such as oak, ivy, and sumac. • Checking for hot- and cold- weather injuries and preventing them. • Ensuring that correct driving practices are followed, to include— ▪ Speed in relation to the weather. ▪ Ground guides. ▪ Seat belts. ▪ Blackout driving. ▪ Gunner positioning (goggles). • Ensuring that equipment is set up, such as camouflage nets or tents. • Ensuring that there are fire-prevention vehicles and tents. • Taking appropriate actions during severe weather. • Using pyrotechnics. • Using the buddy system for safety and during the mission.

Figure E-2. Extensive PCI Checklist

Steps	Actions
	Vehicle commander, squad leader, or team leader. The vehicle commander, squad leader, or team leader ensures that— • There is a map with a current overlay. • There is a current SOI. • Includes the following in the leader packet: ▪ A unit tactical standing operating procedure (TACSOP). ▪ Map markers. ▪ Overlay sheets. ▪ A notebook and a pen or a pencil. ▪ A route reconnaissance GTA card. ▪ A bridge classification GTA card. ▪ A call for-fire GTA card. ▪ A rigging GTA card. ▪ An NBC GTA card. ▪ An *EPW GTA card (GTA 19-7-1).* ▪ Flex cuffs for EPWs. ▪ *A DD Form 1408.* ▪ *A DA Form 4137.* ▪ *A DA Form 3975.* ▪ *A DD Form 2708.* ▪ *A DA Form 3881.* ▪ *A DA Form 2823.* ▪ A compass and a wrist watch. ▪ Batteries for all the equipment. ▪ Binoculars. ▪ NVGs. ▪ A sensitive item serial number list (radio, weapons, NVGs, and so forth).
2	Vehicles. • Tie down all the items according to the load plan, and ensure that— ▪ There is space left for ammunition. ▪ The TA-50 is loaded. ▪ The MOPP chemical protective overgarment (CPOG) is stored for ready access. ▪ The gas tank is topped off. ▪ There are POL package products and weapon oil. ▪ Water cans are full.

Figure E-2. Extensive PCI Checklist (Continued)

Steps	Actions
	▪ Spare fuel cans are full and tied down. ▪ MRE rations are inventoried and stored. ▪ There are weapon-cleaning kits. ▪ The OVM is stored and serviceable. ▪ There are chock blocks. ▪ That a complete first aid kit is available. ▪ A TM for the equipment (-10) on hand. ▪ A vehicle dispatch packet and a *DA Form 2404* are available. ▪ There are tools and tool kits. ▪ There are tire chains. ▪ The spare tire is properly stowed. • Perform a premovement inspection and ensure that— ▪ There is a current TM (-10) on hand. ▪ There are no fuel leaks. ▪ The fire extinguisher is sealed, tagged, and dated. ▪ The engine coolant level is correct. ▪ The oil level is correct. ▪ The transmission fluid level is correct. ▪ The battery level is correct and the cables are tight. ▪ The air filter is clean and secure. ▪ The lights are operational. ▪ There are no exhaust leaks. ▪ The fuel filter is drained. ▪ Warning triangles are on hand. ▪ The canvas is attached and tied down. ▪ The gauges are functional. ▪ The tire pressure is correct. ▪ The vehicle service is current. ▪ The interior of the vehicle is clean and orderly.
3	Crew-served weapons. Ensure that— • All mounting equipment is present. • The TM for the weapon (-10) is available. • There is a spare barrel and BII (such as gloves, a T&E mechanism, a wrench, a blank adapter, a tripod mount with an adapter, and a cleaning kit). • The NVDs are operational. • The tripod is clean and serviceable. • The weapons are mounted with pins.

Figure E-2. Extensive PCI Checklist (Continued)

Steps	Actions
4	Communications equipment. Ensure that— • Radios (vehicles or manpacks)— ▪ Are operational. ▪ Are secure. ▪ Have the proper frequency setting. ▪ Have a matching unit operational/set. ▪ Have an antenna with the tip protector tied down. ▪ Have connectors that are clean and serviceable. ▪ Have batteries available. ▪ Have the TM for the equipment (-10) is on hand. ▪ Have cleaning materials available (such as cotton swabs, alcohol, steel wool, and so forth). • The field telephone (TA 312) is operational. • The digital nonsecure voice terminal (DNVT) with wire is operational (communication). • The appropriate phone listing is available. • Radio antennas (OE 254s) are complete. • Communication test sets, including tools, are on hand. • Frequencies, call signs, challenges, and passwords are disseminated. • Each team has one SINCGARS manpack for quick-dismount operations.
5	NBC equipment. • Ensure that protective clothing is complete and serviceable. • Ensure that the M40 is complete, such as— ▪ M13 deconnaissance kit. ▪ M258A1 deconnaissance kit. ▪ M256 detector kit. ▪ M8 detector paper. ▪ M9 detector paper. ▪ Antifogging kit. ▪ Hood. ▪ Optical inserts, if needed. ▪ Waterproof bag.

Figure E-2. Extensive PCI Checklist (Continued)

Steps	Actions
	• Ensure that the mask and hood are serviceable. • Implement the proper MOPP level. • Brief the threat condition. • Ensure that nerve agent antidotes are available. • Issue and mount M11 deconnaissance apparatus, to include the following: ▪ Two DS2 cans per vehicle. ▪ An M8 alarm. ▪ An M229 refill kit for each M8. • Ensure that NBC defense equipment is available, such as— ▪ An M256 agent detector kit. ▪ An M 174 RADIAC meter. ▪ Two sets of batteries for the IM 174. ▪ An IM 93/147 RADIAC meter (dosimeter). ▪ A PP 157A RADIAC charger 1/IM 93/147. ▪ NBC contamination signs (marking kits).

Figure E-2. Extensive PCI Checklist (Continued)

Appendix F

Fratricide Avoidance

Fratricide is defined as the unforeseen or unintentional death, injury, or damage to friendly personnel or equipment. It is committed during the employment of friendly weapons and munitions with the intent to kill the enemy or destroy his equipment or facilities. Fratricide prevention is the commander's responsibility, and all leaders assist him across all operating systems to accomplish this mission.

FRATRICIDE

F-1. Friendly forces operating in the sustainment area are particularly vulnerable to fratricide due to the high concentration of soldiers. Most MP missions are conducted in the sustainment area in support of these friendly forces. During MP operations, it is critical that MP know where other friendly elements are operating. With this knowledge, they anticipate dangerous conditions and take steps to either avoid or mitigate them. The platoon leader is always aware of changes and developments in the situation that may place friendly forces in danger from MP long-range weapon systems.

F-2. It is important that other friendly forces operating in the AO know the location of MP to prevent them from becoming victims of fratricide. MP leaders constantly report the location of their forces to higher HQ. Higher HQ disseminates this information to other friendly forces. When the platoon leader perceives potential for fratricide, he quickly responds to prevent it.

FRATRICIDE EFFECTS

F-3. Fratricide results in unacceptable losses and increases the risk of mission failure; it usually affects the unit's ability to survive and function. Units experiencing fratricide suffer the following consequences:

- Loss of confidence in the unit's leadership.
- Increased self-doubt among the leaders.
- Hesitancy to employ deadly force.
- Oversupervision of units.
- Hesitancy to conduct night operations.
- Loss of aggressiveness.
- Loss of initiative.
- Disrupted operations.
- General degradation of unit cohesiveness, morale, and combat power.

FRATRICIDE CAUSES

F-4. The following paragraphs discuss the primary causes of fratricide. MP leaders must identify any of the factors that may affect their soldiers and strive to eliminate or correct them.

TARGET IDENTIFICATION

F-5. The primary cause of fratricide is the lack of positive target identification. MP must positively identify targets before engaging.

LAND NAVIGATION FAILURES

F-6. MP may stray into unknown locations, report wrong locations, and become disoriented. In this situation, they may unexpectedly encounter an errant unit, that may fire their weapons at the friendly force.

REPORTING AND COMMUNICATIONS FAILURES

F-7. Units at all levels may fail to generate timely, accurate, and complete reports as locations and tactical situations change. This distorts the tactical picture available at each level and can lead to erroneous clearance of supporting fire.

WEAPON ERRORS

F-8. Lapses in individual discipline can result in fratricide. These incidents include charge errors, accidental discharges, mistakes with explosives and hand grenades, and incorrect gun data use.

OPERATION HAZARDS

F-9. A variety of explosive devices and materiel may create danger on the battlefield, such as UXO, unmarked or unrecorded minefields (including scatterable mines), and booby traps. Failure to mark, record, remove, or otherwise anticipate these threats will lead to casualties.

FRATRICIDE RISK ASSESSMENT

F-10. *Table F-1, page F-4,* shows an example of a work sheet for evaluating fratricide risk in the context of mission requirements. The work sheet lists six mission accomplishment factors that affect the fratricide risk, along with related considerations for each factor. Platoon leaders assess the potential risk in each area (low, medium, or high), and assign a point value to each (one point for low risk, two for medium risk, three for high risk), and add the point values to calculate the overall fratricide assessment score.

F-11. They use the resulting score only as a guide. The platoon leader's final assessment is based on observable risk factors, such as those listed on the work sheet, and on his feel for the intangible factors affecting the operation. Note that descriptive terms are listed only in the low- and high-risk columns of the work sheet. The assessment of each factor determines whether the risk matches one of these extremes or lies somewhere between them as a medium risk.

Table F-1. Fratricide Risk Assessment Work Sheet

Factors Affecting Fratricide	Potential Risk Categories (With Variable Conditions and Point Values)		
	Low Risk (1 point)	Medium Risk (2 points)	High Risk (3 points)
Understanding of the Plan			
Commander's intent	Clear	⟷	Vague
Complexity	Simple		Complex
Enemy situation	Known		Unknown
Friendly situation	Clear		Unclear
ROE/ROI	Clear		Unclear
Environmental Factors			
Intervisibility	Favorable	⟷	Unfavorable
Obscuration	Clear		Obscured
Battle tempo	Slow		Fast
Positive target ID	100 percent		None
Control Measures			
Command relationships	Organic	⟷	Joint/ combined
Audio communications	Loud/clear		Jammed
Visual communications	Easily seen		Obscured
Graphics	Standard		Not understood
SOP	Standard		Not used
Liaison personnel	Proficient		Not trained
Location/navigation	Sure		Unsure

Table F-1. Fratricide Risk Assessment Work Sheet (Continued)

Factors Affecting Fratricide	Potential Risk Categories (With Variable Conditions and Point Values)		
	Low Risk (1 point)	**Medium Risk (2 points)**	**High Risk (3 points)**
Equipment (Compared to US Equipment)			
Friendly	Similar	⟷	Different
Enemy	Different		Similar
Training			
Individual proficiency	MOS qualified		Untrained
Unit proficiency	Trained		Untrained
Rehearsals	Realistic	⟷	None
Habitual relationship	Yes		No
Endurance	Alert		Fatigued
Planning Time (Based on the 1/3–2/3 Rule)			
Higher HQ	Adequate		Inadequate
Own unit	Adequate	⟷	Inadequate
Subordinate elements	Adequate		Inadequate
Overall risk assessment (by total point value)	26-46 points	42-62 points	58-78 points
NOTE: Point values alone may not accurately reflect fratricide risk. The commander tailors the assessment to the unit's requirements. Refer to *FM 100-14*.			

PREVENTIVE MEASURES

F-12. These measures provide the platoon leader with a guide to actions he can take to reduce and prevent fratricide risk. These guidelines are not directive in nature, nor are they intended to restrict initiative. All MP leaders must learn to apply them as appropriate based on the specific situation and METT-TC factors.

Platoon leaders must educate their soldiers on the following reduction and prevention of fratricide principles:

- Identifying and assessing the potential fratricide risks in the estimate of the situation. Express the risks in the OPORD and applicable FRAGOs.
- Maintaining situational awareness. Focusing on such areas as current intelligence, unit locations and/or dispositions, denial areas (minefields and scatterable mines; contaminated areas, such as improved conventional munitions [ICM] and NBC), SITREPs, and METT-TC factors.
- Ensuring positive target identification. Review vehicle and weapons ID cards. Become familiar with the characteristics of potential friendly and enemy vehicles, including their silhouettes and thermal signatures.
- Establishing a command climate that emphasizes fratricide prevention. Enforce fratricide prevention measures, placing special emphasis on the use of doctrinally sound TTP. Ensure supervision in the execution of orders and in the performance of all tasks and missions to standard.

F-13. Additional considerations for fratricide reduction and prevention include the following:

- Reporting MP patrol locations accurately. The TOC and CP track the patrols' locations in relationship to friendly forces.
- Recognizing the signs of battlefield stress. Maintain unit cohesion by taking quick, effective actions to alleviate the battlefield stress.
- Conducting individual, leader, and collective (unit) training covering fratricide awareness.
- Giving complete and concise mission orders.

- Simplifying mission orders by using SOPs that are consistent with the doctrine. Periodically review and update SOPs as needed.
- Striving to provide maximum planning time for leaders and subordinates.
- Planning and establishing effective communications.
- Ensuring that ROE are clear and understood.
- Conducting rehearsals when the situation allows.
- Being in the right place at the right time. If an MP patrol becomes lost or misoriented, leaders must know how to contact higher HQ immediately for instructions and assistance.
- Including a discussion of fratricide incidents in all AARs.

FRIENDLY FIRE INCIDENTS

F-14. MP patrol's could become involved in a friendly fire incident in one of several ways: as the victim of the fire, as the firing element, or as an observer observing an attack of one friendly element on another. If a patrol encounters such a situation, it should follow the procedures for the various situations listed below:

FALL VICTIM TO FRIENDLY FIRES

F-15. When falling victim to friendly fire, react to contact until recognizing friendly fire, then—

- Cease fire.
- Take immediate actions to protect soldiers and vehicles.
- Use a visual recognition signal directing the firing unit to cease fire.
- Report the following on the next higher unit net:

- Announce that friendly fire has been received.
- Request medical assistance as needed.
- Give the location and direction of the firing vehicles.
- Warn the higher unit not to return fire if the firing unit has been identified as friendly.

ENGAGE FRIENDLY FORCES

F-16. When engaging friendly forces, cease fire and report the following on the next higher unit net:

- The ID of the engaged friendly force (if the unit is unidentified, report the number and type of vehicles).
- The location of the incident.
- The direction and distance to the engaged force.
- The type of fire.
- The target effects.

OBSERVE A FRIENDLY FIRE INCIDENT

F-17. When observing a friendly fire incident, seek cover for protection, then—

- Use a visual recognition signal directing the firing unit to cease fire.
- Report the following on the next higher unit net:
  - The ID of the engaged friendly force (if the unit is unidentified, then report the number and the type of vehicles).
  - The location of the incident.
  - The direction and distance to the victim and the firing unit.
  - The type of fire.
  - The target effects.

- Provide assistance as needed (when safe to do so).

LEADER RESPONSIBILITIES

F-18. In all situations involving the risk of fratricide and friendly fire, leaders must be prepared to take immediate actions to prevent casualties and equipment damage or destruction. Recommended actions in fratricide situations include—

- Identifying the incident and ordering the parties involved to cease fire.
- Conducting an in-stride risk assessment.
- Identifying and implementing the controls to prevent the incident from recurring.

Appendix G

MK19 Qualification Tables

The tables in this appendix are designed to enable the MK19 gunner and assistant gunner to qualify with the weapon.

NOTE: These tables are not for ASV qualification. The ASV tables are under development.

PRIMARY GUNNER, MK19 QUALIFICATION AND ZERO/PRACTICE TABLES

G-1. *Table G-1, page G-2,* is a zero/practice table for the MK19 primary gunner.

Table G-1. Primary Gunner Zero/Practice Table

Task	Condition	Standard	Ammunition	Time
Mount the MK19 on a vehicle or a M3 tripod	Given a mounted MK19, an M3 tripod, an M64 gun mount, a T&E mechanism, an organic carrier, and a selected firing position	The gun crew must be able to mount the MK19 on a vehicle or a tripod.	0	2 minutes
Zero the MK19 using a 400-meter vehicle target	Given a mounted MK19, a 400-meter vehicle target, a selected firing position, and 4 40-millimeter training practice (TP) rounds	The gunner must impact on the target with at least one of the 4 40-millimeter TP rounds.	4	No time limit
Engage a stationary vehicle target at 1,100 meters	Given a mounted MK19, a vehicle target at 1,100 meters, a selected firing position, and 4 40-millimeter TP rounds	The gunner must impact on the vehicle target with at least 1 of the 4 rounds.	4	No time limit
Engage a stationary vehicle target at 1,500 meters	Given a mounted MK19, a vehicle target at 1,500 meters, a selected firing position, and 4 40-millimeter TP rounds	The gunner must impact on the vehicle target with at least 1 of the 4 rounds.	4	No time limit

Table G-1. Primary Gunner Zero/Practice Table (Continued)

Task	Condition	Standard	Ammunition	Time
Engage a stationary vehicle target at 600 meters	Given a mounted MK19, a vehicle target at 600 meters, a selected firing position, and 4 40-millimeter TP rounds	The gunner must impact on the vehicle target with at least 1 of the 4 rounds.	4	No time limit
Dismount the MK19 from the vehicle or the tripod	Given a mounted MK19 and a selected firing position	The gunner must take the MK19 out of operation from the carrier or tripod.	0	1 minute

PRIMARY GUNNER, INDIVIDUAL MK19 QUALIFICATION (DAY), DISMOUNTED

G-2. *Table G-2* is the MK19 qualification table for the primary gunner, dismounted. The commander may decide to fire using this table (mounted) based on the METT-TC.

Table G-2. Primary Gunner (Day), Dismounted

Task	Condition	Standard	Ammunition	Time
Mount the MK19 on an M3 tripod or a vehicle	Given a mounted MK19, an M3 tripod, an M64 gun mount, a T&E mechanism, an organic carrier, and a selected firing position	The gun crew must be able to mount the MK19 on a vehicle or a tripod.	0	120 seconds
Engage a stationary vehicle target at 800 meters	Given a mounted MK19, a selected firing position, 6 40-millimeter TP rounds, and a vehicle target at 800 meters	The gunner must impact on the vehicle target with at least 2 of the 6 rounds.	6	90 seconds
Engage a stationary vehicle target at 600 meters	Given a mounted MK19, a selected firing position, 4 40-millimeter TP rounds, and a vehicle target at 600 meters	The gunner must impact on the vehicle target with at least 1 of the 4 rounds.	4	90 seconds
Engage a stationary vehicle target at 1,100 meters	Given a mounted MK19, a vehicle target at 1,100 meters, a selected firing position, and 6 40-millimeter TP rounds	The gunner must impact on the vehicle target with at least 2 of the 6 rounds.	6	120 seconds

Table G-2. Primary Gunner (Day), Dismounted (Continued)

Task	Condition	Standard	Ammunition	Time
Engage a stationary vehicle target at 1,500 meters	Given a mounted MK19 vehicle target at 1,500 meters, a selected firing position, and 8 40-millimeter TP rounds	The gunner must impact on the vehicle target with at least 2 of the 8 rounds.	8	180 seconds
Dismount the MK19 from the vehicle or the tripod	Given a mounted MK19 and a selected firing position	The gunner must take the MK19 out of operation from the carrier or tripod.	0	60 seconds

PRIMARY GUNNER, MK19 QUALIFICATION, MOUNTED

G-3. *Table G-3, page G-6,* is the primary gunner's individual MK19 qualification table, mounted.

Table G-3. Primary Gunner, Mounted

Task	Condition	Standard	Ammunition	Time
Mount the MK19 on a vehicle or a M3 tripod	Given a mounted MK19, an organic carrier, an M3 tripod, an M64 gun mount, a T&E mechanism, and a selected firing position	The gun crew must be able to mount the MK19 on a vehicle or a tripod.	0	120 seconds
Engage a stationary vehicle target at 400 meters	Given a mounted MK19, a selected firing position, 4 40-millimeter TP rounds, and a vehicle target at 400 meters	The gunner must impact on the vehicle target with at least 2 of the 4 rounds.	4	90 seconds
Engage a stationary vehicle target at 1,100 meters	Given a mounted MK19, a selected firing position, 8 40-millimeter TP rounds and a vehicle target at 1,100 meters	The gunner must impact on the vehicle target with at least 2 of the 8 rounds.	8	120 seconds

Table G-3. Primary Gunner, Mounted (Continued)

Task	Condition	Standard	Ammunition	Time
Engage troops at 600 meters	Given a mounted MK19, a selected firing position, 4 40-millimeter TP rounds, and a troop target at 600 meters	The gunner must impact on the target with at least 1 of the 4 rounds	4	60 seconds
Engage a stationary vehicle target at 1,500 meters	Given a mounted MK19, a selected firing position, 6 40-millimeter TP rounds, and a vehicle target at 1,500 meters	The gunner must impact on the vehicle target with at least 1 of the 6 rounds	6	120 seconds
Engage a stationary vehicle target at 1,100 meters and troops in the open at 600 meters	Given a mounted MK19, a selected firing position, 10 40-millimeter TP rounds, a vehicle target at 1,100 meters, and a troop target at 600 meters	The gunner must impact on the vehicle and troop targets with at least 2 rounds per target	10	180 seconds
Dismount the MK19 from a vehicle or a tripod.	Given a mounted MK19 and a selected firing position	The gunner must take the MK19 out of operation from the carrier or tripod	0	60 seconds

PRIMARY GUNNER, MK19 QUALIFICATION (NIGHT)

G-4. *Table G-4* is the primary gunner's MK19 qualification table for night firing.

Table G-4. Primary Gunner (Night)

Task	Condition	Standard	Ammunition	Time
Mount the AN/TVS-5 to the MK19	Given a complete AN/TVS-5, an MK19, an M3, and a T&E mechanism or organic carrier	The gunner must mount the AN/TVS-5 on the MK19.	0	120 seconds
Use a zeroed MK19 with an AN/TVS-5 mounted for observation and engage a 400-meter vehicle target	Given a complete AN/TVS-5, a mounted MK19, a selected firing position, 6 40-millimeter TP rounds, and a vehicle target at 400 meters.	The gunner must impact the target with at least 2 of the 6 rounds.	6	60 seconds

Table G-4. Primary Gunner (Night) (Continued)

Task	Condition	Standard	Ammunition	Time
Engage a stationary vehicle target at 600 meters	Given a mounted MK19, a selected firing position, 8 40-millimeter TP rounds, and a vehicle target at 600 meters	The gunner must impact the vehicle target with at least 4 of the 8 rounds.	8	120 seconds
Engage troops in the open at 800 meters	Given a mounted MK19, a selected firing position, 8 40-millimeter TP rounds, and troops in the open at 800 meters	The gunner must impact the troop target with at least 4 of the 8 rounds.	8	120 seconds
Dismount the AN/TVS-5 from the MK19	Given an MK19 and an AN/TVS-5	The gun crew must take the AN/TVS-5 out of operation.	0	120 seconds

ASSISTANT GUNNER, MK19 FIRING TABLE, MOUNTED

G-5. *Table G-5* is for familiarization only.

Table G-5. Assistant Gunner, Mounted

Task	Condition	Standard	Ammunition	Time
Mount the MK19 on a vehicle	Given a zeroed, mounted, MK19; a T&E mechanism; an MK64 gun mount; an organic carrier; and a selected firing position	The gun crew must be able to mount the MK19 on a vehicle or a tripod	0	120 seconds
Engage a stationary vehicle target at 400 meters	Given a mounted MK19, a vehicle target at 400 meters, a selected firing position, and 4 40-millimeter TP rounds	The gunner must impact the vehicle target with at least 2 of the 4 rounds	4	90 seconds
Engage troops in the open at 600 meters	Given a mounted MK19, a troop target at 600 meters, a selected firing position, and 4 40-millimeter TP rounds	The gunner must impact the troop target with at least 1 of the 4 rounds.	4	60 seconds

Table G-5. Assistant Gunner, Mounted (Continued)

Task	Condition	Standard	Ammunition	Time
Engage a stationary vehicle target at 1,100 meters	Given a mounted MK19, a vehicle target at 1,100 meters, a selected firing position, and 8 40-millimeter TP rounds	The gunner must impact the vehicle target with at least 2 of the 8 rounds.	8	120 seconds
Engage a stationary vehicle target at 1,100 meters and troops at 600 meters	Given a mounted MK19, a vehicle target at 1,100 meters, troops in the open at 600 meters, a selected firing position, and 10 40-millimeter TP rounds	The gunner must impact the vehicle and troop targets with at least 2 rounds per target.	10	180 seconds
Dismount the MK19 from a vehicle or a tripod	Given a mounted MK19 and a selected firing position	The gunner must take the MK19 out of operation from the carrier or tripod.	0	NA

ASSISTANT GUNNER, MK19 FIRING TABLE, DISMOUNTED

G-6. *Table G-6* is a dismounted firing table. The commander may decide to fire using *Table G-5* based on the METT-TC.

Table G-6. Assistant Gunner, Dismounted

Task	Condition	Standard	Ammunition	Time
Mount the MK19 on a vehicle or an M3 tripod	Given a mounted MK19, an M3 tripod, an M64 gun mount, a T&E mechanism, an organic carrier, and a selected firing position	The gun crew must be able to mount the MK19 on a vehicle or a tripod.	0	120 seconds
Engage a stationary vehicle target at 800 meters	Given a mounted MK19, a vehicle target at 800 meters, a selected firing position, and 6 40-millimeter TP rounds	The gunner must impact the vehicle target with at least 2 of the 6 rounds.	6	60 seconds
Engage a stationary vehicle target at 600 meters	Given a mounted MK19, a vehicle target at 600 meters, a selected firing position, and 4 40-millimeter TP rounds	The gunner must impact the vehicle target with at least 1 of the 4 rounds.	4	90 seconds

Table G-6. Assistant Gunner, Dismounted (Continued)

Task	Condition	Standard	Ammunition	Time
Engage a stationary vehicle target at 1,100 meters	Given a mounted MK19, a vehicle target at 1,100 meters, a selected firing position, and 6 40-millimeter TP rounds	The gunner must impact the vehicle target with at least 2 of the 6 rounds.	6	90 seconds
Engage a stationary vehicle target at 1,500 meters	Given a mounted MK19, a vehicle target at 1,100 meters, a selected firing position, and 6 40-millimeter TP rounds	The gunner must impact the vehicle target with at least 2 of the 6 rounds.	6	120 seconds
Engage a stationary vehicle target at 1,500 meters	Given a mounted MK19 vehicle target at 1,500 meters, a selected firing position, and 8 40-millimeter TP rounds	The gunner must impact the vehicle target with at least 2 of the 8 rounds.	8	120 seconds
Dismount the MK19 from a vehicle or a tripod	Given a mounted MK19 and a selected firing position	The gunner must take the MK19 out of operation from the carrier or tripod.	0	60 seconds

ASSISTANT GUNNER, MK19 FIRING TABLE (NIGHT)

G-7. *Table G-7* is for night firing familiarization only.

Table G-7. Assistant Gunner (Night)

Task	Condition	Standard	Ammunition	Time
Mount the AN/TVS-5 to the MK19	Given a mounted AN/TVS 5 and a T&E mechanism	The gunner must mount the AN/TVS-5 on an MK19.	0	120 seconds
Engage a stationary vehicle target at 600 meter	Given a mounted MK19, a vehicle target at 600 meters, a selected firing position, and 6 40-millimeter TP rounds	The gunner must impact on the vehicle target with at least 1 of the 6 rounds.	6	120 seconds
Engage troops in the open at 800 meters	Given a mounted MK19, troops in the open at 800 meters, a selected firing position, and 6 40-millimeter TP rounds	The gunner must impact on the vehicle target with at least 1 of the 6 rounds.	6	120 seconds
Dismount the AN/TVS-5 from the MK19	Given an MK19 and an AN/TVS-5	The gun crew must take the AN/TVS-5 out of operation within 2 minutes.	0	120 seconds

MK19 SCORECARD

G-8. *Figure G-1* is the scorecard for primary and assistant gunner qualification.

MK19 Qualification Scorecard

1. Name (Last, First, Middle Initial)	2. SSN	3.Grade	4. Unit	5. Date

Mounted Table

Rounds	Range (Meters)	Target	Time (Seconds)	Required Hits	Hits
4	400	Vehicle	90	2	
8	1,100	Vehicle	120	2	
4	600	Troops	60	1	
6	1,500	Vehicle	120	1	
10	1,100/600	Vehicle/ troops	180	2/2	
				Total hits	

Rating:
Distinguished = 5 of 5 qualified engagements.
Qualified = 4 of 5 qualified engagements.
Unqualified = 3 and below.

Dismounted Table

Rounds	Range (Meters)	Target	Time (Seconds)	Required Hits	Hits
6	800	Vehicle	90	2	
4	600	Vehicle	90	1	
6	1,100	Vehicle	120	2	
8	1,500	Vehicle	180	2	
				Total hits	

Rating:
Distinguished = 4 of 4 qualified engagements.
Qualified = 3 of 4 qualified engagements.
Unqualified = 2 and below.

Night-Firing Table

Rounds	Range (Meters)	Target	Time (Seconds)	Required Hits	Hits
6	400	Vehicle	60	2	
8	600	Vehicle	120	4	
8	800	Troops	120	4	
				Total hits	

Rating:
Distinguished = 3 of 3 qualified engagements.
Qualified = 2 of 3 qualified engagements.
Unqualified = 1 and below.

Standards

All standards achieved within table = distinguished.
Missed 1 standard within table = qualified.
two or more missed standards = unqualified.

Overall Rating:
Three distinguished tables = distinguished.
Two distinguished tables = qualified.
One distinguished table = unqualified.

Figure G-1. MK19 Qualification Tables

Appendix H

Countermine Operations

Military operations are often conducted in areas where warring factions have left unrecorded mines and minefields scattered across the landscape. In these environments, emphasize mine awareness at all levels of command and plan countermine operations into every military operation.

OVERVIEW

H-1. Undertake countermine operations to breach or clear a minefield. All the tasks fall under breaching or clearing operations and include detecting, reporting, reducing, proofing, and marking.

H-2. MP efforts in countermine operations are in detecting, reporting, and marking. They conduct countermine operations in conjunction with their other missions. Countermine operations include–

- Denying infiltrators, insurgents, and terrorist groups the opportunity to use mines.
- Using mine detection equipment.
- Detecting mines visually or by probing.
- Reporting the location of suspected mines and UXO.
- Marking the location of these devices.

H-3. MP patrols look for suspicious persons along MSRs; watch approaches to critical points like bridges or defiles; and maintain surveillance of MSRs, key terrain, and critical facilities. Check with the HN police and local nationals for information on unusual activity in a particular AO. Mines are usually emplaced at night. Use night vision and early warning devices to maintain surveillance and detect enemy activity.

H-4. Mine and UXO awareness involve soldier and leader skills. Soldier skills are a mix of individual and collective tasks that are required for an element to maintain its combat effectiveness in and around a mined environment. Soldier skills involve individual and collective tasks that are required for basic survival in a mined environment. They include minefield indicators, probing techniques, mine detector operations, extraction drills, survival rules, casualty treatment, and evacuation drills. The soldier's basic mine awareness skills are critical to his and the unit's survival. Leader skills involve planning missions, assessing situations, and tracking and disseminating mine information. Soldiers must be proficient in all mine awareness skills to effectively operate in a mined environment (refer to *FM 20-32*).

DETECT

H-5. Detection is the actual confirmation and location of the mines and may be accomplished through reconnaissance or unintentionally (such as a vehicle running into a mine). Use mine detection in conjunction with intelligence-gathering operations, minefield bypass reconnaissance, and breaching and clearing operations. Use the following three methods to detect mines and UXO:

- Visual.
- Physical (probing).
- Electronic (mine detector).

VISUAL INDICATORS

H-6. Mine and UXO indicators are part of all combat operations. Understanding and recognizing mine indicators could determine whether or not a soldier becomes a casualty. The following may indicate the presence of mines and UXO:

- Trip wires.
- Signs of road repair (such as new fill or paving, road patches, ditching, or culvert work).
- Signs, usually understood only by local populace, placed on trees, posts, or stakes. Threat forces mark their minefields to protect their own forces.
- Dead animals.
- Vehicles that are damaged.
- Disturbances in previous tire tracks or tracks that stop unexplainably.
- Wires leading away from the side of the road. They may be firing wires that are partially buried.

H-7. Check for odd features in the ground or patterns that are not present in nature. Plant growth may wilt or change color, rain may wash away some of the cover, the cover may sink or crack around the edges, or the material covering the mines may look like mounds of dirt. Civilians may know where mines or booby traps are located in the residential area. Civilians staying away from certain places or out of certain buildings are good indications of the presence of mines or booby traps. Question civilians to determine the exact location of these devices.

H-8. Pieces of wood or other debris on a road may be indicators of pressure or pressure-release firing devices. These devices may be on the surface or partially buried. The enemy uses mines that are fired by command, so search road shoulders and areas close to the objects.

PHYSICAL INDICATORS

H-9. Physical detection (probing) is very time-consuming. Use it primarily for clearing operations, self-extraction, and covert breaching operations. Detection of mines by visual or electronic methods should be confirmed by probing. Use the following procedures and techniques when probing for mines:

- Roll up your sleeves, and remove any jewelry to increase sensitivity. Wear a Kevlar helmet, with the chin strap buckled, and a protective fragmentation vest.
- Stay close to the ground, and move in a prone position to reduce the effects of an accidental blast. When moving into a prone position–
  - Squat down without touching your knees to the ground.
  - Scan forward up to 2 meters and to the sides up to 3 meters for mine indicators.
  - Probe the area around your feet and as far forward as possible.
  - Kneel on the ground after the area is found to be clear. Continue probing forward until you are in a prone position.
- Use sight and touch to detect trip wires, fuses, and pressure prongs.
- Use a slender, nonmetallic object as a probe, and—
  - Probe every 5 centimeters across a 1-meter front.
  - Push the probe gently into the ground at an angle that is less than 45 degrees.
  - Apply just enough pressure on the probe to sink it slowly into the ground.
  - Check the probe for resistance. If the probe encounters resistance and does not go into the ground freely, carefully pick the soil

away with the tip of the probe and remove the loose dirt by hand. Take care to prevent functioning the mine.

- Stop probing when you touch a solid object, and use two fingers from each hand to carefully remove the surrounding soil and identify the object. If the object is a mine, remove enough soil to show the mine type and mark its location.

DANGER
Do not attempt to remove or disarm the mine.

H-10. Probing is extremely stressful and tedious. The senior leader sets a limit to the time a prober can actually probe in the minefield. To determine a reasonable time, the leader considers the METT-TC factors, weather conditions, the threat level, the unit's stress level, and the prober's fatigue level and state of mind. As a rule, 20 to 30 minutes is the maximum amount of time that an individual can probe effectively.

DANGER
Use extreme caution when probing. If the probe is pushed straight down, its tip may detonate a pressure fuse.

ELECTRONIC INDICATORS

H-11. Electronic detection is effective for locating mines, but this method is time-consuming and exposes personnel to enemy fire. Confirm suspected mines by probing.

H-12. The AN/PSS-12 mine detector can only detect metal, but most mines have metal components in their design. The detector locates and identifies plastic or wooden mines by a slight metallic signature. Refer to *TM 5-6665-298-10* for more information about the employment and operation procedures for the AN/PSS-12. The detector is handheld and identifies suspected mines by an audio signal in the headphones.

H-13. As in probing, take consideration for the maximum amount of time an individual can operate the detector. The leader considers the METT-TC factors, weather conditions, the threat level, the unit's stress level, and the individual's fatigue level and state of mind. As a rule, 20 to 30 minutes is the maximum amount of time an individual can use the detector effectively.

REPORT

H-14. Intelligence concerning enemy mines and UXO is reported by the fastest means available. Report sightings of these devices using a SPOTREP format. SPOTREPs originate from patrols that have been sent on specific reconnaissance missions or from MP patrols that discover mine information in the course of their normal route operations.

MARK

H-15. When mines and UXO are detected, mark the location to prevent friendly follow-on forces and local nationals from accidentally encountering them. Mark the mines and UXO with standard North Atlantic Treaty Organization (NATO) markers (a red triangle with white letters). Use concertina wire to construct a perimeter around the mined area and place markers on the fencing about waist-high.

Appendix I

Route Classification and Signing System

The military route classification system helps in planning and executing battlefield movements. Military engineers develop route classifications using a route classification formula. This formula consists of a series of numbers and letters that express, in standard sequence, the route width, the route type, the lowest MLC, overhead clearance, obstructions to the traffic flow, and special conditions on a given route. Findings are based on information extracted from route reconnaissance reports.

OVERVIEW

I-1. The first step in completing a route reconnaissance overlay is to understand what information must be included on it. As a minimum, include the following information on the overlay:

- The route classification formula.
- The name, rank, and social security number of the person preparing the overlay.
- The unit conducting the reconnaissance.
- The DTG that the overlay was conducted.
- The map name, edition, and scale.
- Any remarks necessary to ensure the complete understanding of the information contained on the overlay.

I-2. This appendix focuses on the route classification system and the methods necessary to determine the classification of a road. Refer to *FM 5-170* for more information about route classification and reconnaissance.

ROUTE CLASSIFICATION FORMULA

I-3. The route classification is derived from the information gathered during the route reconnaissance. The formula is recorded on the route reconnaissance overlay *(Figure I-1)* and consists of the following:

- The route width (in meters).
- The route type (based on the ability to withstand weather).
- The lowest MLC.
- The lowest overhead clearance (in meters).
- Any obstructions (OB) to traffic flow, if applicable.
- Any special conditions, such as snow blockage (T) or flooding (W).

I-4. The following are examples of route classification formulas:

- 6.1/Z/40/∞. A fair-weather route (Z) with a minimum traveled way of 6.1 meters and an MLC of 40. Overhead clearance is unlimited (∞), and there are no obstructions to traffic flow. This route accommodates both wheeled and tracked, single-flow traffic without obstruction.
- 7/Y/50/4.6 (OB) (W). A limited, all-weather route (Y) with a minimum traveled way of 7 meters, an MLC of 50, an overhead clearance of 4.6 meters, and an obstruction. This route width is not suitable for double-flow traffic (wheeled or tracked). This route is subject to regular, recurrent flooding.

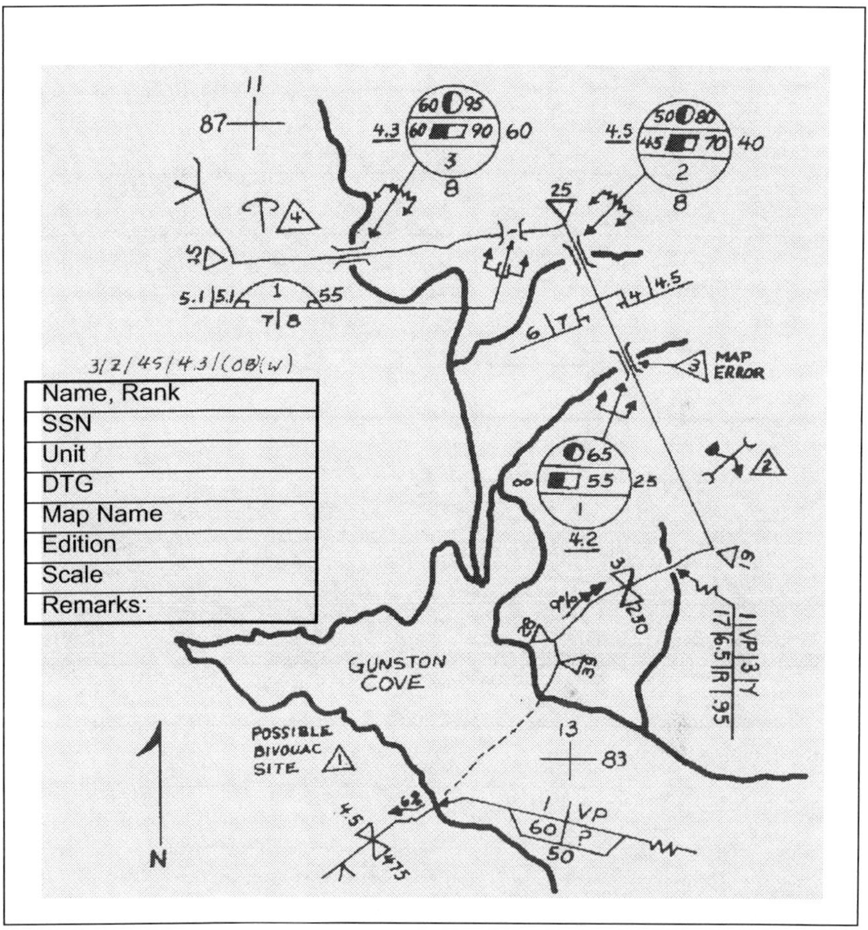

Figure I-1. Route Reconnaissance Overlay

I-5. *Table I-1, page I-4,* gives route reconnaissance symbols used on a route reconnaissance overlay.

Table I-1. Route Reconnaissance Symbols

Explanation	Symbol	Remarks
1. Abbreviated bridge symbol	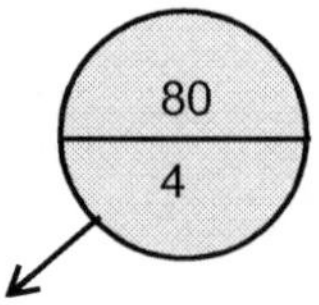	Use this symbol only when the map scale does not permit the use of the full NATO bridge symbol. If this symbol is used, submit DA Form 1249. Draw an arrow to the map location of the bridge. Show the bridge's serial number in the lower portion of the symbol and the MLC for single-flow traffic in the upper portion. If there are separate load classifications for tracked or wheeled vehicles, show the lesser classification. Underline the classification number if the width or overhead clearance is below the minimum standard.
2. Axial route	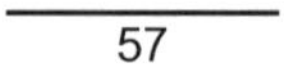	Use a solid line and identify the route by an odd number.
3. Bypass difficult	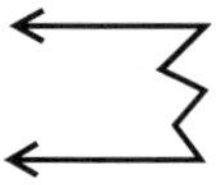	Use this symbol when the obstacle can be crossed in the immediate vicinity, but some work is necessary to improve the bypass.
4. Bypass easy	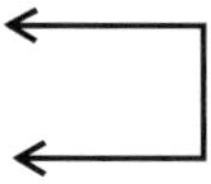	Use this symbol when the obstacle can be crossed in the immediate vicinity by a US 2 1/2-ton truck (or NATO equivalent) without work to improve the bypass.
5. Bypass impossible	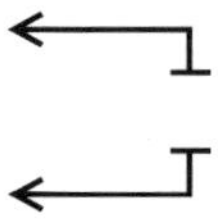	Use this symbol when the obstacle can be crossed only by repairing or constructing a feature or by detouring around the obstacle.
6. Civil or military route designation		Write the designation in parentheses along the route.

Table I-1. Route Reconnaissance Symbols (Continued)

Explanation	Symbol	Remarks
7.Concealment		Show roads lined with trees by a single line of circles for deciduous trees and a single line of inverted Vs for evergreen trees. Show woods bordering a road by several rows of circles for deciduous trees and several rows of inverted Vs for evergreen trees.
8. Critical points		Number, in order, and describe critical points on *DA Form 1711-R*. Use critical points to show the features not adequately covered by other symbols on an overlay.
9. Damage or destruction	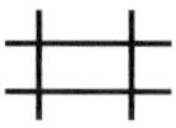	Damage or destruction that prevents movement along the route.
10. Ferry *Ferry type* P–pedestrian V–vehicular	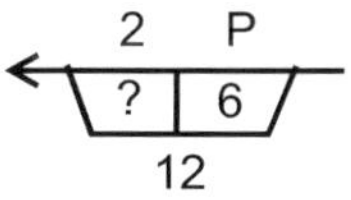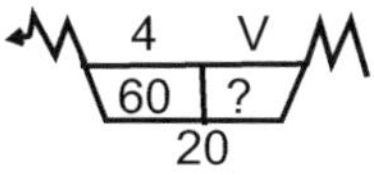	Draw an arrow to the ferry location. The data above the symbols shows, in order, the left approach, the ferry's serial number, the ferry type, and right approach. The data inside the symbol shows, from left to right, the MLC and the dead weight capacity in tons. The number below the symbol shows the turnaround time, in minutes. A question mark indicates unknown information. Show the different approaches by a zigzag line, and show easy approaches by a straight line.

Table I-1. Route Reconnaissance Symbols (Continued)

Explanation	Symbol	Remarks
11. Ford *Ford type* P–pedestrian V–vehicular	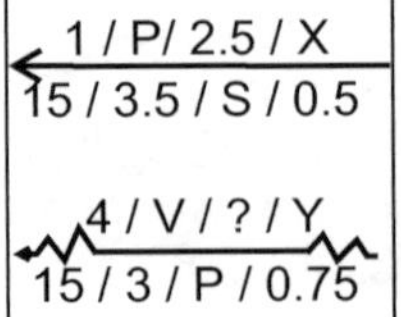	Draw an arrow to the ford location. The data above the line shows, in order, the left-bank approach, the ford serial number, the ford type, the stream velocity (in meters per second), seasonal limitations, and the right-bank approach. A zigzag line corresponding in position to the shore where the approach is located represents difficult approaches. Straight lines identify an easy approach. The left and right banks are determined by looking downstream. The data below the line shows, in order, the length, the width, the bottom type, and the depth. All measurements are in meters. Seasonal limiting factors: X – none; Y – significant; ? – unknown information. Bottom type: M – mud; C – clay, S – sand; G – gravel; R – rock; P – artificial paving.
12. Full NATO bridge symbol	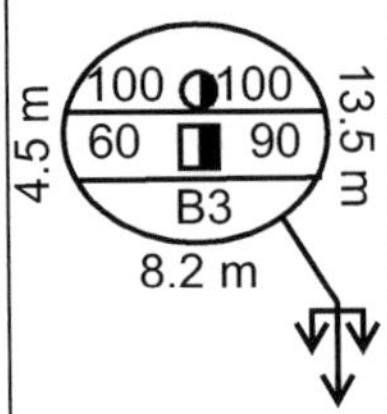	Indicate wheeled vehicles in the upper third of the symbol with the two-way wheeled classification on the left and the one-way wheeled classification on the right. Show tracked vehicles in the center third of the symbol with the two-way tracked classification on the left and the one-way tracked classification on the right. Place the bridge serial number in the lower third of the symbol. Draw an arrow to the bridge location and show the bypass conditions on the arrow shaft. Place the traveled-way width below the symbol, the overhead clearance to the left of the symbol, and the overall length to the right of the symbol.

Table I-1. Route Reconnaissance Symbols (Continued)

Explanation	Symbol	Remarks
13. Grades	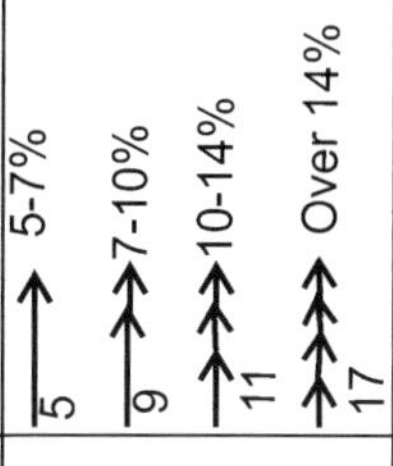	Show the actual percent of grade to the right of the symbol. Any grade of 7 percent or more is an obstruction. Include the obstruction in the route classification formula. Arrows should point uphill. The length of the arrow represents the length of the grade if the map scale permits.
14. Lateral route	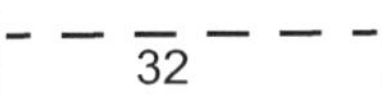	Use a broken line, and identify the route by an even number.
15. Limits of sector		Show the beginning and ending of a reconnoitered section of a route or road with this symbol.
16. MSR		Label the route as MSR and assign a code name.
17. Obstacle a. Proposed block b. Prepared but passable c. Completed block	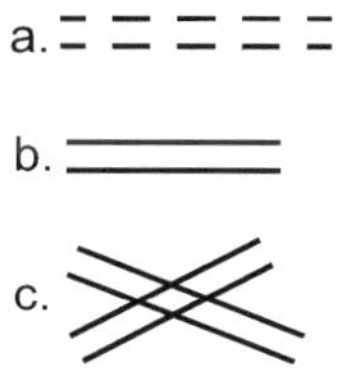	Place the center of the symbol over the location of the blocked part of the route. Use parallel broken lines for a proposed block, parallel lines for a prepared but passable block, and crossed lines for a completed block.
18. Clearance unlimited		Use this symbol when overhead clearance is unlimited.
19. Parking area		Designate parking areas.
20. Railroad crossing	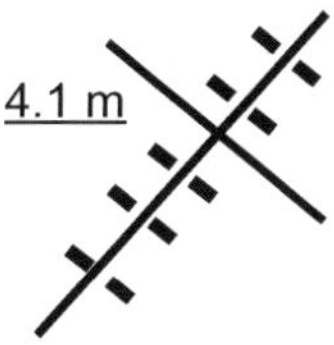	Use this symbol to show a level crossing where passing trains would interrupt the traffic flow. If there is a power line present, show its height, in meters, from the ground. Underline the overhead clearance if it is less than 4.3 meters.

Table I-1. Route Reconnaissance Symbols (Continued)

Explanation	Symbol	Remarks
21. Railway bridge symbol	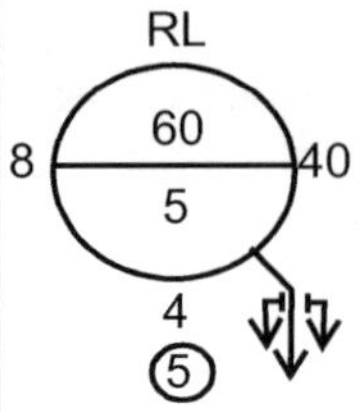 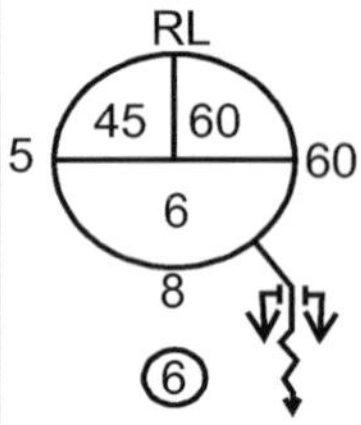 RL = railroad bridge	Place RL above the symbol to indicate a railway bridge. At the left of the symbol, show the overhead clearance. Show the overall length of the bridge on the right of the symbol. Indicate the traveled-way width below the symbol, and underline it if it is below standard for the classification. Inside the symbol, show the bridge classification in the upper half. If the class is different for single- and double-flow traffic, show single flow on the left and double flow on the right. Place the railway bridge serial number in the lower half of the symbol. Draw an arrow to the bridge location. On the arrow shaft, indicate the ease of adapting the bridge for road vehicle use. A zigzag line means it would be difficult to adapt, and a straight line means it would be easy to adapt. Place the bypass symbol on the arrow shaft to indicate the bypass conditions.
22. Route classification formula	6 / Z / 30 / 4 / (OB)	Express the formula in the order of the route width, the route type, MLC, minimum clearance, obstructions (if present), and special conditions. All measurements are in meters. Route types: X – all weather; Y – limited all-weather route; Z – fair-weather route. Special conditions: (T) – regular snow blockage; (W) - regular flooding.
23.Series of sharp curves		Write the number of curves and the radius of the sharpest curve of the series to the outside of the triangle.

Table I-1. Route Reconnaissance Symbols (Continued)

Explanation	Symbol	Remarks
24. Sharp curve		Point the vertex of the triangle to the curve location and indicate the radius of the curve, in meters, outside the triangle. A curve of 45 meters or less must be reported on the overlay, and a curve of 25 meters or less is an obstacle.
25. Traffic control HQ	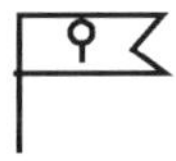	None
26. Traffic control post		Manned traffic control post.
27. Tunnel	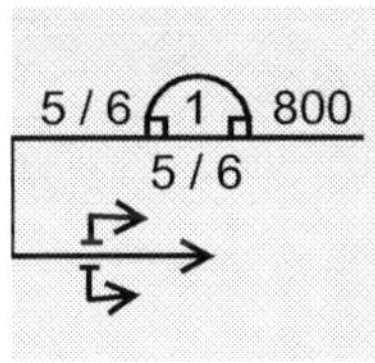	Draw an arrow to the tunnel location. Place the bypass condition symbol on the arrow. Show the minimum and maximum overhead clearances to the left of the symbol, the tunnel serial number inside the symbol, and the total tunnel length to the right of the symbol. Below the symbol, show the traveled-way width. If sidewalks are present, follow with a slash and the total traveled way, including sidewalks. Underline the traveled way if the road entering the tunnel is wider than the traveled way of the tunnel. Use a question mark to show unknown information.

Table I-1. Route Reconnaissance Symbols (Continued)

Explanation	Symbol	Remarks
28. Turnout; the symbol may be amplified as follows: a. Wheeled vehicle b. Tracked vehicle c. A length of road exceeding 1 kilometer	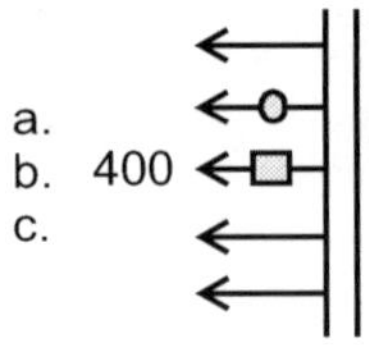	Use this symbol to show the possibility of driving off the road. Draw the arrow in the direction of the turnout (right or left of the road). For wheeled vehicles, draw a small circle on the shaft of the arrow. For tracked vehicles, draw a small square on the shaft of the arrow and place the length of the turnout, in meters, at the tip of the arrow. When the turnout is longer than 1 kilometer, use double arrows.
29. Underpass constrictions (arched or rectangular)	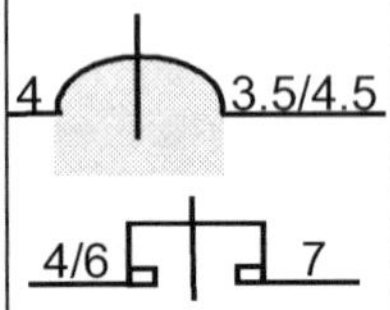	Draw the symbol over the road. Place the width of the traveled way, in meters, to the left of the symbol. If sidewalks are present, follow the traveled-way width with a slash and the total width, including sidewalks. Underline the traveled-way width if the road entering the underpass is wider than the underpass traveled way. Show the overhead clearance, in meters, to the right of the symbol. Show both minimum and maximum overhead clearances, if different.
30. Unknown information	?	Use a question mark with almost any symbol when information required for the symbol is unknown or doubtful.
31. Width constriction	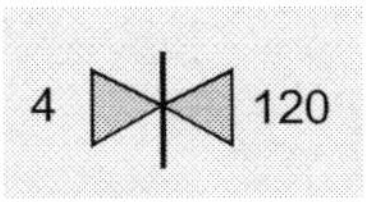	The minimum width is to the left of the symbol. The length of the minimum width is to the right of the symbol.

ROUTE WIDTHS

I-6. The route width is the narrowest width of a traveled way on a route *(Figure I-2)* and is expressed in meters. It may be the width of a bridge, a tunnel, an underpass, or any other constriction that limits the traveled-way width. The establishment of the traveled-way width will help determine the number of lanes of traffic that the route may accommodate.

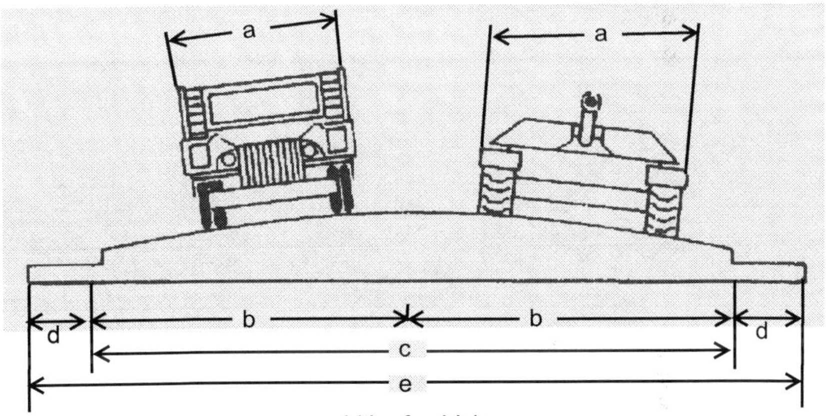

a = width of vehicle
b = width of lane
c = width of traveled way
d = width of hard shoulder
e = width of grading

Figure I-2. Road Width

I-7. Based on the number of lanes of traffic, categorize the route as either—

- Limited access. Limited access permits passage of isolated vehicles of the appropriate width in one direction only.

- Single lane. A single lane permits use in only one direction at any one time. Passing or movement in the opposite direction is impossible.
- Single flow. Single flow permits the passage of a column of vehicles and allows isolated vehicles to pass or travel in the opposite direction at predetermined points only.
- Double flow. Double flow permits two columns of vehicles to proceed simultaneously.

ROUTE TYPES

I-8. Route types are designated by their ability to withstand the effects of weather, and are determined by the worst section of the route. The three types of routes are—

- Type X. Type X is an all-weather route that, with reasonable maintenance, is passable throughout the year to its maximum volume of traffic. This type of route is normally formed of roads having waterproof surfaces, is only minimally affected by precipitation or temperature changes, and is never closed from the effects of weather.
- Type Y. Type Y is a limited, all-weather route that, with reasonable maintenance, is passable throughout the year, but at times having a volume of traffic considerably less than maximum capacity. This type of route is normally formed of roads that do not have waterproof surfaces and are affected by weather or temperature changes. This type of route is closed for short periods (up to one day at a time) by adverse weather conditions during which heavy use of the road would probably lead to complete collapse.
- Type Z. Type Z is a fair-weather route that may become impassable in adverse weather. It may

be closed for long periods, and may require major maintenance or construction efforts to reopen it.

MILITARY LOAD CLASSIFICATION

I-9. A route MLC is a class number representing the safe load-carrying capacity and indicating the maximum vehicle class that can be accepted under normal conditions. Usually, the lowest bridge MLC (regardless of the vehicle type or conditions of traffic flow) determines the route MLC. If there is not a bridge on the route, the worst section of road determines the route overall classification.

I-10. Whenever possible, the basic military road network is composed of average routes and includes a number of heavy traffic routes and a few very heavy traffic routes. Individual routes are grouped and identified in broad categories, such as—

- Average traffic routes are Class 50.
- Heavy traffic routes are Class 80.
- Very heavy traffic routes are Class 120.

OVERHEAD CLEARANCE

I-11. Overhead clearance is the vertical distance between the road's surface and any overhead obstacle (power lines, overpasses, tunnels, and so forth) that denies the use of the route or road to all vehicles or loads that exceed this height. This distance is represented in the route classification formula, in meters. Use the infinity symbol (∞) in the formula for unlimited overhead clearance. Any point along the route where the minimum overhead clearance is less than 4.3 meters is considered an obstruction.

ROUTE OBSTRUCTIONS

I-12. Route obstructions are conditions that restrict the type, amount, or speed of the traffic flow. They are indicated in the route classification formula as OB. If an obstruction is encountered and reported, depict its exact nature on the route classification overlay. Reportable obstructions include the following:

- Overhead obstructions, such as tunnels, underpasses, power lines, and so forth, with a clearance of less than 4.3 meters.
- Reductions in traveled-way widths that are below the standard minimums prescribed for the type of traffic flow *(Table I-2)*. This includes reductions caused by bridges, tunnels, craters, and lanes through mined areas, protruding buildings, or rubble.
- Slopes (gradients) of 7 percent or greater.
- Curves with a radius of 25 meters or less. Curves with a radius of 25.1 to 45 meters are not considered obstructions; however, report them on the reconnaissance overlay.
- Ferries.
- Fords.

Table I-2. Traffic Flow

Type of Vehicles	Limited Access (In Meters)	Single Lane (In Meters)	Single Flow (In Meters)	Double Flow (In Meters)
Wheeled	At least 3.5	3.5 to 5.5	5.5 to 7.3	Over 7.3
Tracked or combination vehicles	At least 4.0	4.0 to 6.0	6.0 to 8.0	Over 8.0

SPECIAL CONDITIONS (SNOW BLOCKAGE AND FLOODING)

I-13. Conditions, such as snow and water, are not normally classified as obstructions, except where these conditions are regular, recurrent, and serious. In cases where snow accumulation is excessive and is either blocking traffic or has the potential to block traffic, the symbol following the route classification formula is T. Similarly, where flooding poses the same problem, the symbol following the formula is W.

CALCULATIONS

I-14. Make calculations for the various curves, slopes, and fords that may affect vehicle movement along the route.

CURVES

I-15. The speeds that vehicles move along the route and security planning measures are affected by sharp curves in the roadway. As previously stated, curves with a radius of 25 meters and less are obstructions to traffic and are indicated by OB in the route classification formula and identified on *DA Form 1248*. There are several ways to measure the radius of a curve, but the two easiest are the tape measure and formula methods.

Tape Measure Method

I-16. The quickest way to estimate the radius of a sharp curve is by using a tape measure to find the radius *(Figure I-3, page I-16)*. Imagine the outer edge of the curve as the outer edge of a circle. Find (estimate) the center of this imaginary circle; then measure the radius using a tape measure. Start from the center of the circle and measure to the outside edge of the curve. The length of the tape measure from the center of the imaginary circle to its outer edge is the curve radius. This method is practical for curves having a radius up to 15 meters and located on relatively flat ground.

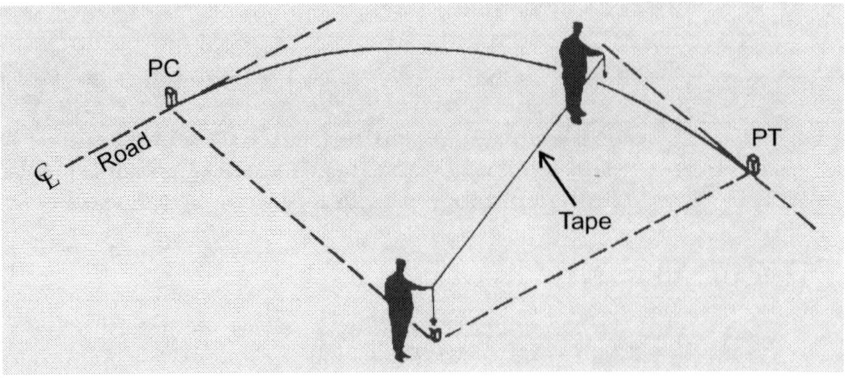

Figure I-3. Find the Radius with a Tape Measure

Formula Method

I-17. The other, and often more practical, method of determining a curve radius *(Figure I-4)* is based on the following formula:

$R = (C^2/8m) + (M/2)$

where—

R = radius of the curve

C = distance from the centerline of the road to the centerline of the road at the outer extremities (beginning and end) of the curve

M = perpendicular distance from the center of the tape to the centerline of the road

I-18. When conditions warrant, set M at 2 meters from the centerline, then measure C at 2 meters from the centerline. Use this method when there is a time limitation or because natural or man-made restrictions prevent proper measurements.

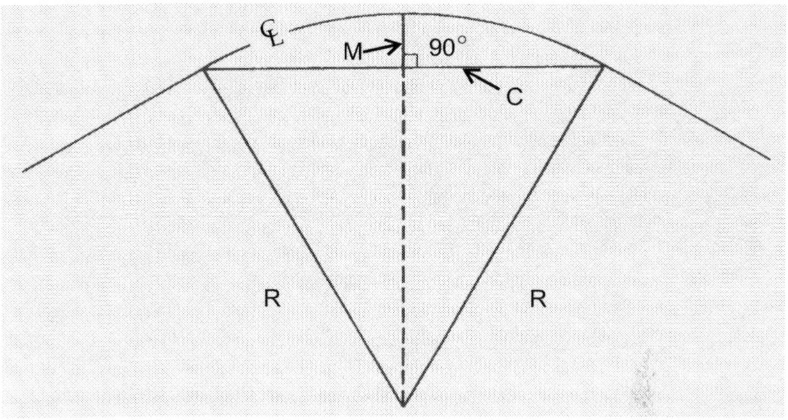

Figure I-4. Determine a Curve Radius

I-19. If C is 15 meters and M is fixed at 2 meters, the formula becomes:

$R = (15^2/16) + 2/2$

I-20. The result of this calculation would be an obstruction to traffic flow, and OB would be placed in the route classification formula.

SLOPES

I-21. The rise and fall of the ground is known as the slope or gradient (grade). Roadways with a slope of 7 percent or greater affect the movement speed along the route and are considered an obstruction. The percent of slope is the ratio of the change in elevation (the vertical distance to the horizontal ground distance) multiplied by 100 *(Figure I-5, page I-18)*. It is important to express the vertical and horizontal distance in the same unit of measure. Although not an obstruction, report all hills with a slope greater than 5 percent on the reconnaissance overlay.

I-22. There are numerous methods and instruments available for determining the percent of slope, but the

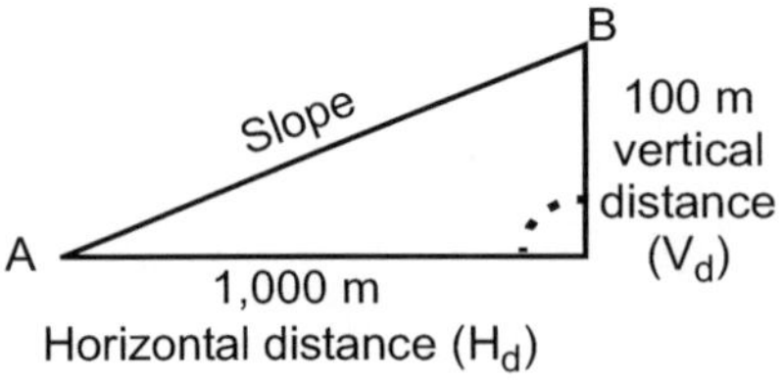

$$\textit{Percent of slope} = \frac{V_d}{H_d} \times 100$$

$$= \frac{100\ m}{1,000\ m} \times 100$$

$$= +10\ \textit{percent}$$

Figure I-5. Calculate the Percent of Slope

two hasty and most practical methods are the map and pace methods.

Map Method

I-23. Use a large-scale map (such as 1:50,000) to estimate the percent of slope quickly. After identifying the slope on the map, find the difference in elevations between the top and bottom of the slope by reading the elevation contours or spot elevation. Then, measure and convert the horizontal distance (usually road distance) to the same unit of measurement as the elevation difference. Substitute the vertical and horizontal distances in the percent-of-slope formula and compute the percent of slope *(Figure I-6)*. This is a hasty method and is the less accurate of the two methods.

Pace Method

I-24. The pace method is another quick way to estimate the percent of slope. Before using this method, accurately determine the height and pace of each member of the reconnaissance team. As a general rule of thumb, the eye level of the average soldier is 1.75 meters above the ground *(Figure I-7, page I-20)*. The pace of the average soldier is 0.75 meters. Perform the following steps to determine the percentage of slopes:

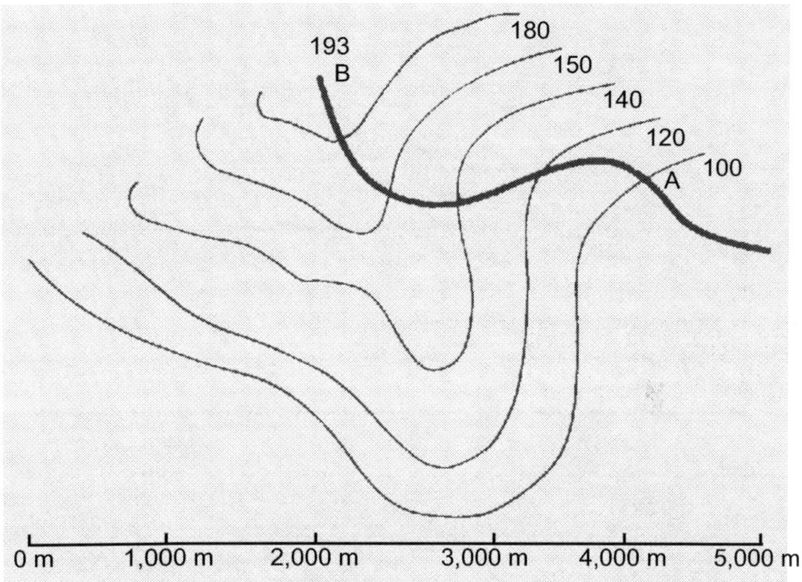

Find the difference in elevation (V_d)

Spot elevation B	=	193 m
Spot elevation A	=	100 m
Difference in elevation (V_d)	=	93 m

Find the horizontal distance (H_d)

Road distance A to B (use a piece of paper and a graphic scale as shown in FM 21-26) = 3,720 m

Use the percent-of-slope formula.

$$\frac{V_d}{H_d} \times 100$$

$$= \frac{93}{3,720} \times 100$$

$$+ 2.5 \text{ percent}$$

Figure I-6. Determine the Slope with a Map

Find:	Percent of slope	=	
Given:	Eye-level height	=	1.75 m
	Pace	=	0.75 m
Answer:	Vertical distance	=	2 x 1.75 m
		=	3.5 m
	Horizontal distance	=	75 paces +125 paces x 0.75
		=	150 m

$$Percent\ of\ slope = \frac{V_d}{H_d} \times 100 = \frac{3.5}{150} \times 100$$

Figure I-7. Determine the Slope Using the Pace Method

Step 1. Stand at the bottom of the slope with your head and eyes level.

Step 2. Sight an easily identifiable spot on the slope, horizontal to eye-level.

Step 3. Walk forward and stand on the spot, recording the number of paces. Repeat this procedure until you reach the top of the slope. Estimate fractions of eye level.

Step 4. Compute the vertical distance by multiplying the number of sightings by the eye-level height of 1.75 meters. Compute the horizontal distance by totaling the number of paces and converting them to meters by multiplying by the known pace-to-meter conversion factor of 0.75 meters.

Step 5. Calculate the percent of slope by substituting the values into the percent-of-slope formula *(Figure I-7)*. Because this method considers horizontal ground distance and incline distance as equal, the reasonable accuracy is only for slopes less than 30 percent. This method requires repeated practice to achieve reasonable accuracy.

FORDS

I-25. A ford is a location in a water barrier where the current, bottom, and approaches allow personnel and vehicles to cross and remain in contact with the bottom during the crossing. While engineers are fully trained and equipped in this area, MP are also capable of performing this task. In the course of a route reconnaissance, MP identify potential fording sites as a contingency for bridges that may become unusable. When identifying a ford on the reconnaissance overlay, include the following information in the ford symbol:

- The ford location, as indicated by an arrow from the ford symbol to the site on the overlay.
- A serial number assigned to each ford for reference. Follow the unit SOP in assigning serial numbers, ensuring that the same number is not duplicated on any one overlay.
- The type of ford, as determined by bottom composition, width, and water depth. Use V for vehicle, P for pedestrian, or VP for both. Approaches are not considered in determining the ford type.
- The stream's normal current velocity is expressed in meters per second. Seasonal limiting factors follow the stream velocity notation and are shown by the following letters:
  - X - No seasonal limitations except for potential sudden flooding of limited duration (flash flooding).
  - Y - Serious, regular, or recurrent flooding or snow blockage.

NOTE: If the Y symbol is used, the route type in the route classification formula automatically becomes type Z.

- The length of the ford (distance from nearshore to far shore), in meters.
- The nature of the bottom surface, using the most appropriate letter symbol—
  - M – mud.
  - C – clay.
  - S – sand.
  - G – gravel.
  - R – rock.
  - P – artificial paving.
- The normal stream depth at its deepest point, expressed in meters.

I-26. Separate all elements of the ford symbol by slashes. Substitute a question mark for any item of information that is unknown. Record all the information on the route reconnaissance overlay.

Stream Width

I-27. There are many methods of determining (estimating) the stream's width. The most common and practical is the compass method. Begin by using a compass to take an azimuth from a point on the nearshore and close to the water's edge to a point on the opposite shore and close to the water's edge *(Figure I-8)*. On the nearshore, establish another point that is on a line and at a right angle to the azimuth selected. The azimuth to the same point on the far shore is plus or minus 45 degrees (800 mils) from the previous azimuth. Measure the distance between the two points on the nearshore. This distance is equal to the distance across the stream.

Current Velocity

I-28. Current velocities vary in different parts of a stream. Velocity is usually slower near the shore and

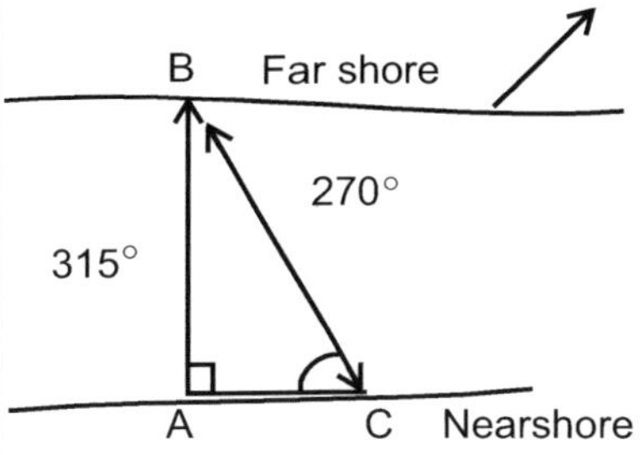

A - Point on nearshore
B - Point on far shore
AB - Distance to be measured

Azimuth of line AB is 315° (moving left, add 45°; moving right, subtract 45°).
Azimuth of line CB is 270°.
Difference between azimuth AB and azimuth CB is 45°.
Distance along AC equals distance along AB.

Figure I-8. Determine Stream Width

faster in the main channel of the stream. To estimate stream velocity—

- Measure a distance along the nearshore.
- Throw a light, floating object that will not be affected by the wind into the stream.
- Record the time that is takes for the object to travel measured distance.
- Repeat this procedure at least three times to establish an accurate average.
- Use the average time of the test in the following formula *(Figure I-9, page I-24)* to determine the stream's velocity.

Velocity (in meters per second) = measured distance (in meters) / average time (in seconds)

TEMPORARY ROUTE SIGNING

I-29. A military route sign system, like the US highway sign system, can enable road users to reach their destinations by following route signs and road markings displayed along the roadside. MP patrols monitor signs on a routine basis, checking specific signs before critical

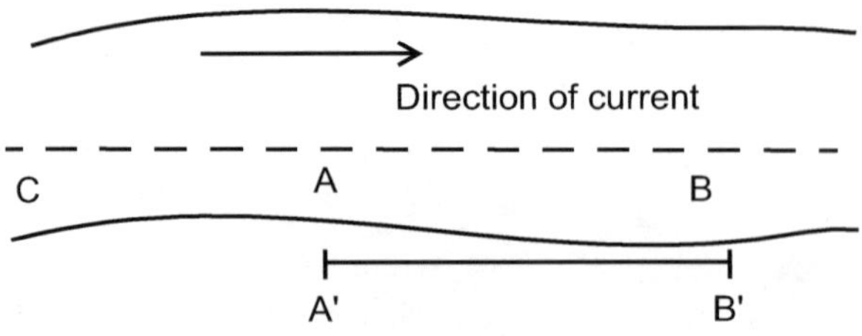

$$Velocity = \frac{A'B' \text{ (meters or feet)}}{\text{Time to float from } A \text{ to } B \text{ (seconds)}}$$

Measure distance A'B'.
Throw a floating object into the stream at C.
Determine the time required for the object to float to distance from A to B.

NOTE: This should be done three times, with the final velocity being the average of the three.

Figure I-9. Determine Stream Velocity

moves. Engineers erect permanent signs, but these signs can be damaged, destroyed, or moved by weather, saboteurs, and battle.

I-30. When MP on patrol encounter immediate and temporary MSR obstructions, like blown bridges or NBC contamination, they construct and erect signs quickly to guide vehicles around the obstruction. Prepare and post temporary signs to—

- Identify routes.
- Reroute traffic around problem areas.

- Help convoys and units move quickly and easily to their destinations even on an unfamiliar route.
- Show the drivers the locations of staging areas, tactical AAs, detours, key units, and facilities.
- Give directions, distances, and general information.
- Help lost military personnel find their way to the closest MP element.

TEMPORARY SIGNING

I-31. Obtain signing materials, signs, paint, and wire through the Army supply system. In an emergency, use boards, shingles, or cardboard. Use a portable sign-making kit to prepare signs when the signs will not conflict with standardization agreements (STANAGs).

I-32. Preplanned route signing for convoys and units traveling long stretches of MSRs and link routes must be planned well in advance. It is a time-consuming and manpower- and material-intensive operation. Signs used for this type of route are built by engineers or by MP battalion or company supply personnel. If tasked to take part in such an activity, ensure that after the signs have been constructed and coded for a particular route or operation, they are stored in a secure place to avoid compromise.

TEMPORARY ROUTE SIGNING

I-33. Place signs where they will support the traffic control plan and the traffic circulation plan. Specific sign locations are shown on the traffic control plan overlay prepared at the PM operations section. Changes in sign locations are reflected on the traffic control plan overlay as signs are added and deleted.

I-34. **Plan**. When MP encounter immediate and temporary MSR obstructions, like blown bridges or NBC contamination, they use the squad route sign kit or construct and erect hasty signs quickly to guide vehicles around the obstructions. Planning considerations include the following:

- Identifying routes.
- Rerouting traffic around problem areas.
- Helping convoys and units move quickly and easily to their destinations, even on unfamiliar routes.
- Showing drivers the locations of staging areas, tactical AAs, detours, key units, and facilities.
- Giving directions, distances, and general information.

I-35. **Execute**. Place signs where they will support the traffic control plan and the traffic circulation plan. Specific sign locations are shown on the traffic control plan overlay. Changes in sign locations are reflected on the traffic control plan overlay as signs are added or deleted.

I-36. Often, one three-man team can place signs along a route. To post a sign, a team member dismounts and walks 50 to 100 meters up the road. The other team members provide security and check and confirm the sign's placement to ensure that the drivers will be able to see the sign.

I-37. MP squads use the route signing kit to place signs along a route within the squad's AO. One team posts the signs while the other teams provide security. The squad leader checks and confirms the sign placement by traveling the entire route. He develops an overlay with the location of each sign in his squad's AO. Squad overlays are consolidated and contribute to a platoon traffic control plan.

I-38. More teams, up to a squad, may be needed for signing tasks in urban areas, in areas where a thorough reconnaissance has not been conducted, and in areas where the threat is unknown or is thought to be great. One team erects signs while following teams provide overwatch security. Use the following guidelines when placing signs:

- Place hazard signs about 150 meters before the hazard.
- Place regulatory signs where a regulation takes effect.
- Place all signs on the side of the road facing the traffic flow, about 1 meter off the traveled roadway. Conceal them from air view. If no cover is available, slant the sign stake forward.
- Place signs 1 to 2 meters above road level. Place all the signs at the same height if possible. Sign height is governed by roadside foliage, by whether the route is in an urban or a rural area, and by day or night use. In urban areas, place signs so that they are not hidden by vehicles or pedestrians, do not hinder pedestrians, and can be seen at night with street lighting or vehicle headlights. In open country, a good sign height is between thigh and knee height. This usually makes signs visible by day and night. Be sure that signs are not obscured by foliage.
- Use the least number of signs needed to be effective. Ensure that every sign is necessary and specific.
- Use more signs in urban areas than in rural areas.
- Use more signs on night routes than on routes primarily used during the day.
- Use signs to inform drivers to follow the common route when one road is used for two signed routes. Use signs to inform drivers when the routes diverge.

I-39. Conceal all the signs so that they are seen only from the direction from which they are approached.

There is no exact rule stating the distance from which a sign should be visible. However, the distance should be no greater than security allows and not less than is reasonable for those receiving directions.

I-40. Carefully conceal illuminated signs. Ensure that the light source is just strong enough to light the sign, but not strong enough to be seen from the air. This entails masking and covering the light sources. Consider placing chemical light sticks on top of the signs.

I-41. Place temporary route signs where they will provide warning and reaction time for the drivers. Do not block existing civilian signs. Place warning panels at convenient distances from where a route regulation takes effect. This distance can be shown on the panel (for example, BLACKOUT 500 METERS). In areas where blackout drive is the rule and drivers are using night-vision equipment, employ IR emitters on signs along routes to help guide the units to their destination. Just like any other light source, ensure that the IR emitters cannot be seen from the air.

I-42. Place guide signs at road junctions to prevent confusion. Put signs on both sides of the road if needed. Place confirmation signs *(Table I-3)* 150 meters beyond the critical road junctions to let drivers know that they are on the correct route.

Table I-3. Signs Identifying Military Routes and Locations

Type	Examples	Description	Purpose	Placement
Direction indicator		White, 30-centimeter diameter disk with black directional arrow (ID number or name is mounted below the disk) or White disk on black rectangular board with the identification number above	Indicates the direction and identifies the route or the name	At intersections where roads merge and where routes separate

Table I-3. Signs Identifying Military Routes and Locations (Continued)

Type	Examples	Description	Purpose	Placement
Guide signs (use together at important road junctions)		Rectangular with the symbols in white on black background; includes directional arrow and route number, name, and/or symbol	Indicates the locations, distances, directions, and routes	Where needed
Warning signs (use together at important road junctions)		Same as above	Shows the correct direction to take at route junctions	A sufficient distance (50 to 100 meters) before a junction to allow drivers to make the turn safely (On roads where speed is restricted, place the signs 25 meters before the junction.)

Table I-3. Signs Identifying Military Routes and Locations (Continued)

Type	Examples	Description	Purpose	Placement
Confidence signs		Same as above	Reassures the drivers that they are still on the correct route (Use them in urban areas to assure drivers that they are following the correct route. Use them on long stretches of road where it is unnecessary to use warning and confirmation signs for a considerable distance.)	Where needed
Confirmation signs		Same as above	Lets drivers know that they are on the correct route after changing direction	Just after turns, but visible while making the turn, if possible

Table I-3. Signs Identifying Military Routes and Locations (Continued)

Type	Examples	Description	Purpose	Placement
Countdown signs		Same as above	Warns of significant locations, such as SPs, RPs, starts and ends of routes, link routes, MSR junctions, and blackout areas—anything requiring a major change to the movement	Series of three signs at 100-meter intervals before the designated location
Regulatory signs		Military equivalent of civilian signs like STOP and YIELD	Regulates and controls the traffic on a route	Where needed (posted by engineers and considered permanent)

Table I-3. Signs Identifying Military Routes and Locations (Continued)

Type	Examples	Description	Purpose	Placement
Hazard signs		Yellow, diamond-shaped background with information printed in black	Indicates traffic hazards, such as dangerous corners, steep hills, or crossroads (rarely used in COMMZ, as civilian signs usually suffice. Use military signs in combat zones.)	Where needed (posted by engineers and considered permanent)
Military casualty evacuation route signs	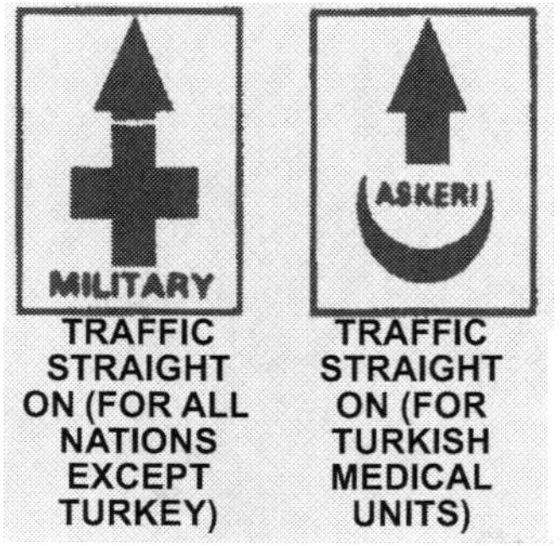	Rectangular white background with a red directional arrow, cross, or crescent; the word MILITARY; the unit or subunit designation; and other information like national markings or A directional disk with four segments cut out to form a cross or a directional disk with a crescent cut out (include the same information as above.)	Indicates evacuation route for military casualties	Where needed

Table I-3. Signs Identifying Military Routes and Locations (Continued)

Type	Examples	Description	Purpose	Placement
Civilian casualty evacuation route signs	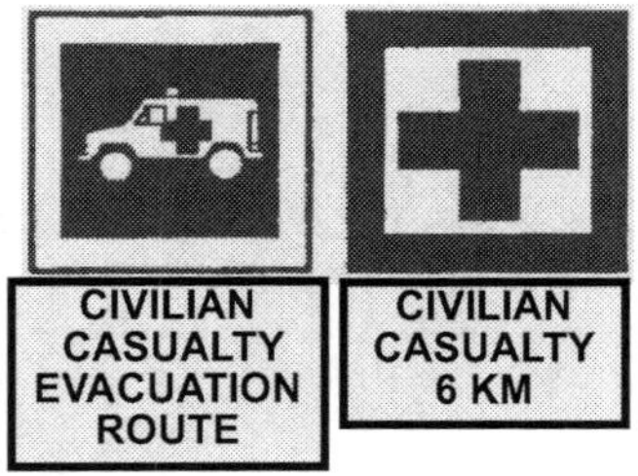	Blue Geneva Conventions information sign (includes ambulance in white with a red cross or crescent and the words CIVILIAN CASUALTY EVACUATION ROUTE beneath the sign in the HN language)	Indicates civilian casualty evacuation routes	Along routes for civilian traffic (designated by the HN)
Blackout warning signs		Based on the Geneva Conventions hazard warning sign (Place the legend and the distance on a rectangular plaque beneath the warning sign.)	Indicates the beginning of a blackout area	Same as warning signs

Table I-3. Signs Identifying Military Routes and Locations (Continued)

Type	Examples	Description	Purpose	Placement
Blackout enforcement signs		Geneva Conventions prohibitory sign; plaque beneath says VEHICLE LIGHTS FORBIDDEN	Indicates that a blackout is in effect	Every 100 meters along the blackout route
Blackout relaxation signs		Same as a blackout warning sign	Indicates the end of a blackout area	At the end of the blackout route

I-43. Place detour signs next to general traffic signs to identify the detour. Place the detour sign to the side (left or right) of the general sign that corresponds to the new direction to be taken.

I-44. Use signs to mark the entrance to a HQ or an installation along with a halt sign or other regulatory signs. Signs marking turnoffs and roads or tracks should include a directional disk or a directional arrow. Place signs to—

- Indicate where vehicles leave a signed route to get to the HQ or the installation.
- Mark the road or track leading to the HQ or the installation.
- Mark the entrance to the HQ or the installation.

SIGNING FOR KEY POINTS ON ROUTES

I-45. Place countdown signs *(Table I-3, page I-29)* at the beginning and end of a route. Clearly mark SPs and RPs with the appropriate countdown signs. Feeder routes to the SP may require signs to help the convoys find it.

I-46. Mark route detours with countdown signs showing the distance to where the detour begins. The signs clearly indicate the route to be detoured.

I-47. Place countdown signs so that they give clear warning of the end of the detour. Mark the end of the detour with a sign reading DETOUR END. Erect a warning sign at the end of the detour to show how to return to the original route.

I-48. Position detour signs ahead of time where terrain will require a defile or the like. Position the signs off the road, facedown on the ground. To implement the detour, erect the signs.

NOTE: The point where a link route meets the circuit is an ideal place for MP control. It is the point where the circuit and the link route begin and end.

I-49. Place signs at junctions of axial and lateral MSRs from all four directions. Place countdown signs 300, 200, and 100 meters before the junction.

I-50. To keep the number of signs to a minimum, battalion-size and smaller groups use a directional sign. The space for the military symbol may contain the UIN or the unit map-marking symbol. Print the UIN or the symbol so that it can be read when the arrow is vertical. This allows the sign to be used as a disk direction.

I-51. Therefore, only one type of sign is needed for all purposes. Put unit signs in place immediately before a unit movement. Remove them as soon as possible after the move.

MAIN SUPPLY ROUTE SIGNS

I-52. MSR signs identify MSRs by number, a pictorial symbol, or a name. Names and pictures—

- Are easier to identify and to remember than numbers.
- Prevent confusion with link-route signs marking the routes of units having three-figure identification numbers.
- Prevent security compromises by removing the chance use of an identification number already used on a map overlay.

I-53. There are two types of MSRs, axial and lateral, in a theater of operations. On the theater of operation's traffic circulation plan—

- Axial MSRs run to and from the forward edge of the battle area (FEBA) and are identified by odd numbers, like 87 or 215. Axial MSRs are shown as solid lines on the traffic circulation plan overlay. Axial routes are represented by a pictorial symbol on the route sign and are marked on the overlay as pictorial; for example,

MSR CLUB (PIC). On axial MSRs, *up* is toward the FEBA. It is shown on overlays and signs as a plain arrow. *Down* is away from the FEBA. It is shown as an arrow with a bar on the tail end.

- Lateral MSRs run parallel to the FEBA and are identified by even numbers, like 86 or 214. Lateral MSRs are shown as broken lines on the traffic circulation plan overlay. A named route sign represents lateral routes. The names are short, three- or four-letter words like *fox*, *ant*, or *hen*. *Up* and *down* on lateral MSRs show only general directions of travel. The general direction shown by *up* or *down* varies with the theater of operation and with the location of the FEBA. *Up* is usually to the north or east; d*own* is usually to the south or west. A plain arrow on an overlay or sign indicates *up*; a barred arrow shows *down*. To avoid confusion on lateral MSRs, use the letters N, E, S, W, NE, SE, NW, and SW on route signs to show the general direction of movement.

This paragraph implements STANAG 2010.

I-54. Ensure that all route signs are large enough to be read easily in poor light. Size is not specified, but—

- Signs for international use cannot be less than 40 centimeters by 33 centimeters.
- Bridge classification signs must conform to STANAG 2010.

LINK ROUTE SIGNS

I-55. There is no set requirements for the design of link route signs *(Figure I-10)*. (Sometimes showing direction with a white arrow on a black background is enough.) A link route connects a unit or an activity to an MSR. A departing convoy follows the link route signs to the MSR. The convoy follows the MSR until guided off the MSR by signs warning of the need to exit and again follow link route signs until arriving at its destination. That part of a traveled route coinciding with an MSR will not have link route signs.

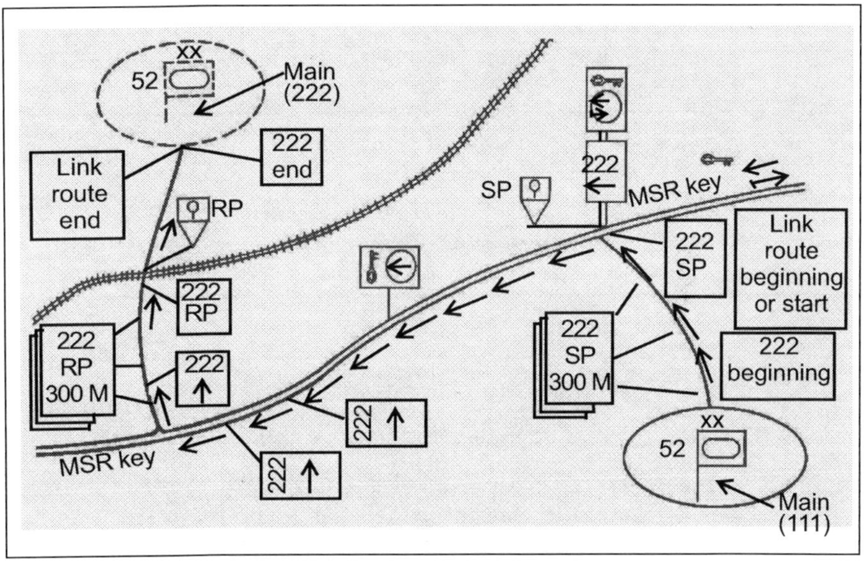

Figure I-10. Link Route Signing

HEADQUARTERS AND INSTALLATION SIGNS

I-56. All signs must be used in HQ and parts of HQ, down to battalion or equivalent level. Examples are—

- Medical installations.
- Water points.
- Ammunition depots.
- TCPs.
- Decontamination stations.
- Fuel installations.

This paragraph implements STANAG 1059.

I-57. HQ and installation signs show the military symbol for the HQ or the installation. For HQ at brigade level and above, include the national distinguishing letters given in STANAG 1059. The sizes of the sign and the symbol are not standardized. The symbol and the background must be clearly contrasting colors. The actual colors used are left to the discretion of each nation.

DIRECTIONAL SIGNS

I-58. A directional sign *(Table I-3, page I-29)* must be—

- A white disk displaying a black directional arrow accompanied by a route identification number or name that can be mounted beneath the disk.
- A white disk mounted or superimposed on a black rectangular board with the number or name on the board above the disk.

I-59. Directional disks cannot be less than 30 centimeters in diameter. Drill eight holes in the disks at equal intervals around the circumference so that they can be erected with the arrow pointing in the correct direction.

I-60. Directional signs show directions for axial and lateral routes on hastily prepared routes or very temporary routes like detours. The initial sign shows the use of disk directional signs and orders the drivers to follow the disk direction.

I-61. An MSR may require hundreds of signs. Make the signs in quantity without the black arrow. Stencil or stick the arrows on later. The superimposed signs are easiest to store and use. It is hard to secure the two elements of the circular two-piece signs, and it is hard to position that sign for display.

I-62. Use yellow instead of white for signs used during prolonged snowfall conditions or for signs permanently erected where there can be prolonged snowfalls.

PORTABLE SIGN-MAKING KIT

I-63. The new portable sign-making kit (national stock number [NSN] 9905-01-478-3723) is lightweight and simple to use and contains tools and materials that, when used in conjunction with vehicle BII and locally available materials, produce temporary signs in support of MP operations. Since the kit is not fully self-contained, the user must plan ahead. The new, lighter kit gives MP units the ability to pack according to the METT-TC. Even though a significant difference exists between signs made with the kit and specifications in STANAGs (STANAG route signs generally have a black background with white letters, and the kit's sign background material is white), it does not differ from the kit it replaces. The soft-sided sign kit comes with white plastic bags that can be used to make signs up to 21 inches wide by 23 inches high. *Figure I-11,* shows a sample sign; and *Table I-4, page I-42,* shows a component listing for sign making.

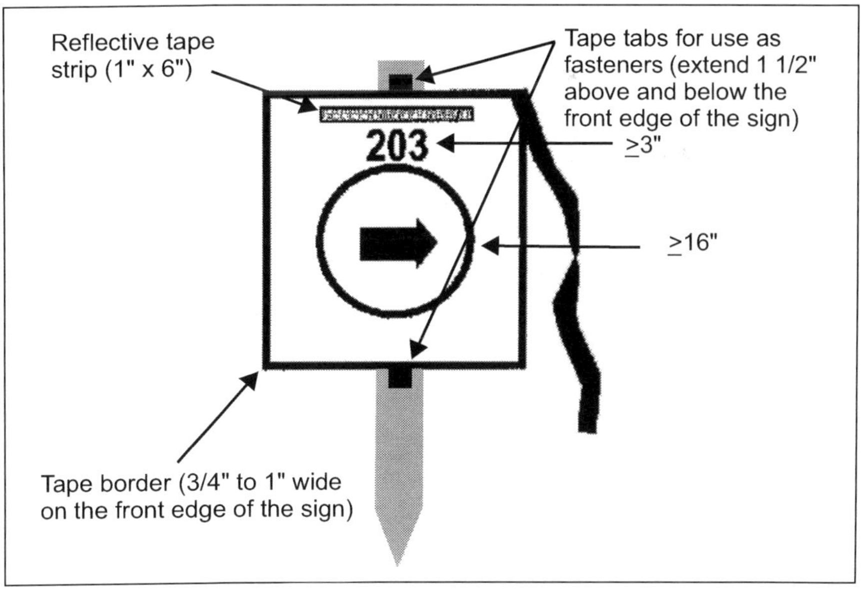

Figure I-11. Sample Sign

Table I-4. Component Listing

Tools	NSN	
Tool bag	5140-01-481-4460	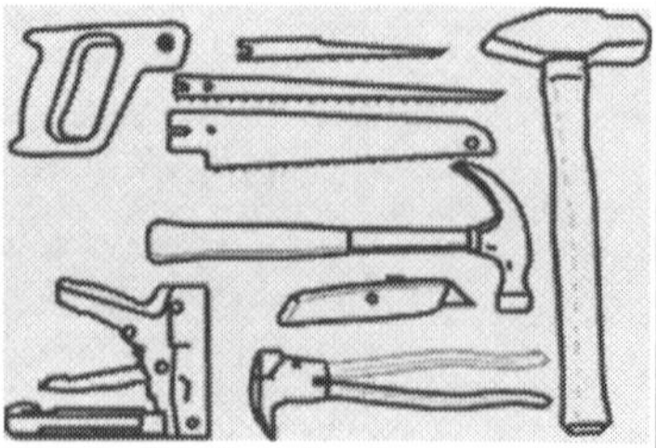
Carpenter's hammer	5120-01-112-8350	
Utility knife Replacement blades	5110-00-892-5071 5110-00-293-2865	
Linemans tool	5120-00-293-3485	
Blacksmith's/engineer's hammer (3 lb)	5120-00-900-6103	
Nested saw set	5110-00-293-0090	
Staple gun	5120-00-889-1796	
Fasteners	**NSN**	
One pound of 1.75" steel screw-shank nails (concrete)	5315-00-543-3023	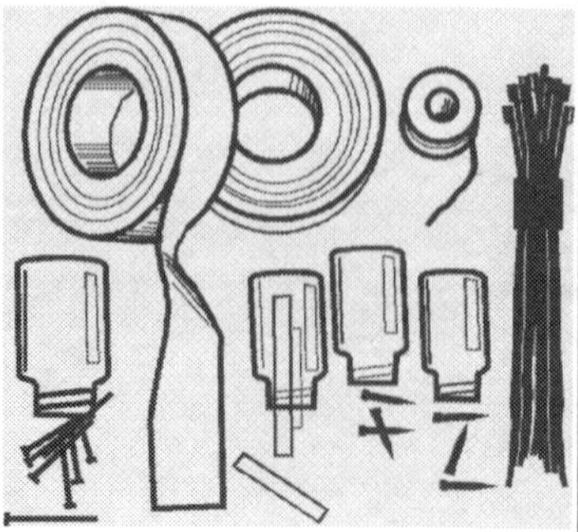
Five pounds of 3.5" (10-12d) common nails	5315-00-753-3885	
One-half pound of 1.5" steel barbed-shank nails (roofing)	5315-01-381-9609	
Tape, pressure sensitive (2" x 160', 100 miles an hour)	7510-00-074-5124	
Staples, 1/2", box of 5,000	5315-00-889-2605	
Cable ties, pkg of 100	5975-00-156-3253	
Wire, nonelectrical, 1-lb roll	9505-00-293-4208	

Table I-4. Component Listing (Continued)

Marking Materials	NSN	
Black tube-type poster markers (2 each)	7520-01-336-2856	
Red tube-type poster marker	7520-01-336-2857	
Blue tube-type poster marker	7520-01-336-2858	
Bags, white plastic, trash, 8 gal, roll/box of 20 to 30 (2 each)	8105-01-481-4225	
Reflective tape (white 1" x 150')	9390-00-949-8044	
Ribbon, flagging, surveyors (arctic-glo pink)	9905-00-194-9703	

Appendix J

Nuclear, Biological, Chemical Defense

The technology to produce and deliver chemical and biological agents and weapons of mass destruction is widely available. This technology provides military and nonmilitary leaders in unstable regions with a means to deter more technically advanced nations. Weapons of mass destruction can radically change the flow of battle, shift the balance of combat power, and place a superior force at risk. Areas targeted with these weapons may remain hazardous for days, even weeks after an attack. Additionally, nuclear reactor and chemical production facility accidents or sabotage can release radiological or chemical contaminants that can threaten friendly forces and military operations.

OVERVIEW

J-1. Enemy forces may threaten sustainment areas to seize and maintain the initiative, while degrading or eliminating a unit's flexibility and capability to support operations. Sustainment area activities are lucrative targets for enemy NBC attacks. Attacking sustainment nodes weakens the main battle area's force effectiveness without adversely affecting an enemy's immediate maneuver. To achieve these aims, threat activities in sustainment areas target key critical support and logistic facilities. These facilities include—

- Special-weapons storage sites and delivery systems.

- C^2 facilities.
- Air defense artillery sites.
- Air bases.
- Seaports.
- MSRs.

HAZARD DETECTION AND REPORTING

J-2. The best units to perform NBC reconnaissance are those units with the M93 NBC Reconnaissance System (FOX). Planners consider the disadvantages of conducting NBC reconnaissance with HMMWV-equipped units. The disadvantages of using a HMMWV include the following:

- The crew conducts operations in MOPP 4 while in contamination.
- The vehicle stops to conduct survey and detection.
- The vehicle has difficulty keeping up with the maneuver force over rugged terrain.
- The readings obtained using an M256A1 kit, detector paper, and a chemical agent monitor (CAM) are not as reliable as readings from the M93 (FOX).

J-3. Planners use the IPB and the METT-TC to determine the type of NBC reconnaissance technique and the type of equipment to be used for the NBC reconnaissance.

J-4. Early detection of NBC hazards is required for timely warning of units and personnel in the affected areas. NBC reconnaissance is an active contamination avoidance measure that provides commanders with information on NBC hazards in an AO.

J-5. NBC reconnaissance contributes to the overall intelligence collection effort and consists of the following five critical tasks: detecting, identifying, marking, reporting, and sampling. The four types of NBC reconnaissance include search, survey, sample, and surveillance.

J-6. MP may be tasked to conduct NBC reconnaissance while conducting other reconnaissance operations. They are equipped to detect and identify radiological and chemical hazards, mark the contaminated area, and report the hazard. They are not equipped to collect or handle samples.

J-7. NBC surveillance activities are planned to support the intelligence collection effort. MP conduct NBC surveillance by monitoring MSRs, critical facilities, and key terrain. They employ NBC detector equipment and measures that include the following:

- M8A1 automatic chemical agent alarm system.
- AN/VDR-2 RADIAC set.
- CAM.
- Visual observation for unexplained artillery or spray.
- M256A1, M8, and M9 paper.

J-8. If MP observe the indications of an NBC attack, they may be ordered to conduct an NBC survey to define the boundaries of the contamination, locate and mark clean bypass routes, and occupy OPs to observe the designated area. MP conduct an NBC survey, on order, to obtain detailed information concerning a contaminated area, including the location of the general boundaries of the contaminated area and the intensity of the contamination (nuclear) or the type of agent (chemical).

J-9. MP locate the boundaries and place warning markers around the contaminated area at all the entry points. They report information using an NBC 4 report and inform the road users of the contaminated area, and direct the traffic to alternate routes to avoid the contaminated area.

J-10. At all echelons, MP provide continuous information to higher HQ using NBC 1 and NBC 4 reports *(Figure J-1)*. These report formats provide a rapid means of disseminating information. Use the NBC 1 report to record the initial use and subsequent data concerning enemy NBC attacks. The initial NBC 1 report precedence is flash and all others are immediate. *Table J-1, pages J-6 through J-10,* gives an explanation of the various lines in an NBC report.

J-11. Use the NBC 4 report for radiation dose rate measurements. Usually, the unit submits two reports—one on initial contact and another for peak dose rate. Measure radiation dose rates in the open, 1 meter above the ground. If radiation dose rates are taken inside a vehicle or a shelter, at least one outside reading is necessary to determine the correlation factor. Other items of the report are optional. Refer to *FM 3-3* for more information about NBC reports.

J-12. As MP move throughout an AO, they monitor for the presence of an NBC hazard, mark contaminated areas, send NBC reports through operational channels, and direct traffic around or through hazards to ensure that the troops and the supplies get where they are needed on the battlefield.

CONTAMINATION MARKING

J-13. Mark contamination so that unsuspecting personnel will not be exposed to it. When detecting or suspecting NBC hazards, mark all likely entry points into the area and report the contamination to higher HQ. The only exception to this policy occurs when marking the area would help the enemy. If the commander makes this exception, report the hazard to protect friendly units.

NBC 1 Report (Observer's Report)		
NBC 1 (Nuclear) Report	**NBC 1 (Biological) Report**	**NBC 1 (Chemical) Report**
B. NB062634	B. LB206300	B. LB200300
C. 90-degree grid	D. 200410Z	D. 201405Z
D. 201405Z	E. 200414Z	E. 201412Z
G. Aircraft	F. LB206300, actual	F. LB206300, estimated
H. Surface	G. Aerial spray	G. Bomblets
J. 60	H. Unknown	H. Nerve
L.15 degrees		V. Airburst
NOTE: Line items B, D, H, and either C or F should always be reported. Use other line items if the information is known.		

NBC 4 Report (Reconnaissance, Monitoring, and Survey Results)	
NBC 4 (Nuclear) Report	**NBC 4 (Chemical) Report**
Q. LB123987	H. Nerve V
R. 35	Q. LP200300, liquid
S. 201535Z	S. 170610Z
NOTES: **1. Line items H, Q, R, and S may be repeated as often as necessary.** **2. In line R, descriptive words, such as initial, peak, increasing, decreasing, special, series, verification, or sumary may be added.**	

Figure J-1. Example of NBC 1 and NBC 4 Reports

Table J-1. Explanation of Line Items in NBC Report

Line	Nuclear	Chemical and Biological	Remarks
A	Strike serial number	Strike serial number	Assigned by the NBC center
B	Position of the observer	Position of the observer	Grid coordinates (universal transverse mercator [UTM] or place).
C	Direction of the attack from the observer, to include the unit of measure	Direction of the attack from the observer	Nuclear: Use degrees magnetic north (DGM), degrees true north (DGT), degrees grid north (DGG), mils magnetic north (MLM), mils true north (MLT), or mils grid north (MLG) Chemical: Measure clockwise from the grid or magnetic north (in degrees or mils)
D	DTG of detonation	DTG for the start of the attack	Nuclear: Use Zulu time. Chemical: Designate time zone used.
E	NA	DTG of the end of the attack	Designate time zone used.
F	Location of the area attacked	Location of the area attacked	Use grid coordinates (or place). State whether the location is actual or estimated.
G	Suspected or observed event and the means of delivery or kind of attack	Kind of attack	State whether the attack was by artillery, mortars, rockets, missiles, bombs, or spray.

Table J-1. Explanation of Line Items in NBC Report (Continued)

Line	Nuclear	Chemical and Biological	Remarks
H	Type of burst	Type of agent and type of burst P = persistent NP = nonpersistent	Nuclear: Specify air, surface, or subsurface. Chemical: State whether by air, ground, or spray attack.
I	NA	Number of munitions or aircraft	If known.
J	Flash-to-bang time	NA	Use seconds.
K	Crater present or absent and the diameter	Description of terrain and vegetation	Nuclear: Sent in meters. Chemical: Sent in NBC 6.
L	Cloud width at H+5 minutes	NA	State whether measured in degrees or mils.
M	Stabilized cloud top or cloud bottom angle or cloud top or bottom height at H+10 minutes	NA	Nuclear: State whether the angle is cloud top or cloud bottom and whether it is measured in degrees or mils. Chemical: State whether the height is cloud top or cloud bottom and whether it is measured in meters or feet.
N	Estimated yield	NA	Sent as kilotons.
O	Reference DTG for estimated contour line not H+1	NA	Use when the contours are not plotted at H+1.
P	For radar purposes only	NA	None

Table J-1. Explanation of Line Items in NBC Report (Continued)

Line	Nuclear	Chemical and Biological	Remarks
PA	NA	Predicted hazard area (coordinates)	If the wind speed is 10 kilometers per hour or less, this item is 010 (the radius of the hazard area in kilometers).
PAR	Coordinates of external contours of the radioactive cloud	NA	Six-digit coordinates. Letter R identifies RADAR set.
PB	NA	Duration of the hazard in the attack and hazard area	State in days, hours, minutes, and so forth
PBR	Downwind direction of the radioactive cloud and the unit of measure	NA	DGM, MLM, DGT, MLT, DGG, or MLG. The letter R identifies a RADAR set.
Q	Location of the reading	Location of the sampling and the type of sample	Nuclear: UTM or the place. Chemical: UTM or the place. State whether the test was air or liquid.
R	Dose rate or actual value of decay exponent	NA	State the dose rate in centigray per hour (cGyph).
S	DTG of the reading	DTG contamination detected	State the time of the initial identification test sample or when the reading was taken.

Table J-1. Explanation of Line Items in NBC Report (Continued)

Line	Nuclear	Chemical and Biological	Remarks
T	H+1 DTG	DTG of the latest contamination survey of the area	NBC 5 and NBC 6 reports only.
U	1,000-cGyph contour line	NA	Plot in red.
V	300-cGyph contour line	NA	Plot in green.
W	100-cGyph contour line	NA	Plot in blue.
X	20-cGyph contour line (30 cGyph contour line is used by other NATO forces)	Area of actual contamination	Nuclear: Plot in black. Chemical: Plot in yellow.
Y	Direction of the left and right radical lines	Downwind direction of the hazard and the wind speed	Nuclear: Direction measured clockwise from grid north (GN) to the left and then right radial lines (degrees or mils, state which), four digits each. Chemical: Direction is four digits (degrees or mils) and the wind speed is three digits (kilometers per hour).

Table J-1. Explanation of Line Items in NBC Report (Continued)

Line	Nuclear	Chemical and Biological	Remarks
Z	Effective wind speed Downwind distance of Zone 1 Cloud radius Unit of measure for each category	NA	Three digits—effective wind speed (kilometers per hour) Three digits—downwind distance of Zone 1 (kilometers or nautical miles) Two digits—cloud radius (kilometers or nautical miles) (the wind speed is less than 8 kilometers per hour, this line contains only a three-digit radius of Zone 1 [kilometers])
ZA	NA	Significant weather phenomena	Air stability: Two digits Temperature in centigrade: Two digits Humidity: One digit Significant weather phenomena: One digit Cloud cover: One digit
ZB	Used to transmit correlation factors or transmission factors	Remarks for additional chemical and biological attacks only.	Include any additional information.

PROCEDURES FOR MARKING

J-14. Face markers away from the contamination. For example, if markers are placed 50 meters from the outside edge of a contaminated area to mark a radiological hot spot, they face away from the point of the highest contamination reading. Place markers at roads, trails, and other likely points of entry. When time and mission permit, use additional markers. The distance between the signs varies. In open terrain, place them farther apart than in hilly or wooded areas. Soldiers should be able to stand in front of a marker and see the markers to the left and right of it.

J-15. Units discovering a marked, contaminated area do not have to conduct elaborate, time-consuming surveys. The new unit checks the extent of the contamination and alters its plans if necessary. If the size of the hazard has expanded or decreased, relocate the signs. If the hazard is gone, remove the signs and report changes to higher HQ.

TYPES OF MARKERS

J-16. US forces use standard NATO markers to make it easier for allies to recognize the hazards. These markers are in the standard NBC marking set. Colors and inscriptions on a marker indicate the type of hazard. Other contamination information is written on the front of the sign.

RADIOLOGICAL CONTAMINATION DETECTION AND MONITORING

J-17. Conduct radiological monitoring using an AN/VDR2, and start on the order of the commander or as set by the SOP. Mark the area with radiological contamination markers. Report radiation dose rates and the times and locations of the readings to higher HQ using the NBC 4 nuclear report. All units start continuous monitoring when–

- Moving from one area to another on the battlefield.

- Getting a fallout warning.
- Knowing that a nuclear burst was reported, seen, or heard.
- Detecting radiation of 1 centigray per hour by periodic monitoring.
- Being ordered by the unit commander.

J-18. Continuous monitoring stops on order from the higher HQ or when the dose rate falls below 1 centigray per hour, except for units on the move.

TECHNIQUES FOR MONITORING

J-19. Use both direct and indirect techniques when conducting radiological monitoring. Refer to *FM 3-3* for more information on radiological monitoring.

Direct Monitoring

J-20. The direct technique is the simplest and the most precise. The soldier must—

- Stand at the desired location.
- Hold the RADIAC meter waist high and turn it slowly 360 degrees. Refer to *TM 11-6665-251-10*.
- Record the highest reading on *DA Form 1971-R*.
- Take the reading in the open, at least 10 meters from buildings or large structures, if possible.

Indirect Monitoring

J-21. Use the indirect technique inside shelters or vehicles. The soldier must—

- Stand at the center of the shelter.
- Hold the RADIAC meter 1 meter above the floor and slowly rotate it 360 degrees.
- Record the highest reading.
- Take all readings from one selected position when monitoring from inside a vehicle or shelter.

SURVEY TEAM

J-22. MP may be tasked to perform as part of a survey team. When essential contamination information cannot be obtained from monitoring reports, a radiological survey may be required. Radiological surveys are directed efforts to learn the extent and intensity of radiological contamination. A survey requires a control team and one or more survey teams. The HQ directing the survey usually provides the control team. In a team assigned to a survey—

- The team leader monitors the RADIAC meter and provides C^2.
- The driver operates the vehicle.
- The gunner provides security.

J-23. The control team instructs the survey teams on what to do. The MP team leader ensures that each team member receives the following information from the control team:

- Purpose. The survey determines the presence and the level of fallout or induced radiation near ground zero, in a specific area, along a route, at a specific location, or any combination of these.
- Start and finish times. The times during which a survey will be conducted.
- Survey execution. The route to be followed and the locations where readings are to be made or the distances needed between the readings are given.
- Radiation safety precautions. The control team tells the survey team the turn-back dose. The dose rate is monitored on the AN/VDR2. If the turn-back dose rate is reached as the survey team travels forward, the team immediately leaves the fallout area by the same route used to enter it, unless instructed otherwise. If the dose rate decreases as the survey team moves ahead, the team continues to perform its mission.

- Recording limitations. The survey team will not convert inside readings to outside readings. Record readings in the dose rate column only.
- Communications. The method and times to be used to communicate survey data to the chain of command.
- Special instructions. Any special instructions for marking the contaminated areas or special security precautions.

CHEMICAL AGENT DETECTION

J-24. During routine operations, MP may detect chemical agents while making periodic checks for contamination. Or as previously discussed, teams may be tasked to conduct a chemical reconnaissance along an MSR or in a given area after an enemy attack. Some chemical agents are odorless, colorless, tasteless, and invisible. Teams must use chemical agent detection equipment and materials to be certain that all chemical agents are detected.

PREPARATION

J-25. When assigned a mission or a task to locate and mark a suspected contaminated area, MP leaders must ensure that they prepare properly for the mission. Preparation for any NBC reconnaissance mission begins with the inspection of personnel and equipment. As a minimum, each squad must have the following equipment on hand:

- M8 paper.
- M9 paper.
- M256 and M256A1 detector kits.
- M8A1 alarm.
- CAM.
- Marking kit.
- M13 decontamination apparatus (DAP).

- M291 decontamination kit.
- MOPP gear.
- Mark 1 nerve agent auto injector kit (NAAK).
- VS-17 marking panels.

J-26. In addition to ensuring that the proper equipment is on hand, leaders must ensure that alarms and paper are properly mounted and functional and that all external equipment is stowed. The platoon leader includes a rehearsal of NBC reconnaissance techniques when preparing the mission. The platoon leader coordinates with the unit chemical officer for any special instructions, ensuring that thorough decontamination support is available at the conclusion of the mission.

RECONNAISSANCE

J-27. When detecting for chemical contamination during NBC reconnaissance, the MP leader—

- Ensures that all members of the party wear MOPP 4 with M9 paper attached to their protective clothing and vehicle.
- Drives along the route until the chemical agent alarm sounds. By the time the alarm sounds, the team may be some distance from where the agent was first encountered. Keep this in mind when selecting the area to be searched.
- Gives special attention to shell craters, low-lying patches of woods, defiles, ravines, streambeds, and areas covered with high grass or underbrush.
- Has one MP remain in each vehicle to provide security and to monitor the radio.
- Has other patrol members dismount and check for contaminants.
- Ensures that one person uses M8 detector paper on the tip of a stick.

- Ensures that one person operates the M256 chemical agent detector kit. (Refer to *TM 3-6665-307-10* for detailed operating instructions for the M256, and refer to *FM 3-6* for more information.)

J-28. When MP find chemical contamination, they mark the area with NBC markers. They service the automatic chemical agent alarm to ready it for further operation. MP must be outside the contaminated area when the alarm is serviced. *TM 3-6665-312-12&P* provides instructions for operators and organizational personnel. MP remount the vehicles and continue the reconnaissance. While moving, they send an NBC 4 report by radio and follow this procedure each time a chemical agent is detected. The goal of the reconnaissance is to define the contaminated area only to the degree necessary to provide friendly forces with the information needed to avoid the area.

BIOLOGICAL AGENT DETECTION

J-29. For warning of biological attacks or contamination outside a unit's immediate area, the unit relies on warnings relayed through its chain of command from division level or higher units. MP help provide warning of biological hazards by their reports of area activity passed on from their route, zone, and area reconnaissance.

J-30. Be alert to any sign that biological agents are being used. Promptly report any unusual occurrences of sickness in troops or civilians. Report any unusual actions of animals, birds, or large numbers of dead animals or birds not likely to have been caused by combat. Observe the types of weapons and munitions used during attacks. Compare them to the known characteristics of enemy biological attacks. Refer to *FM 3-3* for detailed information on biological agent detection, identification, and defense.

J-31. Biological attacks are most likely to occur at night or during extended twilight to avoid toxin exposure to direct sunlight. (Direct sunlight degrades biological agents.) Cloudy and foggy days are also favorable for the use of biological agents.

J-32. Watch for the following indications that a biological attack has occurred:

- A low-flying aircraft that appears to be producing a mist or a spray.
- The use of any type of spray device.
- The use of ammunition that does not seem to have any immediate effect, such as a bomblet.
- Unusual types of bomblets.
- Insect swarms that suddenly appear after an aircraft drops containers that do not seem to have any immediate effects.

SELF-DEFENSE MEASURES

J-33. MP not only provide early warning for units in the sustainment area, but they also employ NBC defense measures to protect their own forces. NBC defense is a balance of the following three principles:

- Avoidance.
- Protection.
- Decontamination.

AVOIDANCE

J-34. Avoiding NBC attacks and hazards is the key to NBC defense. Avoiding contamination involves bypassing contaminated areas and avoiding detection by the enemy. If forces avoid detection, they are less likely to be targeted for attack and will not need to apply protection and decontamination. *FM 3-3* provides detailed guidance on conducting contamination avoidance procedures.

J-35. NBC contamination avoidance is especially critical for light forces. Leaders conduct risk assessments to determine what protective measures they will direct. Refer to *FM 3-4* for detailed information on conducting a risk assessment.

PROTECTION

J-36. A unit may become contaminated because of direct NBC attack or because the mission causes them to enter a contaminated area. In either case, NBC protection is an integral part of operations. Avoidance and protection are closely related. Techniques that work for avoidance also provide protection. There are four broad groups of activity that comprise protective measures and include the following:

- Hardening positions and protecting personnel.
- Assuming MOPP.
- Reacting to attack.
- Using collective protection.

J-37. Refer to *FM 3-4* for detailed guidance on conducting NBC protective procedures.

DECONTAMINATION

J-38. The extent and timing of decontamination depend on the tactical situation, the mission, the degree, the type of contamination, and the resources available. Refer to *FM 3-5* for detailed guidance on conducting decontamination operations.

J-39. The primary purpose of decontamination is to stop erosion of combat power and to reduce the number of casualties. Combat power drops as soon as personnel don MOPP gear. The mask restricts the field of vision and reduces observation and target acquisition ability. Communication is more difficult. Mobility is reduced because personnel reduce their physical work rate to avoid heat stress. The longer a unit remains contaminated, the greater the chance of NBC casualties. The three levels of decontamination are immediate, operational, and thorough.

Immediate Decontamination

J-40. Immediate decontamination is a basic soldier survival skill carried out by soldiers as soon as possible after they discover they are contaminated. Its basic purposes are to minimize casualties, save lives, and limit the further spread of contamination. Personnel treat any contact between chemical or biological agents and bare skin as an emergency. Some agents can kill if they remain on the skin for longer than a minute. The best technique for removing or neutralizing these agents is to use the M291 skin decontamination kit. Leaders must ensure that their soldiers are trained to execute this technique automatically, without waiting for orders.

J-41. Begin personal wipe down within 15 minutes of contamination. The wipe down removes or neutralizes contamination on the hood, mask, gloves, and personal weapon. For chemical and biological contamination, use mitts from the M295 individual equipment decontamination kit (IEDK). For radiological contamination, wipe off the contamination with a cloth or simply brush or shake it away.

J-42. Begin the operator's spray down of the equipment immediately after completing the personal wipe down. The spray down removes or neutralizes contamination on the surfaces that operators must touch frequently to perform their mission. For chemical and biological contamination, use on-board decontamination apparatuses, such as the M11, the M13, or the M295 IEDK to decontaminate surfaces where DS2 cannot be applied.

NOTE: Wash decontamination solution number 2 (DS2) off surfaces no more than 30 minutes after application. If necessary, use 5-gallon water cans or other water sources to assist in removing DS2.

J-43. For radiological contamination, brush or scrape away the contamination with whatever is at hand or flush it with water and wipe it away.

Operational Decontamination

J-44. Operational decontamination allows a force to continue fighting and sustain its mission after being contaminated. It limits the hazard of transferring contamination by removing most of the gross contamination on equipment and nearly all the contamination on soldiers. This speeds the weathering process and allows clean areas (people, equipment, and terrain) to stay clean. Following operational decontamination, soldiers who have removed sources of vapor contamination from their clothing and equipment unmask temporarily in hazard-free areas to eat, drink, and rest.

J-45. Accomplish operational decontamination using the assets of the parent unit. Make use of the following decontamination techniques:

- Vehicle wash down.
- MOPP gear exchange.

J-46. Conduct these procedures at the same time. They are best performed at squad level. Vehicles and personnel that are not contaminated should not go through either technique. Conduct vehicle wash down as far forward as possible. It is normally performed using the power-driven decontamination equipment (PDDE) crew with assistance from the squad decontamination crew. It is most effective if started within 1 hour after contamination.

J-47. The two steps in vehicle wash down are as follows:

Step 1. Button up the vehicle and secure the equipment.

Step 2. Wash down the vehicle and the equipment with hot, soapy water for 2 to 3 minutes.

J-48. Because speed is important, do not check vehicles for contamination after vehicle wash down. Remove only gross contamination.

Thorough Decontamination

J-49. Thorough decontamination reduces contamination to negligible risk levels. It restores combat power by removing nearly all contamination from the unit and individual equipment. This allows troops to operate equipment safely for extended periods at reduced MOPP levels. A contaminated unit conducts detailed troop decontamination (DTD) under the supervision of the chemical unit. Ordinarily, the chemical unit selects a site, sets it up, and performs the detailed equipment decontamination (DED) with assistance from the contaminated unit. A small risk from residual contamination remains, so make periodic contamination checks after this operation.

J-50. Conduct thorough decontamination as part of an extensive reconstitution effort in brigade, division, and corps support areas. Support sites at lower levels cannot provide the quantities of decontamination resources (water, decontaminants, and time) required for such an extensive process. In some cases, a contaminated unit could conduct a thorough decontamination operation with organic decontamination assets, but support from a chemical unit is usually required.

J-51. After thorough decontamination, the unit moves into an adjacent AA for reconstitution. Support elements from the brigade, division, or corps support areas replenish combat stocks, refit equipment, and replace personnel and equipment. The newly reconstituted unit leaves the AA fully operational and fit to return to battle.

J-52. Thorough decontamination does the best job of getting rid of contamination and its hazards, but it is often not possible. It requires large quantities of valuable resources that may not be immediately available. The next best solution is to decontaminate only what is necessary to sustain the force and continue to fight.

MILITARY POLICE LEADERS' RESPONSIBILITIES

J-53. MP leaders ensure that all soldiers understand the concepts of NBC defense and the skills necessary to survive an attack. Individual NBC skills include–

- Recognizing hazards and taking cover.
- Using protective measures and MOPP gear.
- Knowing avoidance measures, first aid, and decontamination.

J-54. MP commanders also designate and train special-purpose teams. These special-purpose NBC teams include the following:

- Chemical agent detection teams.
- Radiological survey and monitoring teams.
- Decontamination teams.

J-55. To assist with NBC defense, MP commanders rely on the unit NBC officer and the NBC NCO. They—

- Provide technical assistance to the commander on NBC defense training and operations.
- Coordinate unit NBC defense activities.
- Give NBC defense instructions to achieve basic operating standards of proficiency for the unit and individuals.
- Plan and supervise NBC defense training aspects of operational training exercises and maneuver.
- Supervise the preparation of unit NBC defense SOPs.
- Supervise the operation and maintenance of NBC material.
- Determine by dosimetry or calculation (as appropriate) the total dose of radiation and the time spent in radiologically contaminated areas to avoid exceeding command exposure guidance.
- Prepare fallout prediction patterns.

- Plan NBC reconnaissance and advise commanders on the best routes to cross or by-pass an NBC contaminated area.
- Plan and coordinate decontamination within the unit.
- Maintain records of the unit's radiation exposure.
- Estimate downwind hazards for chemical attacks.
- Report NBC data to the next higher HQs.
- Analyze the unit's vulnerability to an NBC attack.
- Evaluate individual and unit competence in NBC defense, and advise the commander on the unit's ability to survive and continue operations in an NBC environment.
- Advise the commander on all matters pertaining to unit NBC defense.

MISSION-ORIENTED PROTECTION POSTURE LEVELS, ALARMS, AND SIGNALS

J-56. Soldiers on the integrated battlefield face a combination of NBC and conventional attacks. Individual and unit protection against chemical attack or contamination hinges on effective use of MOPP gear and on individual proficiency in basic NBC skills. All MP must be familiar with the standard MOPP levels shown in *Figure J-2, page J-24,* and *Table J-2, page J-25*.

J-57. When an NBC attack is recognized, every soldier receives the warning and assumes the appropriate MOPP level. Those in immediate danger need warnings they can see or hear. The alarm or signal must be simple and unmistakable for quick and correct reaction. Units not immediately affected need the information to prepare for the hazard or to change plans.

Figure J-2. Illustrated MOPP Levels

J-58. Mark the contaminated area if an NBC hazard has been located. The NBC Warning and Reporting System (NBCWRS) and the contamination markers contribute to the warning procedures for follow-on forces. In the immediate area of contamination, several methods (or a combination of methods) will allow quick reaction by all platoon members.

J-59. These methods include vocal alarms (shout of *GAS*), the M8A1 alarm, nonvocal alarms (horn blast or banging of metal-to-metal objects), and visual alarms (most commonly, hand-and-arm signals). The tactical situation may not allow for audible alarms; therefore, the platoon SOP should clearly detail the visual signals for contamination.

BIOLOGICAL DEFENSE

J-60. The key protective measure against a biological attack is maintaining a high order of health, personal

Table J-2. Standardized MOPP Levels

MOPP Equipment	MOPP Levels						Command
	MOPP Ready	MOPP Zero	MOPP 1	MOPP 2	MOPP 3	MOPP 4	Mask Only
Mask	Carried	Carried	Carried	Carried	Worn[1]	Worn	Worn
Overgarment	Ready[3]	Available[4]	Worn[1]	Worn[1]	Worn[1]	Worn	Available
Vinyl overboots	Ready[3]	Available[4]	Available[4]	Worn	Worn	Worn	Available
Gloves	Ready[3]	Available[4]	Available[4]	Available[4]	Available[4]	Worn	Available
Helmet protective cover	Ready[3]	Available[4]	Available[4]	Worn	Worn	Worn	Available
Chemical protective undergarment[2]	Ready[3]	Available[4]	Worn[2]	Worn[2]	Worn[2]	Worn[2]	Available

[1]In hot weather, the coat or hood can be left open for ventilation.

[2]The chemical protective undergarment (CPU) is worn under the battle dress uniform (BDU) (applies to the special operation forces and armor vehicle crewman).

[3]Must be available to the soldier within 2 hours, and the second set must be available within 6 hours.

[4]Within arms reach of the soldier.

NOTE: Refer to *Figure J-2* for an illustration of the MOPP levels.

hygiene, and sanitation discipline. Biological attacks are hard to detect. If an attack occurs, chances of survival are better if soldiers are healthy, physically fit, keep all immunizations up to date, and maintain good personal hygiene. Keep your body clean to prevent ingestion of biological agents. Keep small cuts or scratches covered and germ-free by using soap, water, and first aid measures. Since insects carry biological agents, keep clothes buttoned and your skin covered to prevent insect bites.

NUCLEAR ATTACK DEFENSE

J-61. MP can defend themselves against nuclear attack before, during, and after the event.

DEFENSE BEFORE A NUCLEAR ATTACK

J-62. The best defense against a nuclear attack is to dig in. Prepare unit defensive positions, which vary from individual foxholes to improved positions, whenever the tactical situation permits. Keep individual weapons, equipment, clothing, and other issue items in vehicles. Secure equipment because the blast wave will convert unsecured items into lethal missiles. Disperse and protect supplies, explosives, and flammable materials. Turn off all electronic equipment and remove antennas from vehicles.

J-63. Reverse slopes of hills and mountains give some nuclear protection. The initial radiation, heat, and light from the fireball of a nuclear blast tend to be absorbed by hills and mountains. Using of gullies, ravines, ditches, natural depressions, fallen trees, and caves can reduce nuclear casualties.

DEFENSE DURING A NUCLEAR ATTACK

J-64. Defense during a nuclear attack is either dismounted or mounted. When taking dismounted defensive actions, never run for cover. Immediately drop flat on the ground (face down) or to the bottom of a foxhole, with your head toward the blast. Cover exposed skin as much as possible and close your eyes. Remain down until the blast wave has passed and debris has stopped falling. Stay calm, check for injury, check weapons and equipment for damage, and prepare to continue the mission.

J-65. MP in vehicles should exit and move away from the vehicle and take dismounted defensive actions.

DEFENSE AFTER A NUCLEAR ATTACK

J-66. Once the attack ends, forward an NBC 1 nuclear report, organize survivors, secure and organize equipment, assist casualties, improve protection against possible fallout, and begin continuous monitoring. If the radiation dose rate reaches a hazardous level after fallout is complete, be prepared to move, on order, to a less hazardous area.

Fallout Warning

J-67. The first person to detect the arrival of fallout is usually the radiological monitor operating a RADIAC meter. As soon as he notes a dose rate of 1 centigray per hour or higher, he warns unit personnel. All personnel hearing the warning relay it to others. If the mission allows, move into a shelter with overhead cover and stay there until given an *all clear* signal or until directed to move. If the mission does not allow the unit to take cover, decontamination becomes more important and perhaps more difficult.

Supervision of Radiological Monitoring

J-68. Perform radiological monitoring routinely with a AN/VDR-2 RADIAC meter to determine the presence and intensity of a radiation hazard. MP leaders must ensure that their teams are properly trained on this equipment. The two types of monitoring is periodic and continuous.

J-69. **Periodic Monitoring**. Periodic monitoring (once every hour) assures the platoon that the area is not contaminated or, if applicable, provides a warning when contamination is detected after the platoon arrives. Periodic monitoring is initiated under the following conditions:

- After the first use of nuclear weapons in the theater of war.
- When the platoon is out of contact with higher HQ.
- When ordered by higher HQ.
- When the platoon stops continuous monitoring.

J-70. **Continuous Monitoring**. Continuous monitoring is the surveillance for radiation in the platoon's area or position. Initiate continuous monitoring when any of the following situations occur:

- When a nuclear detonation is observed or reported in the area of operations.
- When an NBC 3 report is received and the platoon is in the predicted area of contamination.
- When ordered by higher HQ.
- When a dose rate of 1 centigray per hour is recorded in periodic monitoring.

Supervision of Tactical Dosimetry Operations

J-71. Each MP team is issued a dosimeter. Selected soldiers wear them. Check all dosimeters to be used for the operation. Turn in dosimeters for recharging if they do not read zero. If a charger is not available, note the original reading. Ensure that the dosimeter readings are reported accurately. Collect readings at least once daily. Average these readings, round to the nearest 10, and report the average and the radiation exposure status (RES) to higher HQ.

J-72. Use individual actions for nuclear defense when operating in or crossing radiologically contaminated areas. Close vehicles tightly and cover cargoes with tarps or tenting. Mission permitting, keep the speed down to prevent dust. Vehicles should maintain adequate following distances to stay out of the dust raised by preceding vehicles.

J-73. After the unit exits a contaminated area, check personnel, equipment, and cargo for contamination and decontaminate them, if necessary. Monitor dose rates closely to ensure compliance with operational exposure guidance (OEG). Update the RES if appropriate.

CHEMICAL ATTACK DEFENSE

J-74. MP must be prepared to defend themselves before, during, and after a chemical attack.

DEFENSE BEFORE A CHEMICAL ATTACK

J-75. The best defense before a chemical attack is preparation. Use protective procedures. Ensure that all personnel have protective masks available, and ensure that each mask fits and functions properly. All personnel should wear the proper protective clothing according to the MOPP level designated by the commander. Protect all equipment and supplies from liquid chemical contamination by keeping them organized and covered.

J-76. Emplace an M8A1 automatic chemical agent alarm. The M8A1 is the primary means of detecting an upwind chemical attack. The system provides two essential elements of survival: detection of a toxic agent cloud and early warning to the troops in the monitored position. The platoon leader decides where to place the chemical alarm. Place the detector units no more than 150 meters upwind from the platoon's perimeter or position. Space the available detector units about 300 meters apart, and ensure that each detector unit is connected to the alarm unit by telephone cable (WD-1).

Position the alarm units near radiotelephone assets; this makes it easy to alert the unit of an attack. Blowing sand or dust, rain, sleet, snow, temperatures below 40 degrees Fahrenheit, and tropical conditions can affect the operation of the alarm.

J-77. If an NBC hazard cannot be avoided, soldiers must be prepared to protect themselves and equipment from the effects of exposure. Base the type and degree of protection required on the unit's mission and the hazard. Note that the line between contamination avoidance and protection is not distinct; many actions contribute equally to both.

DEFENSE DURING A CHEMICAL ATTACK

J-78. Give the alarm. All unmasked soldiers put on their protective masks and other MOPP gear. All personnel move inside their vehicles and close all doors and hatches, if applicable. This aids in the protection from gross liquid contamination. The platoon leader directs the use of M256A1 detector kits and the CAM to determine the type of agent, and he submits an NBC 1 report. The platoon then continues the mission.

DEFENSE AFTER A CHEMICAL ATTACK

J-79. Forward an NBC 1 chemical report, treat casualties, perform immediate decontamination as required, and mark the contaminated area.

SYMPTOMS AND TREATMENT OF CASUALTIES

J-80. Potential adversaries may have access to a wide variety of biological and chemical agents on the modern battlefield. These agents can be dispensed alone or with other carriers or agents. Casualties resulting from exposure to biological or chemical agents require medical treatment as quickly as possible.

J-81. The first step in the treatment process is usually appropriate self-aid and buddy-aid measures. These vary depending on the agent. Soldiers mask to prevent inhaling or ingesting additional agents and remove agents from exposed skin by washing with soap and water or using the M291 kit. They use buddy-aid procedures to observe each other for early symptoms of toxic exposure and provide medical assistance.

J-82. The platoon leader selects separate casualty collection points for contaminated and noncontaminated casualties to prevent cross contamination. Decontaminate all contaminated casualties as thoroughly as the situation allows before being evacuated. The platoon includes the number of contaminated patients in its casualty evacuation request. This allows the evacuation team to send the proper number of vehicles for pickup.

J-83. Chemical agents fall into four major categories: nerve, blister, blood, and choking. The primary routes of attack on the body are through the respiratory system and the skin. These agents are especially dangerous because they can kill or incapacitate quickly. The first, and most important, step in dealing with them effectively is to recognize symptoms so that proper treatment can be administered. *Table J-3, page J-32,* shows protection and detection measures, symptoms, and treatment and decontamination procedures for chemical agents.

UNMASKING PROCEDURES

J-84. Soldiers unmask as soon as possible except when a biological or chemical attack is expected. Use the procedures outlined in the following paragraphs to determine if unmasking is safe.

J-85. If an M256 or M256A1 detector kit is available, use it to supplement unmasking procedures. The kit

Table J-3. NBC Protection and Detection Chart

Agent Type	Protection	Detection	Symptoms	Effects	First Aid	Decontamination
Nerve	Protective mask and suit	M8A1, M256A1, CAM, M8 paper, or M9 paper	Difficult breathing, drooling, nausea, vomiting, convulsions, or blurred vision	Incapacitates	Use Mark 1 NAAK and convulsant antidote for nerve agents (CANA).	Use an M291 kit and flush eyes with water.
Blister	Protective mask and suit	M256A1, M8 paper, M9 paper, or CAM	Burning eyes, stinging skin, and irritated nose (no symptoms with mustard or nitrogen mustard)	Blisters the skin and damages the respiratory tract	Use the same procedures as for second- and third-degree burns. Refer to FM 21-11.	Use an M291 kit and flush eyes with water.
Blood	Protective mask	M256A1	Convulsions and coma	Incapacitates	None	None
Choking	Protective mask	Odor only (resembles newly-mown hay or green corn).	Coughing, choking, nausea, headache, and tightness in the chest	Floods and damages the lungs	Avoid movement and keep warm.	None

does not detect all agents; therefore, use the proper unmasking procedures, which take about 15 minutes. If all the tests with the kit (including a check for liquid contamination using M8 detector paper) have been performed and the results are negative, use the following procedures:

- The senior person selects one or two soldiers to start unmasking procedures. If possible, they move to a shady place. Bright, direct sunlight can cause pupils to constrict, giving a false symptom.
- The selected soldiers unmask for 5 minutes and then clear and reseal their masks.
- The leaders observe the unmasked soldiers for 10 minutes. If no symptoms appear, request permission from higher HQ to signal *all clear*.

J-86. The leaders watch the soldiers for possible delayed symptoms. Always have first aid treatment immediately available in case it is needed.

J-87. If an M256 or M256A1 kit is not available, unmasking procedures take about 35 minutes. When a reasonable amount of time has passed after the attack, find a shady area and use M8 paper to check the area for possible liquid contamination. Conduct unmasking using the following procedures:

- The senior person selects one or two soldiers. They take a deep breath and break their mask seals, keeping their eyes wide open.
- The soldiers clear and reseal their masks after 15 seconds. The leaders observe them for 10 minutes.
  - If no symptoms appear, the same soldiers break the seals, take two or three breaths, and clear and reseal their masks. Observe them for 10 minutes.

- If no symptoms appear after 10 minutes, the same soldiers unmask for 5 minutes and then remask.
- If no symptoms appear after 10 minutes, the leaders request permission from higher HQ to signal *all clear*. He continues to observe soldiers in case delayed symptoms develop.

J-88. Give the all-clear signal by word of mouth through the chain of command. Higher HQ initiates the all-clear signal after testing for contamination proves negative. If required, use standard sound signals, such as a continuous, sustained blast on a siren, a vehicle horn, or a similar device. When *all clear* is announced on the radio, it must be authenticated before compliance. The commander designates the specific all-clear signal and includes it in the SOP.

Appendix K

Civil-Disturbance Measures

This appendix discusses the crowd control measures used when there is a civil disturbance.

OVERVIEW

K-1. The mission of the military forces in a civil disturbance is to apply the minimum force necessary to help local or HN authorities restore law and order. Most often, military personnel disperse unauthorized assemblages and patrol disturbed areas to prevent unlawful acts. Other missions assigned to MP or other military forces include—

- Maintaining the mechanics of essential distribution, transportation, and communications systems.
- Making a show of force.
- Setting up roadblocks.
- Cordoning off areas.
- Dispersing crowds.
- Releasing riot control agents (RCAs).
- Serving as security forces or reserves.
- Initiating needed relief measures, such as distributing food or clothing or establishing emergency shelter.
- Employing nonlethal munitions and equipment.

K-2. Military control force commanders must know what options are available to them. They select the option that is best for the specific physical and psychological environment. They must be able to reduce the intensity of the confrontation and restore order.

K-3. The commitment of military forces to civil-disturbance control operations does not automatically give these forces police power. There are legal and commonsense reasons to restrict the police power of military forces. All military leaders and planners must be familiar with laws, regulations, and policies that govern military involvement in civil disturbances.

K-4. In all contacts with the civilian population and the participants of the disturbance, military forces must display fair and impartial treatment and must adhere to the principle of minimum force. Whenever possible, civil police apprehend, process, and detain civil law violators. Military forces perform these functions only when necessity dictates and to the minimum extent required. The return these functions to civil authorities as soon as possible.

K-5. As the disturbance subsides, the commander takes steps to restore control to civil authorities. The control force gradually reduces the number and scope of its operations. It takes care not to give the impression that all controls have been removed. Withdrawal is not immediate. The disturbance may flare up again if the participants get the impression that authorities have abandoned the operations. The control force gradually withdraws in a phased return of control to civil authorities.

CIVIL DISTURBANCES ON DEPARTMENT OF DEFENSE INSTALLATIONS OUTSIDE THE CONTINENTAL UNITED STATES

K-6. On DOD installations overseas and at US embassies and consulates, HN forces generally control disturbances targeted at US facilities because of the possible international political ramifications of foreign civil disturbances. Status of Forces Agreements define the legal considerations that guide and constrain actions by US military commanders. Commanders must have an effective liaison with HN authorities. Through close coordination with HN authorities, US commanders can determine the level of visibility and the involvement, if any, of US forces.

CROWD BEHAVIOR

K-7. The presence or absence of social factors like leadership, moral attitudes, and social uniformity may influence crowd behavior. Leadership has a profound effect on the intensity and direction of crowd behavior. When blocked from expressing its emotions in one direction, a crowd's frustration and hostility can be redirected elsewhere. The first person to give clear orders in an authoritative manner is likely to be followed. Agitators can exploit a crowd's mood and convert a group of frustrated, resentful people into a vengeful mob. Skillful agitators using television, radio, and other communications media can reach large portions of the population and incite them to unlawful acts without having direct personal contact. In a civil-disturbance environment, any crowd can be a threat to law and order because it is open to manipulation.

K-8. Additionally, crowd behavior may be affected by emotional contagion or panic.

- Emotional contagion provides the crowd psychological unity. The unity is usually temporary, but it may be long enough to push a crowd to mob action. When emotional contagion prevails, normal law and authority are suppressed, increasing the potential for violence.
- Panic can occur during a civil disturbance when—
  - Crowd members perceive their safety is at risk and attempt to flee the area.

- Crowd members cannot disperse quickly after exposure to RCAs.
- Escape routes are limited.
- Escape routes are blocked or congested.

K-9. Control force members are also susceptible to crowd behavior. They may become emotionally stimulated during a tense confrontation. Commanders must counteract this. Control force members must exercise restraint individually and collectively. Rigorous training, firm and effective supervision, and complete awareness and understanding of ROE and ROI are necessary to offset the effect of crowd contagion on the control force.

CROWD TACTICS

K-10. In civil disturbances, crowds employ any number of tactics to resist control or to achieve their goals. Tactics may be unplanned or planned and nonviolent or violent. The more purposeful the disturbance, the more likely the possibility of well-planned tactics.

K-11. Nonviolent tactics may range from name-calling to building barricades. Demonstrators may converse with control force members to distract them or to gain their sympathy. They may try to convince control force members to leave their posts and join the demonstrators. Demonstrators may use verbal abuse. Expect obscene remarks, taunts, ridicule, and jeers. Crowd members want to anger and demoralize the opposition. They want authorities to take actions that later may be exploited as acts of brutality.

K-12. Sometimes women, children, and elderly people are placed in the front ranks. This plays on a control force's sympathy to try to discourage countermeasures. When countermeasures are taken, agitators take photographs to stir public displeasure and embarrass the control force. Demonstrators may form human blockades to impede traffic by sitting down in roads or at the entrances to buildings. This can disrupt normal activity, forcing control personnel to physically remove the demonstrators. Demonstrators may lock arms, making it hard for the control force to separate and remove them. It also makes the control force seem to be using excessive force.

K-13. Groups of demonstrators may trespass on private or government property. They want to force mass arrests, overwhelm detainment facilities, and clog the legal system. Demonstrators may resist by going limp and forcing control force members to carry them. They may chain or handcuff themselves to objects or to each other. This prolongs the demonstration. Agitators may spread rumors to incite the crowd and try to force the control force to use stronger measures to control or disperse the crowd. The agitators want to make the control force appear to be using excessive force. Terrorist groups may try to agitate crowds as a diversion for terrorist acts. They also try to provoke an overreaction by the control force.

K-14. Violent crowd tactics, which may be extremely destructive, can include physical attacks on people and property, fires, and bombings. Crowd use of violent tactics is limited only by the attitudes and ingenuity of crowd members, the training of their leaders, and the materials available to them. Crowd or mob members may commit violence with crude, homemade weapons, or they may employ sophisticated small arms and explosives. If unplanned violence occurs, a crowd will use rocks, bricks, bottles, or whatever else is at hand. If violence is planned, a crowd can easily conceal makeshift weapons or tools for vandalism. They may carry—

- Balloons filled with paint to use as bombs.
- Bolt cutters to cut through fences.
- Picket signs to be used as clubs.

- Pipes wrapped in newspapers to throw as deadly missiles.
- Firecrackers dipped in glue and covered with BBs or small nails to use as deadly grenades.
- Plywood shields and motorcycle helmets to protect against riot batons.
- Safety goggles to protect against tear gas.

K-15. A crowd may erect barricades to impede troop movement or to prevent a control force from entering certain areas or buildings. They may use vehicles, trees, furniture, fences, or any other material. In an effort to breach barriers, rioters may throw grapples into wire barricades and drag them. They may use grapples, chains, wire, or rope to pull down gates or fences. Long poles or spears may be used to keep control forces back while removing barricades or to prevent the use of bayonets. They also may crash vehicles into gates or fences to breach them.

K-16. Rioters can be expected to vent their emotions on individuals, troop formations, and control-force equipment. Rioters may throw rotten fruits and vegetables, rocks, bricks, bottles, improvised bombs, or other objects from overpasses, windows, and roofs. In the past, troops, firefighters, and utility workers on duty during a civil disorder have been beaten, injured, or killed. Vehicles have been overturned, set on fire, or otherwise damaged.

K-17. Rioters may direct dangerous objects like vehicles, carts, barrels, and liquids at troops located on or at the bottom of a slope. On level ground, they may drive wheeled vehicles at the troops, jumping out before the vehicles reach the target. This tactic is also used to breach roadblocks and barricades.

K-18. Rioters may set fire to buildings and vehicles to block the advance of troops. Fires are also set to create confusion or diversion, destroy property, and mask looting and sniping. Rioters may flood an area with gasoline or oil and ignite it. On the other hand, they may pour gasoline or oil down a slope or drop it from buildings and ignite it.

K-19. Weapons fire against troops may take the form of selective sniping or massed fire. The fire may come from within the ranks of the rioters or from buildings or other adjacent cover. The weapons used can vary from homemade one-shot weapons to high-powered rifles. Snipers may try to panic control force members into firing a volley into the crowd. Innocent casualties make a control force appear both undisciplined and oppressive.

K-20. Explosives may be used to breach a dike, levee, or dam. Bombs can be exploded ahead of troops or vehicles so rubble blocks a street. They can be used to block an underpass by demolishing the overhead bridge. In extremely violent confrontations, bombs placed in buildings may be timed to explode when troops or vehicles are near. Demolition charges can be buried in streets and exploded as troops or vehicles pass over them. Explosive-laden vehicles can be rolled or driven at troops. Rioters may attach explosives to animals and force them toward troops. They then detonate the explosives by remote control. Even harmless looking objects like cigarette lighters and toys can be loaded with explosives and used as weapons.

COMPANY LEVEL OPERATIONS

K-21. Normally, civil-disturbance operations are conducted at company level. The company operations section coordinates for special equipment that includes the following:

- Kevlar and face shield.
- Body armor.
- Shields, work gloves, and batons.
- Protective masks.

- Elbow pads.
- Shin guards.
- Weapons and bayonets.
- CS/OC spray.
- Smoke grenades.
- Breaching ramps.
- Scaling ladders.
- Portable radios.
- Portable lighting.
- Bullhorns.
- Video home system (VHS) cameras to videotape individuals in the crowd for identification.
- Transportation assets to move people to detention cells or detention areas.
- Hand and leg irons and flex cuffs.
- NVDs.

K-22. The MP operations center also coordinates for support, to include medical and emergency operations center (EOC) personnel, SRTs, hostage negotiators, MWD teams, PSYOP, civil affairs, local or HN law enforcement personnel, and apprehension or detention facilities for civilian personnel.

K-23. Rehearsals are imperative to the overall success of suppressing civil disturbances even before they occur. Rehearsals should cover the use and employment of nonlethal munitions, practice situations in which ROE or the use of force is applied, the use of riot dispensers, hand and arm signals or voice commands, and formations.

K-24. MP use appropriate crowd control measures that include formations such as—

- Using the line formation to push or drive a crowd straight back.
- Using the echelon formation to turn or move crowds away from buildings, fences, or walls.

- Using the wedge formation penetrate and split a crowd.
- Using the diamond formation to enter a crowd.

K-25. When a small unit, normally a squad, moves a dignitary through a crowd or an apprehension team into a crowd, they generally use the diamond formation. The other formations are trained at squad level and above but are normally performed by a platoon or company. The line, echelon left or right, and wedge formations are the basis for platoon and company formations. A squad must be skilled in the basic formations before practicing in platoon-size or larger formations.

K-26. Refer to *FM 19-15* for more information about civil-disturbance techniques and procedures.

NONLETHAL MUNITIONS

K-27. Nonlethal munitions are explicitly designed and primarily employed to incapacitate personnel or material, while minimizing fatalities, permanent injury to personnel, and undesired damage to property and the environment. Unlike conventional lethal munitions that destroy their targets through blast, penetration, and fragmentation, nonlethal munitions employ means other than gross physical destruction to prevent the target from functioning.

K-28. The doctrine and concepts of operation for nonlethal weapons are designed to reinforce deterrence and expand the range of options available to commanders. They enhance the capability of US forces to accomplish the following objectives:

- Discourage, delay, or prevent hostile actions.
- Limit escalation.
- Take military action in situations where the use of lethal force is not the preferred option.
- Protect our forces better.

- Disable equipment, facilities, and personnel temporarily.

K-29. Nonlethal munitions are not required to have a zero probability of producing fatalities or permanent injuries, and complete avoidance of these effects is not guaranteed or expected. When properly employed, nonlethal weapons should significantly reduce these effects when compared with using lethal weapons to physically destroy the same target.

K-30. Nonlethal capabilities provide a wider range of options that augment but do not replace traditional means of deadly force. The option to resort to deadly force must always remain available when the commander believes it is appropriate to the mission.

K-31. MP normally use the lowest level of force necessary to control most incidents that involve noncombatants. RCAs, such as CS and MWD, and crowd control devices, such as the riot baton, are examples of nonlethal means currently used by MP forces. These options are currently listed in *AR 190-29*.

K-32. Training for the use of nonlethal munitions requires leaders and soldiers to understand the limited use of these systems in environments with restrictive ROE. Training will be continuous at all levels to ensure that nonlethal munitions are properly employed.

K-33. The countermeasures for thwarting virtually all nonlethal options are usually apparent, quickly learned, and readily available. Because they do not kill, nonlethal options teach an adversary what to avoid in the future. Commanders must be prepared to stay one step ahead of motivated belligerents. Many nonlethal options have both maximum effective and minimum safety ranges. Individuals struck short of the minimum safety range often suffer severe injuries or death, while the effects of most nonlethal devices are greatly mitigated at longer ranges. In order to be effective, an adversary must be engaged within the effective zone (beyond the minimum safety range and short of the maximum effective range).

K-34. Nonlethal munitions and equipment currently available are shown in *Appendix L*. Refer to *Figure 3-2, page 3-11,* for the range of the munitions.

NONLETHAL MUNITIONS EMPLOYMENT CONSIDERATIONS

K-35. Nonlethal munitions employment must be well documented in ROE. Leaders must constantly ensure that soldiers understand when and how to effectively employ them. Incorrect application of these munitions can have significant operational and political ramifications. Employment considerations include the following:

- **Individual.** When possible, do not change individual weapons. Designate individuals as nonlethal shooters. Nonlethal shooters carry lethal munitions only for personal protection. Carry lethal rounds in a place to avoid confusing nonlethal rounds with lethal rounds.
- **Squad.** Squad leaders carry stun grenades and maintain their weapon loaded with lethal ammunition. Ideally, the squad does not change their task organization to accommodate the addition of nonlethal equipment. They should designate nonlethal shooters instead.
- **Patrols.** Commanders do not plan a nonlethal patrol, but they plan a combat and security patrol with a nonlethal attachment when the mission dictates. Carrying a shotgun limits the flexibility an individual has because of the time it takes to transition from nonlethal to lethal. Use shotguns only from a fixed position where adequate coverage is available.
- **Static position.** Individuals on static positions have their weapons loaded with lethal

ammunition, and additionally they carry nonlethal munitions. Both shotguns and M203s work well from static positions and are used together when possible.

- **Contact teams.** Teams of personnel whose primary responsibilities are to make personal contact with the adversary are established before deployment. Train team members in unarmed self-defense, open-hand control, handcuffing, and flex cuffing. Equip contact teams with personal protection gear, 9-millimeter pistols, and flex cuffs. Due to the physical nature of contact teams, do not arm contact team members with long rifles. If security for the contact team is an issue, attach a security element. MWD teams may also augment contact teams to help locate the adversary. Refer to *FM 90-40* for more information on nonlethal techniques and procedures.

Appendix L

Weapons and Equipment

MP must be rapidly deployable, versatile, and as lethal as the force that they support. MP are organized and equipped to provide functional battlefield capabilities that range from the deliberate attack (area security) to civil-disturbance control operations (L&O). Although not all inclusive, the listed data is intended to aid in planning MP operations. In any tactical situation, it is important that MP be able to properly identify threat weapons, vehicles, and aircraft. When threat is observed, MP report their location, activity, and direction of travel to higher HQ by the fastest means available.

FRIENDLY VEHICLES

L-1. *Table L-1, page L-2,* gives MP leaders a quick look at the capabilities of the various vehicles in the Army inventory.

Table L-1. Vehicle Capabilities

Vehicle Capabilities	M998	M1025/1026	M1114	M1117	Light Medium Tactical Vehicle (LMTV)
Weight (gross vehicle weight [GVW]) (in pounds)	4,950	5,250	12,100	29,500	16,500
Height (in inches)	69	73/73	73	102	112
Length (in inches)	180	180/185	190.5	237	251
Width	85	85/85	85	101	96
Fuel capacity (in gallons)	25	25	25	50	58
Maximum speed (in miles per hour)	55	55	55	63	58
Range (in miles)	350	300	275	440	400+

Table L-1. Vehicle Capabilities (Continued)

Vehicle Capabilities	M998	M1025/1026	M1114	M1117	LMTV
Fording Capabilities					
With kit (in inches)	60	60	60	NA	60
Without kit (in inches)	30	30	30	60	60
Pay load (maximum pounds)	2,500	2,500	NA	3,360	5,000

FRIENDLY WEAPONS

L-2. *Table L-2* gives MP leaders a quick look at the capabilities of the weapons used by MP units.

Table L-2. Weapon Capabilities

Weapon Capabilities	M2	M4	M9	M16A2	MK-19	M136 (AT4)	M203	M249
Weight (in pounds)	84	7.5	2.6	8.7	76	14.8	11	15.5
Length (in inches)	66	29.75 closed, 33 open	8.5	39	43	40	39	41.1
Range								
Maximum (in meters)	6,765	3,600	1,800	3,600	2,212	2,100	400	3,600
Arming (in meters)	NA	NA	NA	NA	18 to 30	10	14	NA
Minimum safe (in meters)	NA	NA	NA	NA	28	NA	31	NA

Table L-2. Weapon Capabilities (Continued)

Weapon Capabilities	M2	M4	M9	M16A2	MK-19	M136 (AT4)	M203	M249
Effective Range								
Area target (in meters)	1,830	600	NA	800	2,212	NA	350	800
Point target (in meters)	1,200	500	50	580	1,500	NA	160	600
Moving target (in meters)	NA	NA	NA	200	NA	NA	NA	NA
Rate of Fire (in Rounds Per Minute)								
Cyclic	500	700 to 970	NA	700 to 800	375	NA	NA	800
Rapid	40*	NA	NA	NA	60	NA	35	200*
Sustained	40*	12/15	60	16	40	NA	35	85
*With barrel change.								

FRIENDLY NONLETHAL EQUIPMENT AND MUNITIONS

L-3. The following descriptions and illustrations describe nonlethal equipment and munitions that are currently available.

NONLETHAL EQUIPMENT

L-4. The following nonlethal equipment provides bodily protection for soldiers involved in a nonlethal operation and allows soldiers to capture the threat with a minimum of force.

Nonballistic and Ballistic Riot Face Shields

L-5. The riot face shield *(Figure L-1)* provides soldiers with improved facial protection from thrown objects. The face shield is lightweight, adaptable to the current helmet, transparent, and scratch-resistant. It is adjustable to up and down positions. The mechanism for attaching and removing the shield from the helmet is robust and simple and requires no tools in the field.

Figure L-1. Riot Face Shield

L-6. The ballistic riot face shield is similar in design, but it is heavier and provides facial protection up to a 9-millimeter full metal jacket (FMJ) or 124 grains at 1,400 feet per second. This face shield is primarily used by SRTs in force-entry scenarios and can also be used for MOUT operations.

Nonballistic and Ballastic Riot Body Shields

L-7. The nonballistic riot body shield *(Figure L-2)* is 24 inches by 48 inches by .157 inches. It provides soldiers with improved protection from frontal, side, and overhead assaults. This shield is lightweight, transparent, and scratch-resistant. The ballistic riot body shield *(Figure L-3, page L-8)* is 24 inches by 36 inches with a 4- by 16-inch window for viewing the threat. It is similar in design to the nonballistic shield but is heavier. It provides ballistic protection up to a 9-millimeter FMJ or 124 grains at 1,400 feet per second. This shield is primarily used by SRTs in forced-entry scenarios and can also be used for selected MOUT operations.

Figure L-2. Nonballistic Riot Body Shield

Figure L-3. Ballistic Riot Body Shield

Nonballistic and Ballistic Shin Guards

L-8. Nonballistic shin guards *(Figure L-4)* provide soldiers with improved protection from thrown objects. They are lightweight and black in color.

Figure L-4. Nonballistic Shin Guards

L-9. Ballistic shin guards are similar in design to nonballastic guards, but they are heavier and provide protection up to a 9-millimeter FMJ or 124 grains at 1,400 feet per second. They are primarily used by SRTs in forced-entry scenarios and can also be used in selected MOUT operations.

Baton

L-10. The standard 36-inch wooden riot baton *(Figure L-5)* is currently in use for riot control. It is used for self-defense and to keep rioters out of arm's reach of the soldiers conducting crowd control tactics.

Figure L-5. Standard 36-inch Wooden Riot Baton

Portable Bullhorn

L-11. The portable bullhorn *(Figure L-6, page L-10)* is a critical communication device when conducting crowd control tactics. The bullhorn can facilitate communication with the crowd in conjunction with linguist or PSYOP support. It also assists in communicating commands to troops engaged in crowd control by projecting over the crowd's noise.

Individual Voice-Amplification System

L-12. The individual voice-amplification system *(Figure L-7, page L-10)* is a critical communication device for conducting crowd control tactics when using RCAs and wearing a protective mask. This device facilitates oral communications and increases the user’s ability to be heard on radios and other devices.

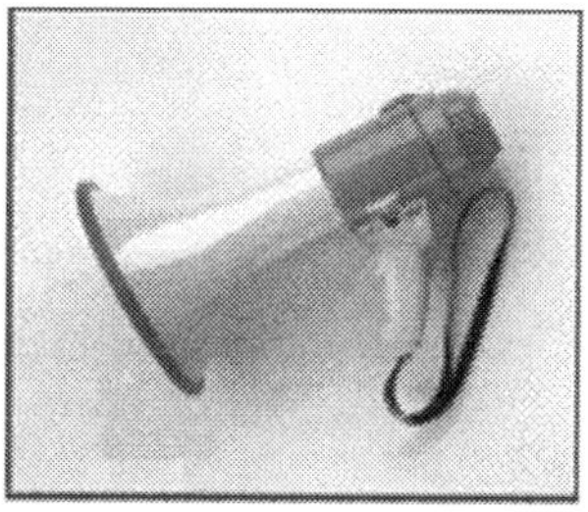

Figure L-6. Portable Bullhorn

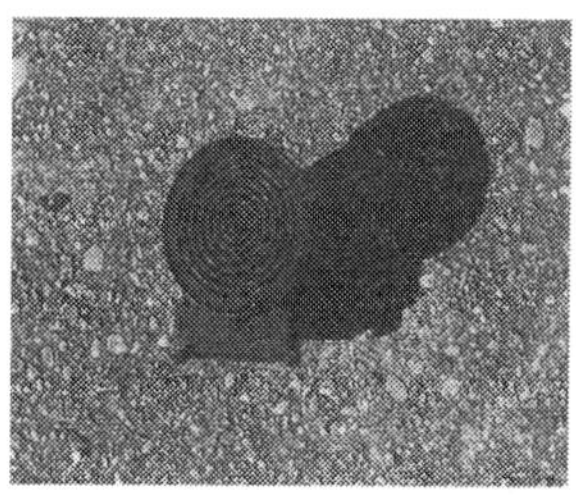

Figure L-7. Individual Voice-Amplification System

Individual Oleoresin Capsicum and M36 Individual Chlorobenzyl Malononitrile Dispersers

L-13. These individual RCA dispersers *(Figure L-8)* are used primarily for self-defense and to keep rioters out of arm's reach of soldiers conducting crowd control tactics or engaged in missions were noncombatant threat exists.

Midsize Riot Control Dispersers of Oleoresin Capsicum and Chlorobenzylidene Malononitrile

L-14. This RCA disperser *(Figure L-9)* is primarily used by formations conducting crowd control tactics, law

Figure L-8. Individual OC and CS Dispersers

enforcement, and I/R operations. It is lightweight, can be operated by one person, and is easily refilled and pressurized with available maintenance equipment or 2½- or 5-ton truck compressors. It is intended to provide a small unit with self-defense capabilities from large crowds out to 10 meters (more range is possible based on the wind) or offensively to clear crowds from critical areas (toward preplanned escape routes).

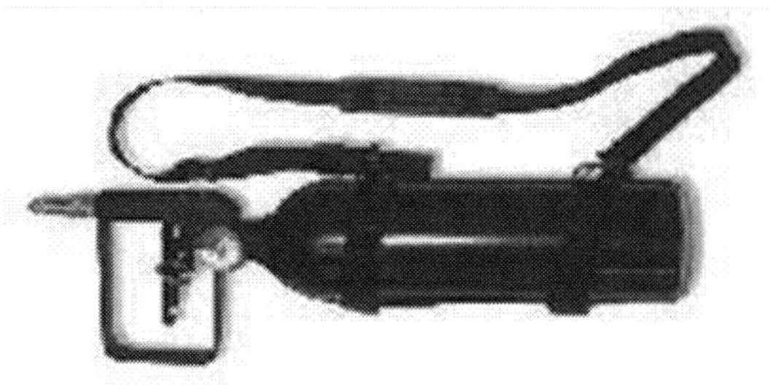

Figure L-9. Midsize Riot Control Dispersers of OC and CS

M33A1 Riot Control Disperser Filled With Chlorobenzylidene Malononitrile or Dibenz (B, f)-1, 4-Oxazepine (CR)

L-15. This RCA disperser *(Figure L-10)* is primarily used by formations conducting crowd control tactics. It is twice as heavy as a midsized disperser, carries a larger payload, and dispenses powdered CS or liquid CR. One individual can operate it, and it can be refilled and pressurized with available M4 compressors. It is intended to provide a small unit with self-defense capabilities from large crowds out to 15 meters (100 meters is possible based on wind speed and direction). Use the M33A1 offensively to clear crowds from critical areas.

Figure L-10. M33A1 Riot Control Disperser Filled With CS or CR

L-16. Some training is required to operate the dispenser. The unit NBC NCO may conduct this training. The device requires up to 2,000 pounds per square inch of pressure to function properly, and it is maintained with an M254 maintenance kit.

High-Intensity Xenon Searchlight

L-17. Use this individual, handheld searchlight *(Figure L-11)* for illumination in crowd control operations during darkness. Use it for general illumination of the operational area; to pinpoint agitators or threat, to reduce the ability of rioters to see troop formations and actions, and to enhance tactical deception techniques for units conducting crowd control operations.

Figure L-11. High-Intensity Xenon Searchlight

Lightweight Disposable Restraints

L-18. Lightweight disposable restraints *(Figure L-12, page L-14)* are also known as flex cuffs. Individual soldiers can carry large quantities of these restraints to immobilize individuals being detained. When freeing a detainee, cut the restrains off him with the safe cutting device supplied with the restraints or with utility shears. A reusable, red restraint training device is available.

Shotgun Munitions Carrier

L-19. The shotgun munitions carrier *(Figure L-13, page L-14)* is a 12-gauge ammunition carrier that straps to the stock of the M12 shotgun. It allows the firer to carry nonlethal ammunition that is readily available.

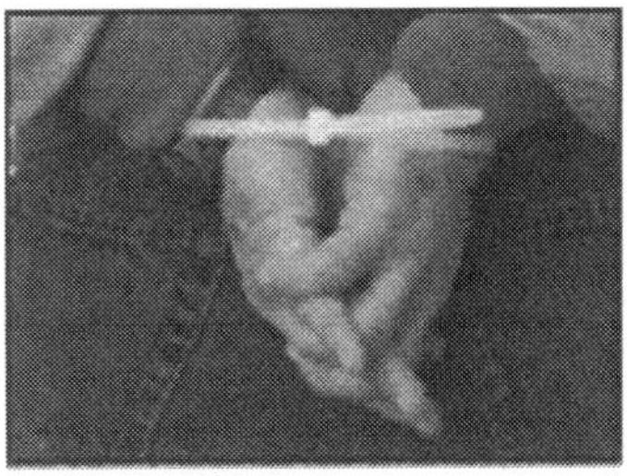

Figure L-12. Lightweight Disposable Restraints

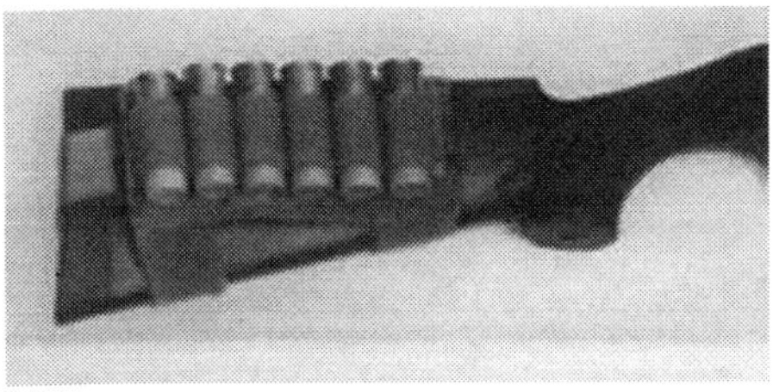

Figure L-13. Shotgun Munitions Carrier

Portable Vehicle-Arresting Barrier (PVAB)

L-20. The PVAB *(Figure L-14)* is designed to assist with short-term physical security of critical facilities and for use at checkpoints. Use the PVAB to augment a vehicle checkpoint. It adds the ability to stop any light vehicle (up to 7,500 pounds) that attempts to flee without killing the occupants. This device may be set up by 2 or 3 soldiers in less than an hour.

L-21. When emplaced, the PVAB resembles a standard speed bump. When armed, the device operates within 1.5 seconds and deploys a high-tensile net that catches the vehicle and slows it to a stop with internal braking

Figure L-14. Vehicle Entrapped in a PVAB

mechanisms. The net and other features prevent occupants from fleeing the vehicle. The PVAB is best suited for mobile, short-term vehicle checkpoints in areas under US control where there is a threat of terrorist-type activity and where the threat uses the cover of noncombatants to infiltrate US AOs.

NONLETHAL MUNITIONS

L-22. The following munitions provide soldiers with a nonlethal way to break contact, enforce a buffer zone, or stun an individual.

12-Gauge Nonlethal Point Target Cartridge Round (M1012)

L-23. The point target cartridge round *(Figure L-15, page L-16)* stuns individuals by delivering a strong blow to the body without penetrating it. This munition allows soldiers to enforce a buffer zone (standoff distance) with a violent crowd, break contact, or stun an individual target for possible detention by snatch teams. Fire the round at the center mass of an adult subject at ranges of 10 to 30 meters. Beyond 30 meters, the projectile loses accuracy and may no longer have the velocity required to stun an individual. This round has applications in law enforcement, I/R facilities, and US military detention facilities.

DANGER
Shots fired at subjects closer than 10 meters or shots to the head or groin may cause serious injury or even death.

Figure L-15. 12-Gauge Nonlethal Point Target Cartridge Round

12-Gauge Aerial Diversionary-Device Round

L-24. The aerial diversionary-device round *(Figure L-16)* provides multishot nonlethal capability to distract individuals or crowds. In crowd control, it delivers a flash bang projectile over the heads of a violent or potentially violent crowd, and is used to distract the crowd (in combination with other distraction devices and troop maneuvers). It allows other troop formations to maneuver to positions that are more advantageous

L-25. The round is designed to be fired at ranges of 75 to 100 meters and is placed about 5 meters above the crowd.

DANGER
Shots fired directly at subjects or in enclosed areas may cause serious injury.

Figure L-16. 12-Gauge Aerial Diversionary-Device Round

12-Gauge Nonlethal Area Target Cartridge Round (M1013)

L-26. The area target cartridge round *(Figure L-17, page L-18)* provides the capability to stun or deter two or three threats by delivering a strong blow to the body without penetrating it. This round has a wide range of capabilities for tactical, law enforcement, I/R, and US military detention operations.

L-27. The round is designed to be fired at the center mass of an adult threat at ranges of 10 to 30 meters.

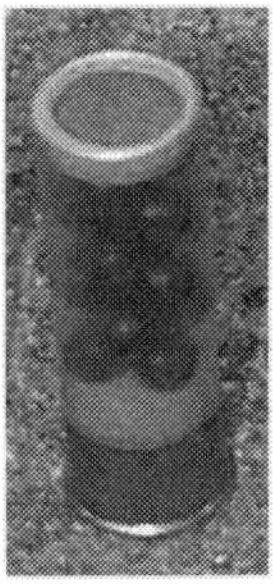

Figure L-17. 12-Gauge Nonlethal Area Target Cartridge Round

Shots fired closer than 10 meters may cause serious injuries. The projectile loses accuracy when shot beyond 30 meters and may no longer be effective.

40-Millimeter Sponge Round (Point) (M1006)

L-28. The 40-millimeter sponge round *(Figure L-18)* delivers a strong, stunning blow to a threat's body without penetrating it. This round has a wide range of capabilities for tactical, law enforcement, I/R, and US military detention operations.

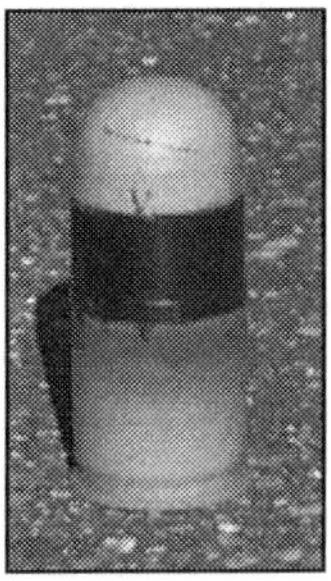

Figure L-18. M1006 40-Millimeter Sponge Round

L-29. The round is designed to be fired at the center mass of an adult threat at ranges of 10 to 50 meters.

WARNING

Shots fired closer than 10 meters may cause injuries.

40-Millimeter Crowd Dispersal Round (Area) (XM1029)

L-30. The 40-millimeter crowd dispersal round *(Figure L-19)* delivers a strong, stunning blow to a threat's body without penetrating it. This round has a wide range of capabilities for tactical, law enforcement, I/R, and US military detention operations. In crowd control, it provides a nonlethal capability that can be used to break contact and enforce a buffer zone (standoff distance) with a violent crowd.

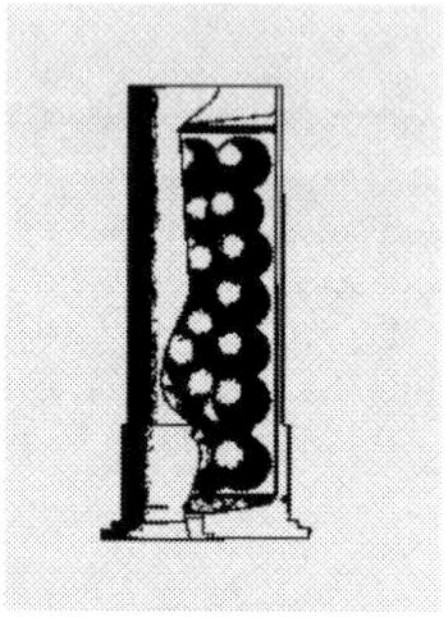

Figure L-19. 40-Millimeter Crowd Dispersal Round (Area)

L-31. The round is designed to be fired at the center mass of an adult threat at ranges of 10 to 30 meters.The projectile loses accuracy when shot beyond 30 meters and may not stun or deter the threat.

WARNING
Shots fired closer than 10 meters may cause injuries.

40-Millimeter Carrying Pouch

L-32. The 40-millimeter carrying pouch is slung over the shoulder of a soldier. It provides the ability to carry nonlethal ammunition separate from lethal rounds.

M84 Stun Grenade (Diversionary Device, Hand-Thrown)

L-33. The M84 *(Figure L-20)* is a hand-thrown, flash bang, stun device used primarily by SRTs in forced-entry scenarios. It is used for selected MOUT or crowd control operations.

Figure L-20. M84 Stun Grenade

L-34. The M84 is designed to be thrown into a room (through an open door, a standard glass window, or other opening). It delivers a loud bang and a brilliant flash that temporarily disorients and detracts the occupants. Because of its reusable metal body, do not throw it into a crowd, as it may be returned to friendly troops in the form of a projectile. Instead, throw it into a controlled area in conjunction with other deception and distraction techniques.

NONLETHAL, TURRET-MOUNTED, 66-MILLIMETER GRENADE LAUNCHER (M315) AND LIGHT VEHICLE OBSCURATION SMOKE SYSTEM, 66-MILLIMETER, NONLETHAL GRENADE (L96/97 XM98/99)

L-35. The LVOSS/M315 *(Figure L-21)* is a 66-millimeter smoke grenade-launching platform designed to give HMMWVs ASV M1117 the ability to obscure their position in the same manner as armored vehicles. These launchers can launch any of the smoke or nonlethal 66-millimeter munition.

M315 LVOSS

Figure L-21. LVOSS and M315

L-36. The M315 turret-mounted, 66-millimeter, multipurpose, adjustable grenade launcher installation kit was developed to provide a vehicle-mounted nonlethal platform. The system is capable of delivering nonlethal payloads to support a variety of mission requirements and can be mounted on various vehicles (M1025, M1026, M966, M114, or ASV M1117) equipped with a machine gun or a TOW II missile system mount. The system electronically fires 66-millimeter cartridges from four adjustable firing tubes.

L-37. The LVOSS, 66-millimeter, nonlethal grenade (L96/97 XM98/99) is an area target munition that can be fired from the standard LVOSS/M315 launcher *(Figure L-22)*. It can be mounted on selected armament carrier HMMWVs and can be fired from any 66-millimeter, smoke-launching system found on most armored vehicles.

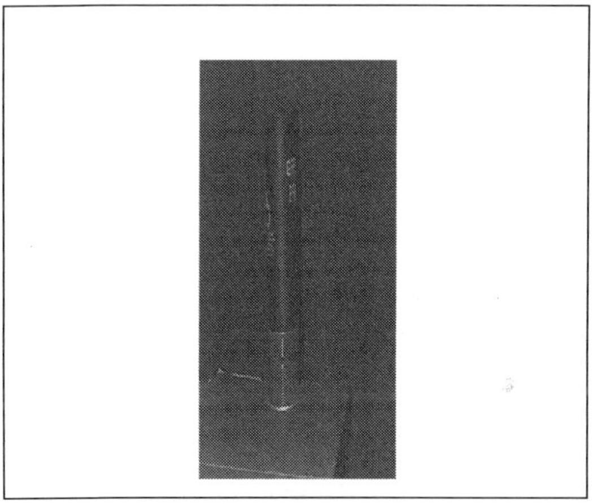

Figure L-22. LVOSS/M315 66-Millimeter Nonlethal Grenade

L-38. These rounds are designed to be fired from 80 to 100 meters. They deliver a flash bang diversionary warning (M98), a payload of rubber nonpenetrating projectiles (M99) (affecting a 10- to 20-meter circular area), or CS/cinnamic acid (CA) practice below the point of burst (L96A1/L97A1). These munitions lack the ability to be precision-delivered and are meant to affect a large number of people at long standoff ranges. This provides convoys and crowd control formations a long range support weapon to affect crowds beyond the range of shoulder-fired nonlethal weapons. Use the rounds to provide supporting nonlethal fires to crowd control formations.

M5 Modular Crowd Control Munitions

L-39. The MCCM *(Figure L-23)* munition is similar in operation to a claymore mine, but it delivers nonlethal effects to the threat by delivering a strong, nonpenetrating blow to the body with multiple submunitions (600 rubber balls). This round has a wide range of capabilities for tactical, law enforcement, I/R, and US military detention operations. In crowd control, it provides a nonlethal counterpersonnel capability that can be used to break contact, enforce a buffer zone (standoff distance), or demonstrate a show of force.

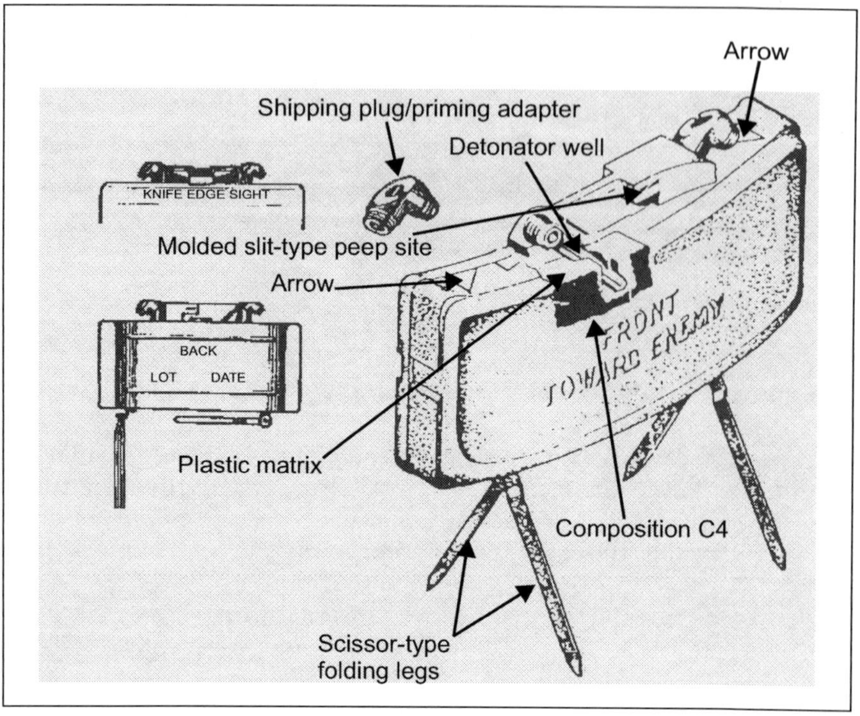

Figure L-23. M5 MCCM

L-40. This round is designed to be fired at the center mass of adult threats at ranges of 5 to 15 meters. It has a shot arc covering between 60 to 80 degrees (laterally).

WARNING

Shots fired at subjects closer than 5 meters may cause injury.

FRIENDLY COMMUNICATION, SINGLE-CHANNEL, GROUND-TO-AIR RADIO SYSTEM (SINCGARS)

L-41. SINCGARS is the primary communication system for MP. It is a series of 2-way FM radio sets that use an 18-element keypad for tuning 2,320 channels. This includes 8 preset channels in the single-channel mode and 6 preset channels in the jam-resistant, frequency-hopping mode. *Table L-3* and *Table L-4* provide the planning ranges for SINCGARS.

Table L-3. Voice Transmission Maximum Planning Ranges

Type of Radio	RF Switch Position	Planning Ranges
Manpack or vehicular	LO M HI	200 to 400 meters 400 meters to 5 kilometers 5 to 10 kilometers
Vehicular only	PA	10 to 40 kilometers

L-42. To increase the transmission range of SINCGARS, connect the OE-254 antenna group. It is an omnidirectional, biconical antenna designed for broadband operation, without field adjustment, from 30 to 88 megahertz, up to 350 watts. The following is the tabulated data transmission range of the OE-254:

- Between two OE-254 antenna groups–
  - Average terrain: 36 miles (57.9 kilometers).

- Difficult terrain: 30 miles (48.3 kilometers).
- Between an OE-254 antenna group and a vehicular whip antenna–
  - Average terrain: 30 miles (48.3 kilometers).
  - Difficult terrain: 25 miles (40.3 kilometers).

Table L-4. Data Transmission Maximum Planning Ranges

Type of Radio	Baud Rate Used	RF Switch Position	Planning Ranges* (In Kilometers)
Manpack/ vehicular	600 to 4,800 baud per second (BPS)	HI	3 to 5
(short range)	16,000 BPS (16 kilobauds per second [KBPS])	HI	1 to 3
Vehicular	600 to 2,400 BPS	PA	5 to 25
(long range)	4,800 BPS	PA	5 to 22
	16,000 BPS (16 KBPS)	PA	3 to 10

***Ranges are based on the line of sight and are average for normal conditions. Ranges depend on the location, the sighting, the weather, and the surrounding noise level, among other factors. The use of the OE-254 antenna increases the range for both voice and data transmissions. Enemy jamming and mutual interference conditions degrade these ranges. In data transmission, the use of a lower baud rate increases the range.**

THREAT WEAPONS AND EQUIPMENT

L-43. In any tactical situation, it is important that MP be able to properly identify threat weapons, vehicles, and aircraft. MP report threat location, activity, and direction of travel to higher HQ by the fastest means available.

L-44. *Tables L-5* through *L-12, pages L-27* through *L-52,* outline the most common features of threat weapons, equipment, and vehicles that MP may encounter in a hostile environment. Special operations forces and airborne assault, reconnaissance, and insurgent units often use these weapons.

Table L-5. Small Arms

Small Arms Characteristics	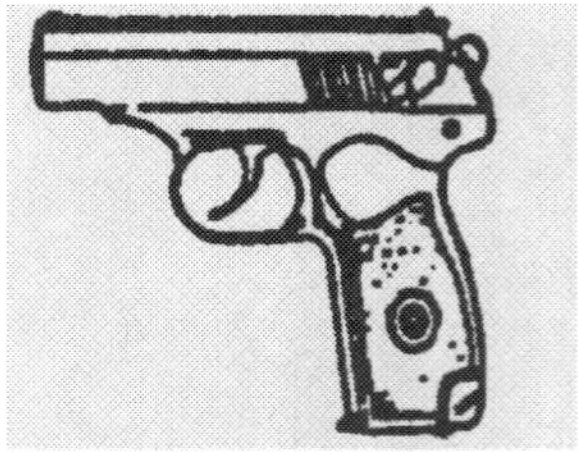 **9-Millimeter PM Pistol**	 **5.45-Millimeter AK-74 Assault Rifle**
Range, effective and maximum (in meters)	50	500/1,000
Rate of fire, practical and cyclic (rounds per minute)	30	100/600
Ammunition type	9- by 18-millimeter ball	5.45- by 39-millimeter ball, ball tracer, incendiary T
Fire mode	Semiautomatic	Selective semiautomatic or fully automatic

Table L-5. Small Arms (Continued)

Small Arms Characteristics	 **5.45-Millimeter Squad Machine Gun (RPK)-74 Light MG**	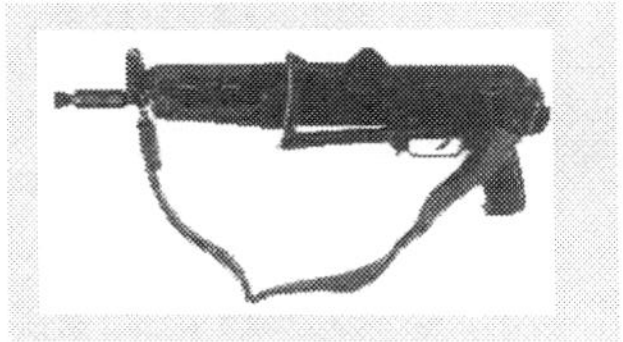 **5.45-Millimeter AKSU-74 Submachine Gun (SMG)**	 **7.62-Millimeter Soviet Sniper Rifle (SVD)**
Range, effective and maximum (in meters)	800/2,500	300/1,500	1,300 with scope, (800 without)/ 3,800
Rate of fire, practical and cyclic (rounds per minute)	150/600	150/600	30
Ammunition type	5.45- by 39-millimeter ball, ball tracer, incendiary T	5.45- by 39-millimeter ball, ball tracer, incendiary T	7.62- by 54R-millimeter light or heavy ball, steel core, tracer, AT, incendiary "rose" sniper bullet
Fire mode	Selective semiautomatic or fully automatic	Selective semiautomatic or fully automatic	Semiautomatic

Table L-5. Small Arms (Continued)

Machine Gun Characteristics	 RPK-74	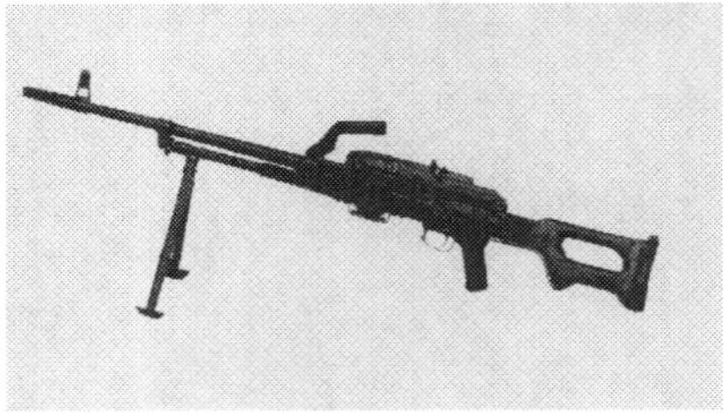 PKM
Range, effective and maximum (in meters)	800/2,500	1,000/3,800
Rate of fire, cyclic and practical (rounds per minute)	600/150	650/250
Ammunition type	5.45- by 39-millimeter, rimless	7.622- by 54-millimeter ball, ball tracer, API-T, incendiary
Fire mode	Selective	NA
Armor penetration (in millimeters)	NA	8 millimeters at 500

Table L-6. Grenade Launchers

Grenade Launcher Characteristics	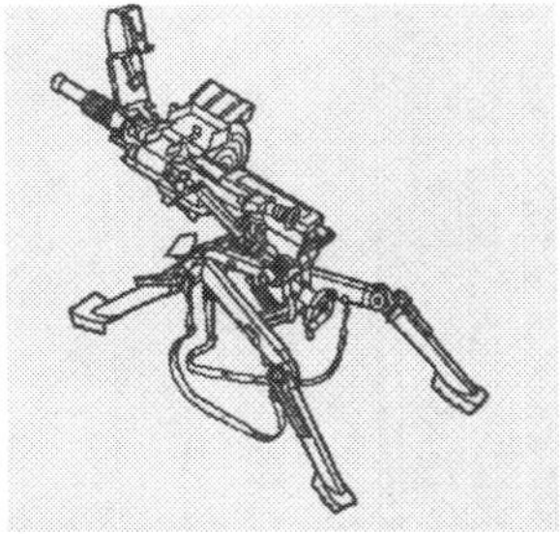 Armored Gun System (AGS)-17	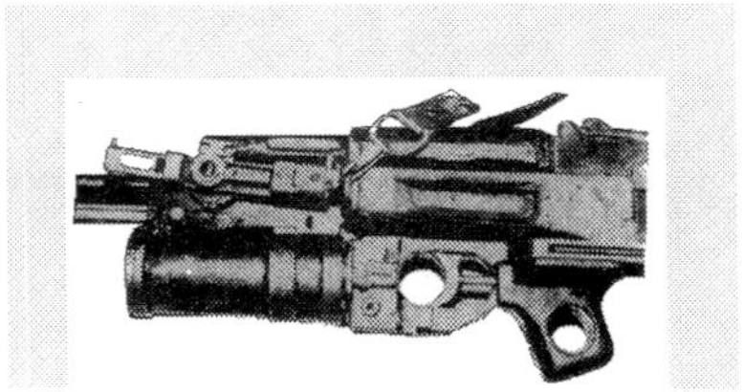 BG-15/GP-25
Range, effective and maximum (in meters)	1,200 indirect fire, 700 direct fire/1,730	4 to 400/400
Rate of fire, practical and cyclic (rounds per minute)	60 to 100/100 to 450	5
Elevation (in degrees)	7 to 87	NA
Traverse (in degrees)	30	NA
Platform	Tripod, vehicle, or helicopter	Under barrel grenade launcher
Fire mode	Selective, semiautomatic, or fully automatic	Single shot
Feed	29-round belt in drum magazine	Muzzle loaded
Ammunition type	30-millimeter, HE fragment	40-millimeter, HE fragment, bounding HE fragment
Caliber (in millimeters)	30	40
Crew	3	1
Remarks	15-meter lethal area of burst	Bounding HE fragment strikes the ground and "bounds" up 1.5 to 2 meters before exploding

Table L-7. Rocket-Propelled Grenades

AT Grenade Launcher Characteristics	 Rocket-Propelled Grenade (RPG)-7V					 RPG-16D
Projectile	PG-7	PG-7M	PG-VR	PG-VL	OG-7/7V	PG-16
Range, moving and stationary target (in meters)	300/500	NA	NA	NA	NA	500/800
Armor penetration (in millimeters)	330	330	>750	>PG-7M <PG-VR	NA	375
Ammunition type	High-explosive, antitank (HEAT)	HEAT	Tandem HEAT	HEAT	HE fragment	HEAT
Warhead caliber (in millimeters)	85	72	105	93	NA	58.3
Rate of fire (rounds per minute)	6	6	6	6	6	6
Crew	NA	NA	NA	NA	NA	2
Tube caliber (in millimeters)	40	40	40	40	40	58.3

Table L-7. Rocket-Propelled Grenades (Continued)

AT Grenade Launcher Characteristics	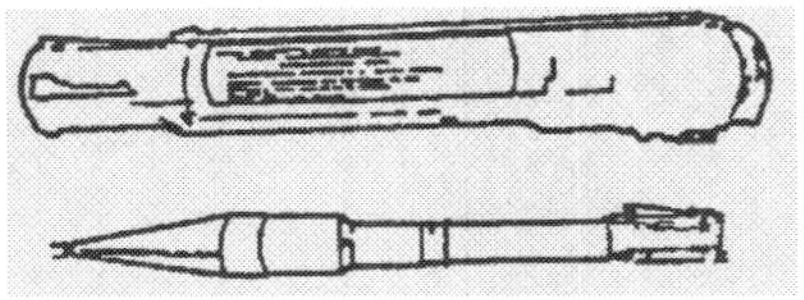 **RPG-18 and 22**		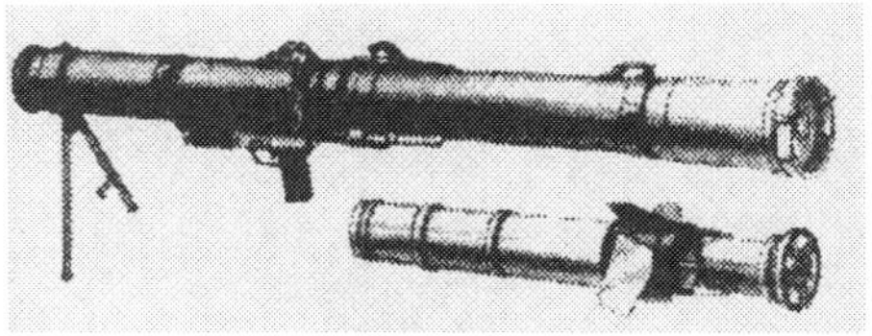 **RPO-A**
	RPG-18	RPG-22	RPO-A
Projectile			
Range, moving and stationary target (in meters)	200	250	200 direct fire 1,000 indirect fire
Armor penetration (in millimeters)	360	390	NA
Ammunition type	HEAT	HEAT	FAE
Rate of fire (rounds per minute)	NA	NA	2
Crew	1	NA	NA
Tube caliber (in millimeters)	64	73	93

Table L-8. Recoilless Rifles

AT Gun and Rifle Characteristics	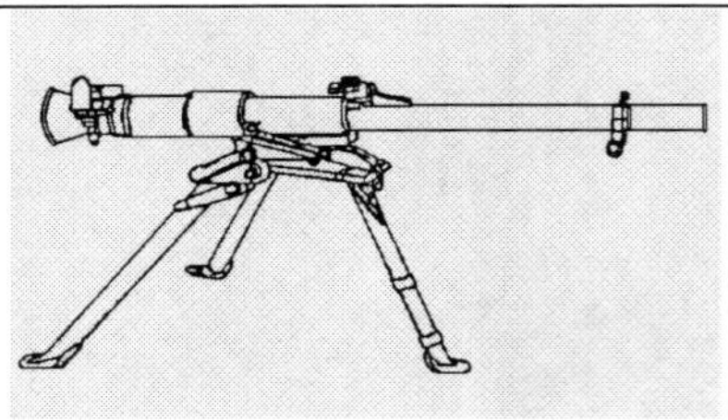 Self-Propelled Gun (SPG-9) Recoilless Rifle	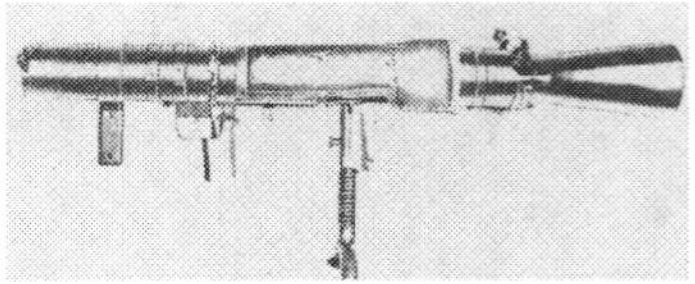 84-Millimeter Carl Gustaf Recoilless Rifle
Range, HEAT (in meters)	1,000	500
HEAT, rocket assist (RA)	NA	250
HE	NA	500
Indirect fire	NA	1,000 HE, 1,300-meter smoke, 2,300-meter illumination
Armor penetration (in millimeters)	400	500 HEAT, 900 HEAT, RA
Rate of fire, maximum and sustained (rounds per minute)	6	6
Fire control	Direct-view optics II (DVO)	3x DVO
Ammunition type	HEAT, HE, rocket-assisted projectile (RAP), armor piercing (AP) AT hollow charge	HEAT, HEAT RA, illumination, smoke, high explosive, dual purpose (HEDP), HE
Crew	3 or 4	

Table L-8. Recoilless Rifles (Continued)

Infantry Fighting Vehicle (IFV) Characteristics	 **Bronevaya Maschina Piekhota, Armored Vehicle, Infantry (BMP)-1P IFV**	 **BMP-2 IFV**	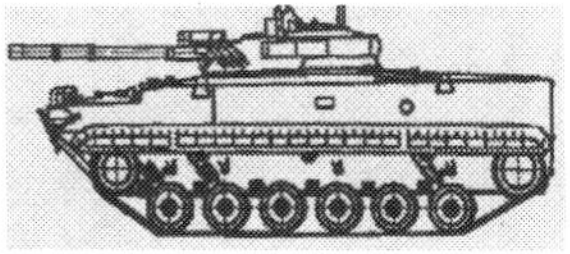 **BMP-3 IFV**
Main Armament (Caliber and Model)	73-millimeter 2A38 gun	30-millimeter 2A42 gun	100-millimeter 2A70
Ammunition type	HEAT FS, HE fragment	High explosive incendiary (HEI), high explosive tracer (HET), APT	HE fragment, HE
Range, effective (in meters)	800	1,500 light armor, 3,000 air, 2,500 ground	4,000 HE fragment
Rate of fire (rounds per minute) sustained/maximum	10	300/500	10/15

Table L-8. Recoilless Rifles (Continued)

IFV Characteristics	BMP-1P IFV (Continued)	BMP-2 IFV (Continued)	BMP-3 IFV (Continued)
Secondary Armament	AT-4a/5a ATGM	AT-4a/5a ATGM	AT-10 ATGM
Range, effective (in meters)	2,000/4,000	2,000/4,000	5,000
Rate of fire, cyclic and practical (rounds per minute)			
Penetration (millimeters at meter range)	600/650	600/650	660
Auxiliary Armament	7.62-millimeter PKT MG		30-millimeter 2A72 gun
Range, effective (in meters)			
Ammunition type			APT, HE, HET
Auxiliary Armament			
Model		7.62-millimeter PKT MG	3- by 7.62-millimeter PKT MG
Commander	IR	IR	IR
Speed, road and off road (kilometers per hour)	65/45	65/50	70/50

Table L-8. Recoilless Rifles (Continued)

IFV Characteristics	BMP-1P IFV (Continued)	BMP-2 IFV (Continued)	BMP-3 IFV (Continued)
Vehicle Characteristics (Continued)			
Range, road and off road (in kilometers)	600/570	600	600
Ground clearance (in millimeters)	390	420	190 to 510
Armor, hull/turret (in millimeters)	19/23	19/23	
Dimensions (length by width by height, in meters)	6.735 by 2.94 by 2.068	6.86 by 3.13 by 2.45	7.2 by 3.2 by 2.6
Crew and passengers	3/8	3/7	3/7

Table L-8. Recoilless Rifles (Continued)

IFV Characteristics	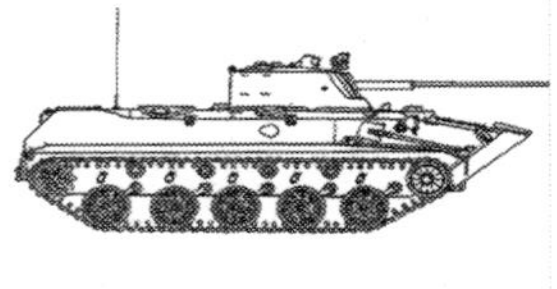 BMD-2	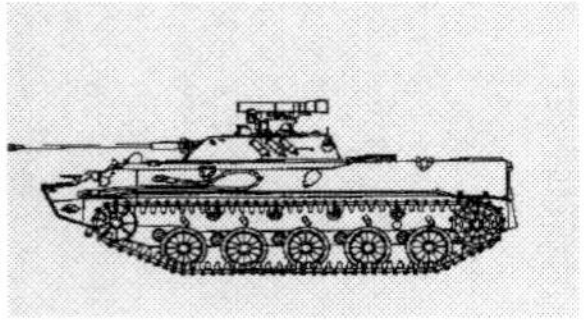 BMD-3	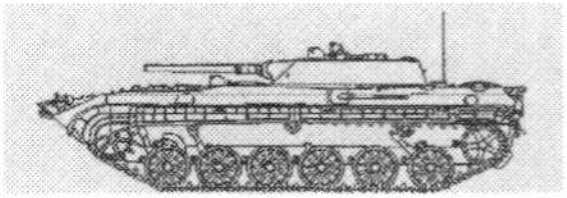 BRM
Main Armament (Caliber and Model)	30-millimeter 2A42 gun	30-millimeter 2A42 gun	73-millimeter 2A38 gun
Ammunition type	APT, fragment T, HEI	APT, fragment T, HEI	HEAT FS, FE fragment
Range, effective (in meters)	1,500 light armor, 2,000 air, 4,000 soft skin	2,000 AP, 4,000 HE	800
Rate of fire (rounds per minute) sustained and maximum	240/600	240/600	10
Secondary Armament	AT-4/5 ATGM	AT-4a/5a ATGM	7.62-millimeter PKT MG
Range, effective (in meters)	2,000/4,000	2,000/4,000	1,000
Rate of fire, cyclic and practical (rounds per minute)			650/250
Penetration (millimeters at meter range)	600 at 650	600 at 650	8 at 500 meters

Table L-8. Recoilles Rifles (Continued)

IFV Characteristics	BMD-2 (Continued)	BMD-3 (Continued)	BRM (Continued)
Auxiliary Armament		30-millimeter, AG-17 grenade launcher	
Model	2- by 7.62-millimeter PKT MG	7.62-millimeter PKT MG	
Vehicle Characteristics			
Night sights		Active IR	Second generation
Speed, road and off road (kilometers per hour)	60/35/10	70/45/10	70/10
Range, road and off road (in kilometers)	500/350	500	500
Armor, hull and turret (in millimeters)		16	10
Dimensions (length by width by height, in meters)	5.4 by 2.63 by 1.615	6.1 by 3.134 by 2.25	6.75 by 2.97 by 1.98
Crew and passengers	2/5	2/5	3/6

Table L-9. Armored Personnel Carriers (APCs)

APC Characteristics	 **BTR-60PB**	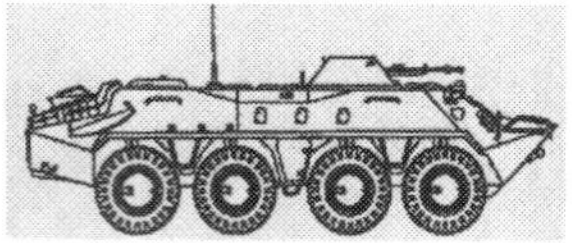 **BTR-70**	 **BTR-80**
Main Armament (Caliber and Model)	14.5-millimeter Soviet 14.5-millimeter, heavy, MG (KPVT)	14.5-millimeter MG (KPVT)	14.5-millimeter MG (KPVT)
Range, effective (in meters)	2,000 ground, 1,400 air	2,000 ground, 1,400 air	2,000 ground, 1,400 air
Rate of fire (rounds per minute) sustained and maximum	80/600	80/600	80/600
Secondary Armament	7.62-millimeter PKT MG	7.62-millimeter PKT MG	7.62-millimeter PKT MG
Range, effective (in meters)	1,000	1,000	1,000
Rate of fire, cyclic and practical (rounds per minute)	650/250	650/250	650/250
Penetration (millimeters at meter range)	8 at 500 meters	8 at 500 meters	8 at 500 meters

Table L-9. APCs (Continued)

APC Characteristics	BTR-60PB (Continued)	BTR-70 (Continued)	BTR-80 (Continued)
Vehicle Characteristics			
Commander	IR	IR	IR
Driver	IR	IR	IR
Range, road and off road (in kilometers)	500	450	800
Armor, hull and turret (in millimeters)		10/7	
Dimensions (length by width by height, in meters)	7.5 by 2.8 by 2.23	7.85 by 2.8 by 2.45	7.62 by 2.9 by 2.35
Crew and passengers	3/7	2/8	3/7

Table L-9. APCs (Continued)

APC Characteristics	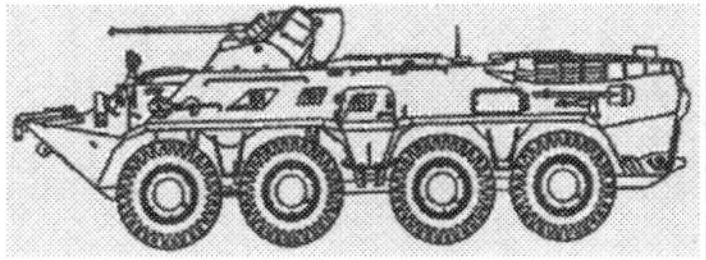 BTR-80A	 BTR-D	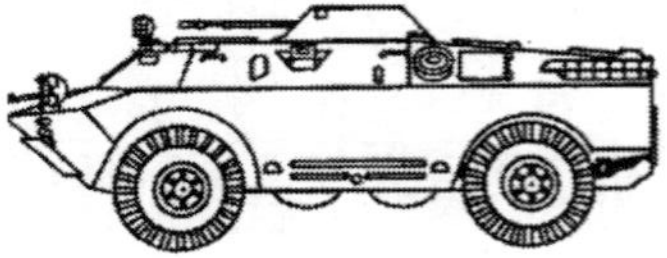 BRDM-2 Reconnaissance
Main Armament (Caliber and Model)	30-millimeter 2A72 gun	2 by 7.62 PKT MG	14.5-millimeter KPVT MG
Range, effective (in meters)	2,000 APT, 4,000 HEI, 800 night	1,000	2,000 ground, 1,400 air
Rate of fire (rounds per minute) sustained and maximum		250/650	80/600
Secondary Armament	7.62-millimeter PKT MG		7.62-millimeter PKT MG
Range, effective (in meters)	1,000		1,000
Rate of fire, cyclic and practical (rounds per minute)	650/250		650/250
Penetration (millimeters at meter range)	8 at 500 meters		8 at 500 meters

Table L-9. APCs (Continued)

APC Characteristics	BTR-80A (Continued)	BTR-D (Continued)	BRDM-2 Reconnaissance (Continued)
Vehicle Characteristics			
Night sights			
Range, road and off road (in kilometers)	800	500/350	750
Armor, hull and turret (in millimeters)			
Dimensions (length by width by height, in meters)		5.883 by 2.63 by 1.67	5.75 by 2.262 by 2.31
Crew and passengers	2/8	3/10	4
Remarks		Airborne assault. Can have AGS-17 grenade launcher. BTR-RD has 5 AT-4 antitank guided missiles (ATGMs), dismountable to 2 tripods.	

Table L-9. APCs (Continued)

APC Characteristics	 MTLB
Main Armament (Caliber and Model)	7.62 PKT MG
Range, effective (in meters)	1,000
Rate of fire (rounds per minute) sustained and maximum	250/650
Vehicle Characteristics	
Night sights	
Range, road and off road (in kilometers)	
Armor, hull and turret (in millimeters)	7/7
Dimensions (length by width by height, in meters)	6.45 by 2.85 by 1.87
Crew and passengers	2/11
Remarks	

Table L-10. ATGM

ATGM Characteristics	AT-3c SAGGER 9K11 Malyutka	AT-4a/b SPIGOT 9K111/9K111M Fagot/Faktoriya	AT-5 a/b SPANDREL 9K133/9K133m Konkurs/ Konkurs-M	AT-6 a/b/c SPIRAL 9K114 Shturm
Range (in meters)	500 to 3,000	70 to 2,000/75 to 3,500	70 to 4,000/75 to 4,000	400 to 5,000/400 to 5,000/400 to 6,000
Flight time to maximum range (in seconds)	23 to 26	11/19.5	19	13.3/16/18.6
Guidance and command link	Wire MCLOS, 9M14-2: wire SACLOS	Wire SACLOS	Wire SACLOS	Radio freguency (RF) SACLOS
Warheads	HEAT, 9M14-2: tandem HEAT	HEAT	HEAT/tandem HEAT	HEAT or FAE
Remarks	The AT-3 continues to be improved and proliferated worldwide. The M14-2 missile is effective against reactive armor.	Thermal sight detection range is 3,600 meters, with identification at 2,000 meters.	Thermal sight detection range is 3,600 meters, with identification at 2,000 meters.	FAE ranges 400 to 5,000 meters.

Table L-10. ATGM (Continued)

ATGM Characteristics	AT-7 SAXHORN 9K115 Metis	AT-8 SONGSTER 9K112 Kobra	AT-9 9K120 Ataka	AT-10 STABBER 9K116-1/9K116 Bastion/Kastet
Range (in meters)	40 to 1,000	100 to 4,000	400 to 6,000	100 to 5,000
Flight time to maximum range (in seconds)	6	10	14.5	15
Penetration (rolled hardened armor (RHA)/behind extended range artillery (ERA), in millimeters)	500	800	1,000/800	660
Guidance and command link	Wire SACLOS	RF (30 gigahertz) SACLOS		Laser beam rider SACLOS
Warheads	HEAT	HEAT	Tandem HEAT, blast, FAE, antihelicopter	HEAT
Remarks	Thermal sight detection range is 3,200 meters, with identification at 1,600 meters. Emplace and displace times are 12/20 seconds.	Fired through the main gun tube.		Fired through the main gun tube. Off BMP-3 range is 100 to 4,000 meters, but it can fire on the move.

Table L-10. ATGM (Continued)

ATGM Characteristics	AT-11a/b SNIPER 9K119/9K119M Svir/Invar	AT-12 STABBER 9K116-2 Sheksna	AT-13 9K115-2 Metis-M	AT-14 Kornet
Range (in meters)	100 to 5,000	100 to 5,000	80 to 1,500	100-5,500
Flight time to maximum range (in seconds)	15	12 to 4,000	8.4	22
Penetration (RHA/ behind ERA in millimeters)	770 AT-11a, 1,050 AT-11b	800	1,000/800	1,200/980
Guidance and command link	Laser beam rider SACLOS	Laser beam rider SACLOS	Wire SACLOS	Laser beam rider SACLOS
Warheads	HEAT/tandem HEAT	HEAT	Tandem HEAT, FAE	Tandem HEAT, FAE
Remarks	Fired through the main gun tube. T-72 variants have a 4,000-meter maximum range.	Fired from the halt through the main gun tube.	FAE has greater power than 152-millimeter artillery round.	Thermal sight range is 4,000-meter detection and 3,500-meter identification. Maximum altitude of engaged helicopters is 3,000 meters.

Table L-10. ATGM (Continued)

ATGM Characteristics	AT-16 9K121 VIKhR	Milan (France [FR]) 1/2/2T/3	Tow (US) 1/ITOW/ 2a/2b	HOT (FR/Germany [GE])
Range (in meters)	1,000 to 10,000	20 to 2,000	3,750/3,750/65-3,750/200 to 3,750	75 to 4,000
Flight time to maximum range (in seconds)	23	12.5	2a: 20, 2b: 21	16.5
Penetration (RHA/ behind ERA in millimeters)	1,050/900	600/1,060/1,200/ 1200	500, ITOW:800, 2:920, 2a 1,000	850, 2:950
Guidance and command link	Laser beam rider SACLOS (lock on before launch)	1 and 2: SACLOS wire SACLOS IR	SACLOS wire	SACLOS wire
Warheads	Tandem HEAT			
Remarks	Maximum airborne target speed is 800 kilometers per hour.	Milan 3 medium range ATGM has IR command link resistant to jamming and tandem warhead.	TOW 2b is fly over, shoot down, top attack, with self-forming penetrant 2nd generation forward looking infrared (FLIR).	HOT 3 long-range ATGM has a tandem warhead, bispectral day and night sights, and may be mounted on posts, vehicles, or helicopters.

Table L-11. Rotary Wing Aircraft

Rotary Wing Aircraft Characteristics	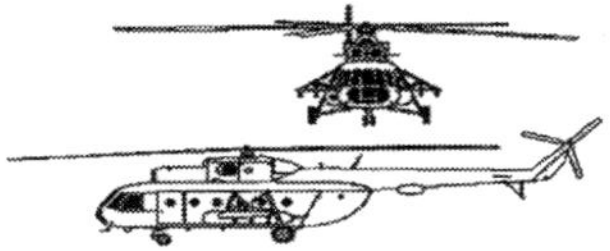 Mi-8 HIP	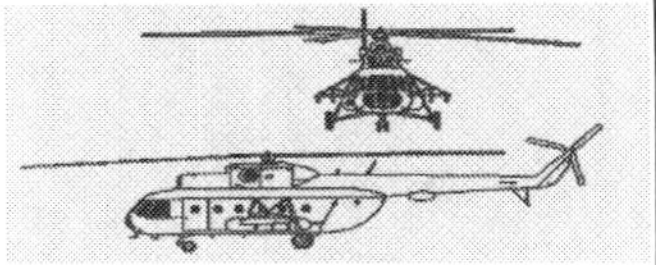 Mi-17 HIP H	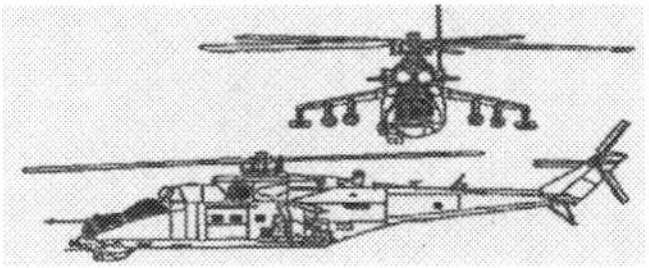 Mi-24 HIND D/E/F
Mission	Utility, transport	Utility, transport	Attack, close support
Gun	12.7 millimeter	12.7 millimeters	30-millimeter HIND F 12.7-millimeter HIND D/E
ATGM	AT-2c	AT-2c, AT-3c	4-16 AT-2c HIND D AT-6c HIND E/F
Rockets	57 millimeters	57 millimeters	57 millimeters or 80 millimeters
Bombs	250 kilograms, 500 kilograms	250 kilograms, 500 kilograms	250 kilograms, 500 kilograms

Table L-11. Rotary Wing Aircraft (Continued)

Rotary Wing Aircraft Characteristics	Mi-8 HIP (Continued)	Mi-17 HIP H (Continued)	Mi-24 HIND D/E/F (Continued)
Range, normal payload (in kilometers)	460	495	480
Crew	3	3	2
Passengers	24	24	8
Remarks	Troops can fire personal weapons through windows. It has four external hard points.	Improved version of Mi-8MT, have upgraded engines and six external hard points.	
	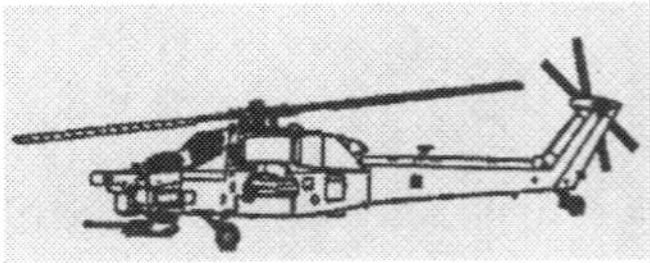 **Mi-28N HAVOC**	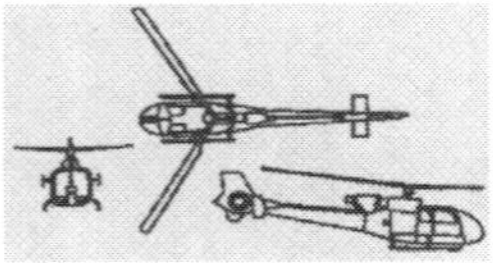 **Gazelle (FR)**	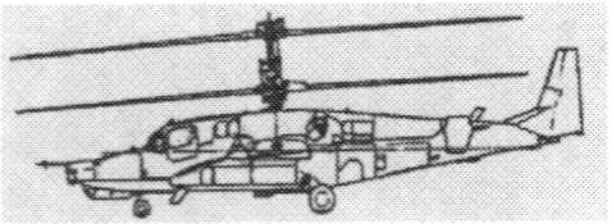 **Ka-50 HOKUM "Black Shark" or "Werewolf"**
Mission	Attack, close support	Light attack utility	Attack, close support
Gun	23 millimeters 30 millimeters	2 by 7.62 millimeters	2A42 30-millimeter cannon

Table L-11. Rotary Wing Aircraft (Continued)

Rotary Wing Aircraft Characteristics	Mi-28N HAVOC	Gazelle (FR)	Ka-50 HOKUM
ATGM	16 AT-6/9/16	AT-3, HOT	AT-16
Rockets	57 millimeters	5 millimeters 68 millimeters 2.75 inches	80 millimeters
Bombs	NA	NA	250 kilograms, 500 kilograms
Range, normal payload (in kilometers)	470 1,100 with drop tanks	710	455
Crew	2 or 3	1 or 2	1
Passengers	NA	2	NA
Remarks	This system is not operationally fielded in any armed force.	NA	The armored cockpit with-stands 23-millimeter fire and the windscreen withstands 12.7-millimeter fire.

Table L-11. Rotary Wing Aircraft (Continued)

Rotary Wing Aircraft Characteristics	MD 500 (US)	UH-1H (US)	BO-105 (GE)	Lynx (UK)
Mission	General purpose, light attack	Utility	General purpose, light attack	General purpose, attack
Gun	7.62 millimeters	7.62 millimeters	20 millimeters	2 by 20 millimeters
ATGM	TOW	NA	HOT TOW	HOT Hellfire TOW
Rockets	70 millimeters	NA	NA	Sura, 80 millimeters
Range, normal payload (in kilometers)	540	465	318	540
Crew	2	2	2	2
Passengers	2	11	3	10

Combat Aircraft Characteristics	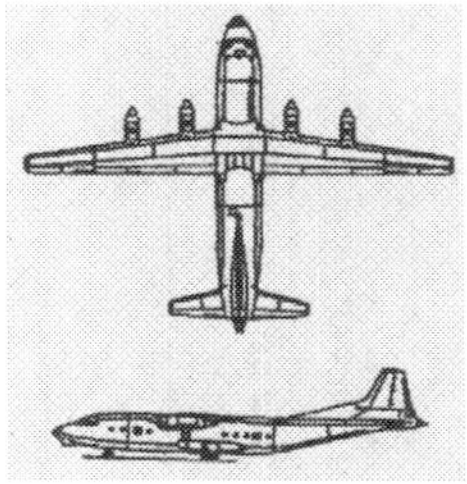 An-12 CUB	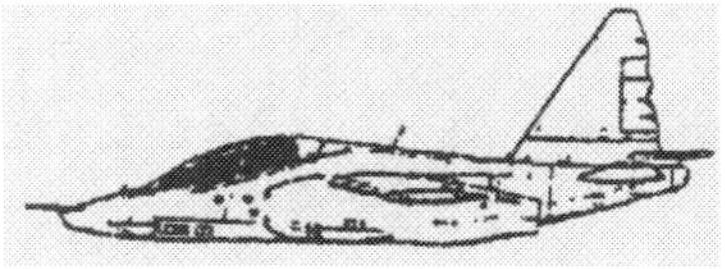 SU-25 FROGFOOT
Mission	Medium transport aircraft	CAS
Gun(s)	NR-23 23 millimeters in the tail turret	GSh-30-2, 30 millimeters (250 rounds), AO-17a, 30-millimeter pods
ATGMs		16x AT-16 or AT-9
Rockets		57-millimeter S-5 pod, 80-millimeter S-8 pod (up to 8 pods)
Air-to-surface missiles		AS-7, AS-10, AS-11, AS-14, AS-17
Bombs		100 kilograms, 350 kilograms, 500 kilograms (up to 4,000 kilograms)
Combat radius (in kilometers)	1,500 to 1,800	495
Payload, paratroops	90	
Remarks	It can operate from dirt strips.	SU-25M: AT, SU-25UB: trainer, ceiling 10,000 meters, 4,344 kilograms maximum payload, 6.5 gram limit.

Table L-12. Combat Aircraft (Continued)

Combat Aircraft Characteristics	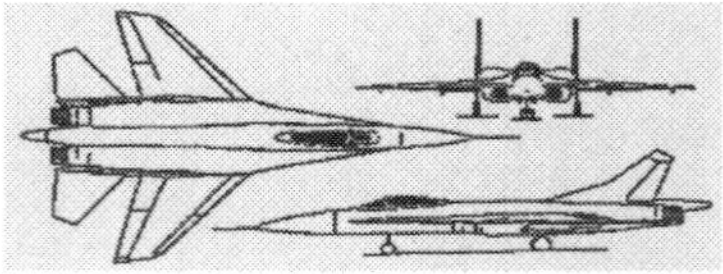 SU-27 Flanker B
Mission	Fighter/intercepter, fighter bomber variant
Gun(s)	GSh-301, 30-millimeter cannon
ATGMs	
Rockets	80-millimeter S-8 or 120-millimeter S-25
Air-to-air missiles	R-27, R-73, AA-10, AA-11
Bombs	100 kilograms, 250 kilograms, 500 kilograms
Combat radius (in kilometers)	1,125/1,950 with tanks

Glossary

∞	infinity symbol
"	inch(es)
'	foot; feet
<	more than
>	less than
°	degree
?	unknown information
AA	assembly area
AAR	after-action review
AB	A map's reference point.
ABD	air base defense
ACE	ammunition, casualty, and equipment
ADC	area damage control
AFJI	Air Force joint instruction
AFM	Air Force manual
AFSF	Air Force security force
AGS	armored gun system
AI	area of interest
alt	alternate
ammo	ammunition
ANCD	automated net control device

AO	area of operation
AOE	Army of Excellence
AOR	area of responsibility
AP	armor piercing
APC	armored personnel carrier
APL	antipersonnel land mines
APT	armor piercing tracer
AR	Army regulation; automatic rifle (SAW)
AS	area security
ASAP	all-source analysis product
ASCC	Army service component command
ASI	addition skill identifier
ASP	ammunition supply point
ASV	armor security vehicle
AT	antitank
AT/FP	antiterrorism/force protection
ATGM	antitank guided missiles
ATTN	attention
AWOL	absent without leave
BCC	battlefield circulation control
BCOC	base cluster operations center
BDOC	base defense operations center
BDU	battle dress uniform

BHL	battle handover line
BII	basic-issue item
bldg	building
BMP	Bronevaya Maschina Piekhota, armored vehicle, infantry
BPS	baud per second
BSA	brigade support area
BT	bomb trench
c	distance of the centerline of the road
C	clay; distance from the centerline
C^2	command and control
C4	composition C4
CA	cinnamic acid; civil affairs
CAC	crossing area commander
CAM	chemical agent monitor
CANA	convulsant antidote for nerve agents
CAS	close air support
CASEVAC	casualty evacuation
CCIR	commander's critical information requirements
cdr	commander
CERTEX	certification exercise
cGyph	centigray per hour
CI	civilian internee

CID	criminal investigation division
℄	centerline
cm	centimeter
co	company
COA	course of action
COMMEX	communication exercise
COMMZ	communications zone
CONEX	container express
CONUS	continental United States
CP	command post
CPOG	chemical protective overgarment
CPU	chemical protective undergarment
CR	Dibenz (B, f)-1, 4-oxazepine
CS	combat support; chlorobenzul-malononitrile
CSC	crossing site commander
CSS	combat-service support
CTOC	corps tactical operations center
CTUS	customs territory of the United States
DA	Department of the Army
DAP	decontamination apparatus
DC	dislocated civilian
DCG	degrees grid north

DD	Department of Defense
Dec	December
DED	detailed equipment decontamination
DFAS-IN	Defense Finance and Accounting Service-Indianapolis
DFC	defense force commander
DGG	degrees grid north
DGM	degrees magnetic north
DGT	degrees true north
DM	designated marksman
DNVT	digital nonsecure voice terminal
DOD	Department of Defense
DS	direct support; decontamination solution
DS2	decontamination solution number 2
DSA	division support area
DTD	detailed troop decontamination
DTG	date-time group
DTO	division transportation office
DVO	direct-view optics
DZ	drop zones
E	east
EA	engagement area
EAC	echelon above corps

EEFI	essential elements of friendly information
EEP	engineer equipment park
EIC	end item code
ENDEX	end of exercise
EO	executive order
EOC	emergency operations center
EPW	enemy prisoner of war
ERA	extended range artillery
ERP	engineer regulating points
FA	field artillery
FAE	fuel air explosive
FCF	field confinement facility
FDC	fire direction center
FEBA	forward edge of the battle area
FFIR	friendly force information requirements
FIST	fire support team
FLIR	forward looking infrared
FM	field manual
FMFM	fleet Marine force manual
FMJ	full metal jacket
FN	foreign nation
FO	forward observer

FOX	M93 NBC Reconnaissance System
FP	force protection
FPF	final protective fire
FPL	final protective line
FR	France
FRAG	fragment
FRAGO	fragmentary order
FSO	fire support officer
G	gravel; M203
G2	Assistant Chief of Staff, G2 (Intelligence)
G5	Assistant Chief of Staff, G5 (Civil Affairs)
gal	gallon
GB	guard bunker
GE	Germany
GM	grid to magnetic
GMG	grenade machine gun
GN	grid north
GP	general purpose
GPS	Global Positioning System
GS	general support
GTA	graphic training aid
GVW	gross vehicle weight

HAVOC	Soviet mi-28 aircraft
H_d	horizontal distance
HE	high explosive
HEAT	high explosive antitank
HEDP	high explosive, dual purpose
HEI	high-explosive incendiary
HET	high-explosive tracer
HHC	headquarters and headquarters company
HI	high
HIND	Soviet helicopter
HIP	Soviet medium lift helicopter
HMMWV	high mobility multipurpose wheeled vehicle
HN	host nation
HOKUM	Soviet Ka-? aircraft
HOT	air-to-ground antitank missile, also call Euromi
HTS	highway traffic section
hwy	highway
HQ	headquarters
IBCT	initial brigade combat team
ICITAP	International Criminal Investigative Training Assistance Program

ICM	improved conventional munitions
ID	identification
IEDK	individual equipment decontamination kit
IFV	infantry fighting vehicle
IPB	intelligence preparation of the battlefield
IPC	interpersonal communication
I/R	internment and resettlement
IR	infrared
IRIC	Internment Resettlement Information Center
JRA	joint rear area
Jul	July
KBPS	kilobaud per second
km	kilometers
KPVT	Soviet 14.5-millimeter heavy machine gun
L	location
L&O	law and order
LAW	light antiarmor weapon
lb	pound(s)
LC	line of contact
LCE	load-carrying equipment
LD	line of departure

ldr	leader
LMTV	light medium tactical vehicle
LO	low
LOC	lines of communication
LOGSEC	logistics security
LOTS	logistics over the shore
LP	listening post
LRA	local reproduction authorized
LVOSS	Light Vehicle Obscuration Smoke System
LZ	landing zones
m	meter; perpendicular distance from the center of the tape to the centerline of the road
M	medium; mud
MANSCEN	Maneuver Support Center
MBA	main battle area
MCCM	modular crowd control munitions
MCLOS	manual command line of sight
MCO	Marine Corps orders
MCRP	Marine Corps warfighting publication
MCT	movements control team
MCWP	Marine Corps warfighting publication
MDMP	military decision-making process

MEDEVAC	medical evacuation
MELIOS	mini eyesafe laser infrared observation set
MEVA	mission-essential or vulnerable area
METL	mission-essential task list
METT-TC	mission, enemy, troops, terrain, time available, and civilian considerations
MG	machine gun
MHZ	mega hertz
MI	military intelligence
MIJI	meaconing; intrusion; jamming; interference
mil	military
MILES	multiple-integrated laser engagement system
MILVAN	military-owned demountable container
min	minute
MLC	military load classification
MLG	mils grid north
MLM	mils magnetic north
MLT	mils true north
mm	millimeter
MMS	maneuver and mobility support
MOPP	mission-oriented protection posture

MOUT	military operations on urbanized terrain
MP	military police
mph	miles per hour
MPI	military police investigation
MRE	meal, ready-to-eat
MRO	medical regulating officer
MSR	main supply route
MTMC	military traffic management command
MTF	medical treatment facility
MTP	mission training plan
MWD	military working dogs
N	north
NA	not applicable
NAAK	nerve agent auto injector kit
NAI	named area of interest
NATO	North Atlantic Treaty Organization
NBC	nuclear, biological, chemical
NBCWRS	Nuclear, Biological, Chemical Warning and Reporting System
NCO	noncommissioned officer
NCOIC	noncommissioned officer in charge
NE	northeast
NEO	noncombatant evacuation operations

NGO	nongovernmental organization
NLP	nonlethal procedure
NLW	nonlethal weapons
no	number
NP	nonpersistent
NSN	national stock number
NTC	National Training Center
NVD	night vision device
NVG	night vision goggle
NW	northwest
NWP	Naval warfare publication
OB	obstruction
OC	observer/controller; oleoresin capsicum
OCOKA	observation and fields of fire, cover and concealment, obstacles, key terrain, and avenues of approach
OCONUS	outside the continental United States
Oct	October
OD	other detainees
OEG	operational exposure guidance
OF	observed fire
OIC	officer in charge
OP	observation post
OPCON	operational control

OPFOR	opposing forces
OPLAN	operation plan
OPNAVINST	Chief of Naval Operations Instruction
OPORD	operation order
OPSEC	operations security
ORP	objective rally point
OVM	on-vehicle material
P	pedestrian; artificial paving; persistent
PA	power amplifier
PAO	Public Affairs Office
PC	point of curvature
PCI	precombat inspection
PCIR	police and criminal information requirements
PDDE	power-driven decontamination equipment
PDF	principal direction of fire; portable document format
PERSCOM	personnel command
PEWS	Platoon Early-Warning System
PIAP	police information assessment process
PIO	police intelligence operations
PIR	priority intelligence requirements

PKT	The standard vehicle with a mounted light machine gun and is the basic Soviet coaxial AFV weapon.
PL	phase line
PLGR	precision lightweight global positioning system receiver
plt	platoon
PM	provost marshal
PMCS	preventive-maintenance checks and services
PMO	Provost Marshals Office
POE	port of embarkation
POL	petroleum, oils, and lubricants
PRC	populace and resource control
PSG	platoon sergeant
PSYOP	psychological operations
PT	point of tangency
PVAB	portable vehicle-arresting barrier
PVO	private volunteer organization
pyro	pyrotechnics
R	radio; rock; radius of curve; radar set
R&S	reconnaissance and surveillance
RA	rocket assist
RADAR	radio detection and ranging

RADIAC	radiation detection, indication, and computation
RAOC	rear-area operations center
RAP	rocket-assisted projectile
RCA	riot control agents
RES	radiation exposure status
RF	radio frequency
RHA	rolled hardened armor
RL	release line; railroad bridge
ROE	rules of engagement
ROI	rules of interaction
RORO	roll-on/roll-off
RP	release point; retained person
RPG	rocket-propelled grenade
RPK	squad machine gun
RTO	radio telephone operator
S	size; sand; south
S2	Intelligence Officer (US Army)
S3	Operations and Training Officer (US Army)
S4	Supply Officer (US Army)
S5	Civil Affairs Officer (US Army)
S&I	standardization and interoperability
SACLOS	semiautomatic command line of sight

SAGGER	An AT-3 ATGM.
SALUTE	size, activity, location, unit, time, and equipment
SAM	surface-to-air missile
SAW	squad automatic weapon
SAXHORN	A nickname for the Soviet AT-7-ATGM.
SE	southeast
SF	security forces
SINCGARS	Single-Channel, Ground-to-Air Radio System
SIR	serious incident report
SITREP	situation report
SJA	Staff Judge Advocate
SM	soldier's manual
SMCI	senior military customs inspector
SMG	submachine gun
SNAP	size, nature, activity, protection
SONGSTER	The nickname for the Soviet AT-8 ATGM.
SPANDREL	An AT-5 ATGM.
SPIGOT	An AT-4 ATGM.
SPIRAL	A Soviet AT-6 ATGM.
SOC	security operations center
SOFA	Status of Forces Agreement

SOI	signal operating instructions
SOP	standard operating procedure
SP	start point
SPG	self-propelled gun
SPOTREP	spot report
sqd	squad
SRT	special-reaction team
SSCO	small-scale contingencies operations
SSN	social security number
STABBER	A nickname for the Soviet AT-10 ATGM.
STANAG	standardization agreement
STX	situational training exercise
SVD	Soviet sniper rifle
SW	southwest
T	time; snow blockage
T&E	traversing and elevating
TA	training area
TACSOP	tactical standing operating procedure
TAI	traffic accident investigation
TC	training circular
TCF	tactical combat force
TCMD	transportation control and movement document

TCP	traffic control post
TEM	training execution model
TL	team leader
TLP	troop-leading procedures
TM	technical manual
tm	team
TO	table of organization
TOC	tactical operations center
TOE	table(s) of organization and equipment
TOW	tube-launched, optically tracked, wire-guided missile
TP	training practice
TRADOC	United States Army Training and Doctrine Command
TRL	traffic regulating line
TRP	target reference point
TRS	transportation railway service
TSC	theater support command
TTP	tactics, techniques, and procedures
U	unit
UCMJ	Uniform Code of Military Justice
UIN	unit identification number
UK	United Kingdom
UO	urban operation

US	United States
USACIDC	United States Army Criminal Investigation Division Command
USAF	United States Air Force
USAPA	United States Army Publishing Agency
USATSC	United States Army Training Support Center
USCG	United States Coast Guard
USDA	United States Department of Agriculture
USDOJ	United States Department of Justice
UTM	universal transverse mercator
UXO	unexploded ordnance
V	vehicular
V_d	vertical distance
VHS	video home system
VP	both vehicle and pedestrian
W	flooding; west
WO	warning order
Y	limited, all-weather route, significant, serious, regular, or recurrent flooding or snow blockage
X	all-weather route
Z	Zulu; fair-weather route

Bibliography

AR 190-8. *Enemy Prisoners of War, Retained Personnel, Civilian Internees and other Detainees {OPNAVINST 3461.6; AFJI 31-304; MCO 3461-1}.* 1 October 1997.

AR 190-11. *Physical Security of Arms, Ammunition and Explosives.* 30 September 1993.

AR 190-12. *Military Police Working Dogs.* 30 September 1993.

AR 190-14. *Carrying of Firearms and Use of Force for Law Enforcement and Security Duties.* 12 March 1983.

AR 190-29. *Misdemeanors and Uniform Violation Notices Referred to US Magistrates or District Courts.* 1 March 1984.

AR 190-30. *Military Police Investigations.* 1 June 1978.

AR 190-40. *Serious Incident Report.* 30 November 1993.

AR 190-47. *The Army Corrections System.* 15 August 1996.

AR 190-51. *Security of Unclassified Army Property (Sensitive and Nonsensitive).* 30 September 1993.

AR 525-13. *Antiterrorism Force Protection (AT/FP): Security of Personnel, Information, and Critical Resources.* 10 September 1998.

AR 700-93. *Processing and Shipping DOD Sponsored Retrograde Materiel Destined for Shipment to the United States, Its Territories, Trusts, and Possessions.* 15 May 1980.

DA Form 1156. *Casualty Feeder Report.* 1 June 1966.

DA Form 1248. *Road Reconnaissance Report.* 1 July 1960.

DA Form 1249. *Bridge Reconnaissance Report.* 1 July 1960.

DA Form 1251. *Ford Reconnaissance Report.* 1 January 1955.

DA Form 1355-1-R. *Hasty Protective Row Minefield Record (LRA)* 00 October 1997.

DA Form 1594. *Daily Staff Journal or Duty Officer's Log.* 1 November 1962.

DA Form 1711-R. *Engineer Reconnaissance Report (LRA).* 00 May 1985.

DA Form 1911. *Altitude and Azimuth (SIN-COS)* 00 February 1957.

DA Form 1971-R. *Radiological Data Sheet-Monitoring or Point Technique (LRA).* 00 September 1994.

DA Form 2028. *Recommended Changes to Publications and Blank Forms.* 1 February 1974.

DA Form 2404. *Equipment Inspection and Maintenance Worksheet.* 1 April 1979.

DA Form 2823. *Sworn Statement.* 1 December 1998.

DA Form 3881. *Rights Warning Procedures/Waiver Certificate.* 00 November 1989.

DA Form 3975. *Military Police Report.* 00 December 1998.

DA Form 4137. *Evidence/Property Custody Document.* 01 July 1976.

DA Form 5367-R. *Personnel Status Report (LRA).* 00 December 1984.

DA Form 5517-R. *Standard Range Cards (LRA).* 00 February 1986.

DA Pamphlet 190-12. *Military Working Dog Program.* 30 September 1993.

DA Pamphlet 190-51. *Risk Analysis for Army Property.* 30 September 1993.

DD Form 1265. *Request for Convoy Clearance.* 1 January 1959.

DD Form 1408. *Armed Forces Traffic Ticket (Book, Consisting of 25 Three-Part Sets).* 00 December 1987.

DD Form 1911. *Material Courier Receipt.* 00 May 1982.

DD Form 2707. *Confinement Order (Available on DOD Website).* 00 November 1999.

DD Form 2708. *Receipt of Inmate or Detained Person* 00 November 1999.

DD Form 2745. *Enemy Prisoner of War (EPW) Capture Tag.* 00 May 1996.

DFAS-IN 37-1. *Finance and Accounting Policy Implementation.* January 2000.

Geneva Convention Relative to the Treatment of Prisoners of War. August 1949.

FM 100-14. *Risk Management.* 23 April 1998.

FM 101-5. *Staff Organization and Operations.* 31 May 1997.

FM 101-5-1. *Operational Terms and Graphics*

{MCRP 5-2A}. 30 September 1997.

FM 101-5-2. *US Army Report and Message Formats.* 29 June 1999.

FM 17-95. *Cavalry Operations.* 24 December 1996.

FM 17-98. *Scout Platoon.* 10 April 1999.

FM 19-10. *The Military Police Law and Order Operations.* 30 September 1987.

FM 19-15. *Civil Disturbances.* 25 November 1985.

FM 19-20. *Law Enforcement Investigations.* 25 November 1985.

FM 19-25. *Military Police Traffic Operations.* 30 September 1977.

FM 20-32. *Mine/Countermine Operations.* 29 May 1998.

FM 2-11. *First Aid for Soldiers.* 27 October 1988.

FM 21-26. *Map Reading and Land Navigation.* 7 May 1993.

FM 21-60. *Visual Signs.* 30 September 1987.

FM 21-75. *Combat Skills of the Soldier.* 3 August 1984.

FM 23-25. *Light Antiarmor Weapons.* 17 August 1994.

FM 25-100. *Training the Force.* 15 November 1988.

FM 25-101. *Battle Focused Training.* 30 September 1990.

FM 27-10. *The Law of Land Warfare.* 18 July 1956.

FM 3-19.1. *Military Police Operations.* 22 March 2001.

FM 3-19.30. *Physical Security.* 8 January 2001.

FM 3-19.40. *Military Police Interment/Resettlement Operations.* 1 August 2001.

FM 3-3. *Chemical and Biological Contamination Avoidance {FMFM 11-17}.* 16 November 1992.

FM 3-4. *NBC Protection {FMFM 11-9}.* 29 May 1992.

FM 3-5. *NBC Decontamination {MCWP 3-37.3}.* 28 July 2000.

FM 3-6. *Field Behavior of NBC Agents (Including Smoke and Incendiaries) {AFM 105-7; FMFM 7-11-11}.* 3 November 1986.

FM 3-34.2. *Combined-Arms Breaching Operations.* 31 August 00.

FM 34-2. *Collection Management and Synchronization Planning.* 8 March 1994.

FM 34-130. *Intelligence Preparation of the Battlefield.* 8 July 1994.

FM 41-10. *Civil Affairs Operations.* 14 February 2000.

FM 44-8. *Combined Arms for the Air Defense.* 1 June 1999.

FM 5-170. *Engineer Reconnaissance.* 5 May 1998.

FM 55-10. *Movement Control.* 9 February 1999.

FM 55-20. *Rail Transport in a Theater of Operations.* 1 June 2000.

FM 6-20. *Fire Support in the Airland Battle.* 17 May 1988.

FM 6-30. *Tactics, Techniques, and Procedures for Observed Fire.* 16 July 1991.

FM 7-8. *Infantry Rifle Platoon and Squad.* 22 April 1992.

FM 7-10. *The Infantry Rifle Company.* 14 December 1990.

FM 90-10-1. *An Infantryman's Guide to Combat in Built-Up Areas.* 12 May 1993.

FM 90-40. *Multiservice Procedures for the Tactical Employment of Nonlethal Weapons (NLP) (S&I, CDR, USATSC, ATTN: ATIC-TMSD-T (Bldg {MCRP 3-15.8; NWP 3-07-31; USCG Pub 3-07.31}.* 6 October 1998.

Geneva Convention Relative to the Treatment of Prisoners of War. August 1949.

GTA 19-7-1. *Enemy Prisoner of War (EPW) Basic Commands.* 1989.

Section 660 of the Foreign Assistance Act.

STANAG 1059. *Distinguishing Letters for Geographical Entities for Use in NATO.* 17 April 2000.

STANAG 2010. *Military Load Classification Markings.* 18 November 1980.

TC 25-10. *A Leader's Guide to Lane Training.* 26 August 1996.

TM 11-5860-202-10. *Operator's Manual for Mini Eyesafe Laser Infrared Observation Set (MELIOS) AN/PVS-6 (NSN 5860-01-35-8551) (EIC: N/A).* 1 September 1995.

TM 11-6665-251-10. *Operator's Manual for Radiac Set AN/VDR-2 (NSN 6665-01-222-1425).* 1 March 1988.

TM 3-6665-312-12&P. *Operator's and Organizational Maintenance Manual Including Repair Parts and Special Tools List for M8A1 Automatic Chemical Agent Alarm (NSN 6665-01-105-5623) and*

Auxiliary Equipment M10 Power Supply (6665-00-859-2225); M10A1 Power Supply (6665-00-093-2739); M228 High Profile Mounting Kit (6665-00-859-2212) and M182 Low Profile Mounting Kit (6665-00-110-9492) {TO 11H2-17-1}. 17 March 1985.

TM 5-6665-298-10. *Operator's Manual for AN/PSS-12 Mine Detecting Set {TM 5-6665-298-10}*. 28 February 1995.

Index

Foreword

We remain a nation at war in an era of persistent conflict, but we do not stand alone. Our nation has many multinational partners, equally committed to freedom, rule of law and stability. It is clear that we are stronger when we act with partners in today's operating environment. Therefore, security force assistance is no longer an "additional duty." It is now a core competency of our Army.

As distinctions separating war and peace blur and challenges to security increase, we must seek to help our multinational partners successfully confront their security challenges. Security force assistance builds our multinational partners' capability to defeat regular, irregular, and hybrid threats prevalent in an era of persistent conflict.

The two pillars of security force assistance are the modular brigade and Soldiers acting as advisors. Their tactical efforts build partner capability and improve the security situation. Over time, U.S. forces and partners foreign security forces collectively set the conditions to defeat common threats and ultimately achieve strategic success.

This manual addresses common characteristics and considerations for conducting security force assistance and clarifies what units and individual advisors must understand to work "by, with, and through" their counterparts. Since every situation and foreign security force is unique, units and individuals conducting security force assistance must carefully analyze the operational environment, especially the relationships of foreign security forces and their population. Conducting successful security force assistance requires adaptive units led by well-informed, culturally astute leaders. FM 3-07.1 provides commanders, staffs, and advisors with the common doctrine for achieving this goal.

Conducting foreign security assistance requires great skill in building relationships and "leading from behind." We must all increase our understanding of this important mission.

MARTIN E. DEMPSEY
General, U.S. Army
Commanding General
U.S. Army Training and Doctrine Command

Field Manual
No. 3-07.1

Headquarters
Department of the Army
Washington, DC, 1 May 2009

Security Force Assistance

Contents

Distribution Restriction: Approved for public release; distribution is unlimited.

Figures

Preface

This field manual (FM) is the Army's doctrinal publication for security force assistance (SFA). It provides doctrinal guidance and direction for how U.S. forces contribute to SFA. It focuses on the brigade combat team (BCT) conducting SFA and advising foreign security forces. It is based on lessons learned from previous advising efforts and recent combat operations with a view to the future. It supports the Army Education System instruction on the theory and conduct of SFA.

The two primary audiences for this manual are leaders in BCTs conducting SFA and Soldiers assigned as advisors. The BCT is the largest fixed tactical organization and the key formation of the Army's modular organization. Army modularity allows commanders to add selective units to assist the BCT as it conducts SFA. The BCT and higher echelons provide the framework for advisors to function and accomplish the mission—developing foreign security forces and, when appropriate, working by, with, and through foreign security forces to achieve the desired end state of the host nation's internal defense and development. Thus, Soldiers assigned as advisors are the key element of SFA.

This FM provides the conceptual framework for conventional forces to conduct SFA within the construct of full spectrum operations, across the spectrum of conflict. It addresses SFA at operational and tactical levels.

Army doctrine is consistent and compatible with joint doctrine. FM 3-07.1 expands on fundamental principles for SFA expressed in FM 3-07. FM 3-07.1 uses text and concepts developed with North Atlantic Treaty Organization partners and ABCA (America, Britain, Canada, Australia, and New Zealand program armies).

Terms that have joint or Army definitions are italicized and the number of the proponent manual follows the definition. FM 3-07.1 will be the proponent manual for foreign security forces as a term.

FM 3-07.1 applies to the Active Army, Army National Guard/U.S. Army National Guard, and U.S. Army Reserve unless otherwise stated.

Headquarters, US Army Training and Doctrine Command is the proponent for this publication. The preparing agency is the Combined Arms Doctrine Directorate, US Army Combined Arms Center. Send written comments and recommendations on Department of the Army (DA) Form 2028 (Recommended Changes to Publications and Blank Forms) directly to: Commander, US Army Combined Arms Center and Fort Leavenworth, ATTN: ATZL-CD (FM 3-07.1), 201 Reynolds Avenue (Building 285), Fort Leavenworth, KS 66027-1352. Send comments and recommendations by e-mail to leav-cadd-web-cadd@conus.army.mil. Follow the DA Form 2028 format or submit an electronic DA Form 2028.

Introduction

> *Our strategy emphasizes building the capacities of a broad spectrum of partners as the basis for long-term security. We must also seek to strengthen the resiliency of the international system to deal with conflict when it occurs.*
>
> National Defense Strategy, June 2008

In an era of persistent conflict, the United States supports the internal defense and development of international partners, regardless of whether those partners are highly developed and stable or less developed and emerging. While many of these partners are nations, they can also include alliances, coalitions, and regional organizations. U.S. support to these partners ranges from providing humanitarian assistance to major combat operations. U.S. support includes conducting conflict transformation, bolstering partner legitimacy, and building partner capacity. A vital part of these three aspects of U.S. support is assisting partner security forces.

Security force assistance (SFA) is not new for Army forces. In fact, General George Washington's Inspector General of the Army acted as an advisor for Army forces. Baron Friedrick Wilhelm von Steuben instilled discipline and professionalism into an army that previously lacked formalized training. His 1779 *Regulations for the Order and Discipline of the Troops of the United States*, adapted from the Prussian army, formed the doctrinal backbone of the Continental Army throughout the Revolutionary War. Additionally, the lineage of the Army's operations field manual, FM 3-0, can be traced to this document. As a benefactor of advisors such as von Steuben, the Army has since undertaken what is called SFA several times throughout its history.

Army doctrine defines *security force assistance* as the unified action to generate, employ, and sustain local, host-nation, or regional security forces in support of a legitimate authority (FM 3-07). SFA is part of the FM 3-0 construct of full spectrum operations. Similarly, it is conducted across the spectrum of conflict or in any of the operational themes. SFA is normally part of a larger security sector reform effort, while in other instances, SFA is not tied to reform but to building partner capacity.

Three general situations exist in which SFA may occur: an internally focused bilateral relationship, an externally focused bilateral relationship, and a multilateral relationship. Consequently, SFA supports the appropriate partner's plans. When SFA supports a host nation, it also supports that host nation's strategy. If SFA supports a host nation's externally focused efforts, it must support the host nation's national security strategy. SFA may support regional security forces, such as those of the African Union or the Organization of American States. In these cases, SFA supports that organization's plans.

ORGANIZATION

This FM primarily focuses on the brigade combat team and advisors. It also provides a wider doctrinal framework for SFA. It is organized in the following manner:

- **Chapter 1 focuses above the brigade combat team**. It provides the overall doctrinal context for SFA.
- **Chapters 2 through 6 focus on the brigade combat team level**. These chapters provide a framework for SFA, its considerations for the brigade combat team operations process, considerations for augmenting the modular brigade for security force assistance, its considerations for the unit employment, and its sustainment considerations.
- **Chapters 7 through 10 focus on the individual advisor**. These chapters discuss the advisor, advisor cultural and communication considerations, the advisor working with counterparts, and cross-cultural influencing and communication.
- The appendices discuss Army special operations forces imperatives, legal considerations, and media considerations.

Chapter 1

The Strategic Context

This chapter places security force assistance in strategic context. First, it explains security force assistance and its relationship to full spectrum operations, spectrum of conflict, and operational themes (see FM 3-0). A discussion of U.S. national strategy and defense policies that impact security force assistance follows. This chapter then examines operations and programs related to security force assistance. Finally, it discusses the relationship of security force assistance and related operations and programs.

SECURITY FORCE ASSISTANCE AND THE STRATEGIC CONTEXT

1-1. *Security force assistance* is the unified action to generate, employ, and sustain local, host-nation or regional security forces in support of a legitimate authority (FM 3-07). Security force assistance (SFA) improves the capability and capacity of host-nation or regional security organization's security forces. These forces are collectively referred to as foreign security forces. ***Foreign security forces*** **are forces—including but not limited to military, paramilitary, police, and intelligence forces; border police, coast guard, and customs officials; and prison guards and correctional personnel—that provide security for a host nation and its relevant population or support a regional security organization's mission**.

1-2. SFA occurs within the framework of full spectrum operations (see FM 3-0). In most situations involving this assistance, there is relatively little weight on offensive and defensive operations from a U.S. perspective. However, when U.S. forces accompany foreign security forces (FSF) in combat, the weight of offensive and defense operations will change to address the situation and align with the foreign security force's efforts. SFA is not just a stability operation, although it is a key contributor to the primary stability tasks of establish civil security and establish civil control.

1-3. SFA can be conducted across the spectrum of conflict, from stable peace to general war. This assistance could focus on improving the security forces of a host nation that is currently under no immediate threat, on paramilitary forces to counter an insurgency, or on advising FSF in major combat operations against an external threat. Figure 1-1 illustrates how SFA spans the spectrum of conflict.

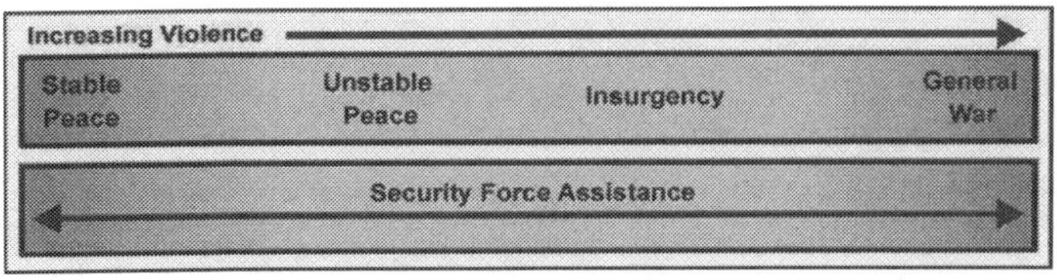

Figure 1-1. The spectrum of conflict and security force assistance

1-4. SFA can occur as part of any operational themes—peacetime limited engagement, limited intervention, peace operations, irregular warfare, and major combat operations. SFA often occurs during peacetime military engagement, peace operations, and irregular warfare. SFA involving paramilitary forces will often be limited to the irregular warfare operational theme, although paramilitary forces should eventually be absorbed, disbanded, or maintained in a long-term relationship with special operations forces.

1-5. While Army units are vital contributors, SFA is part of a comprehensive approach. This comprehensive approach should include close collaboration with military and civilian joint and multinational forces. The host-nation or regional security organization is the key actor within the comprehensive approach. It is vital that SFA is based on an objective, continuous assessment that examines the organization, training, equipment, rebuilding, and advising of the forces involved.

1-6. Law enforcement, military, intelligence, and border forces operate and cooperate within the security sector. U.S. forces understand how these units are intended to operate in the host-nation scheme, not necessarily using a U.S. model. U.S. forces should plan to help develop the respective capabilities so that these units can carry out their security functions.

1-7. Fully developed and institutionalized FSF plan, prepare, and execute at all the appropriate levels. Thus, U.S. forces must integrate SFA into the operations process at all levels. SFA should be designed to work seamlessly with the host-nation government at all levels, from ministries addressing the security sector to initial entry-level FSF training.

NATIONAL STRATEGY AND DEFENSE POLICIES

1-8. U.S. strategy and foreign policy affects all SFA. More importantly, SFA is critical to these policies as it is a primary tool for building partnership capacity with other nations. Key strategy documents for SFA include the National Security Strategy, the National Defense Strategy, the National Military Strategy, the Quadrennial Defense Review, and DODD 3000.05.

1-9. The National Security Strategy outlines the President's vision for providing enduring security for the American people in a volatile, uncertain, complex, and ambiguous strategic environment. The National Security Strategy of 2006 addresses stability operations within the broad engagement strategy for regional conflict. It supports SFA by "tailoring assistance and training of military forces to support civilian control of the military and military respect for human rights in a democratic society." This strategy also states that "[t]he world has found through bitter experience that success often depends on the early establishment of strong local institutions such as effective police forces and a functioning justice and penal system. This governance capacity is critical to establishing the rule of law and a free market economy, which provide long-term stability and prosperity."

1-10. The National Defense Strategy encompasses the Secretary of Defense's vision for the Department of Defense role in protecting the American people and U.S. interests. This strategy stems from the National Security Strategy. The 2008 National Defense Strategy highlights SFA. It states that "[o]ur strategy emphasizes building the capacities of a broad spectrum of partners as the basis for long-term security.... [l]ocal and regional conflicts in particular remain a serious and immediate problem...we will help build the internal capacities of countries at risk...[b]y helping others to police themselves and their regions, we will collectively address threats to the broader international system."

1-11. The National Military Strategy supports the National Security Strategy, National Defense Strategy, and Quadrennial Defense Review. The 2004 National Military Strategy states "[t]he Department will also work to strengthen alliances and partnerships by helping other nations increase their ability to defend themselves and protect common security interests." Working with other nations and strengthening partnerships appears throughout the National Military Strategy, which supports the importance of SFA.

1-12. The Quadrennial Defense Review is a congressionally mandated Department of Defense (DOD) review of strategy, programs, and resources. The 2006 Quadrennial Defense Review directed that joint ground forces "possess the ability to train, mentor and advise FSF and conduct counterinsurgency campaigns." This review also stated that conventional forces need to be capable of training, equipping, and advising local forces; deploying and engaging with partners; and supporting stability operations. Additionally, the review envisioned conventional ground forces continuing to perform more of the tasks traditionally performed by special operations forces. Conventional forces must understand foreign cultures and societies as well as be able to train, mentor, and advise FSF.

1-13. DODD 3000.05 provides guidance on stability operations. DODD 3000.05 establishes that stability operations are a core military mission and equal in importance to combat operations. DODD 3000.05 also reinforces the importance of SFA by stating "[t]he immediate goal often is to provide the local populace with security, restore essential services, and meet humanitarian needs. The long-term goal is to help develop host-nation capacity for securing essential services, a viable market economy, rule of law, democratic institutions, and a robust civil society."

RELATED OPERATIONS AND PROGRAMS

1-14. Several operations or programs directly or potentially relate to SFA. These operations and programs provide the context for SFA. They include security cooperation, security assistance, foreign internal defense, internal defense and development (IDAD), and security sector reform. For more information on operations related to SFA, see DOD 5105.38M, DOD security cooperation guidance, JP 3-07.1, FM 3-07, FM 3-05.137, and AR 11-31.

SECURITY COOPERATION

1-15. *Security cooperation* is all Department of Defense interactions with foreign defense establishments to build defense relationships that promote specific U.S. security interests, develop allied and friendly military capabilities for self-defense and multinational operations, and provide U.S. forces with peacetime and contingency access to a host nation (JP 3-07.1). Finally, security cooperation occurs across the spectrum of conflict and is not exclusively a peacetime activity. Security cooperation includes security assistance programs administered by the DOD as well as activities that enhance interoperability and the collective capability of combined forces (using Title 10 or exercise funding and authority). Security cooperation consists of a focused program of bilateral and multilateral defense activities conducted with foreign countries to serve mutual security interests and build defense partnerships. Security cooperation is governed by various sections of Title 10 and specific public laws addressing DOD interactions with other nations.

1-16. AR 11-31 establishes Department of the Army policy and prescribes responsibilities and procedures for the planning, integration, programming, budgeting, and execution of Army security cooperation activities. It is the policy of the U.S. Army to conduct security cooperation activities in compliance with higher-level guidance and in the execution of Army responsibilities under Title 10 and Title 22 which governs the transfer, exchange, conduct, and development of articles and services via a variety of U.S. Government programs.

1-17. Army security cooperation consists of official, cooperative, and noncombat interactions among any Army elements, Active Army, or Reserve Components. This includes any U.S. Government or nongovernment entity supporting the military and civilian joint and multinational organizations.

1-18. Commanders distinguish security cooperation from SFA based its emphasis in building relationships and capacities by using programmatic activities. Thus, security cooperation may support or be supported by SFA.

SECURITY ASSISTANCE

1-19. *Security assistance* is a group of programs authorized by the Foreign Assistance Act of 1961, as amended, and the Arms Export Control Act of 1976, as amended, or other related statutes by which the United States provides defense articles, military training, and other defense-related services by grant, loan, credit, or cash sales in furtherance of national policies and objectives (JP 3-57). Security assistance is a specific subset of security cooperation and may focus on external or internal threats.

1-20. DOD 5105.38-M describes the scope of security assistance programs in detail. Security assistance allows the transfer of military articles and services to friendly foreign governments. These transfers may be carried out via sales, grants, leases, or loans. If these transfers are essential to the security and economic well-being of allied governments and international organizations, they are equally vital to the security and economic well-being of the United States. U.S. forces can use security assistance to deliver defense weapon systems to foreign governments; to train international students; to advise other governments on improving internal defense capabilities; and to provide guidance and assistance in establishing infrastructures and economic bases for regional stability. Finally, security assistance cannot be conducted using Title 10 or exercise funds.

1-21. The military component of security assistance, implemented by DOD with policies established by the Department of State, has four principal components. They are international military education and training, foreign military sales, foreign military financing program, and peace operations. The foreign military financing program and international military education and training fall inside the military assistance budget process of the Department of State. Security assistance can also include funding peace operations.

International Military Education and Training

1-22. International military education and training contributes to internal and external security of a country by providing training to selected foreign militaries and related civilian personnel on a grant aid basis (see JP 3-07.1). These programs help to strengthen foreign militaries through training for the proper functioning of a civilian-controlled, apolitical, professional military. International military education and training serves as a foreign policy tool where the United States shapes doctrine; promotes self sufficiency in maintaining and operating U.S. acquired defense equipment; encourages the value of rule of law; and occasionally has a marked effect on the policies of the recipient governments. Foreign students, many of who occupy the middle and upper echelons of their country's military and political establishments, are taught U.S. doctrine and weapons systems employment resulting in greater cooperation and interoperability. All international military education and training activities are SFA.

Foreign Military Sales

1-23. Foreign military sales is a nonappropriated program through which foreign governments can purchase defense articles, services, and training from the United States (see JP 3-07.1). Eligible nations use this program to help build national security infrastructures. However, nations that require assistance are often unable to finance their needs. SFA supports foreign military sales when these sales include U.S. forces providing equipment and related training. It does not support foreign military sales when these sales provide equipment without related training.

Foreign Military Financing Program

1-24. The foreign military financing program provides funding to purchase defense articles and services, design and construction services, and training through foreign military sales or commercial channels (see JP 3-07.1). This program can assist nations with weak economies that would otherwise be unable to afford U.S. assistance. The foreign military financing program can fund foreign military sales. SFA and the foreign military financing program may be separate activities, although they can also be complementary.

FOREIGN INTERNAL DEFENSE

1-25. *Foreign internal defense* is the participation by civilian and military agencies of a government in any of the action programs taken by another government or other designated organization to free and protect its society from subversion, lawlessness, and insurgency (JP 3-05). U.S. foreign internal defense efforts involve all instruments of national power to support host-nation IDAD programs. The military instrument of foreign internal defense includes indirect support, direct support (not involving U.S. combat operations), and combat operations. Foreign internal defense can occur across the spectrum of conflict and within any operational theme. Foreign internal defense is often a part of irregular warfare or peacetime military engagement operational theme. SFA may support the military instrument of foreign internal defense. JP 3-07.1 discusses foreign internal defense in detail.

Note: It is a national policy decision for U.S. forces to accompany FSF on combat operations.

Military Instrument of Foreign Internal Defense

1-26. The military instrument of foreign internal defense supports the other instruments of national power through activities ranging from peacetime military engagement to major combat operations. Although foreign internal defense may include capacity building across the host nation, U.S. forces focus on combating internal threats to help the host nation maintain legitimacy and influence over the population. As such, foreign internal defense may be part of irregular warfare.

1-27. SFA and foreign internal defense support the host nation's IDAD. Consequently, forces conducting SFA must have proper mindset—they are part of a developmental and supporting effort. This applies even if SFA is part of working with a potential adversary or an existing enemy. SFA and foreign internal defense focus on working by, with, and through FSF and not conducting unilateral efforts to achieve objectives.

Indirect Support

1-28. Indirect support efforts are foreign internal defense operations that emphasize the principle of host-nation self-sufficiency. Indirect support focuses on building strong national infrastructures through economic and military capabilities that contribute to self-sufficiency. For example, indirect support could include U.S. assistance in improving a host-nation's intelligence infrastructure. Contributions from U.S. forces to indirect support come from security cooperation guidance through security assistance. They are supplemented by multinational exercises, exchange programs, and selected joint exercises. Indirect support that is focused on assisting security forces is SFA.

1-29. Although they indirectly assist those involved, military exchange programs are not SFA. Military exchange programs foster mutual understanding between forces; familiarize each force with the organization, administration, and operations of the other; and enhance cultural awareness. These exchange programs often have long-term implications for strengthening democratic ideals and respect for human rights among supported governments. However, such exchange programs (funded with Title 10 monies) do not themselves become vehicles for security assistance training, as they do not intend to assist security forces but to foster mutual understanding and enhance cultural awareness.

1-30. Joint and multinational exercises strengthen U.S. and host-nation relations and interoperability of forces, but they are not SFA. They complement security assistance and civil-military operations by validating host-nation needs and capabilities and by providing a vehicle for the conduct of humanitarian and civic assistance programs (see JP 3-07.1). Strict legal restrictions clarify the type of support that U.S. forces can provide and on the monetary limits of such support. These efforts are not SFA as they do not intend to directly assist security forces; instead, they intend to improve relations and interoperability. See appendix B.

Direct Support

1-31. Direct support does not involve combat operations but includes the use of U.S. forces providing direct assistance to the host-nation civilians or military. Direct support often focuses on civil-military operations (primarily, providing services to local civilians), psychological operations, intelligence and communications sharing, and logistic support. Sometimes direct support will be SFA, such as psychological operations, military training support, logistic support, and intelligence and communications sharing. SFA is not a civil affairs activity such as foreign humanitarian assistance, humanitarian and civic assistance, or military civic action.

1-32. Direct support differs from security assistance in that direct support is joint- or Service-funded. Direct support does not usually involve transferring arms and equipment and does not often include training local military forces. Direct support operations are normally conducted when the host nation has not attained or regained self-sufficiency and when it still faces social, economic, or military threats beyond its capability.

Combat Operations

1-33. The introduction of U.S. combat forces into foreign internal defense operations requires a Presidential decision. These operations are a temporary solution until host-nation forces can stabilize the situation and provide security for civilians. In all cases, U.S. combat operations support the host-nation IDAD program and remain strategically defensive in nature. When U.S. forces are advising, mentoring, partnering, or augmenting host-nation forces in foreign internal defense combat operations, this is SFA. When Army units conduct independent combat operations alongside FSF, this is not SFA.

1-34. U.S. tactical participation in host-nation internal conflicts requires judicious and prudent rules of engagement and guidelines for applying force. Inappropriate destruction and violence attributed to U.S. forces may easily reduce the legitimacy and sovereignty of the supported government. Adversaries may also use such incidents to fuel anti-American sentiments and assist their cause.

SECURITY RELATIONSHIPS

1-35. A complex relationship exists among security cooperation, security assistance, and the military instrument of foreign internal defense. The left side of figure 1-2 depicts this relationship, including how aspects of foreign internal defense and security assistance overlap. The right side of figure 1-2 illustrates how foreign internal defense focuses on internal threats to a host nation and how security assistance focuses on external threats. The column depicts how security (military, intelligence, and civilian), economic, and governance are considerations common to both foreign internal defense and security assistance.

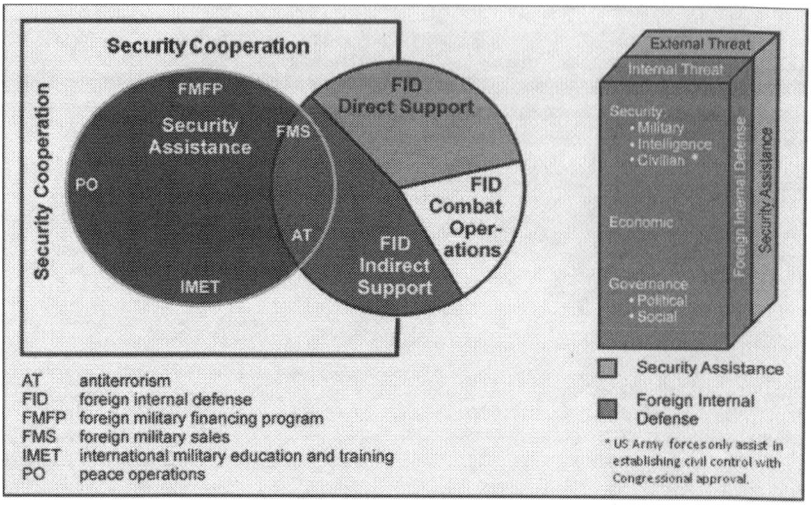

Figure 1-2. Security cooperation, security assistance, and foreign internal defense

INTERNAL DEFENSE AND DEVELOPMENT

1-36. *Internal defense and development* is the full range of measures taken by a nation to promote its growth and to protect itself from subversion, lawlessness, and insurgency. It focuses on building viable institutions (political, economic, social, and military) that respond to the needs of society (JP 3-07.1). IDAD aims to forestall or defeat the threat and to correct core grievances that prompt violence. Ideally it is a preemptive strategy. However, if an insurgency or other threat develops, it becomes an active strategy to combat that threat. To support the host nation effectively, U.S. forces, especially planners, consider the host-nation's IDAD strategy.

1-37. IDAD focuses on building viable institutions that respond to the needs of society. IDAD is a U.S. doctrinal term; it is not used universally by all countries. Internal defense and development blends four interdependent functions to prevent or counter internal threats: balanced development, security, neutralization, and mobilization. Finally, IDAD involves a cyclic interaction of execution, assessment, and adaptation. JP 3-07.1 discusses IDAD and its functions in detail.

Security Sector Reform

1-38. *Security sector reform* is the set of policies, plans, programs, and activities that a government undertakes to improve the way it provides safety, security, and justice (FM 3-07). Security sector reform aims to provide an effective and legitimate public service that is transparent, accountable to civil authority, and responsive to the needs of the public. It may include integrated activities to support defense and armed forces reform; civilian management and oversight; justice, police, corrections, and intelligence reform; national security planning and strategy support; border management; disarmament, demobilization, and reintegration; and concurrent reduction of armed violence. SFA is a vital component of security sector reform when security sector reform includes U.S. assistance to FSF. See chapter 6 of FM 3-07 for an additional discussion of security sector reform.

Security Force Assistance Relationship

1-39. Paragraph 1-35 explained and figure 1-2 supported the complex relationship among security cooperation, security assistance, and the military instrument of foreign internal defense. Figure 1-3 explains the relationship of SFA to security cooperation, security assistance, and the military instrument of foreign internal defense. SFA supports the military instrument of foreign internal defense, much of security assistance efforts, and some security cooperation efforts. The right side of figure 1-3 also shows that SFA addresses security issues, but not economic or governance issues.

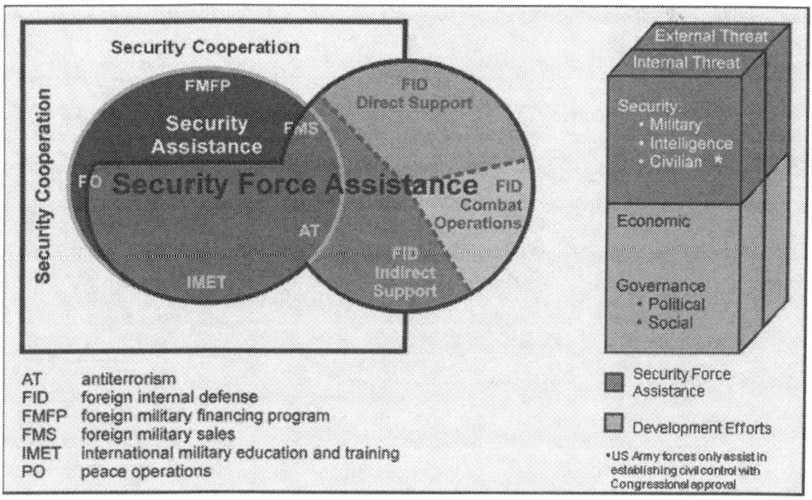

Figure 1-3. Relationship of security force assistance with security cooperation, security assistance, and foreign internal defense

COUNTRY TEAM

1-40. The country team will play a vital role with respect to SFA. The country team is the senior in-country U.S. coordinating and supervising body headed by the U.S. chief of mission. The chief of mission leads the country team and integrates U.S. efforts in support of the host nation. The chief of mission is the highest U.S. civil authority in a foreign country. As the senior U.S. Government official permanently assigned in the host nation, the chief of mission is responsible to the President for policy oversight of all U.S. Government programs. The Foreign Service Act assigns the chief of mission to a foreign country with responsibility for the direction, coordination, and supervision of all government executive branch employees in that country except for Service members and employees under the command of a U.S. area military commander.

1-41. The country team is composed of the senior member of each represented department or agency. As permanently established interagency organizations, country teams represent a priceless resource. They often provide deep reservoirs of local knowledge and interaction with the host-nation government and population. See FM 3-07 and JP 3-07.4 for more information on the country team.

THEATER COMMAND STRUCTURES

1-42. SFA is inherently multinational. Nations do not relinquish their national interests by participating in multinational operations. As in any multinational operation, commanders conducting SFA prepare to address issues related to legality, mission mandate, and prudence early in the planning process. In multinational operations, consensus often stems from compromise.

1-43. Political considerations heavily influence the ultimate shape of the command structure in which U.S. forces will conduct SFA. However, participating nations should strive to achieve unity of command for the operation to the maximum extent possible. All participants must understand missions, tasks, responsibilities, and authorities. While well defined in U.S. doctrine, command relationships are not necessarily part of the doctrinal lexicon of nations with which the United States may operate in coalition or combined operations.

1-44. Command and control relationships of higher echelons take precedence when conducting SFA. As a result, the brigade combat team (BCT) may be a supported command from a U.S. perspective as well as acting as a supporting command for FSF in the BCT's area of operations. For example, a BCT may support a joint interagency task force conducting international police training. Another example is a joint command supporting a BCT with joint fires; intelligence, surveillance, and reconnaissance; or other capabilities while the BCT supports FSF in training. The basic structures for multinational operations fall into one of three types: lead nation, integrated, or parallel command. Alliance and coalition commands use these structures.

Lead Nation Command Structure

1-45. A lead nation command structure exists when all member nations place their forces under the control of one nation. The lead nation command has a dominant lead nation command and staff arrangement with subordinate elements retaining strict national integrity. This command may also have an integrated staff and multinational subordinate forces. Integrating the staff allows the commander to draw upon the expertise of allied or coalition partners in areas where the lead nation may have less experience. This is the most desirable command structure for SFA.

Integrated Command Structure

1-46. Multinational commands organized under an integrated command structure provide unity of effort in a multinational setting. An integrated, combined command has several distinguishing features. Besides a designated single commander, the staff consists of representatives from all member nations. Subordinate commands and staffs are integrated into the lowest echelon necessary. Although effective for SFA, the integrated command structure works less effectively than the lead nation command structure.

Parallel Command Structure

1-47. Under a parallel command structure, no single force commander is designated. The coalition leadership develops a means for coordination among the participants to attain unity of effort. They can accomplish this by using coordination centers. Nonetheless, because of the absence of a single commander, the use of a parallel command structure should be avoided if at all possible. This is the least desirable command structure for SFA.

Alliance Command Structures

1-48. In combined commands, national political objectives are addressed and generally subsumed within multinational force objectives at the alliance treaty level. Combined command relationships often reflect either an integrated command structure or a lead nation command structure. In combined operations such as SFA, these structures should be used to the maximum practical extent. Combined command and force structures often mirror the degree of allied member participation. Senior military officers from member nations often lead subordinate commands. Effective operations within an alliance, including SFA, require senior political and military authorities to agree on the type of command relationships that will govern the operations of the forces. Potential political sensitivities associated with SFA can impact command relationships and operating procedures.

Coalition Command Structures

1-49. Coalitions often form in response to unforeseen crises that occur outside the area of an established alliance, which is often the case for SFA. Coalition command relationships routinely evolve as a coalition develops. Coalitions often have one of two basic structures: lead nation or parallel. For SFA, all participating coalition members should augment the host-nation headquarters staff with representatives. These representatives might include designated deputies or assistant commanders, planners, and logisticians. Such staffing provides the coalition commander with representative leadership and accessible expertise on the capabilities of the coalition members as well as facilitates the planning process. This integration of the multinational command elements into the coalition planning process should occur as early as practical.

Integration of Conventional and Special Operations Forces

1-50. Many criteria determine when conventional forces, special operations forces, or a combination are appropriate to conduct SFA. Both force levels and force characteristics suggest optimal, acceptable, and undesirable force package options in planning and resourcing SFA. Options for the deployment of a modular brigade augmented for SFA, a select number of conventional military transition teams, or special operations forces depend on conditions of the operational environment, priorities of the IDAD strategy, overall U.S. national policy, and forces available.

1-51. Rarely will U.S. forces conducting SFA be homogenous in terms of conventional or special operations forces, but rather contain both. FM 3-0 and other Army doctrine define command and control relationships and support relationships when both forces operate in the same area. Recent operational experience has shown that, in improving the effectiveness of FSF, the supported commander conducting SFA may be either a conventional force or special operations force commander. Subordinate commanders and staffs of both conventional and special operations forces may act in the supporting role to the U.S. ambassador. Typically this is done through the security assistance officer or the defense attaché at the American Embassy.

1-52. If U.S. forces are to train a large number of FSF in a short time, then the requirement may exceed the capacity of available special operations forces. Large SFA missions occur at the geographic command level or higher headquarters. Planners at these levels determine the number of conventional and special operations forces as part of their mission analysis aimed at training enough FSF in the given time frame. To best prepare for a large SFA, U.S. forces might deploy one or more augmented modular brigades supported by Army special operations forces. These forces conduct complementary foreign internal defense tasks under Title 22 authorities. Another option involves using special operations forces training FSF over several years while using a broad mix of funding and authorities. Another option uses a modular brigade to support a select number of company- or battalion-sized foreign units.

1-53. A force analysis of how best to integrate conventional and special operations forces to accomplish SFA considers host-nation and U.S. national policies. Skill sets among conventional and special operations forces trainers, advisors, and mentors avoid universally excluding or including final, integrated force packages for SFA. Planners leverage skill sets by integrating and coordinating conventional and special operations units with assigned tasks within SFA. Force analysis of the FSF to be trained often suggests a two-phase approach to training. This analysis also suggests a transition to the FSF by either conventional or special operations forces using basic and advanced training. The type of training depends on the nature of the foreign security force being trained. Additionally, integrating the advisor may be a phased process using either conventional or special operations forces as the initial advisor. This method lends itself to the classic crawl-walk-run approach to assisting FSF.

Chapter 2
Framework

This chapter discusses a doctrinal framework for the conduct of security force assistance. This security force assistance framework is composed of the mindset required of units and Soldiers, imperatives for success, inherent tasks and activities, and the three types of security force assistance.

MINDSET

2-1. Conducting successful security force assistance (SFA) requires a specific mindset. This mindset focuses on working by, with, and through foreign security forces (FSF) to support the host nation's internal defense and development (which includes local security requirements) or regional organization's charter. Soldiers conducting SFA must also understand that legitimacy is vital. The relevant population must perceive FSF as legitimate for long-term success. Those conducting SFA must understand that the military instrument of national power is only one part of a comprehensive approach. The imperatives of SFA provide the foundation for proper mindset.

Note: Host nation includes partner organization.

2-2. SFA focuses on how FSF organizations, institutions, systems, capabilities, and limitations—not U.S. organizations, systems, procedures, and equipment—can be enabled to address the host-nation or partner organization problems. Local forces have advantages over outsiders. They inherently understand the local culture and behavior that outsiders simply lack. To tap into those advantages, advisors must resist blatant military solutions. To overcome the temptation to do what they know and do best, whether relevant or not to the situation, advisors must accept that they are bound by unique situations.

IMPERATIVES

2-3. The imperatives of SFA come from the historical record and recent experience. These imperatives do not replace the principles of war or the principles of joint operations. Rather, they provide focus on how to successfully conduct SFA. The six imperatives apply to SFA at every level of war, for any echelon, and for any Soldier.

2-4. Since there is a close relationship between SFA and Army special forces conducting foreign internal defense, SFA planners should also consider the twelve Army special operations forces imperatives, especially at the tactical level (see appendix A).

Understand the Operational Environment

2-5. An in-depth understanding of the operational environment—including available FSF, opposing threats, and civil considerations—is critical to planning and conducting effective SFA. Units and Soldiers conducting SFA must clearly understand the theater, population, and FSF with which they are working, especially FSF capabilities. Diplomatic, informational, military, economic, sociological, psychological, and geographic research and understanding are essential prerequisites for successful SFA. Tactically, successful SFA requires identifying the friendly and hostile decisionmakers, their objectives and strategies, and the ways they interact. Further, the conditions of the operational environment can rapidly change, and those conducting SFA must anticipate these changes and exploit potential and possibly fleeting opportunities.

PROVIDE EFFECTIVE LEADERSHIP

2-6. Leadership, a critical aspect of any application of combat power, proves important in the dynamic and complex environments associated with SFA. The operational environment in which SFA occurs places a premium on effective leadership at all levels, from the most junior to the most senior general officer and agency director. Leading FSF or a combined group of U.S. and foreign security forces is inherently more challenging due to differences in culture, language, training, and other aspects. Leaders focus on transitions; their decisions move the foreign security force toward independent operations.

BUILD LEGITIMACY

2-7. Ultimately, SFA aims to develop security forces that contribute to the legitimate governance of the local populace. Significant policy and legal considerations may apply to SFA activities. Legitimacy is the most crucial factor in developing and maintaining internal and international support. The United States cannot sustain long-term SFA without legitimacy. Legitimacy is a concept that goes beyond a strict legal definition; it includes the moral and political legitimacy of a host-nation government or partner organization. Local civilians and the international community determine the government's legitimacy based on collective perception of the credibility of its cause and methods.

MANAGE INFORMATION

2-8. Successful SFA disseminates timely and protected relevant information, integrates it during planning, and leverages it appropriately during execution. Effective and efficient information management supports decisionmaking throughout capacity building. Managing information encompasses the collection, analysis, management, application, and preparation of information. Maintaining continuity between rotating SFA elements is critical. For long-term SFA efforts, FSF expect more professional abilities from U.S. advisors. To meet expectations and success in general, units conducting SFA establish and integrate lessons learned. Capturing and incorporating evolving tactics, techniques, and procedures and aspects of the operational environment is vital; with each successive rotation, planners strive to maintain continuity and momentum. Finally, planners synchronize effective information engagement with the entire mission. See appendix C. SFA operations inherently affect the dynamic operational environment.

ENSURE UNITY OF EFFORT

2-9. SFA often includes many actors, making unity of effort essential for success. SFA will include U.S. and foreign security forces, including conventional forces, special forces, or a combination. Other civilian and military joint and military organizations are often involved in SFA. Planners integrate them into one cohesive effort.

SUSTAIN THE EFFORT

2-10. Sustainability consists of two major components: the ability to sustain SFA effort throughout the operation and the ability of the FSF to sustain their operations independently. While each situation will vary, Army personnel conducting SFA must avoid assisting FSF in techniques and procedures beyond the FSF's capability to sustain. U.S. tactics, techniques, and procedures must be modified to fit the culture, educational level, and technological capability of the FSF. Those involved in SFA must recognize the need for programs that are durable, consistent, and sustainable by both the U.S. and FSF. They must not begin programs that are beyond the economic, technological, or cultural capabilities of the host nation to maintain without U.S. assistance. Such efforts are counterproductive.

TASKS

2-11. SFA aims to establish conditions that support the partner's end state, which includes legitimate, credible, competent, capable, committed, and confident security forces. This requires a force capable of securing borders, protecting the population, holding individuals accountable for criminal activities, regulating the behavior of individuals or groups that pose a security risk, and setting conditions in the operational area that enable the success of other actors.

2-12. This section addresses the six SFA tasks: organize, train, equip, rebuild and build, and advise and assist. When supporting the development of FSF, commanders understand their role with the host-nation or regional security organization. These tasks facilitate SFA planners to assess and allocate resources based on conditions.

ORGANIZE

2-13. Organize is a SFA task that encompasses all measures taken to assist FSF in improving its organizational structure, processes, institutions, and infrastructure. U.S. forces must understand the existing security organizations of FSF to better assist them. Subsequently, SFA personnel may help the host nation organize its security forces to meet the needs of its security environment.

2-14. Organizing a foreign security force depends on the host nation's social and economic conditions, cultural and historical factors, and security threats. SFA aims to create an efficient organization with a command, intelligence, logistic, and operations structure viable for the host nation. Conventional forces with limited special purpose teams (such as explosive ordnance disposal) are preferred. Doctrine or standing operating procedures should apply across the force, as should unit structures. The organization must facilitate the collection, processing, and dissemination of intelligence across and throughout all security forces. As the foreign security force strengthens, U.S. leaders and trainers should expect more independent organizational decisions.

2-15. The host nation determines the structure of its military forces, to include approving all organizational designs. These may include changing the numbers of forces, types of units, and internal organizational designs.

2-16. Organization should address all FSF, from the ministerial level to the patrolling police officer and soldier. Building a competent FSF infrastructure—including civilian information systems—is critical for success. Commanders work with ministries responsible for national and internal security, including the ministry of national defense, the interior ministry, and the justice ministry. Commanders assess strengths and weaknesses of the ministerial organization as well as training requirements of their employees. The joint and multinational advisor team at the ministry level helps the host nation develop a procurement and management system to meet its requirements.

2-17. Organizing a foreign security force requires resolving issues related to—

- Recruiting.
- Promotion screening and selection.
- Pay and benefits.
- Leader recruiting and selection.
- Personnel accountability.
- Demobilization of security force personnel.

Recruiting

2-18. Recruiting is critical when establishing FSF. The recruiting program assimilates local culture and themes that resonate with the populace. The program ensures that FSF have members from all major demographic groups. Joint and multinational partners should encourage and support efforts to recruit from minority populations. Mobile recruiting can target specific areas, ethnic groups, or tribes. Forces should encourage moderate groups and factions within hostile or potentially hostile ethnic groups to join the foreign security force. Often host-nation governments resist recruiting disaffected ethnic groups. However, even moderate success in recruiting from these groups provides enormous payoffs. It builds the security forces' legitimacy and often quiets legitimate fears of such groups regarding their relationship to the host-nation government. Effectively disarming, demobilizing, and reintegrating former enemies or other armed groups should be part of the overall plan. These programs should be mentioned in recruiting efforts.

2-19. A proper recruiting program specifies appropriate behavioral, physical, and moral standards. Ideally, recruits are centrally screened and inducted. Recruiting centers reside in safe areas safe. All recruits undergo a basic security check and are vetted against lists of suspected enemies. Foreign agencies and personnel perform this screening. Membership in illegal organizations needs to be carefully monitored. Past membership need not preclude a person from joining the foreign security force; however, any ongoing relationship with an illegal organization requires constant monitoring. FSF personnel limits the members of an illegal unit, tribal militia, or other militant faction to a single military or police unit.

Promotion Screening and Selection

2-20. Selection for promotion must stem from proven performance and aptitude for increased responsibility. Objective evaluations ensure promotion is by merit and not through influence or family ties. One method of selection identifies the most competent performers, trains them, and recommends them for promotion. The second method identifies those with social or professional status within the training group, then trains and recommends them for promotion. The first method may lead to more competent leaders but could be resisted for cultural reasons. The second method ensures the new leader will be accepted culturally but may sacrifice competence. The most effective solution comes from combining the two methods.

Pay and Benefits

2-21. Appropriate compensation precludes a culture of corruption in FSF. Spending the money needed for adequate wages and producing quality security forces is less costly than ending up with corrupt and abusive forces that alienate the populace. Paying the police adequately is especially important; their duties and contact with civilians often expose them to opportunities for corruption.

2-22. Payroll procedures and systems are also vitally important. They must be transparent and accountable, so FSF members receive their full pay and entitlements. Centralized administration of compensation through secure, automated systems minimizes the risk for corruption and manipulation; however, the supporting infrastructure will have to be available or developed over time.

2-23. Effective FSF can help improve host-nation social and economic development through the benefits each member receives. Every recruit should receive a basic education, job training, and morals and values inculcation.

Leader Recruiting and Selection

2-24. Officer candidate standards should be high. Candidates should have good health and pass an academic test with higher standards than the test for enlisted recruits. Recruiters carefully vet officer candidates to ensure that they do not have close ties to any radical or enemy organization.

2-25. Noncommissioned officers should be selected from the best enlisted FSF members. Objective standards, including proficiency tests, should be established and enforced to ensure that promotion to their ranks comes from merit. Many armies lack a professional noncommissioned officer corps; establishing one for a foreign security force may prove difficult. In the meantime, adjustments will have to be made, placing more responsibility on commissioned officers.

Personnel Accountability

2-26. Host-nation leaders carefully track and account for FSF. Proper personnel accountability reduces corruption, particularly in countries with manual banking systems where soldiers are paid in cash. In addition, many personnel failing to report for duty can indicate possible attacks, low unit morale, or enemy and militia influences on the foreign security force.

Demobilization of Foreign Security Force Personnel

2-27. Host nations develop programs to keep a class of impoverished and disgruntled former officers and soldiers from forming. As the foreign security force mature, officers who perform poorly or fail to meet the standards will need to be removed. Providing some form of government-provided education grants or low-interest business loans enables discharged personnel to earn a living outside the military. All soldiers who serve for several years and are then removed should receive a lump-sum payment or pension to ease their transition to civilian life. These programs should not apply to those guilty of major human rights abuses or corruption. Transition or demobilization planning should start as soon as commanders anticipate the need (it may not be required in all cases). Similar programs may be required when demobilizing nongovernment militias.

TRAIN

2-28. Train is a SFA task to assist FSF by developing programs and institutions to train and educate. These efforts must fit the nature and requirements of their security environment. The imperatives of Army special operations forces for facilitating a comprehensive approach and engaging the threat discriminately apply to the train task (see appendix A).

Training the U.S. Trainers

2-29. Soldiers assigned training missions should receive training on the requirements of developing FSF. The training should emphasize the host nation's cultural background, introduce its language, and provide cultural tips for developing a good rapport. Training should also include protection for troops working with FSF. U.S. trainees must become familiar with FSF organization and equipment, especially weapons not in the U.S. inventory. This training must emphasize the following:

- Sustaining training and reinforcing individual and team skills.
- Using the smallest possible student-to-instructor ratio.
- Developing host-nation trainers.
- Training to standards—not to time.
- Providing immediate feedback; using after action reviews.
- Respecting the culture but learning to distinguish between cultural practices and excuses.
- Learning the language.
- Working with interpreters.

2-30. U.S. forces should show respect for local religions and traditions. They should willingly accept many aspects of the local and national culture, including food (if sanitation standards permit). U.S. forces must stress that they do not intend to undermine or change the local religion or traditions. However, they have to strive to reduce how dysfunctional social practices affect the foreign security force's ability to conduct operations. U.S. trainers and advisors must have enough awareness to identify and stop inappropriate behavior, or at least report it to the appropriate chains of command.

Establishing Training Standards

2-31. Training in SFA involves many of the individual and collective skills performed in conventional military operations. All levels of training for all components should include values training. Metrics for evaluating units should include subjective measures, such as loyalty to the host-nation government, as well as competence in military tasks. Soldiers know how to evaluate military training. However, the acceptance of values, such as ethnic equality or the rejection of corruption, may be a better measure of training effectiveness. Gauging this acceptance is far more difficult than evaluating task performance. While the operational environment varies widely, FSF and trainers can still establish clear measures to evaluate the training of individuals, leaders, and units.

2-32. Effective training programs require clear, detailed individual, leader, and unit performance standards. These standards take into account cultural factors that directly affect the ability of the individual or unit to operate. For example, training a staff or unit to conduct effective operations requires more time in countries where the average soldier is illiterate. Building a security force from the ground up takes far more time than creating one around a trained cadre. Thus, using existing military personnel to form units and cadres for units often proves better than creating novice security forces.

2-33. Often, poorly trained leaders and units commit more human rights violations than well-trained, well-led units. Leaders and units unprepared for the pressure of active operations tend to use indiscriminate force, target civilians, and abuse prisoners. These actions can threaten the popular support and government legitimacy essential for SFA success. Badly disciplined and poorly led FSF can facilitate insurgent recruiting and propaganda efforts.

2-34. Setting realistic measures for FSF and following through on training plans consume time. The pressure is strong to find training shortcuts, employ quick fixes, or train personnel on the job. Trainers should resist such approaches. Such approaches often create more problems than they solve. However, trainers should also avoid the temptation to create long, complex training programs based on unrealistic standards. Effective programs account for the host nation's culture, resources, and short-term security needs. No firm rules exist on how long particular training programs should take. Trainers can use existing and historical training programs to determine how long training should take. To a certain extent, the enemy threat dictates the length of training. As security improves, training programs can expand to facilitate achievement of the long-term end state.

Training Foreign Security Forces

2-35. Members of FSF develop through a systematic training program (individual training and education as well as collective training). The program builds basic skills, then teaches them to work as a team, and finally allows them to function as a unit. Basic military, intelligence, or law enforcement training focuses first on basic skills such as legal considerations, first aid, marksmanship, and fire discipline. Leaders have training in tactics, including patrolling, urban operations, and legal evidence collection. Everyone must master rules of engagement and the law of armed conflict. FSF units should train to standard for conducting the major operations they will face.

Training Leaders

2-36. The effectiveness of FSF directly relates to the quality of their leadership. Building effective leaders requires a comprehensive program of officer, staff, and specialized training. The ultimate success of any SFA effort depends on creating viable FSF leaders able to carry on the fight at all levels and build their nation on their own.

2-37. The standards of leader training reinforce different levels of authority within the foreign security force. Clearly established responsibilities for commissioned and noncommissioned officers specify what is expected of recruits and leaders. To ensure civilian control, subordinate relationships to civilian authorities also are reinforced. In addition, training should establish team dynamics. In some cultures, security forces may need training to understand the vital role of members not in primary leadership positions. For example, noncommissioned officers may be a new or different concept for some FSF.

Employing Newly Trained Forces

2-38. SFA must build the morale and confidence of FSF. Committing poorly trained and badly led forces results in high casualties and invites tactical defeats. While defeat in a small operation may have little strategic consequence in a conventional war, often a small tactical defeat of FSF has serious strategic consequences. If the foreign security force fails, the local populace may begin to lose confidence in the host-nation government's ability to protect them. FSF must prepare for operations so they have every possible advantage. As much as possible, FSF should begin with simpler missions. As their confidence and competence grows, these forces can assume more complex assignments. Collaborating with joint or multinational units can help new forces to adjust to combat stress.

2-39. Newly trained units should enter their first combat operation in support of more experienced foreign units. Operational performance of such inexperienced organizations should be carefully monitored and evaluated to correct weaknesses quickly. The employment plan for FSF should allow enough time for additional training after each operation. Gradually introducing units into combat allows the command to identify poor leaders for retraining or other actions. Competent leaders are also identified and given greater authority and responsibility.

EQUIP

2-40. Equip is a SFA task encompassing all efforts to assess and assist FSF with the procurement, fielding, and sustainment of equipment. All equipment must fit the nature of the operational environment. The SFA principle of ensuring long-term sustainment is a vital consideration for the equip task.

2-41. The strategic plan for security force development should outline equipment requirements. Appropriate equipment is affordable and suitable against the threat. Forces must be able to train on the equipment. Interoperability may be desired in some cases. A central consideration includes the host nation's long-term ability to support and maintain the equipment.

2-42. The initial FSF development plan should use a comprehensive approach with local, regional, national, and other appropriate actors. Fundamentally, it should support the internal defense and development strategy. It should include phases with objectives for FSF to meet over three to four years. Since potential enemies adapt rapidly and situations change, commanders must continually assess the direction and progress of developing FSF.

2-43. The requirement to provide equipment may be as simple as assisting with maintenance of existing formations or as extensive as providing everything from shoes to communications and investigation kits. If the enemy uses heavy machine guns and rocket-propelled grenades, then FSF need comparable or better equipment. This especially applies to police forces, which are often lightly armed and vulnerable to well-armed enemies.

2-44. Primary considerations should include maintainability, ease of operation, and long-term sustainment costs. Few developing nations can support highly complex equipment. In most operations, having many versatile vehicles that require simple maintenance is often better than having a few highly capable armored vehicles or combat systems that require extensive maintenance. Effective maintenance systems for FSF often begin with major maintenance performed by contractors. The program then progresses to arrangements with U.S. forces as they train foreign personnel to conduct the support mission.

2-45. Sources of materials for FSF include U.S. foreign military sales (not technically SFA), multinational or third-nation resale of property, contracts with internal suppliers, or purchases on the international market. The organizations responsible for equipping FSF require flexibility to obtain equipment that meets their needs for quality, timeliness, and cost. As part of their training, FSF also need to learn property accountability to reduce corruption and ensure proper equipment usage. FSF provide equipment the same level of control and protection that U.S. forces provide for similar equipment. (See AR 12-1 and DOD 5105.38M.)

REBUILD AND BUILD

2-46. Rebuild and build is a SFA task to assess, rebuild, and build the existing capabilities and capacities of FSF and their supporting infrastructure. This task requires an in-depth analysis of the capability, capacity, and structures required to meet the desired end state and operational environment. Some FSF may require assistance in building and rebuilding, while other FSF may only need assistance in building.

2-47. FSF need infrastructure support. People need buildings for storage, training, and shelter. Often requirements include barracks, ranges, motor pools, and other military facilities. Construction takes time; the host nation needs to invest early in such facilities if they are to be available when needed. Forces must plan to protect any infrastructures, including headquarters facilities, since these building are attractive targets for insurgents. (See FM 3-34.400 for information on hardening measures to increase infrastructure survivability and improve protection.)

2-48. During some operations, such as counterinsurgency, foreign military and police forces often operate from local bases. Building training centers and unit garrisons requires a long-term force-basing plan. If possible, garrisons should include housing for the host-nation soldiers and their families; government-provided healthcare for the families; and other attractive benefits.

2-49. The host nation may need to make large investments in time and resources to restore or create the nationwide infrastructure necessary to effectively command and control FSF. The host nation will build functional regional and national headquarters and ministries as well as local bases and police stations.

ADVISE AND ASSIST

2-50. Advise and assist is a SFA task in which U.S. personnel work with FSF to improve their capability and capacity. Advising establishes a personal and a professional relationship where trust and confidence define how well the advisor will be able to influence the foreign security force. Assisting is providing the required supporting or sustaining capabilities so FSF can meet objectives and the end state. The level of advice and assistance is based on conditions and should continue until FSF can establish required systems or until conditions no longer require it. Leaders cannot permit FSF to fail critically at a point that would undermine the overall effort. See chapter 6 for a discussion on advisors.

ACTIVITIES

2-51. SFA is inherently a developmental effort. Success is measured by the increase in the foreign security force's capability, capacity, competency, commitment, and confidence in areas in which U.S. forces are assisting them and as a whole. SFA activities include plan and resource, generate, employ, transition, and sustain. Figure 2-1 depicts the SFA activities. These activities blend over time based on the capability and capacity of FSF.

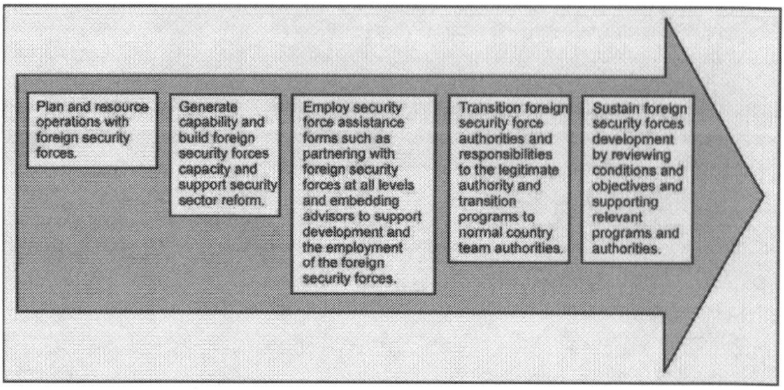

Figure 2-1. Security force assistance activities

PLAN AND RESOURCE

2-52. The plan and resource SFA activity begins as commanders understand the operational environment and determine the requirements of FSF. It also ensures that the United States provides SFA that achieves the objectives and end state of both the host nation and United States. The host nation and United States should then analyze the resource requirements and efforts so that developing FSF have sufficient and appropriate resources. Chapter 4 provides additional detail on planning for facilities and infrastructure.

GENERATE

2-53. The generate SFA activity includes generation of FSF, including the forces, leaders, and warfighting function capabilities (movement and maneuver, intelligence, fires, sustainment, command and control, and protection) based on the desired end state.

EMPLOY

2-54. The employ SFA activity involves FSF transitioning from force generation to mission employment. It does not rely on the maturity of the force or supporting institutions as a whole but is focused on the foreign element in question. Conditions determine when to use an element of FSF. Leaders of U.S. and foreign security forces assess the risk associated with employment and mitigate that risk as much as possible.

TRANSITION

2-55. The transition SFA activity defines the transition between two security forces, when applicable. This could be a transition of authority between U.S. forces to a new foreign security force. Another example could be from a regional foreign security force to a foreign security force with U.S. advisors. It could also be from a host-nation military force to a host-nation police force, with or without U.S. advisors.

SUSTAIN

2-56. The sustain SFA activity occurs when the institutional capacity of the foreign security force has been developed to a point where it is self sustaining. It may continue to have SFA contact through combined exercises, educational opportunity exchange, intelligence sharing, and foreign military sales.

TYPES

2-57. The U.S. or other actors use three types of SFA—augmenting, partnering, and advising—to develop confident, competent, capable, committed, and credible FSF. These types of SFA define the relationship between outside actors and the foreign security force. They may be employed simultaneously, sequentially, or in combination. The progression and types of SFA are determined by the operational environment, the assessment of the FSF, and by resources available. Each of these types requires decidedly different requirements, objectives, and legal considerations.

ADVISING

2-58. Advising is the primary type of security force assistance. Advising is the use of influence to teach, coach, and advise (see chapter 6) while working by, with, and through FSF. This type of SFA relies on the ability of the advisor to provide relevant and timely advice to FSF. Advising helps FSF conduct independent decisionmaking and operations. Advisors may also provide FSF with direct access to joint and multinational capabilities such as air support, artillery, medical evacuation, and intelligence. However, the advisor and advisor team require proper manning and equipment to perform these secondary support functions while staying focused on advising. Multiple sources will pressure advisors as they perform their missions. For example, nearby maneuver units may have other priorities and may focus on accomplishing their missions with or without FSF.

2-59. Advisor teams require a clearly defined and structured chain of command under which to operate. This is not only for logistics and support, but also keeps the advisor focused on developing FSF. Advisor teams will often find themselves answering to their higher military assistance group, the brigade combat team they are attached to, and the host-nation unit with which they are embedded. A well-defined chain of command alleviates confusion concerning who tasks or monitors the team's progress as well as ensures constant and adequate team sustainment.

2-60. Advisors are not partners; U.S. forces act as partners. Advising and partnering are complementary but inherently different activities. Advising requires relationship building and candid discourse to influence development of a professional security force. Partnering incorporates training with combined operations to achieve the same SFA goals. Advisors perform partnership shaping functions, shape discussions with their counterparts, and create opportunities for the partner units.

PARTNERING

2-61. Partnering attaches units at various levels to leverage the strengths of both U.S. and foreign security forces. As a foreign security force's capabilities mature, the echelon and degree of partnering decrease. As the foreign security force conducts more autonomous operations, U.S. forces still provide quick reaction forces and other assistance as appropriate.

2-62. Partner units should establish combined cells for intelligence, operations, planning, and sustainment. While effective coordination is always required and initial efforts may require completely fused efforts, FSF should eventually build the capability and capacity to conduct all efforts autonomously. These combined cells have several functions. They support transparent operations and a comprehensive approach. They also enhance the relationships among U.S. and foreign security forces by demonstrating trust. Finally, they develop the capacity of FSF in key staff areas. However, combined cells are not without risk. Operations and plans always at risk of compromise and prudent precautions should be taken.

2-63. Another partnering technique establishes relationships among command and staff elements of U.S. and foreign security forces. It requires less reorganization. It also allows counterparts to understand each other since foreign personnel can observe U.S. personnel performing their duties. This construct works well when the foreign culture is extremely sensitive to maintaining honor. In this case, foreign personnel can observe without having to expose their lack of knowledge or appear subordinate.

2-64. Unit partnerships do not replace advisor roles or functions. If partnering and advising are used in combination, it forms a three-part relationship amongst FSF, advisors, and the partner units. Partner units should look to the advisor to identify, shape, and facilitate operational partnering opportunities and training events. Advisors support U.S., coalition, and partner unit objectives, but, depending on the operational phase, the partner unit may support advisors or advisors may support the partner unit. Therefore, some level of advisor skills training should be included in the partner unit training program if those units will be conducting SFA activities.

AUGMENTING

2-65. Augmenting is an arrangement where FSF provide individuals or elements to combine with U.S. units, or U.S. individuals or elements combine with FSF. Augmentation improves the interdependence and interoperability of U.S. and foreign security forces. Augmentation can occur at many levels and in many different forms. For example, a U.S. squad can be augmented with host-nation individuals, a U.S. company can be augmented with a host-nation platoon, or a U.S. battalion can be augmented with a company from a foreign security force. Similarly, augmentation can be of short duration for a specific operation or of a longer duration for an enduring mission. Augmenting immerses FSF in a U.S. environment to provide language and cultural awareness to the U.S. unit. U.S. forces can also augment FSF.

COMBINING TASKS, ACTIVITIES, AND ASSESSMENT

2-66. Execution of successful SFA links its activities, tasks, and assessment. Figure 2-2 depicts how to arrange SFA activities, tasks, and assessment in time and purpose. The tasks that occur in one SFA activity will normally continue throughout operations. For example, organize, train, equip, rebuild and build, and advise and assist tasks will continue throughout the SFA activities of employment, transition, and sustain.

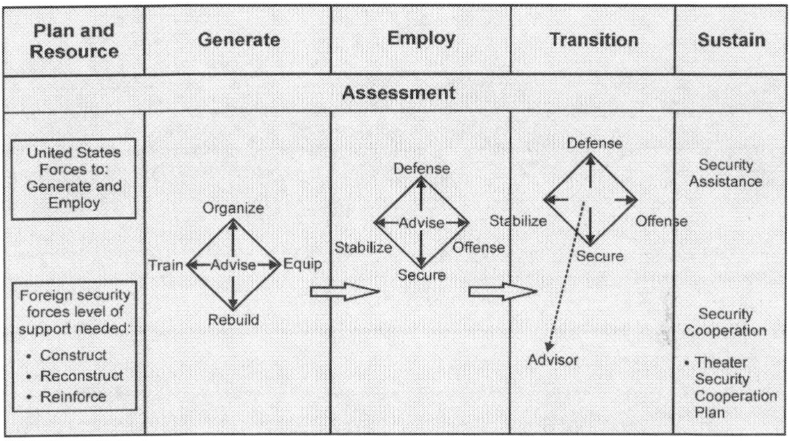

Figure 2-2. Security force assistance tasks, activities, and assessment

Plan and Resource

2-67. The plan and resource SFA activity determines the level of support and number of U.S. forces needed. As all efforts in SFA this activity is part of a comprehensive approach. Planners assess whether they will be constructing a foreign security force from the ground up, reconstructing a foreign security force based on existing capabilities and structure, or merely reinforcing an existing foreign security force. Understanding what capability and capacity the foreign security force should have and what capability and capacity they currently have is vital. Commanders and planners can then determine the size and structure, resources, and funding for the U.S. unit or units that will facilitate or assist the generation of FSF. Chapters 3 and 4 discuss planning and execution considerations for SFA.

Generate

2-68. In the generate SFA activity, U.S. forces focus on organizing, training, equipping, and advising FSF. During generation of FSF, the condition of the existing foreign security force's infrastructure and capabilities determine the assistance needed from U.S. forces. The generate activity adapts to the needs and requirements of the foreign security force. It can be designed to take the individual recruit from initial entry through to an employable unit, or it can augment an existing foreign unit with additional or enhanced capabilities. Like all SFA efforts, generate focuses on bringing a foreign security force's capabilities to a level where it can conduct its mission. With help from advisors in the next phase, FSF can work toward independent and sustained employment. In large SFA efforts, an organization may be required to command and control this activity.

2-69. Once elements of FSF are trained, the focus shifts from organizing, training, equipping, and rebuilding to employing the foreign security force. However, not all FSF units and organizations are ready for employment at the same time. As one unit is employed, others may be part of the generation activity, while still others may be completing the transition activity. FSF may be employed for offensive, defensive, or stability operations while other FSF may be employed against internal or external threats. The primary SFA task for U.S. forces is to advise and assist to facilitate the foreign security force operations.

Transition

2-70. The transition activity focuses on removing the U.S. elements and advisors once FSF are ready to conduct independent operations. All involved actors work in a comprehensive approach to assess when this can be achieved and to coordinate the transition itself. As FSF gradually progress toward the transfer of authority, close relationships forged between the FSF and their partners prove essential to sustainable development and successful transition. Genuine relationships engender trust and confidence, enabling increased responsibility and a well-executed transition process. These relationships also foster a clear understanding of command responsibilities and authorities. Such an understanding ensures FSF approach transition prepared to assume the full weight of their future role in the security sector. Success in developing FSF often depends more on relationships and personalities at the unit level than any other factor.

Chapter 3

Brigade Operations Process

This chapter examines operational considerations for the modular brigade augmented for security force assistance (SFA). It uses battle command and the operations process to examine these considerations.

BATTLE COMMAND

3-1. *Battle command* is the art and science of understanding, visualizing, describing, directing, leading, and assessing forces to impose the commander's will on a hostile, thinking, and adaptive enemy. Battle command applies leadership to translate decisions into actions—by synchronizing forces and warfighting functions in time, space, and purpose—to accomplish missions (FM 3-0). Commanders are central to battle command for brigade combat team (BCT) operations just as they are for any other operation. However, there are key considerations for successful battle command of a BCT conducting SFA. Most importantly, BCT commanders must work with their foreign counterparts, which may require facilitating operations when the foreign security forces (FSF) commander has the lead for the operations process.

UNDERSTAND

3-2. Successful SFA requires that BCT commanders thoroughly understand the operational environment. Conditions of the operational environment affect what SFA commanders provide. Commanders must understand the conditions that make up the current situation, including the relevant population, other actors, and FSF. From this understanding, commanders visualize desired conditions that represent a desired end state. Commanders must understand the capacity and capabilities of foreign security. The host-nation or regional security organization helps to determine desired conditions associated with the foreign security forces. After envisioning a desired end state, commanders then conceptualize how to change the current situation to the desired end state.

3-3. BCT commanders must understand the host-nation's security and political concerns. Security concerns focus on internal and external threats, their capability to affect the operational environment, and the foreign security force's strength and capability to counter those threats. Politically, the considerations associated with the security sector are the legitimacy of the foreign security force, FSF capability and capacity, host-nation or partner organization policy, and the ability to support the host-nation and U.S. Government objectives.

3-4. Commanders use the information obtained from intelligence preparation of the battlefield and intelligence, surveillance, and reconnaissance (ISR) to determine objectives and requirements for assessing and mitigating risk. Risk applies to how well FSF, U.S. Government, host-nation, partner organization, and international organizations can tolerate changes in the operational environment, as well as the challenges and conditions inherent to the operation.

VISUALIZE

3-5. *Commander's visualization* is the mental process of developing situational understanding, determining a desired end state, and envisioning the broad sequence of events by which the force will achieve that end state (FM 3-0). Visualization begins with assignment of the mission, which provides the focus for developing the commander's visualization. However, visualization must be continuous as SFA is dynamic. During planning, commander's visualization provides the basis for developing plans and orders.

3-6. The nature and extent of assistance provided to FSF are determined by this desired end state. SFA is often a slow process that does not lend itself to quick solutions. To ensure long-term success, commanders and host-nation or regional security organizations clarify early what conditions they desire. These conditions include the desired end state for FSF, which often include being—

- **Flexible**. FSF can accomplish broad missions. Flexibility requires an effective command and control organization appropriate for the partner or host-nation. This can include both internal and external threats.
- **Proficient**. The level of proficiency of FSF should include—
  - The capability of working effectively within a comprehensive approach.
  - The capability to integrate their operations with those of multinational partners.
  - Nonmilitary security forces competent in maintaining civil order, enforcing laws, controlling borders, securing key infrastructure, and detaining criminal suspects.
  - Nonmilitary security forces thoroughly trained in modern police ethos and procedures and which understand the basics of investigation, evidence collection, and proper court and legal procedures.
- **Self-sustained**. FSF can manage their own equipment throughout its lifecycle and perform administrative support.
- **Well led**. Leaders at all levels possess sound professional standards and appropriate military values; they are selected and promoted based on competence and merit.
- **Professional**. FSF are—
  - Honest, impartial, and committed to protecting and serving the entire population; they operate under the rule of law and respect human rights.
  - Loyal to the partner organization or host-nation government and serving partner or national interests; they recognize their role as the servants and not masters.
- **Integrated into society**. In the case of a host nation, FSF represent the host nation's major social groups; they are not seen as elements of just one faction.

3-7. These conditions may vary in different countries, but well-trained FSF should—

- Provide reasonable levels of security from external threats without threatening regional security.
- Provide reasonable levels of internal security without infringing on the populace's civil liberties or posing a coup threat to the host-nation government.
- Be founded upon the rule of law, including international principles of the rule of law.
- Be sustainable after joint and multinational forces depart.

3-8. When addressing internal threats, foreign military and police forces often perform unconventional functions. The military may have to fill an internal security role usually reserved for the police, or police may have forces so heavily armed that they would normally be part of the military. In the near term, FSF should—

- Integrate military capabilities with those of local, regional, and national police.
- Maintain the flexibility to transition to more conventional roles of external and internal defense based on long-term requirements.

3-9. To achieve this end state and intermediate objectives, the host nation should develop an internal defense and development (IDAD) strategy (with multinational assistance if required). This strategy should have a section on IDAD that addresses all aspects of SFA.

3-10. U.S. doctrine divides force development into domains: doctrine, organization, training, materiel, leadership and education, personnel, and facilities (DOTMLPF), which may be useful to analyze the IDAD. DOTMLPF elements are tightly linked, simultaneously pursued, and difficult to prioritize. Commanders monitor progress in all domains. Often military forces involved in such programs are tempted to impose their own doctrine and judgment on the host nation. Foreign security force doctrine, like the remaining DOTMLPF domains, must be appropriate to host-nation capabilities and requirements. Paragraphs 3-59 through 3-70 use DOTMLPF as a tool to assess FSF.

DESCRIBE

3-11. After BCT commanders visualize how to conduct SFA, they describe it to their staffs and subordinates to facilitate shared understanding of the mission and commander's intent. Commanders ensure subordinates understand the visualization well enough to begin planning. Commanders express their initial visualization as—

- Initial commander's intent.
- Planning guidance, including an initial concept of operations.
- Commander's critical information requirements.
- Essential elements of friendly information that must be protected.

Commander's Intent

3-12. Like any operation, SFA requires a clear, concise statement of what the force must do and the conditions the force must meet to succeed with respect to the enemy, terrain, and desired end state. In such an operation, the end state should include a description of desired conditions with respect to FSF capability and capacity. The commander's intent links the mission and concept of operations. It describes the key tasks and end state that, along with the mission, form the basis for subordinates' initiative.

3-13. A sample end state could discuss—

- How civilians perceive FSF as credible and legitimate and representing a legitimate authority.
- How FSF have the capability to support stability and security, retain the initiative and freedom of movement, disrupt enemy activities, influence the population, operate with other friendly forces, and sustain themselves.

3-14. Like any operation, how the BCT supports the higher organization's mission is vital. Additionally, the BCT supports FSF. Some sample key tasks are:

- Support the organization of a division of foreign infantry and two battalions of foreign police.
- Train leaders of FSF to create a cadre that can sustain their own training programs.
- Oversee the equipping of FSF and provide new equipment training.
- Support the rebuilding or building of infrastructure required to develop FSF.
- Advise FSF to develop the required capabilities and competencies associated with shared objectives.
- Assist the partner's security forces in meeting their operational requirements.

Planning Guidance

3-15. Commanders develop planning guidance for the staff from the commander's visualization. Planning guidance may be as broad or detailed as circumstances require; however, it must convey to the staff the essence of the commander's visualization. Due to the complex and dynamic nature of FSF, SFA often requires detailed planning guidance. Commanders ensure the staff understands their visualization while still permitting the staff to explore different options. Some considerations are:

- Friendly courses of action (COAs) with differing SFA forms and differing methods on how to augment, partner, or advise.
- Command and control architecture, intelligence architecture, and communications architecture to support dispersed SFA.
- Enemy COAs as they seek to defeat friendly efforts with respect to SFA activities.
- Enemy adaptation given the development of the foreign security force.
- Guidance for security during the conduct of SFA, including the BCT securing its advisor teams. The BCT determines the staging, needs, and ad hoc arrangements of advisor teams.
- ISR guidance to include identifying the focus of effort as it relates to SFA.

Commander's Critical Information Requirements

3-16. Commander's critical information requirements (CCIRs) will be developed for SFA just as for any other operation. These requirements are what commanders need to make decisions or the information they think is important for understanding the operational environment. These CCIRs should be compatible with the FSF commander's information requirements. CCIRs include priority intelligence requirements and friendly force information requirements.

3-17. Priority intelligence requirements are vital to support decisionmaking. During SFA, priority intelligence requirements must be focused on the enemy and the operational environment, not on FSF. U.S. forces do not focus intelligence efforts, such as collection, on the foreign forces they assist.

3-18. Friendly force information requirements provide the information that commanders need to understand the conditions of their own forces, as well as the conditions of the foreign security forces. For example, a BCT commander may direct the staff to better define a foreign security force's critical equipment shortages or maintenance requirements. During execution, commanders may focus on when FSF meet certain objectives or conditions. This information allows the commander and FSF commander to make adjustments, such as changes to the task organization, support requirements, or even the type of SFA.

Essential Elements of Friendly Information

3-19. An *essential element of friendly information* is a critical aspect of a friendly operation that, if known by the enemy, would subsequently compromise, lead to failure, or limit success of the operation, and therefore should be protected from enemy detection (FM 3-0). Essential elements of friendly information should be determined through a comprehensive approach. Information sharing and trust are vital elements of SFA. U.S. commanders must be aware of the inherent risks in this context and employ prudent risk mitigation efforts. Thus, successful SFA requires a balance between sharing information and risk mitigation efforts. Although essential elements of friendly information are not CCIRs, they have the same priority.

Intelligence, Surveillance, and Reconnaissance

3-20. *Intelligence, surveillance, and reconnaissance* is an activity that synchronizes and integrates the planning and operation of sensors, assets, and processing, exploitation, and dissemination systems in direct support of current and future operations. This is an integrated intelligence and operations function. For Army forces, this activity is a combined arms operation that focuses on priority intelligence requirements while answering the commander's critical information requirements (FM 3-0). For the brigade, these efforts will often support FSF. Moreover, the BCT must have means that allow effective intelligence collaboration, such as an intelligence fusion cell with the participation of all appropriate actors.

3-21. ISR efforts focus on the enemy and operational environment within the context of SFA. ISR efforts in SFA are not used on the foreign security force. Rather, U.S. forces assess the foreign security force within a comprehensive approach.

DIRECT

3-22. BCT commanders direct all aspects of operations, although a comprehensive approach, including FSF and other appropriate actors is fundamental to successful SFA. The commander's direction takes different forms during planning, preparation, and execution. Commanders make decisions and direct actions based on their situational understanding, which they maintain by continuous assessment and coordination with FSF. SFA is directed by—

- Preparing and approving plans and orders with FSF.
- Assigning and adjusting missions, tasks, task organization, and control.
- Positioning units to maximize their support to FSF, anticipate threat actions, or create or preserve options.
- Positioning key leaders to ensure observation and supervision of FSF at critical times and places.

- Adjusting SFA and other support priorities and allocating resources based on FSF priorities, as well as opportunities and threats.
- Changing support arrangements, either internally or with respect to FSF.

ASSESS AND LEAD

3-23. Effective battle command requires commanders to continuously assess and lead. Assessment helps commanders to better understand current conditions and determine how the operation is progressing and how best to support FSF. Based on an assessment that is the product of a comprehensive approach, commanders modify plans and orders to better accomplish the mission. This may include changing the type or combination of SFA, support, sustainment, or task organization. If their assessment reveals a significant variance from the original shared commander's visualization, a comprehensive approach must be used to reframe the visualization and develop a new plan.

3-24. Commanders in a BCT conducting SFA lead by example and presence, as Army commanders do in any operation. However, commanders have the additional challenge of leading by, with, and through FSF. Command posts for BCTs conducting SFA have similar challenges in that they should be integrated headquarters as much as practical.

OPERATIONS PROCESS

3-25. As they would in any other mission, BCT commanders conducting SFA balance their time and the staff's time and resources among four major activities in a continuous learning and adaptive cycle called the operations process. The *operations process* consists of the major command and control activities performed during operations: planning, preparing, executing, and continuously assessing the operation. The commander drives the operations process (FM 3-0). While simple in concept, the operations process is complex in execution. The operations process activities may be sequential or simultaneous.

3-26. Figure 3-1 depicts the operations process activities as a cycle that is driven by battle command. This cyclic process applies equally to a BCT conducting major combat operational theme or operations within an irregular warfare operational theme. During the operations process involving SFA, commanders consider the presence and inclusion of other actors, including FSF. Policy, the operational environment, and the forms of SFA determine how U.S. forces, FSF, and other actors interact in the operations process.

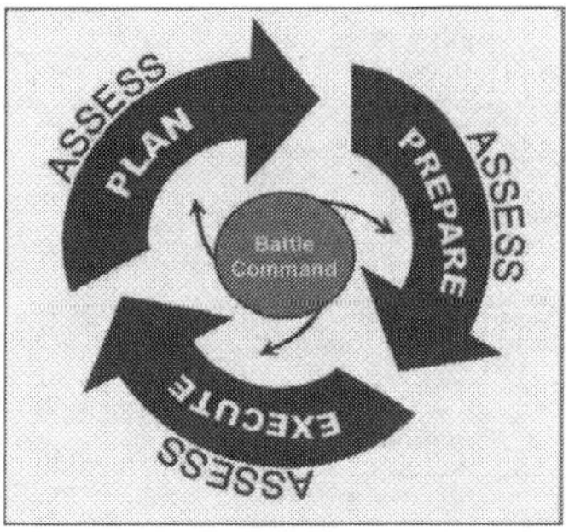

Figure 3-1. The operations process

PLAN

3-27. Planning for SFA, like any other operation, begins either with the anticipation of a new mission or the receipt of mission as part of the military decisionmaking process. *Planning* is the process by which commanders (and the staff, if available) translate the commander's visualization into a specific course of action for preparation and execution, focusing on the expected results (FM 3-0). SFA requires a comprehensive approach to ensure the U.S. commander's visualization synchronizes with policy and the foreign commander's visualization. SFA is the major requirement for a BCT conducting SFA, but the mission may require the brigade to perform other tasks.

3-28. Continuous and open to change, planning for SFA includes identifying how to best assist the FSF and developing a sequence of actions to change the situation. This planning must occur within a comprehensive approach that includes FSF, U.S. forces, and other actors. Similarly, this plan must be nested within policy, IDAD strategy, the campaign plan, and any other higher-echelon plans. It also involves anticipating consequences of actions and developing ways to mitigate them.

3-29. The plan, which includes the commander's intent, provides a common understanding to U.S. and foreign security forces on the actions to take. Plans and orders also provide decision points and branches that anticipate options that enable the force to adapt as the operation unfolds. This is especially important for SFA, as these operations tend to be prolonged efforts. BCTs conducting SFA often rotate before achieving all objectives. As a result, the planning team should establish objectives and milestones that can be achieved during the brigade's mission. These BCT objectives and milestones must support higher-echelon plans, including the campaign plan and IDAD strategy.

Functional Requirements

3-30. The composition of advisor teams is subject to objectives and conditions. For example, security conditions may require three armored vehicles, a crew-served weapon, and nine personnel to conduct independent movement. Another example involves the difference in the number of advisors. Newly generated FSF may require many advisors whereas experienced FSF may only require assistance in a few warfighting functions and consequently need fewer advisors. Understanding the functional requirements provides flexibility to the units assisting FSF. Three examples of determining functional requirements based on the understanding the operational environment include the BCT, a partnering or augmenting unit, and an advisor team.

Brigade Combat Team

3-31. A BCT headquarters has specific functions and requirements. These can include, but are not limited to—

- The functions of—
  - Carrying out SFA tasks and assisting FSF in support of objectives and the end state.
  - Facilitating partnerships between U.S. and foreign security forces.
  - Requesting, coordinating, and synchronizing resources.
  - Providing command and control of subordinate units and assigned advisor teams.
  - Coordinating with other commands and adjacent units.
  - Coordinating with other participants integrated within the comprehensive approach.
  - Developing, implementing, and synchronizing information engagement in assigned areas.
  - Developing relationships and an understanding with appropriate actors involved in SFA.

3-32. The requirements of—

- Having enough staff to handle the communications traffic generated by conditions.
- Being able to analyze information to aid commander and subordinate elements.
- Being able to articulate resource requirements in support of subordinate elements and efforts to conduct SFA tasks.
- Being organized and equipped to conduct operations and support and sustain FSF.
- Including regionally oriented host-nation officers to provide regional expertise.

Partnering or Augmenting Unit

3-33. Functions and requirements depend on the degree and level of partnering or augmenting. No prescriptive way to partner or augment exists; it is based on specific conditions and subject to the needs of policy, higher headquarters guidance, and needs of the FSF. A BCT may designate one battalion to partner or augment while designating other battalions to advise. A partnering or augmenting unit conducting SFA has specific functions and requirements. These can include, but are not limited to—

- The functions of—
  - Partnering with foreign security under the BCT level.
  - Increasing capability, capacity, competency, confidence, and commitment of FSF by conducting combined tactical operations.
  - Providing feedback on performance of FSF and certain shortfalls to embedded advisors and BCT headquarters.
  - Conducting sustainment and medical training with FSF at their home station or on operations.
  - Reporting conditions in the operational environment.
- The requirements of—
  - Having U.S. forces that are capable, competent, committed, and confident.
  - Being interoperable with FSF.
  - Being available for an appropriate period.

Military Advisor Team

3-34. The functions and requirements of a military advisor team are based on conditions—requirements, availability, and footprint. A military advisor team conducting SFA has specific functions and requirements. These can include, but are not limited to—

- The functions of—
  - Under the BCT headquarters, advising, training, and assisting FSF to which it is assigned.
  - Increasing capability, capacity, competency, confidence, and commitment of FSF by providing advice and support during battalion level and higher operations.
  - Assessing partner leaders, staff, and certain shortfalls to BCT headquarters.
  - Conducting sustainment training with FSF at their home or on operations.
  - Reporting on conditions in the operational environment.
- The requirements of—
  - Being manned with trained personnel who are properly equipped so as to be capable of executing their mission in current conditions.
  - Providing personnel with rank, education, and experience corresponding to the echelon and type of unit being advised.

Support and Sustainment Requirements

3-35. Support and sustainment requirements play a vital role in COA development (see chapter 6). The BCT may operate across a broad geographical area when conducting or supporting SFA. Partnering may be on certain levels with FSF and may support numerous embedded advisor teams or other actors. Since FSF often have varied relationships with higher echelons and institutions, the BCT may be involved vertically and spatially. These considerations may require more support and sustainment requirements than an initial analysis might indicate. Support and sustainment considerations include, but are not limited to—

- Distribution of FSF to which U.S. forces are providing SFA.
- The capability of FSF to provide support, sustainment, and medical functions.
- The technical capability and interoperability of U.S. and foreign security forces.
- The infrastructure that supports SFA.
- Overall security conditions where SFA occurs.

- The available contracting support.
- The ability of the environment to accept and absorb the U.S. forces' footprint.

Determining a Course of Action

3-36. COAs for SFA require an understanding of the breadth and width of the security sector. They may have to account for the roles ranging from the individual soldier or policeman to the ministerial level. COA development examines the roles other actors, for example, a gendarme or paramilitary force, in the security sector. Because the United States does not have a gendarme, U.S. forces may not fully appreciate its importance.

3-37. COAs for SFA also need to be flexible and inclusive. Commanders consider the following when examining a partner's security sector:

- The status and challenges of the host-nation security sector or regional security organization.
- The composition and disposition of the partner's security sector.
- Supporting institutions.
- Links among ministries, institutions, and components of the security sector.
- Partner capability and capacity shortfalls and impediments to their development.
- Environmental requirements to enable the building of security capability and capacity.
- Functional requirements to assist building capability and capacity.
- The impact of other actors within the comprehensive approach.

3-38. Once planners analyze the security sector, they can develop COAs that best combine SFA tasks, activities, and types (see chapter 2). While the advising type of SFA can be of any duration, partnering and augmenting tend to be enduring if used.

3-39. All SFA types tend to be more effective if they are long-term efforts. Advising fosters a more personal relationship and infers a stronger commitment to see the partner succeed. Partnering and augmenting include U.S. and foreign security forces conducting operations together. There are levels of partnering. Different advisory relationships use advisor teams embedded within various partner echelons and units, use one advisor team to advise multiple entities, or have a senior leader visit a partner unit to give advice.

3-40. Using any form or combination of forms depends on conditions (such as the level of competency), capability of FSF, and resources available. Policy, guidance, or formal agreements can also help commanders determine which form to use. Historically, a combination of forms appears to be the most effective. However, combining forms often proves more resource intensive and affected by time, geography, and the number of forces. COAs should remain flexible with respect to the form of SFA. Some example changes in conditions are the increased capability of one element in the security sector over another, the desire to increase one element over another based on partner objectives, or the increased targeting of one element over another by the enemy.

3-41. If a desired COA is not feasible given the scope of the security sector or due to other considerations, it may be useful to sequence SFA forms with portions of FSF. For example, a COA that affects smaller portions of the FSF, but in the right area and at the right time, may do more for achieving an objective or end state than a COA that attempts to support all FSF at one time.

3-42. BCTs are task-organized and augmented to accomplish a specific mission. How BCTs are specifically task-organized and augmented is based on the phase and types of SFA that policy and higher headquarters have directed. Chapter 4 describes augmentation in detail.

3-43. Attached units or units under operational control are task-organized under different headquarters as necessary to accomplish assigned and implied SFA missions. Deciding to task-organize specific formations for SFA may reflect previously approved contingency plans or result from current planning efforts. As these missions end, units return to their parent headquarters or are further task-organized by the controlling headquarters.

3-44. As with other operations, gaining commanders task-organize for SFA by designating operational control, attached, or support relationships to another unit or headquarters. Within each BCT, the brigade commander task-organizes all organic, assigned, attached, or operational control units to best suit the operational environment and types of SFA. Commanders may choose to use a support relationship (direct support, reinforcing, or general support reinforcing) among units of the BCT instead of a command relationship. This can also apply to relationships with foreign units.

3-45. SFA is often highly dispersed, as BCTs may have individuals or units assisting FSF over a wide area. This dispersion increases communications and sustainment requirements for the BCT. The nature of SFA also requires BCTs to help FSF meet requests for assets. To do so, the BCT and other units may need to task-organize to provide tactical support or sustainment. Dispersion and the nature of SFA often necessitate a decentralized leadership approach to facilitate effective and rapid support of FSF.

3-46. BCT commanders use the three types of SFA—advising, partnering, or augmenting—to accomplish the mission. The exact nature of a unit's assigned mission, the operational variables, and the mission variables drive modifications to headquarters and maneuver units. For example, a commander may need a unit tasked with advising, a unit tasked with partnering, and a unit tasked with augmenting. In this case, the unit tasked to advise would provide teams for the appropriate FSF while other foreign units continue their traditional missions. The unit that is partnering may provide the host-nation staff to assist the FSF headquarters. The third unit may provide squads to augment foreign platoons. Units may request additional support, sustainment, and medical support. However, BCTs conducting SFA must retain a reserve to conduct full spectrum operations.

3-47. Based on the overall mission analysis, including an analysis of functions and requirements, commanders and staffs begin to determine how to task organize. Due to the level of detail that may be involved, task organization should begin early in the military decisionmaking process, especially during COA development. No standard task organization exists since conditions vary. For example, a squad of infantry and three additional leaders may work as an advisor team for one mission, where another mission may require an advisor team with specialists for different warfighting functions. The art in SFA involves determining where, when, and if to allocate resources; it involves either breaking up or simply moving personnel and units from one foreign unit to another.

3-48. The techniques used by BCT commanders and staffs to allocate other resources and support are just as valid in SFA as in other types of operations. For example, if it is determined that the most important effort is advising, then the BCT commander may weight that effort with additional resources, such as communications equipment or experienced personnel.

3-49. BCTs emphasize unity of effort with FSF. Ideally, interoperability and shared resources extend to the lowest level. This means having one tactical operations center, dining facility, clinic, motor pool, and so on. Not only does this use resources more efficiently, but it also facilitates understanding, trust, and rapport between U.S. and foreign security forces. For Soldiers to effectively conduct SFA work by, with, and through FSF requires a close rapport and high level of trust.

3-50. Part of COA development for SFA includes an analysis of where to focus the brigade's subordinate units. This analysis includes determining which subordinate units support key elements of FSF. The brigade's subordinate units may support provincial reconstruction teams or transition teams that are not organic to the brigade. Another method is for the brigade may provide organic forces to make up transition teams or the military portion of provincial reconstruction teams or transition teams. Transition teams may include, but are not limited to police, border, point of entry, or military transition teams. Figure 3-2 depicts an example for a brigade providing forces as part of a provincial reconstruction team as well as providing the other transition teams. Figure 3-2 is an example of how a brigade may focus its subordinate units. Additionally, the teams are only examples. All of these teams may not be necessary for all brigades, and some brigades may require other types of advisor teams.

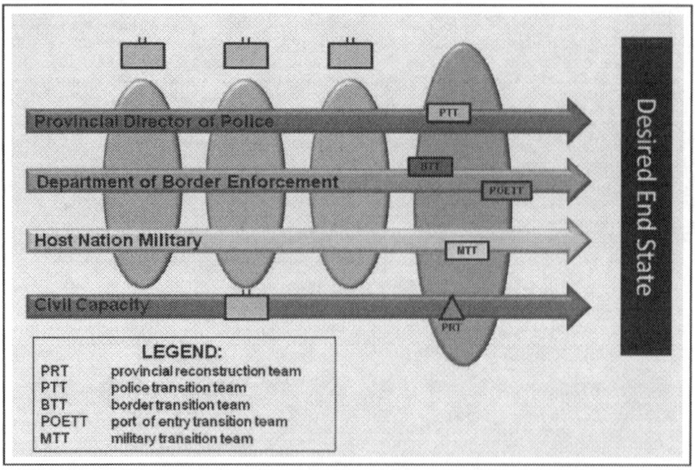

Figure 3-2. Example of brigade focus for subordinate units

PREPARE

3-51. *Preparation* consists of activities performed by units to improve their ability to execute an operation. Preparation includes, but is not limited to, plan refinement; rehearsals; intelligence, surveillance, and reconnaissance; coordination; inspections; and movement (FM 3-0). Like any operation, preparation for SFA creates conditions that improve friendly forces' opportunities for success. It facilitates and sustains transitions, including those to branches and sequels, which are of vital importance for the often dynamic operational environment for SFA. Chapter 5 covers preparation for units conducting SFA, including the BCT.

EXECUTE

3-52. *Execution* is putting a plan into action by applying combat power to accomplish the mission and using situational understanding to assess progress and make execution and adjustment decisions (FM 3-0). Execution of SFA focuses on supporting the needs of the FSF and, like many other operations, is often characterized by phases and transitions between phases. Chapter 4 examines BCT phases for SFA.

Acceptance of Prudent Risk

3-53. SFA, like any other operation, requires prudent, disciplined risk-taking focused on winning rather than preventing defeat. Because uncertainty exists in all operations, every decision involves risk. Among key elements of the art of command are deciding how much risk to accept and minimizing the effects of accepted risk (see FM 6-0). Due to SFA's inherent collaborative nature, determining prudent risk is complex and requires a comprehensive approach.

3-54. Risk reduction measures identified in planning add to the plan's flexibility during execution. A flexible plan can mitigate risk by partially compensating for a lack of intelligence. Again, SFA requires a thorough comprehensive approach to analyzing and agreeing upon risk reduction measures.

Execution Keys to Success

3-55. History and recent operations point to several keys to success for BCTs executing SFA:

- Communication is vital, especially when the BCT shares its operational area with other entities that have cultural differences and lack of communication. It can include FSF or multinational forces, as well as U.S. forces with another mission. For example, an advisor team can develop effective relationships with its foreign security force and local authorities only to have another unit conduct an operation that unhinges those relationships.
- BCTs establish close and continuing relationships with all advisor teams, other actors operating in their operational area, and foreign area officers with local or regional expertise. In addition to advisor teams assigned to military units in direct contact with the BCT, advisors may embed with local police, border, and other forces.
- BCTs establish close and continuing relationships with all foreign units (military, police, and others) operating in their operational areas.
- BCTs establish close and continuing relationships with all political entities and actors in their operational areas.
- Before each operation, as part of the military decisionmaking process, the BCT considers second- and third-order effects of the operation. The team coordinates the commander's intent with advisor teams of FSF and other units in the operational area.
- Unit leaders develop cultural awareness and use this awareness so that operations and relationships achieve the desired end state. In short, the closer a maneuver commander works with advisor teams and the more they interact with local political and cultural leaders, the better the overall chances of mission success.

ASSESS

3-56. The operations process and effective battle command require commanders to continuously assess and lead. Assessment helps commanders to better understand current situations and how SFA is progressing. *Assessment* is the continuous monitoring and evaluation of the current situation, particularly the enemy, and progress of an operation (FM 3-0). Assessment is conducted during planning, preparation, and execution and is evaluated against measures of effectiveness and performances to make decisions and adjustments (see FM 6-0).

3-57. Assessment of SFA requires a comprehensive approach that analyzes the overall operational environment and focuses on FSF and their impact. SFA assessment is performed in close collaboration with the U.S. country team, the host-nation government, and other multinational partners. These partners continually assess the security situation and its influence on other operations. From the assessment, planners develop short-, mid-, and long-range plans. As circumstances change, so do the plans. An insurgency might require employing host-nation forces at earlier stages of development. Some existing security forces may be so dysfunctional or corrupt that the organizations ideally should be disbanded rather than rehabilitated. In some cases, commanders will need to be replaced before their units will become functional. Political realities, however, may make both prospects undesirable. Minimizing and marginalizing particularly corrupt units or leaders while focusing on other units can be effective in this case.

3-58. While every situation differs, BCT leaders assess the following factors throughout planning, preparation, and execution of the operation:

- The operational environment.
- The civil considerations of the operational environment (see appendix B of FM 6-0).
- The core grievances and the threat (possibly including the prerequisites of insurgency).
- FSF elements.
- Methods, successes, and failures of FSF security efforts.
- State of training at all levels, and the specialties and education of leaders.

- Equipment and priority placed on maintenance.
- Logistic and support structure and their ability to meet the force's requirements.
- Extent of acceptance of ethnic and religious minorities.
- Laws and regulations governing the security forces and their relationships to national leaders.

Initial Assessment

3-59. An initial comprehensive assessment offers the BCT a baseline for establishing the scope of effort required. This assessment includes a troop-to-task analysis that determines the type and size of forces needed. FSF may require complete reestablishment or only assistance to increase capacity. They may completely lack a capability (such as internal affairs, corrections, or formal schools for leaders) or they may require temporary reinforcement. As with other military operations, efforts to assist FSF should reinforce success, not failure. For example, instead of building new police stations in every town, improve the good stations and use them as models for other organizations.

3-60. BCT leaders need to make decisions on what shortfalls to address first. The extent of the threats combined with resource limitations inevitably forces commanders to set priorities. However, making efforts to legitimize the host-nation leaders by encouraging them to make the decisions often proves more successful.

3-61. A key effort for SFA is the initial assessment of the DOTMLPF and policy capabilities of a foreign security force. During the early phases of a campaign, a geographic combatant command receives an initial assessment or a unit operating in the area has an assessment of FSF. The incoming BCT updates this assessment as it begins SFA.

3-62. Assessments require objective and subjective reporting, which may require a classification level. At a minimum, a BCT should assess—

- Doctrine.
- Organization.
- Training.
- Materiel and equipment.
- Leadership.
- Personnel.
- Command, control, communications, and intelligence.
- Operational effectiveness.

3-63. The doctrine assessment provides the BCT with an analysis on how the FSF operate. This assessment could focus on formal doctrine or more localized practices.

3-64. The organization portion of the overall assessment provides a perspective on how the FSF are organized. This includes the array of foreign institutions and units.

3-65. The level of training for FSF is a vital consideration. While this assessment is partly subjective, it should objectively examine individual and collective training strategies and mission-essential task lists. In the absence of mission-essential task lists, the BCT should compare FSF to U.S. equivalents.

3-66. The materiel and equipment assessment includes, but is not limited to, weapons, vehicles, equipment, and communications. It should be divided into equipment on hand, equipment readiness, and sustainment. Equipment on hand is quantitative and based on the type of equipment available to security operations. Equipment readiness reflects the operational readiness condition or serviceability of on-hand equipment. Sustainment readiness provides a qualitative and quantitative assessment that reflects how well FSF perform supply, maintenance, transportation, and medical functions. This includes facilities, systems, and structures.

3-67. The leadership assessment analyzes FSF commanders, officers, noncommissioned officers, and other leaders. This includes an assessment of staff members.

3-68. The foreign personnel assessment is quantitative and examines manning, staff manning, and administrative capabilities. It also examines officers, noncommissioned officers, and other personnel.

3-69. The command, control, communications, and intelligence assessment is qualitative and should reflect how well FSF conduct these functions.

3-70. The operational effectiveness assessment examines the foreign security force's ability to execute operations. This portion of the assessment should not be confused with the examination of training, as relatively high levels of training do not necessarily mean that FSF will perform in operations. This assessment is highly subjective.

Assessment During Operations

3-71. Assessment for the BCT conducting SFA should be part of a comprehensive approach. Assessment is directly tied to the commander's decisions throughout planning, preparation, and execution. Critical to effective assessment is developing criteria to evaluate progress toward task accomplishment, achievement of objectives, and attainment of the end state conditions. Assessment criteria are expressed as measures of effectiveness or measures of performance.

3-72. During operations, BCT and FSF commanders and staffs periodically assess the underlying framework of the plan itself. This involves reexamining the logic and assumptions used to develop the original plan, and it requires a deliberately conceived effort to learn as a part of the operation. BCT and FSF collaboration among higher, lower, and adjacent commanders, staffs, and other actors, backed up by qualitative and quantitative assessments, contributes to this learning. Based on this reexamination, commanders update their visualization, direct changes to the order, or develop an entirely new plan to adapt the force to better accomplish the mission.

Chapter 4

The Modular Brigade Augmented for Security Force Assistance

This chapter examines the typical phases for the brigade combat team conducting security force assistance, including a discussion of baseline augmentation by phase. The final section examines some of the subordinate units and elements of the modular brigade augmented for security force assistance.

THE BRIGADE COMBAT TEAM

4-1. The brigade combat team (BCT) is the cornerstone for Army modularity and is designed to operate at the tactical level across the spectrum of conflict. The modular brigade can be augmented, based on the requirements of the operational environment, with enabling assets and capabilities to support distributed security force assistance (SFA). However, the BCT augmented for SFA retains the capability to conduct full spectrum operations—offense, defense, and stability. Any of the three modular BCTs—heavy, infantry, or Stryker—can support SFA. Battalion-sized maneuver, fires, reconnaissance, and sustainment subordinate elements of the BCT support foreign security forces (FSF).

4-2. The gaining theater army force tailors all modular Army forces for use in major combat operations and contingencies. The theater army commander, working for the geographic combatant commander, determines the mix of forces and capabilities required for the BCT. This force tailoring can be done as part of the joint deliberate planning process or as a result of crisis action planning. Force tailoring includes providing BCTs with additional forces, personnel, or capabilities, including the possibility of an embedded provincial reconstruction team. Additional assets and capabilities can include, but are not limited to, command and control, communications, sustainment, engineering, military police, and intelligence.

4-3. In recent operations, the BCT has proven its versatility and capability to learn at the individual and collective levels. It supported other U.S. units conducting SFA and directly increased the effectiveness of FSF. The BCT has operated under varying levels of enemy activity, host-nation capabilities, objectives, and other conditions.

4-4. Because the BCT can operate in nonpermissive and permissive environments, it can conduct SFA across the spectrum of conflict. Thus, it can support civilian and military joint and multinational actors. The support can range from movement security and sustainment to augmentation of reconstruction teams and security for elections.

4-5. BCTs conducting SFA may support FSF development, assist FSF operations, and support and assist the development of host-nation institutions and infrastructure. While providing this support to the host nation, the BCT remains capable of conducting full spectrum operations independently.

BRIGADE COMBAT TEAM PHASES

4-6. If a force lacks the means to accomplish its mission in a single effort, then commanders normally phase the operation. A *phase* is a planning and execution tool used to divide an operation in duration or activity. A change in phase usually involves a change of mission, task organization, or rules of engagement. Phasing helps in planning and controlling and may be indicated by time, distance, terrain, or an event (FM 3-0). Phasing SFA assists in planning and controlling these dispersed and complicated operations. These phases often involve key changes in supporting SFA tasks and types employed.

4-7. BCT phases for SFA are initial response, transformation, and fostering sustainability. These phases mirror the stability operations framework described in FM 3-07 and are based on the operational environment. SFA can start in any phase or may even move to a previous phase due to changes in the conditions of the operational environment.

4-8. Figure 4-1 illustrates an SFA operation that includes all three SFA phases. The horizontal axis depicts an improving security situation, and the vertical axis depicts security capability. The example depicts an ideal situation, although this may not be the case in actual execution. Issues that arise require proper assessment and potentially the use of branches or sequels.

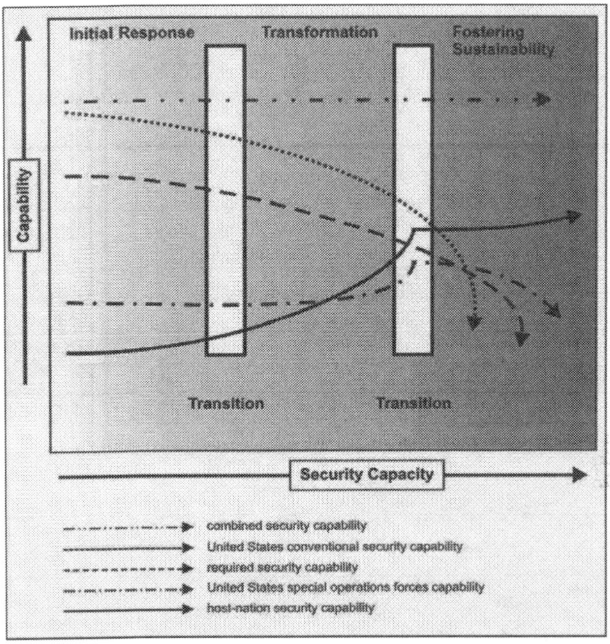

Figure 4-1. Example of phases of security force assistance

4-9. In figure 4-1, the combined security capability represents the overall security capacity of all security forces involved; it is the sum of U.S. conventional security, host-nation security, and the U.S. special operations forces security capabilities. This example assumes that the combined security capability exceeds the required security capability to effectively maintain a secure environment and counter any threat. The U.S. conventional security capability refers to tactical U.S. elements, such as the BCT. It does not refer to higher tactical or operational headquarters or support, sustainment, or medical assets. The U.S. special operations forces capability is depicted as declining in the end of the sustainment phase; as the FSF capability improves, the need for special operations forces declines. The host-nation or regional security organization and U.S. policy determine the end state for special operations.

4-10. For figure 4-1, a major change to enemy resources may require more FSF or additional. Drastic changes to the situation may even require a new operation plan. The figure focuses on capability, not forces, so host-nation security capability may increase without increasing the number or size of host-nation forces.

4-11. Figure 4-2 illustrates how the area of operations for a modular brigade augmented for SFA may change throughout the SFA phases. The differences between the initial response phase and the transformation phase may not change on the surface, but the BCT's relationship with FSF changes drastically. For example, FSF in the north met required conditions so that one of the U.S. brigades conducting SFA was no longer required. The responsibility of the other U.S. brigade thus expanded from providing assistance to one division to two divisions. This brigade's span of control and area of operations similarly expanded. The latter stage of the transformation phase can differ greatly. Areas of operation generally increase in size as they provide SFA to more FSF.

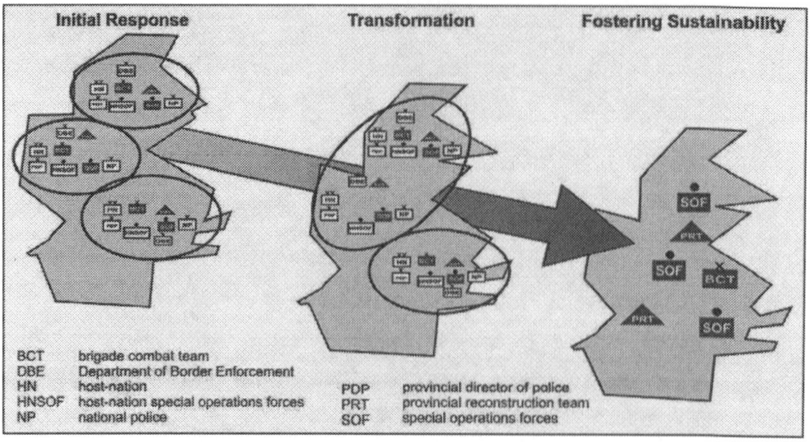

Figure 4-2. Example of changes to a brigade combat team's area of operations

Augmentation By Phase Summary

4-12. BCT augmentation is based on the operational environment and mission. As the three phases are based on the operational environment, they provide a baseline for augmentation. Potential augmentation may require military police, legal, public affairs, civil affairs, psychological operations, engineering, sociocultural experts, sustainment, and military transition team personnel. Table 4-1 illustrates a baseline for augmentation by phase for BCTs.

4-13. An embedded provincial reconstruction team is a key element of a BCT conducting security for assistance. The BCT has the lead on the following tasks: establish civil security, establish civil control (when approved by Congress), and develop and enable foreign security forces. An embedded provincial reconstruction team has the lead for the following tasks: support to economic and infrastructure development, restore essential services, and support to governance. Thus, the BCT and an embedded provincial reconstruction team are able to effectively support FSF and execute all five stability tasks.

Table 4-1. Baseline augmentation for modular brigade

CAPABILITY	INITIAL PHASE (1 HN DIV)	TRANSFORMATION PHASE (2 HN DIV)	TRANSFORMATION PHASE (3 HN DIV)
Police Provincial District	171 - MP Company 2 - MP Team (LTC/SGM) 3 - MP MAJ	171 - MP Company 4 – 2 x MP Team (LTC/SGM) 6 - MP MAJ	171 - MP Company 6 – 3 x MP Team (LTC/SGM) 9 - MP MAJ
Legal HN BDE HN DIV	2 - JAG Team (MAJ/SFC) 2 - JAG Team (CPT/SSG)	4 – 2 x JAG Team (MAJ/SFC) 4 – 2 x JAG Team (CPT/SSG)	6 – 3 x JAG Team (MAJ/SFC) 6 – 3 x JAG Team (CPT/SSG)
Public Affairs	2 - PA (Officer/NCO)	2 - PA (Officer/NCO)	2 - PA (Officer/NCO)
Civil Affairs & Psychological Operations	32 - CA Company = 5 CA Teams 70 – Tactical PSYOP Company = 9 TPTs	32 - CA Company = 5 CA Teams 70 – Tactical PSYOP Company = 9 TPTs	197 - CA Battalion (4 CA Companies) = 20 CA Teams 70 – Tactical PSYOP Company = 9 TPTs
Engineer	8 - FEST-A Team 2 - EOD (CPT/SFC)	8 - FEST-A Team 2 - EOD (CPT/SFC)	8 - FEST-A Team 2 - EOD (CPT/SFC)
Sociocultural Experts	8	8	8
Military Transition Team	56 - MTT Augment	4 - MTT Augment	8 - MTT Augment
TOTAL	358	315	493
EPRT	~15-20	~15-20	~15-20
Sustainment Truck Squad Supply Platoon CCT FMST	 21 personnel 32 personnel 4 personnel 4 personnel	 4 personnel 4 personnel	 4 personnel 4 personnel

CA	civil affairs	LTC	lieutenant colonel
CPT	captain	MAJ	major
CCT	contingency contracting team	MP	military police
DIV	division	MTT	military transition team
JAG	Judge Advocate General	NCO	noncommissioned officer
EOD	explosive ordnance disposal	PA	public affairs
EPRT	embedded provincial reconstruction team	SFC	sergeant first class
FEST-A	forward engineering support team – augmenting	PSYOP	psychological operations team
FMST	financial management support team	SGM	sergeant major
HN	host-nation	SSG	staff sergeant
HTT	human terrain team	TPT	tactical psychological operations team

INITIAL RESPONSE PHASE

4-14. The initial response phase seeks to assist FSF stabilize the operational environment in a crisis state. This phase occurs during or immediately after a conflict where the operational environment prevents civilian personnel from operating effectively. The operational environment is typified as nonpermissive.

Thus, the objective of this phase is to improve the security situation, reducing the threat to the populace and creating the conditions that allow civilian personnel to safely operate.

4-15. SFA in the initial response phase is normally required when FSF lack the capability or capacity to provide the required level of security. This phase often requires SFA efforts to help generate and train or assist new and existing FSF. This phase may require a combination of the types of SFA and considerable support, sustainment, and medical resources. BCT activities during the initial response may have to be conducted with multinational major combat operations, to include providing a safe, secure environment for the local populace.

4-16. The BCT's SFA efforts during this phase focus on improving the foreign security force's capability and capacity so all security forces—U.S., other, and foreign security forces—provide a secure environment and reduce the threat. Figure 4-3 illustrates the BCT's span of control. The figure also depicts three BCTs' operational areas in the initial response phase. As security conditions improve, transition to the transformation phase begins.

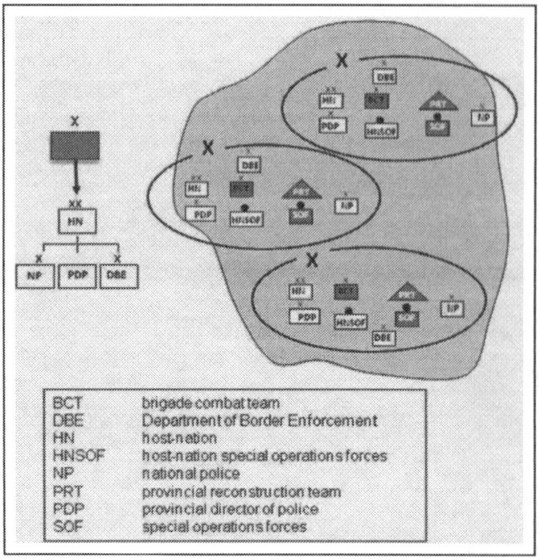

Figure 4-3. Example of initial response span of control and area of operations

4-17. The BCT requires augmentation for the initial response phase. These embedded units provide key enabling capabilities for the brigade, supported FSF, and other supported actors. The embedded provincial reconstruction team is one of the most important of the brigade's augmentation, as it provides interim civilian expertise in developing local institutions to take the lead in national governance, the provision of basic services, fostering economic development, and enforcement of the rule of law. Column two of table 4-1 summarizes augmentation during the initial phase. Although each brigade's mission and conditions in its area of operations are unique, the baseline augmentation for the modular brigade operating in this phase is as follows:

- Embedded provincial reconstruction team.
- Forward engineering support team-augmenting.
- Explosive ordnance disposal team.
- Sociocultural experts.

- Military police company (provincial police transition team).
- Military police team (district police transition team).
- Three military police majors (district police transition team).
- Two judge advocate general teams.
- Civil affairs company.
- Tactical psychological operations company.
- Public affairs team.
- Contingency contracting team.
- Financial management support team.
- Military transition team (specific augmentation personnel include two lieutenant colonels, thirty-one majors, and twenty-three master sergeants).

TRANSFORMATION PHASE

4-18. In the transformation phase, the BCT seeks to assist FSF to stabilize the operational environment in a crisis or vulnerable state. The operational environment in this phase is more permissive than the initial response phase; however, military forces will often be required to provide security to some actors. Activities in this phase normally include a broad range of post-conflict reconstruction, stabilization, and capacity-building efforts, which the embedded provincial reconstruction team is essential for long-term success. Objectives in this phase include continuing efforts to improve the security situation, reducing the threat to the populace, building host-nation capacity across the stability sectors, and facilitating the comprehensive approach to assist FSF.

4-19. The transformation phase represents a broad range of SFA activities that the BCT could perform to support FSF. The initial response phase differs from the transformation phase in the foreign security force's capability to provide for a safe and secure environment. More specifically, FSF have a level of proficiency so they no longer need a permanent U.S. and foreign security forces relationship for tactical operations. However, they may still need full-time advisors and support, sustainment, and medical assistance. The embedded provincial reconstruction team will continue to play a vital role in assisting governance and development efforts throughout this phase.

4-20. Figure 4-4 illustrates the BCT conducting SFA in the transformation phase. In this example, the BCT's span of control expands from assisting one foreign division to assisting two foreign divisions. The figure also depicts how a BCT's operational area has similarly expanded. SFA activities in this phase seek to establish conditions so the host nation's security sector can provide a secure environment with its own security forces.

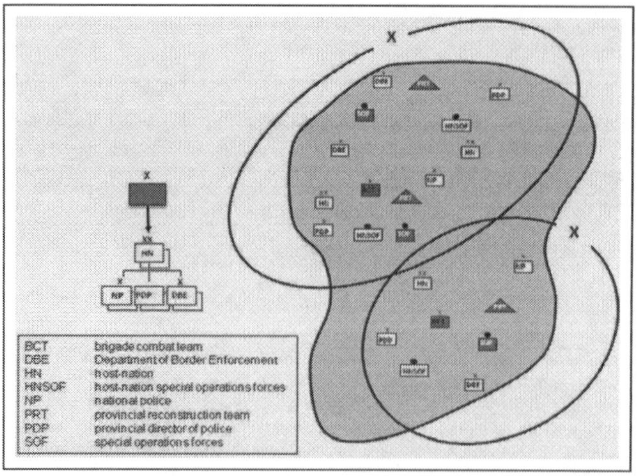

Figure 4-4. Example of transformation phase span of control and area of operations

4-21. The BCT requires augmentation for the transformation phase, like the initial response phase. It often requires similar augmentation due to its greater span of control and area or operations. Column three of table 4-1 summarizes augmentation during the transformation phase with two foreign divisions. The baseline augmentation for the brigade in this phase is as follows:

- Embedded provincial reconstruction team.
- Forward engineering support team-augmenting.
- Explosive ordnance disposal team.
- Sociocultural experts.
- Military police company (provincial police transition team).
- Two military police teams (district police transition team).
- Six military police majors (district police transition team).
- Four judge advocate general teams.
- Civil affairs company.
- Tactical psychological operations company.
- Public affairs team.
- Contingency contracting team.
- Financial management support team.
- Military transition team (specific augmentation personnel include two lieutenant colonels and two master sergeants).

4-22. As conditions of the operational environment improve, the BCT may require a transition to assist three foreign divisions, rather than two divisions as depicted in figure 4-4. This represents a significant expansion of the span of control and area of operations, with a corresponding expansion of required augmentation. Figure 4-5 depicts the BCT supporting three foreign divisions.

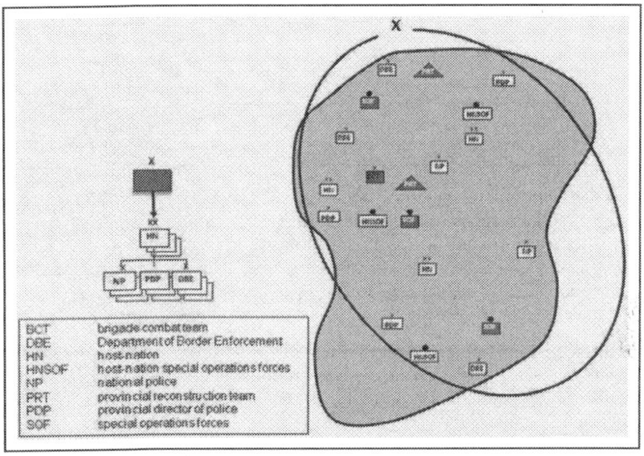

Figure 4-5. Example of a three-division span of control and area of operations

4-23. As in other phases, the BCT requires augmentation to support three divisions. Column four of table 4-1 summarizes augmentation during the transformation phase with three foreign divisions. The baseline augmentation for the brigade in this phase is as follows:

- Embedded provincial reconstruction team.
- Forward engineering support team-augmenting.
- Explosive ordnance disposal team.
- Sociocultural experts.
- Military police company (provincial police transition team).
- Three military police teams (district police transition team).
- Nine military police majors (district police transition team).
- Six judge advocate general teams.
- Civil affairs company.
- Tactical psychological operations company.
- Public affairs team.
- Contingency contracting team.
- Financial management support team.
- Military transition team (specific augmentation personnel include four lieutenant colonels and four master sergeants).

Fostering Sustainability Phase

4-24. In this phase, the focus of SFA continues to shift toward assisting institutions required to sustain FSF. This phase encompasses long-term efforts to assist FSF. In this phase, FSF are capable and viable. They conduct independent operations and can provide a safe, secure internal environment. While the brigade may be initially required during this phase, it is normally withdrawn during this phase. The determination to withdraw the brigade is based on the policy and conditions of the operational environment. Special operations forces, provincial reconstruction teams, and other forces may remain to support a theater security cooperation plan. Thus, remaining SFA missions are normally special operations forces missions, not a BCT mission (special operations forces will normally be involved in all phases of SFA).

TRANSITIONS

4-25. SFA can require a vast array of transitions. These transitions are conducted according to the tactical, operational, and strategic conditions identified during planning (which must support the internal defense and development strategy and using an overall comprehensive approach). Transitions can occur simultaneously or sequentially in different levels or war and in separate echelons. This includes potentially having tactical transitions for different units within one brigade's area of operations.

4-26. There are several major transitions. First, the supported unit may be the U.S. BCT in the beginning of an initial response phase but the foreign security forces will transition to be the supported unit. At this point in the transformation phase, the area in which the BCT conducts SFA will expand. This expansion can occur multiple times during the transformation phase, which is based on conditions, especially the capability and capacity of FSF.

4-27. Operations never proceed exactly as planned. To facilitate flexibility, commanders incorporate branches and sequels into the overall plan. Visualizing and planning branches and sequels are important because they involve transitions. Unless planned, prepared for, and executed efficiently, transitions can reduce the tempo of the operation, slow its momentum, and surrender the initiative to the adversary.

SUBORDINATE UNITS

4-28. The modular brigade augmented for SFA has subordinate units whose sole focus is working with FSF. These advisor teams may be formed from brigade organic resources, external augmentation, or a combination. These teams optimally are embedded with the counterpart unit, or they may reside on a U.S. camp and commute to FSF they support. The method depends on U.S. policy, direction from higher headquarters, the conditions of the operational environment, and capacity of the foreign security force's camps to accommodate the U.S. forces.

4-29. The military transition team (MTT) is a key subordinate unit for the brigade conducting SFA. The MTT mission is to assist FSF units. Figure 4-6 illustrates the relationship between a brigade conducting SFA and a MTT assisting a foreign brigade. In this example, the brigade has a battalion and a division MTT as subordinate units, although it would have other subordinate units. The brigade MTT is subordinate to the battalion and has operational control of one of the parent U.S. battalion's company teams.

4-30. When the brigade provides organic forces to form the basis for the brigade MTT, the company team is the foundation of a brigade MTT. The company team will be augmented with additional personnel and assets. The U.S. brigade must facilitate the MTT operations, such as force protection, transportation, communications, and reachback.

4-31. As companies do not have organic intelligence analysts, personnel must be used to form company intelligence support teams (COISTs). A COIST, an ad hoc organization, assists the MTT and its FSF. Ideally, the COIST should consist of six individuals, which will allow for 24 hour operations and the depth in the team to both integrate with the staff and to successfully support the company's mission. The team should consist of an officer, a noncommissioned officer, and four Soldiers, although the cell can be minimally manned with four personnel. The company intelligence cell works most effectively if its work complements the battalion S-2's work.

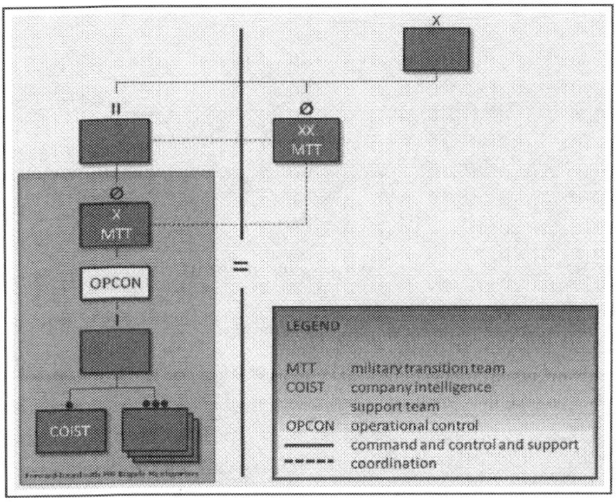

Figure 4-6. Military transition teams

SAMPLE TROOP-TO-TASK MODEL

4-32. The BCT may support assigned to multiple missions, including potentially supporting multiple foreign units in their operational area. Additionally, these FSF organizations may each report through different host-nation government channels and even to different ministries. To synchronize efforts in this case, U.S. forces must achieve unity of command and effort. Similarly, each of the FSF organizational commanders should synchronize their efforts with the host-nation government representatives, as appropriate.

4-33. Figure 4-7 depicts a sample troop-to-task model for a U.S. brigade. It depicts one company team acting as a quick reaction force; thus, the battalion retains the capability to conduct independent full spectrum operations. The figure depicts how U.S. company teams may form the basis of a brigade MTT or a brigade border transition team (and subordinate point of entry transition teams). The figure also depicts how a company team may provide police support. This support is in the form of two platoons supporting police transition teams and a third platoon in a joint security station providing support to a police transition team.

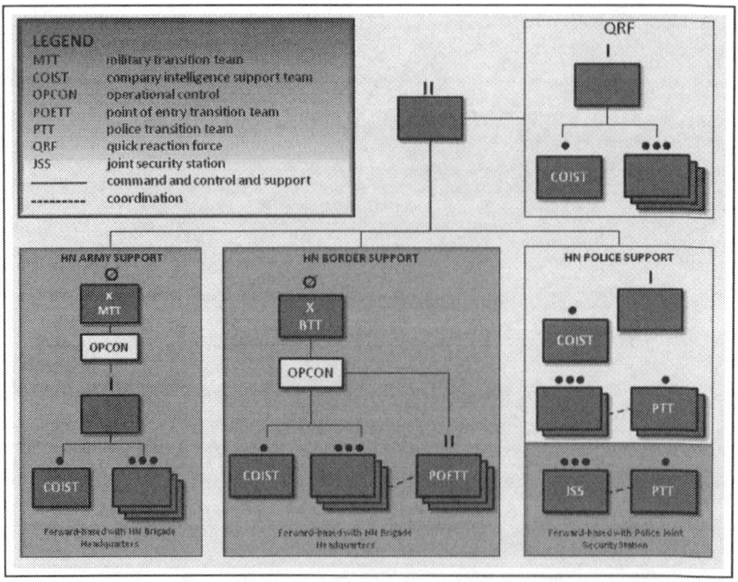

Figure 4-7. Battalions conducting security force assistance

4-34. Subordinate units of BCTs conducting SFA are best located inside a FSF base—not an American base. Co-location facilitates integration with the FSF. It allows the two forces to form mutual understanding and trust. Co-location and the close cooperation it facilitates often improves the population's perception of the legitimacy U.S. and foreign security forces, which can be an essential condition of the overall mission's end state.

4-35. If force protection requires, a U.S. area may be established in the FSF base, although this is not optimal. Key considerations for co-location may include the threat, FSF acceptance, physical space inside the FSF base, sustainment capabilities, medical facilities, and availability of quick reaction forces.

4-36. When U.S. forces are operating out of smaller outposts in an urban environment, the local populace sees the integration and presence of the U.S. and foreign security forces working together. This integration not only enhances overall operational effectiveness and trust, living and working together builds legitimacy of the two forces as well as FSF; it reinforces trust between the FSF and the people they are tasked to protect.

Chapter 5
Unit Operations

Unit operations require continuous activities. Commanders plan for premission, deployment, in-country preparation, employment, redeployment, and postdeployment activities.

PREMISSION

5-1. After receiving a mission, the unit continues detailed preparation activities. The unit obtains supplies, equipment, and training materials and prepares for deployment. Personnel prepare for and rehearse classes to be given in-country. They hone their military skills and conduct extensive area orientation. Paragraphs 5-2 through 5-10 discuss staff actions particular to security force assistance (SFA).

OPERATION PLAN

5-2. The S-3 disseminates the operation plan to unit staff and subordinates. The S-3 ensures predeployment training for Soldiers, to include preparation for training foreign security forces (FSF) and rehearsals for movement. The S-3 reviews the program of instruction for training FSF, to include getting approval from the commander and higher headquarters, if necessary.

CIVIL-MILITARY OPERATIONS

5-3. The civil-military operations section ensures the operation plan minimizes how operations affect the civilian population and addresses ways to mitigate the civilian impact on military operations. Civil-military operations also ensure the operation plan discusses coordination with any augmenting actors. These actors might include civil affairs battalions or brigades, provincial reconstruction teams, or United States Agency for International Development project officers in the operational area. During predeployment training, Soldiers receive training, materials, and briefings on the operational area. This training can cover the history, culture, religion, language, tribal affiliations, local politics, and cultural sensitivities as well as any significant nongovernmental organizations operating in the operational area.

TRAINING

5-4. As discussed in chapter 2, advisors should focus their premission training on the specific requirements of developing FSF. The training emphasizes the host-nation culture and language and provides cultural tips for developing a good rapport with foreign personnel. All premission training for SFA follows the training principles described in FM 7-0.

5-5. Based on the higher commander's training guidance, unit commanders assign missions and approve the draft mission-essential task list that supports SFA. The staff plans, conducts, and evaluates training to support this guidance and the approved mission-essential task list for SFA missions. Commanders prioritize tasks that need training. Since there is never enough time to train in every area, commanders focus on tasks essential for mission accomplishment.

5-6. Once commanders select tasks for training, the staff builds the training schedule and plans on these tasks. The staff provides the training requirements to the commander. After approving the list of tasks to be trained, the commander includes the tasks in the unit training schedule. The staff then coordinates the support and resource requirements with the S-3 and S-4. Finally, the commander ensures standards are enforced during training.

Evaluation

5-7. Evaluations can be either internal or external. Internal evaluations occur at all levels, and they must be inherent in all training. External evaluations are usually more formal and conducted by a headquarters two levels above the unit being evaluated. This subject must be carefully planned and discussed with FSF leaders to account for cultural sensitivities and current capabilities.

5-8. A critical weakness in training is the failure to evaluate each task every time it is executed. Every training exercise provides potential for evaluation feedback. Every evaluation is also a training session. Leaders and trainers must continually evaluate to optimize training. Evaluation must occur as training takes place. Emphasis is on direct, on-the-spot evaluations. However, leaders allow Soldiers to complete the task first. Leaders plan after action reviews at frequent, logical intervals during exercises. This technique allows the correction of shortcomings while they are still fresh in everyone's mind. The after action review eliminates reinforcing bad habits.

Specified Training

5-9. Augmentation elements require area orientation, refresher combat training, field training exercises, and the like. Unit training objectives are for developing capabilities to conduct internal defense and development (IDAD) activities for tactical operations, intelligence operations, psychological operations, populace and resources control operations, and civil affairs and advisory assistance operations in the host-nation language. Units identified for SFA begin intensified training immediately upon deployment notification.

5-10. After deployment to the host nation and before commitment to operations, the unit may receive in-country training at host-nation training centers or at designated training locations. This training helps personnel become psychologically and physically acclimated to the host nation. This training also allows commanders and staff some time to coordinate and plan within their own command and with civilian and military joint and multinational organizations. After commitment, training continues and is stressed between operations, using needed improvements identified in operations as the basis for training.

DEPLOYMENT

5-11. SFA is often conducted in semipermissive environments in which U.S. forces are guests of the host-nation or partner organization. Units move into the operational area by following their deployment standing operating procedures and the operation plan. Paragraphs 5-12 through 5-25 highlight organizations and procedures critical to SFA deployment. This information, however, does not obviate the need for extensive standing operating procedures and an operation plan.

PREDEPLOYMENT SITE SURVEY

5-12. The predeployment site survey unit aims to report accurately to its parent unit the existing FSF mission variables—mission, enemy, terrain and weather, troops and support available, time available, civil considerations (METT-TC). See table 5-1. It also establishes in-country command and control, support, protection, and sustainment relationships for the follow-on unit's mission execution. The site survey coordinates the in-country reception of SFA forces. This coordination wards off most friction during the handoff. This coordination requires all units—outgoing and incoming—to apply a comprehensive approach at all levels of command. Appendix D of FM 3-05.202 discusses site survey procedures in detail.

Table 5-1. Advisor team predeployment site survey checklist

Tasks	Characteristics
Area of operations overview	• Map familiarization. • In sector orientation. • Areas of emphasis.
Graphic control measures	• Unit boundaries. • Routes. • Other United States and multinational support locations. • Maintenance support location. • Medical support locations. • Other points of interest.
United States and multinational unit interviews by specialty staff	• Responsibilities. • Battle rhythm. • Required reports. • Standing operating procedures, continuity book. • Stay-behind equipment.
Briefings on advised unit	• Foreign security forces task organization. • Foreign security forces key leadership. • Intelligence assessment. • Unit operational assessment.
Advised unit introductions	• Among commanders. • Among staffs.

ADVANCE PARTY DEPLOYMENT

5-13. An advance party deploys to the host-nation or partner organization area of operations after the unit completes its mission planning and the proper commander representative declares it deployable. The advance party conducts predeployment activities and coordinates necessary travel documentation and country clearances. The advance party reviews the rules of engagement and any status-of-forces agreements before deploying. If applicable, the proper staff provides an updated threat briefing to the unit. The S-4 section and support company help with out-loading in the unit area. The unit places all supplies and materials on pallets beforehand and prepares the load manifest.

5-14. Upon arrival, the advance party processes through customs and immediately notifies the higher headquarters of its arrival and status. Often the higher headquarters provides point of contacts to the advance party. These contacts help the advance party obtain the proper identification, documentation, and weapons permits.

5-15. The advance party establishes contact with all U.S. and foreign security forces tactical unit commanders and conducts final premission planning activities. The advance party procures secure working, storage, and living areas for the unit and reconnoiters and prepares the training sites. The advance party coordinates for special support equipment to unload heavy supplies and transport the supplies and personnel to the training site.

MAIN BODY DEPLOYMENT

5-16. The unit deploys after completing its planning and the appropriate commander declares the unit ready. The unit follows the same procedures performed by the advance party. Members of the advance party meet the main body on its arrival and brief the main body on any required changes to the operation order.

5-17. After arriving in-country, the unit processes through customs and immediately notifies the higher headquarters of its arrival and status. The unit obtains the proper identification, documentation and weapons permits from U.S. or other sources, if applicable. Often the security assistance organization provides to the unit the necessary point of contacts that expedite this process.

5-18. The unit off-loads equipment and personnel onto vehicles for movement to the training site, or it stores the equipment in a secure area until the processing is completed. FSF secure the equipment but the unit members remember the commander is ultimately accountable for the advisor team's physical and personal security. Supply personnel stay with the equipment and help to guard it.

IN-COUNTRY PREPARATION

5-19. Upon arrival, the commander and S-3 brief the higher headquarters on the planned execution of the mission and reconfirm the required command relationship. Local conditions may require the unit to confirm or establish its in-country and external command and control, support, and sustainment relationships from outside its operational area upon arrival. The unit establishes direct working relationships with its next higher in- or out-of-country supporting element to—

- Determine the limits of the available support and expected reaction time between the initiation of the support request and its fulfillment.
- Confirm or establish communications procedures between the supporting element and the unit, to include alternative and emergency procedures for command and control, all available support, and medical evacuation.

5-20. The unit establishes procedures to promote interagency cooperation and synchronization. The unit—

- Identifies the location of the concerned host-nation, U.S., or other agencies.
- Contacts the concerned agency to establish initial coordination.
- Exchanges information or intelligence.
- Confirms or establishes other coordination protocols as necessary.
- Incorporates the newly established or changed procedures into the plans for mission execution.

5-21. The unit immediately establishes operations security procedures to support its mission execution and identifies rally points incorporated into its defensive, evasion, and personnel recovery plans.

5-22. After receiving a detailed briefing and further guidance from the advance party, unit personnel continue to develop effective rapport with the FSF commander and counterparts. They also assess their working, storage, and living areas for security and verify the location of the training site, communications center, dispensary area, and FSF troop area. With the FSF commander, the unit commander—

- Establishes rapport.
- Conducts introductions in a businesslike, congenial manner.
- Briefs on the unit's mission, its capabilities, and the restrictions and limits imposed on the detachment by the higher U.S. commander.
- Ensures all unit personnel fully support FSF and firmly believes a joint U.S.-FSF effort will succeed. Requests counterpart linkup be made under the mutual supervision of the FSF commander and the unit commander.
- Ensures all current unit plans are tentative and that assistance is needed to finalize them.
- Deduces or solicits the actual estimate of unit capabilities and perceived advisory assistance and material requirements.
- Recommends the most desirable courses of action (COAs) while emphasizing how they satisfy present conditions, achieve the desired training, and meet advisory assistance goals.
- Informs the higher in-country U.S. commander of any significant changes in the unit's plan to assist FSF.

5-23. The S-2 supervises the dissemination of intelligence and other operationally pertinent information within the unit and, as applicable, to higher, lower, or adjacent units or agencies. The S-2 also—

- Monitors the implementation of the intelligence collection plans to include updating the commander's priority intelligence requirements (PIRs), conducting area assessment, and coordinating for additional intelligence support.
- Establishes liaison with FSF intelligence and security agencies (within the guidelines provided by applicable higher authority). Assesses the intelligence threat and resulting security requirements, including coordination with the S-3 on specific security and operations security measures.

5-24. Through the S-2, the commander's PIRs are based on the latest information available and requirements for additional PIRs that arise from modified estimates and plans. The S-2 also—

- Analyzes the foreign unit's status to finalize unit plans for advisory assistance. These plans can include task organization of unit with counterparts, staff functions for planning SFA, and advisory assistance for executing SFA.
- Explains analysis to counterparts and encourages them to help with—and participate in—analyzing, preparing, and briefing the analysis to the foreign unit commander.
- Prepares and briefs the plans for training and advisory assistance.
- Helps the foreign unit inspect the available facilities to identify deficiencies. If the unit finds deficiencies, the S-2 prepares estimates of COAs for the FSF commander to correct them.
- Supervises the preparation of the facilities with their counterparts and informs unit and FSF commanders on the status of the facilities.

5-25. The unit ensures its security is based on the present or anticipated threat. Some recommended actions the unit may take include—

- Hardening its positions based on available means and requirements to maintain low visibility.
- Maintaining unit internal guard system with at least one Soldier who is awake and knows the locations of all other unit personnel. The guard reacts to an emergency by following an internal alert plan and starting defensive actions.
- Maintaining communications with all subordinate unit personnel deployed outside the immediate area controlled by the main body.
- Establishing plans for immediate defensive actions in the event of an attack or a loss of rapport with hostile reaction.
- Discussing visible security measures with foreign counterparts to ensure understanding and to maintain effective rapport. Unit personnel do not divulge sensitive information for the sake of possible rapport benefits.
- Encouraging the foreign unit, through counterparts, to adopt additional security measures identified when analyzing the foreign unit's status and inspecting its facilities.
- Coordinating defensive measures with the foreign unit to develop a mutual defensive plan. Unit personnel obtain from the unit's present reaction and defensive plans for attack. They encourage the foreign unit to conduct mutual full-force rehearsals of defensive plans; if unsuccessful, the unit conducts internal rehearsals of the plans.

EMPLOYMENT

5-26. Once the unit has deployed and arrived in-country, it begins employment. Employment is the conduct of SFA to support the FSF commander. Employment occurs within the advisor team itself. It includes foreign counterparts in the mission planning (preparing the FSF for the mission itself) to increase the capability and capacity of their planning processes, as well as increase the probability of success.

Personnel

5-27. During employment, individual unit personnel perform functions described in paragraphs 5-28 through 5-38.

Commander

5-28. Before the mission, the U.S. commander advises and assists the FSF commander. The latter issues planning guidance for planning the execution of the mission and clarifies commander's intent. The U.S. commander advises and assists the FSF unit commander in the command and control process of tactical operations. By accompanying the FSF unit commander when the mission is received from higher headquarters, the U.S. commander assists any subsequent missions. The U.S. commander monitors how FSF understand the commander's intent and all specified or implied tasks.

5-29. During execution of the mission, the U.S. commander helps the foreign unit commander provide command and control during operations. After monitoring the tactical situation, the U.S. commander recommends changes to the COAs to exploit the situation. After monitoring the flow of information, the U.S. commander recommends improvements to use intelligence collection assets and to keep subordinates reporting required information.

Staff

5-30. Before the mission, the advisor staff advises and assists the foreign counterparts in preparing estimated COAs for essential tasks. The staff helps write tentative plans based on the planning guidance and unit's standing operating procedures. These plans include primary, alternate, contingency, and emergency plans.

5-31. During execution, the staff helps foreign counterparts coordinate the execution of the tasks. They disseminate portions of their plans to other personnel, senior and subordinate staff sections, and supporting elements. They help notify higher, lower, or adjacent staff sections of modified estimates and plans. Finally, the staff—with the S-2 and the S-2 counterpart—helps update the commander's critical information requirements with the latest information and requirements that arise from operations.

Executive Officer

5-32. The executive officer performs the organizational analysis of the unit's coordinating staff sections to ensure efficiency during the planning process according to initial planning guidance. With the foreign counterpart, this officer directs foreign staff sections as they develop estimates and plans. The executive officer monitors the liaison and coordination with FSF higher headquarters, recommending changes to improve efficiency.

Adjutant

5-33. The adjutant provides advice, assists, and makes recommendations to the foreign counterpart for conducting human resources. This includes monitoring the maintenance of foreign unit strength, accountability of casualties, unit morale, and postal activities. This may also include concerns with the foreign pay system, leave procedures, and casualty pay procedures.

Intelligence Officer

5-34. The intelligence officer advises and assists the monitoring of FSF operations security to protect classified and sensitive material and operations and recommends improvements. By helping the foreign counterpart update the situation map, the intelligence officer helps to keep the map current. The intelligence officer recommends improvements to the standing operating procedures of the tactical operations center communications so the intelligence section receives situation reports. This officer helps the counterpart monitor the collection, evaluation, interpretation, and dissemination of information. The intelligence officer assists in the examination of captured insurgent documents and material. This officer helps gather and disseminate intelligence reports from available sources to ensure the exploitation of all assets. The intelligence officer helps the counterpart to brief and debrief patrols operating as a part of reconnaissance and surveillance activities. Finally, the intelligence officer assists, with the advisor operations officer, developing reconnaissance and surveillance plans.

Operations Officer

5-35. The operations officer helps the foreign counterpart to prepare tactical plans using estimates, predictions, and information. This officer monitors command and communications nets, assists in preparing all orders and plans, and helps to supervise the training and preparation for operations. Finally, the operations officer monitors the planning process and makes recommendations for consistency with IDAD goals.

Supply Officer

5-36. The supply officer advises and assists the foreign counterpart in maintaining equipment readiness; monitoring the support provided to the foreign unit, its subunits, and attachments; and recommending improvements. Finally, this officer helps to supervise the use of transportation assets.

Civil-Military Team

5-37. Upon deployment, civil-military teams advise the commander and staff on civil-military considerations and coordinate efforts of any civil affairs units supporting the unit. The civil-military team mentors the counterpart team on the supported foreign element staff on civil-military operations, the legal and moral obligations of military commanders to civilian populations under their control, and the importance of respecting human rights. The civil-military team may introduce the counterpart to relevant nongovernmental organizations, United States Agency for International Development project officers, and provincial reconstruction team staff.

Company-Level Advisors

5-38. Advisors sent to elements below battalion level help foreign counterparts to analyze the mission and commander's intent from higher headquarters. They assist FSF leaders restate the mission, conduct an initial risk assessment, identify a tentative decisive point, and define their own intent. They also assist their foreign counterparts to analyze the mission variables. From these variables, advisors help their foreign counterparts to develop a COA that meets the higher headquarters concept of operations and commander's intent. Finally, they advise and assist in the conduct of operations and the flow of information to the FSF higher commander.

TRAINING

5-39. Training FSF during SFA parallels the training of U.S. forces. Training follows the principles of training described in FM 7-0. This training tailors the objectives to account for cultural differences and focus on the FSF mission and problem set.

Training Assessment

5-40. The commander of the advising unit begins with a training assessment of training plans designed during predeployment and passed on from the preceding U.S. units conducted in coordination with the FSF commander. This assessment is important to both evaluate the foreign security force but also as an exercise to establish and enhance the working relationship between U.S. and foreign security forces. The training assessment should cover all aspects of leadership, training, sustainment, and professionalization. To support an assessment, the advising unit should analyze the following specific foreign unit considerations:

- The unit's mission and mission-essential task list and capability to execute them.
- Staff capabilities.
- Personnel and equipment authorization.
- Physical condition.

- Any past or present foreign influence on training and combat operations.
- Operational deficiencies identified during recent operations or exercises with U.S. personnel.
- Sustainment capabilities, to include sustainment training programs.
- Internal training programs and personnel.
- Training facilities.

5-41. The level of professionalism of FSF, both units and individuals, should also be assessed. Adhering to established rules of engagement, ethics that meet the established laws and regulations of the commanding authority, laws for land warfare, and human rights, promotions based on merit and the security forces' support of civilian leaders and political goals all fall within this assessment.

5-42. The advisor unit, working with FSF leaders, also evaluates current members of the security forces for past military skills and positions. Often military reorganizations arbitrarily shift personnel to fill vacancies outside their knowledge and experience.

Training Plan

5-43. After completing the training assessment, advisors analyze the prepared training plan and determine if changes are necessary. The training plan stresses the deficiencies identified in the training assessment. It also identifies those in the host nation able to help train FSF and to strengthen the legitimacy of the process. Finally, the training plan should consider the unit's eventual self-sustainment. As the foreign security force gains sufficient capacity and capabilities to perform independently, advisors transition from a leading role to a mentoring role.

5-44. The plan for training FSF uses a comprehensive approach, supports the IDAD strategy, and has many echelons. Working with other echelons and actors within the area of operations can provide support and expertise that enhance the training and operations process as well as limit redundancy.

Program of Instruction

5-45. In coordination with the FSF commander and training staff, the advising unit develops a program of instruction. This program incorporates all training objectives that satisfy the training requirements identified during the assessment. The training program must support these requirements. The FSF commander approves the program of instruction prior to execution. When executing the program of instruction, advisors adhere to a training schedule consistent with changes in the mission variables. Advisors ensure through their counterparts and the FSF commander that all personnel receive training. Foreign counterpart trainers rehearse all classes approved on the program of instruction.

Presentation of Instruction

5-46. To present the training material properly, trainers follow the lesson outlines approved in the program of instruction. All training clearly states the task, conditions, and standards desired during each lesson, ensuring the foreign students understand them. Trainers state all warning and safety instructions in the host-nation language. The training to reinforce the concepts includes demonstrations of the execution of each task, stressing the execution as a step-by-step process. Trainers monitor students' progress during instruction and practical exercises, correcting mistakes as they are made.

Training Methods

5-47. An effective method of training used by U.S. forces is the crawl-walk-run method of teaching individual tasks, battle drills, collective tasks, and field exercises. (See FM 7-0.) This method is employed to develop well-trained leaders and units. During all phases, the training must include the mission of the unit in the context of the higher unit's mission to assist with the practical application of the training. Identifying the higher commander's mission and intent, as well as the tasks and purposes of other units in the area, would also add context to the training. This method can also be expanded to include the role of other actors.

5-48. At a minimum, units continue individual training to improve and sustain individual task proficiency while training on collective tasks. Collective training requires interaction among individuals or organizations to perform tasks, actions, and activities that contribute to achieving mission-essential task proficiency. Collective training includes performing collective, individual, and leader tasks associated with each training objective, action, or activity.

Collective Training

5-49. Collective training starts at the squad level. Squad battle drills provide key building blocks that support operations. Trainers can link battle drills and collective tasks through a logical, tactical scenario in a situational training exercise. Although this exercise is mission-oriented, it results in more than mission proficiency. Battle drills and collective tasks support situational training exercises, while these exercises support operations. Advisors must understand the operational environment when training FSF; training incorporates how internal and external threats and civilians affect the environment. Flexibility in using U.S. doctrine in training enhances efforts to make training realistic. Trainers modify U.S. doctrine to fit the FSF level of expertise, their command and control, the tactical situation, and sustainment base. Often the structure and capabilities of FSF differ from that required by U.S. doctrine. When FSF counter an insurgency, these exercises emphasize interplay among psychological and tactical, populace and resources control, intelligence, and civil affairs operations. Similarly, scenarios for each tactical exercise must include interplay among the various possible aspects of IDAD.

Individual Training

5-50. Training individuals of FSF emphasizes physical and mental conditioning, tactical training, basic rifle marksmanship, first aid, combatives, and the operational environment. Individual training also includes general tactics and techniques of security operations and the motivation, operations, and objectives of internal and external threats. Tough and realistic training conditions troops to mentally and physically withstand the strain of continuous offensive operations.

5-51. Foreign personnel cross-train on all types of weapons, communications and other equipment, and skills particular to their unit. Personnel losses must never cause weapons, communications equipment, or essential skills to be lost due to a lack of fully trained replacement personnel.

Small-Unit Leader Training

5-52. SFA frequently entail rapidly changing circumstances; thus, FSF small-unit leaders must be able to plan and execute operations with little guidance. Therefore, trainers stress small-unit leadership training which can occur concurrently with individual training. Tools the trainer uses to train leaders are manuals, previously established training, tactical exercises without troops, battle simulations, and unit missions. This training develops aggressiveness, tactical proficiency, and initiative. Small-unit leader training should include combined arms technical training such as procedures for forward observer and close air support. Leadership training must also include land navigation in difficult terrain and under conditions of limited visibility. Mission readiness and the health and welfare of subordinates are also continuous parts of training.

After Action Review

5-53. In any training, the after action review provides the critical link between training and evaluation. It is a professional discussion that includes the training participants and focuses directly on the training goals. An after action review occurs after all collective foreign unit training. Effective after action reviews review training goals with the concerned FSF unit commander. Advisor personnel ask leading questions, surface important tactical lessons, explore alternative COAs, assist the retention teaching points, and keep the after action review positive.

5-54. The advisor encourages the FSF unit commander to review the training with the entire unit or key subordinate leaders. At this review, the FSF commander stresses how to strengthen the chain of command. During the after action review, advisors avoid criticizing or embarrassing the FSF unit commander or subordinates.

5-55. If possible, after action reviews occur during the field portion of the training when the unit assembles at logical stopping points. These reviews provide feedback from instructors and leaders that increases and reinforces learning. This feedback provides a richer database for key points. During the after action review, the senior evaluator draws information from unit leaders to form possible alternative COAs.

Assessment

5-56. In all operations, including SFA, success is generally measured for three periods: short-, mid-, and long-term. Success in SFA is normally defined within the context of these three periods. In the short-term period, FSF make steady progress in fighting threats, meeting political milestones, building democratic institutions, and standing up security forces. In the mid-term period, FSF lead fighting threats and provide security, have a functioning government, and work towards achieving economic potential. In the long-term period, FSF are peaceful, united, stable, and secure; integrated into the international community; and a full partner in international security concerns.

Measures of Effectiveness

5-57. A *measure of effectiveness* is a criterion used to assess changes in system behavior, capability, or operational environment that is tied to measuring the attainment of an end state, achievement of an objective, or creation of an effect (JP 3-0). A measure of effectiveness provides a benchmark against which the commander assesses progress toward accomplishing the mission.

5-58. The following are tools to assess the effects of operations in meeting expectations defined for the short, mid, and long term. Although all SFA is not focused on a host nation, these tools are based on the four IDAD pillars:

- **Balanced development** attempts to achieve national goals through political, social, and economic programs.
- **Security** includes all activities implemented to protect the populace from the threat and to provide a safe environment for national development.
- **Neutralization** is a political concept that encompasses the physical and psychological separation of the threatening elements from the population.
- **Mobilization** provides organized manpower and resources and includes all activities to motivate and organize popular support of the government.

Measures of Performance

5-59. A *measure of performance* is a criterion used to assess friendly actions that is tied to measuring task accomplishment (JP 3-0). This quantitative analysis determines whether the task or action was performed to standard. The standard may be derived from a procedure or may be time-sensitive. Examples of measures of performance are listed in the appropriate Army tasks in FM 7-15.

5-60. The advisor helps the foreign counterpart to monitor the current situation for unanticipated successes, failures, or enemy actions. As the commanders assess the progress of foreign security operations, they look for opportunities, threats, and acceptable progress. Throughout the operation, the commanders visualize, describe, and direct changes to the operation, feeding the assessments of progress or regression back into the planning process.

REDEPLOYMENT

5-61. Redeployment is the return of forces and materiel to the home or mobilization station. Before redeployment, a unit conducting SFA inventories all supplies and equipment (jointly, if possible) to be passed on to the foreign unit or the relieving U.S. or multinational unit. These actions ensure all items are accounted for and the custody chain for property and equipment is unbroken. The outgoing unit also—

- Prepares a final evaluation report showing its estimate of the foreign unit's capabilities and giving an opinion of its future employment.
- Discusses the foreign unit's performance with its commander. The unit submits a copy of the foreign unit's final evaluation to the next higher FSF commander.

- Passes custody of training schedules, lesson plans, foreign operational records, and the foreign unit's final evaluation to the foreign unit and relieving U.S. or coalition unit.
- Presents debriefings, after action reviews, and the foreign unit's final evaluation to the higher in-country U.S. commander.

5-62. The outgoing unit commander supervises the redeployment. This commander approves prepared redeployment plans or, as necessary, directs changes to the redeployment plans. The commander and operations officer assess and select alternatives or changes to the present redeployment COAs based on the intelligence officer's intelligence estimate. They develop and disseminate a fragmentary order for selected alternatives or changes.

5-63. Unit personnel prepare all accompanying supplies and equipment for shipment. They brief ground support personnel on equipment handling procedures, routes, convoy procedures, and actions to take if a terrorist or insurgent incident occurs. They load personnel and equipment for movement to the departure airfield or airport following the load plan in the unit's operation order. They maintain accountability for all their personnel, supplies, and equipment.

Mission Handoff Procedures

5-64. During long-term SFA, conditions determine the rotation of in-theater units. Time is not the only governing factor. Changes in the operational environment may require reshaping force packages as situations change. In addition, internal administrative concerns might prompt or support a commander's decision to rotate units. Regardless, mission handoff is necessary and defined as the process of passing an ongoing mission from one unit to another with no discernible loss of continuity.

5-65. The overall authority for the handoff and the subsequent transfer of authority lies with the commander ordering the change. The authority for determining the handoff process lies with the incoming commander assuming responsibility for the mission. This changeover process may affect conditions under which the mission will continue.

Redeployment Considerations

5-66. Although intended for a direct handoff between U.S. units, commanders must make specific considerations when making a handoff to a multinational force. The considerations include mission, enemy, terrain and weather, troops and support available, time available, and civil considerations. For units relieved of a function by a government agency, procedures typically entail longer handoff times and more complex coordination. However, the other areas of consideration still apply and may in fact be a greater issue for an agency. Outgoing units that have past, present, or future projects planned with agencies prepare to transfer these projects to responsible agents in the incoming unit.

Mission

5-67. The outgoing unit must prepare a detailed study of their unit's mission statement and detail the current and implied mission tasks. The mission may also require an incoming unit with additional skill sets, such as advanced special operations, near real-time connectivity, civil affairs functional specialists, or complex media production ability. Knowing the mission, commander's concept of the mission, commander's critical information requirement, PIRs, and information requirements helps the incoming unit understand the mission. After a complete in-depth study of the operational area, the outgoing unit should help the incoming staff complete the handoff in a manner that allows for continued, uninterrupted mission accomplishment. The changeover must not allow the adversary to gain any operational advantages.

Enemy

5-68. The outgoing unit must provide the latest available intelligence on all threats that affect the mission. This intelligence includes comprehensive data on terrorists and terrorist-related incidents over the previous several months. In addition to the normal intelligence provided to the incoming unit on a regular basis, the situation may call for a liaison from the outgoing unit. Operations security is critical to prevent the enemy from discovering the impending handoff and then exploiting the fluidity of the change and the concentration of U.S. or multinational forces.

5-69. The in-country unit provides continuous information updates to the incoming unit. PIRs and information requirements were established for the original mission along with operational, strategic, and tactical information. The incoming unit must become familiar with the ongoing and upcoming mission PIRs and information requirements along with their links to planned nonlethal effects.

Troops and Support Available

5-70. To the incoming unit, learning about the friendly forces is as important as knowing the enemy situation. The unit must be familiarized with the command and control structure with which it will work daily. The incoming unit must know all friendly units in adjacent operational environments. It also must be aware of joint and multinational units and the capabilities of their mission support base. The incoming unit must also be aware of other operations, units, and their capabilities.

5-71. Mission handoff is the time for the outgoing unit leadership to introduce the incoming unit leadership to leaders of FSF. This practice helps prevent breaks in intelligence networks already developed by the outgoing unit and reduces the time the incoming unit needs to build trust. The outgoing unit supports the incoming unit's plans to prepare for the transition in counterpart relations. However, potential or anticipated friction between a foreign security force and the incoming unit may cause the relief to occur more slowly. Therefore, incoming and outgoing units need time for overlap to allow for in-country, face-to-face contact with their counterparts before the mission handoff. If possible, outgoing unit members should provide biographical data on their counterparts to include photographs. This information allows unit members to become familiar with their counterparts before deployment and to determine which advisor techniques may need more emphasis. Execution of the mission must continue within the capabilities of the incoming unit, the foreign unit, and the available supporting assets.

Time

5-72. The depth and dispersion of units and the number of operations conducted determines the time required to exchange units. Ideally, an overlap occurs to allow the incoming unit to become familiar with the operational environment and to establish rapport with their foreign counterparts. However, the handoff must occur as quickly as possible. The longer it takes, the more vulnerable become personnel in the operational environment. However, both outgoing and incoming units should not sacrifice continued and uninterrupted execution of ongoing operations for speed. The incoming unit needs enough time to observe training techniques and procedures and to conduct debriefing on lessons learned. Outgoing unit personnel need sufficient time to put incoming unit personnel in contact with agency counterparts.

Civil Considerations

5-73. The outgoing unit provides an in-depth area study, giving close attention to local problems. Often sister units can provide general demographic data that can be expanded upon for unit-specific needs. Popular support for U.S. activities occurring in the operational environment may directly influence changes in the mission statement. The outgoing unit provides this critical information and describes all completed and progressing civic action projects. The incoming unit considers how the host-nation government functions and international civilian or government agencies are involved in or influence the situation in its operational environment.

POSTDEPLOYMENT

5-74. The unit commander debriefs to provide an overview of the mission and all relevant information subsets. Topics can range from military geography, political parties, and military forces to insurgents, security forces, and ongoing agency operations. Concerns specific to a foreign security force and relevant in one SFA operation may not be relevant in another operation. The commander should codify and debrief any information subset that may affect mission success.

Postmission Debriefing

5-75. Redeployment is not the end of the mission. Upon arrival at the redeployment location, the unit undergoes an extensive debriefing. The intelligence staff section officer typically organizes and conducts the debriefing. The intelligence officer coordinates with higher-level intelligence organizations to take part in the returning unit's debriefing, particularly if other organizations tasked the unit to obtain information. All deployed personnel, to include attachments, must be available for the debriefing.

Documentation

5-76. After the debriefing, the unit commander and staff prepare two documents—an after action review and a report of lessons learned.

5-77. The after action review states the who, what, when, where, and how of the operation. It is a permanent record of the major activities of the unit from receipt of mission to debriefing. As such, it is an extremely important template on which past missions may be compared and future missions planned. Within 48 hours of being debriefed, the unit normally submits an after action review through command channels to the higher command. The intelligence and operations officers at each echelon keep copies of unit after action reviews.

5-78. Shortly after completion of the after action review, or simultaneously with its submission, the unit submits a report of lessons learned. This report documents the commander and staff's reflection of the operation and recommendations for the future. Units often prepare the lessons according to the six warfighting functions: movement and maneuver, intelligence, fire support, sustainment, command and control, and protection. This method addresses what worked and did not work during the operation, why it did or did not work, and what changes or substitutions are needed for existing tactics, techniques, and procedures in the unit.

5-79. As a minimum, the lessons learned report is forwarded to appropriate Service- and joint-level lessons learned agencies to allow integration into subsequent unit training and leader education. These agencies might include the Center for Army Lessons Learned and Joint Center for Operational Analysis.

Chapter 6
Sustainment

A brigade combat team may be required to execute operations across the spectrum of conflict. Hence, providing security force assistance (SFA) may require the sustainment staff to analyze the mission to determine support requirements and available capabilities. Since brigade combat teams conducting SFA may be responsible for a large operational area, an accurate and thorough assessment of U.S. assets as well as the foreign security force's assets is critical to mission success. The sustainment planner understands that though the mission can change, methods by which logistics are distributed within a theater of operations do not. Army sustainment enables the commander to execute a mission and sustain the force through full spectrum operations. Sustainment assets are allocated based on those requirements; no standard arrangement fits all situations.

As discussed in chapter 2, one of the principles of SFA is to ensure long-term sustainment. This principle dictates that all policies, strategies, and programs must be durable, consistent, and sustainable by foreign security forces (FSF) and their government once U.S. forces have completed assistance. This can be accomplished by developing the foreign security force's capability and capacity for asset production, management, performance, and maintenance.

SUSTAINMENT IN FOREIGN SECURITY FORCES CAPACITY

6-1. Sustainment is the provision of the logistics, personnel services, and medical support necessary to maintain operations until mission accomplishment. The endurance of SFA forces is primarily a function of their sustainment. Sustainment determines the depth to which the forces can conduct decisive operations, allowing the commander to seize, retain, and exploit the initiative. Endurance is the ability to employ combat power anywhere for protracted periods. Endurance stems from the ability to generate, protect, and sustain a force. It involves anticipating requirements and making the most effective, efficient use of available resources. Sustainment also enables strategic and operational reach. SFA forces require strategic and operational reach to deploy and immediately conduct operations anywhere with little or no advanced notice. FM 4-0 discusses sustainment in detail.

CORE LOGISTIC CAPABILITIES

6-2. Similar to any operation, the core logistic capabilities in SFA provide a framework to facilitate integrated decisionmaking, synchronize and allocate resources, and optimize logistic processes. The challenges associated with support cut across all the core logistic capabilities: supply, maintenance, deployment and distribution, engineering, medical support, logistic services, and operational contract support.

Supply

6-3. Supply operations include identifying requirements; selecting supply sources; scheduling deliveries; receiving, verifying, and transferring products; and authorizing supplier payments. To provide responsive supply operations in SFA to both the advising organization and FSF, logistic planning requires a collaborative environment that includes operations planning, maintenance operations, and the distribution system.

Maintenance

6-4. In SFA, maintenance operations aim to reduce repair cycle times by replacing components, modules, and assemblies as far forward as possible as well as maximizing reliance on parts distribution, visibility, and replacement. Advisors and their foreign counterparts require maintenance capacity for their freedom of action. Force movement and maneuver depends on sufficient readiness levels of transportation equipment and weapons systems. In SFA, maintenance encompasses both intermediate and organizational levels and consists of shop maintenance supporting the supply chain, transportation assets, and weapon systems.

Deployment and Distribution

6-5. The deployment and distribution capability moves forces and logistic support on time by meeting the required delivery date and providing time-definite delivery. By sharing critical information, units create unity of effort among diverse distribution organizations. This unity of effort enhances foreign support efforts and provides end-to-end support to satisfy deployment execution and sustainment operations for advisors and host-nation forces.

Engineering

6-6. Engineering provides capabilities to maximize freedom of action. These include capabilities and activities that support the maneuver of land combat forces; modify, maintain, or protect the physical environment; and contribute to a clear understanding of the physical environment by providing geospatial information and services to commanders and staffs.

Medical Support

6-7. A component of Army health systems oversee operational management of the health service support and force health protection missions for training, predeployment, deployment, and postdeployment operations. Army health system includes all mission support services performed, provided, or arranged by Army Medical Department to support health service support and force health protection mission requirements for civilian and military joint and multinational forces. For more information on Army health systems, see FM 4-02.

Health Service Support

6-8. All support and services performed, provided, and arranged by Army Medical Department promote, improve, conserve, or restore the mental and physical well-being of Soldiers and personnel in other Services, agencies, and organizations as directed. This includes casualty care, medical evacuation, and medical logistics. The health service support mission is a function of the sustainment warfighting function.

Force Health Protection

6-9. Force health protection measures promote, improve, or conserve the mental and physical well-being of Soldiers. These measures enable a healthy and fit force, prevent injury and illness, and protect the force from health hazards. They include some prevention aspects of Army Medical Department functions, combat and operational stress control, dental services, and laboratory services. The force health protection mission is a function of the protection warfighting function.

Logistic Services

6-10. Logistic services comprise the support capabilities that collectively enable the United States to rapidly provide global sustainment for U.S. forces. Logistic services include many disparate activities that are highly scalable capabilities. Included in this area are food, water and ice, base camp, and hygiene services.

Contract Support

6-11. Operational contract support provides acquisition functions to obtain materials, services, and construction of facilities for U.S. and host-nation forces in the local operating areas. The Department of Defense increasingly relies on contractors to perform many functions and tasks and then provides technical oversight. The expeditionary nature of SFA creates an even greater need for this capability.

PRINCIPLES OF SUSTAINMENT

6-12. The principles of sustainment are critical to maintaining combat power, maintaining strategic and operational reach, and providing Army forces with endurance. The principles of sustainment are anticipation, responsiveness, simplicity, economy, survivability, continuity, improvisation, and integration. While these principles are independent, they are also interrelated when used in planning and executing operations. For more information on sustainment principles, see FM 4-0.

MISSIONS

6-13. All brigade support battalions have the same general capability to manage distribution and sustainment operations. The brigade support battalion provides support within an assigned operational area. It is force tailored and task-organized and uses subordinate companies, platoons, and teams to perform specific functions. In the sustainment role, the brigade support battalion is primarily concerned with the continuous management and distribution of stocks and allocation of maintenance in the operational area. Such assistance provides operational reach to maneuver commanders.

6-14. The brigade support battalion contains the following sustainment capabilities:

- Provides supplies, field services, field maintenance, recovery, and field feeding for itself and its assigned subordinates.
- Plans and conducts base and base cluster self-defense; defends against level I threats, assists in destruction of level II threats, and escapes or evades against level III threats.
- Has ability to operate as part of an Army or joint force.
- Coordinates host-nation support.
- Has ability to deploy an advance party to support early entry operations.
- Provides opening, distribution, and sustainment management information and advice to commanders and staff within its operational area.
- Exercises technical supervision over operations for all assigned units.
- Provides limited materiel management for internal stocks—classes I, II, III, IV, V, VII, and IX and maintenance management of internal assets.
- Provides a liaison team to augment other headquarters as necessary.
- Manages and maintains the its property records.
- Maintains data in support of Army equipment status reporting database and Army equipment status reporting system.
- Appoints contracting officer representatives to oversee contractor performance, certify receipt of services, and act as liaisons between the requiring activity and the contracting officer.

6-15. A brigade combat team often has attached companies such as an engineering company, a military intelligence company, an antitank company, or a network support company. Similar to the brigade headquarters and headquarters company, these attached companies are supported by the brigade support battalion regardless of where they are located on the battlefield. If one or part of these companies is task-organized to a maneuver battalion, it remains under the support of the brigade support battalion.

Critical Considerations in Building Host-Nation Sustainment Capacity

6-16. Effective SFA planning begins with accurately assessing and understanding U.S. policy objectives, determining the end state, and identifying resources available. Some possible logistic planning considerations—

- Have a holistic perspective.
- Address the causes, not the symptoms.
- Understand the desired end state.
- Understand the level of FSF senior leader support.

6-17. Logistic planning during SFA uses a holistic perspective. Operational environments can range from peacetime cooperation to counterinsurgency, stabilization to reconstruction. SFA forces determine the foreign security force's logistic and reconstruction requirements based on their internal defense and development strategy; they develop a sustainment end state to support the plan. Rebuilding the host nation's infrastructure and industrial capacity not only provides long-term job opportunities, but also builds trust and legitimacy in FSF, the host-nation government, and partner organizations.

6-18. Logistic planning during SFA addresses the causes, not the symptoms. Operational logistic support problems are the symptoms, not the causes. They result from problems that exist in the supporting systems or in the society's industrial capacity and economic strength. Teaching unit-level inventory management alone will not result in any sustainment improvement.

6-19. Logistic planning during SFA understands the desired end state. Depending on the U.S. strategic interest in a host nation and its security forces, levels of engagement can vary significantly to achieve the desired end state. Sustainment capacity-building activities can be force tailored and scaled to assist in any of the sustainment capability areas.

6-20. Logistic planning during SFA understands the level of FSF senior leader support. These leaders develop and set the policies for resource allocation. They support the various development projects and skills enhancement assistance activities. These senior-level advisors must fully understand the problems identified and an assistance strategy developed by SFA planners. From the beginning, SFA planners work with their counterparts to decide the plans and strategy to promote the vision and ensure continued effort.

6-21. SFA depends on capable logistic support for maneuver and sustainment. Without an improvement in sustainment capacity, FSF cannot significantly improve regardless of how much effort is devoted to advising and training. When planning for sustainment capacity building, planners thoroughly understand the logistic systems to identify causes that lead to functional deficiencies. They must have a holistic perspective and understand the various goals and resources required to formulate an operational approach that effectively and efficiently achieves U.S. objectives.

ASSET MANAGEMENT AND PERFORMANCE

6-22. One tool at the disposal of the commander and planners during an evaluation of foreign infrastructure and industrial capacity is called an asset management plan. This plan, commonly employed in the industrial community, manages an organization's infrastructure and assets to deliver a standard of service. An asset management plan provides information to FSF so they can analyze the capacity, condition, and current and future requirements of their facilities and infrastructure. Physical development should include facilities, transportation networks, area development, utilities, and other infrastructure components.

6-23. A description of the perceived problems of current foreign facilities, or problems created by missing assets, is also important. The description includes how the advisor intends to address deficiencies in foreign facilities and infrastructure. Each organization needs to determine each facility's function and value to their organizational mission by analyzing the current asset performance, costs, benefits, and improvements.

CURRENT ASSET PERFORMANCE

6-24. Asset performance is the most difficult portion of the asset management plan process to capture while assessing conditions of facilities and infrastructure. While historical records may exist, combat operations and possible instability of the host-nation government likely render these records irrelevant. Remaining structures, when measured against U.S. construction standards (designed to protect occupants from man-made or natural disasters), often are deficient, even if untouched by hostilities. Most existing facilities require significant modification to be considered for future use. However, FSF may use them in the interim until new facilities are completed.

6-25. Asset performance may also use gap analysis. Gap analysis defines the present state, the desired end state, and the gap between them. Once a gap has been identified, the commander and staff can develop courses of action to rectify or bridge the gap. Gap analysis alone is not adequate for all problem sets, as goals may evolve and emerge during the course of problem solving. Most problems have many alternative solutions.

COSTS

6-26. Establishing the short, medium, and long-term costs for facilities and infrastructure, from construction to control, is a critical step in prioritizing which facilities get built or refurbished first, and which get support at all. Units develop a forward-looking cost-profile for operating, maintaining, refurbishing, and replacing assets. This profile explains to civilian and military officials the implications and continued costs to maintain assets. SFA requires a plan for U.S. forces to hand off assets to FSF.

6-27. In constructing these models, the ideal cost-profile extends to cover the life of the longest-lived asset in the system. By doing so, the profile estimates the whole-life cost as well as determines the average annual costs. Furthermore, this cost evaluation may need to be revisited and updated annually, forming the basis for an annual bid for funding.

BENEFITS

6-28. Assets should provide some measure of benefit to U.S. and foreign security forces that can be measured or explained during and after SFA. This usually involves translating asset costs and projected performance into a monetary figure. Other benefits may be social or environmental, which may be difficult to quantify in monetary terms. Regardless, some attempt to record all the relevant benefits is important; the more quantitative, the better.

6-29. A benefit evaluation is vital to demonstrating that the ongoing cost of an asset system is reasonable given the scale of the benefits. The expenditure incurred by asset management needs to be justified in some way.

IMPROVEMENTS

6-30. How assets of FSF perform may be improved by acquisition, enhancement, or some other means. An improvement may address the potential reduction of performance and its effect on the rest of the asset system. The improvements represent a potential change to the performance and are typically managed as a project. Such management comes with an appraisal of the additional expenditure, comparing different options and selecting a preferred option based on the commander's guidance and host-nation requirements.

OPERATIONAL CONTRACT SUPPORT

6-31. Contracting is a key source of support for deployed forces across the spectrum of conflict. Because of the importance and unique challenges of operational contract support, the commander and staff need to fully understand their role in planning for and managing contracted support in the operational area. This includes understanding the types of contracted support, contract management, and contract close out.

TYPES OF CONTRACT SUPPORT

6-32. Current doctrine describes three broad types of contracted support: theater of operations support, external support, and systems support.

Theater of Operations Support

6-33. Contracts for theater of operations support provide support to deployed forces under prearranged contracts or contracts awarded from the mission area. They do so by contracting officers under the command and control of the combat sustainment support battalion. Contractors for theater of operations support acquire goods, services, and minor construction support, usually from local commercial sources, to meet the immediate needs of operational commanders. Theater of operations support contracts are the type of contract typically associated with contingency contracting. Sustainment brigades often are the requiring activity for theater of operations support contract actions.

External Support

6-34. Contracts for external support provide a variety of support to deployed forces. External support contracts may be prearranged contracts or contracts awarded during the contingency. These contracts support the mission and may include a mix of personnel including U.S. citizens, third-country nationals, and host-nation subcontractor employees. The largest and most commonly used external support contract is the logistics civilian augmentation program. This Army program is commonly used to provide life support, transportation support, and other support functions to deployed Army and joint forces.

System Support

6-35. System support contracts are prearranged and include, but are not limited to, newly fielded weapons systems, command and control infrastructure, and communications equipment. System support contractors, made up mostly of U.S. citizens, provide support in garrison and may deploy with the force to both training and real-world operations. The unit does not normally have a significant role to play in planning for or coordinating system support contracts other than coordinating and executing support of system support contract-related personnel.

CONTRACT MANAGEMENT

6-36. Contract management requires nominating and tracking contracting officer representatives for every service contract. It will also nominate a receiving official for all supply contracts. To ensure contractors provide the service or item per the contract, they need a quality contracting officer representative and official support. SFA units must identify and coordinate funding requirements in advance with the supporting resource management staff element to secure the correct amount and type of contract funding associated with the mission.

CONTRACT CLOSE OUT

6-37. In long-term operations like SFA, the advising unit ensures direct coordination and transfer of information related to operational contract support before mission handoff. Additionally, when the advance party arrives in the operational area, a designated Soldier actively seeks out current information on local contract support capabilities, policies, and procedures. This individual coordinates the formal handover of existing contract management responsibilities from the redeploying unit. This Soldier must know when recurring service contracts end since it takes 30 to 60 days to obtain funding approval. If the unit waits until the contract is about to expire before requesting additional funds, the sustainment brigade could lose the contracted service until funds are available.

EXTERNAL AGENCIES SUPPORT MANAGEMENT

6-38. Some external agencies help support the management of FSF. External support comes from interagencies, the host nation, and multinational organizations.

Interagency Support

6-39. The Department of Defense performs both supported and supporting roles with other agencies. It often is the lead agency for logistic activities in SFA. When tasked to provide military support to civil agencies, Army forces perform a supporting role. Sustainment forces may be tasked to support the U.S. ambassador that, as the President's representative, serves as senior U.S. Government officer in the country, or may employ resources of other U.S. Government agencies or private firms. Whether supported or supporting, close coordination is the key to efficient and effective interagency operations. In the absence of an American Embassy, or during operations in austere environments or remote locations, the advising unit may be the only organization able to provide supplies, services (including force protection), and life support. More typically, during the early stages of the advising unit deployment, the unit may find the American Embassy to be the primary source of supplies, services, and life support.

6-40. Nongovernmental organizations do not operate in either the military or governmental hierarchy. Their relationship with an advisor organization is neither supported nor supporting. An associate or partnership relationship exists between military forces and engaged nongovernmental organizations, ideally focused on unity of effort. If formed, the focal point where U.S. forces provide coordinated support to nongovernmental organizations would be the civil-military operations center of a joint task force headquarters.

Host-Nation Support

6-41. Host-nation support and local procurement may provide sustainment, operational support, and tactical support. Host-nation support agreements fulfilling the command requirements for support need to be prenegotiated. Such support arrangements must be integrated into the distribution plan and coordinated with joint, allied, and multinational partners to prevent competition for resources and ensure high priority requirements are met. Host-nation support may include functional or area support and may use host-nation facilities, government agencies, civilians, or military units. Preestablished arrangements for host-nation support can reduce the requirement for early deployment of U.S. assets and can offset requirements for early strategic lift by reducing requirements for moving resources to the theater of operations.

Multinational Support

6-42. Multinational support may consist of support provided from one multinational partner to another. One or more of the following organizational and management options facilitates multinational support:

- National support elements provide national support.
- Individual acquisition and cross-servicing agreements provide limited support.
- A lead nation provides specific support to other contributing nation forces.
- A role-specialist nation provides a specific common supply item or service.
- A multinational integrated sustainment unit provides limited common supply and support.
- A multinational joint sustainment center manages common-user logistic support.

6-43. In all cases, the multinational force commander directs specific multinational support within the applicable laws and regulations of the host nation. When operating within a formal alliance, the advising unit executes support following applicable standardization agreements.

FACILITATING THE REDEPLOYMENT PROCESS

6-44. Sustainment assets facilitate the redeployment and the retrograde process through its command and control of the distribution system. Many of the procedures are used to deploy forces, draw pre-positioned stocks, conduct reception staging and onward movement, and distribute supplies within the theater of operations apply to the redeployment process. Two factors in particular complicate redeployment operations: supporting operations and meeting requirements.

6-45. The same elements that operate and manage the theater of operations distribution system during deployment and sustaining operations perform similar roles during redeployment. When redeployment,

deployment, and sustaining operations occur simultaneously, the theater sustainment command may find it necessary to rebalance its forces or change the missions of subordinate units to effectively support ongoing operations and redeployment.

6-46. Requirements vary widely depending on the nature and scale of redeployment operations, infrastructure, and other factors. For example, redeployment operations could range from personnel only to entire units and their equipment. Depending on the political and military strategy, unit rotations may occur while decisive operations continue unabated or during operational transitions. Key considerations include, but are not limited to, several characteristics. These characteristics include the size of the force redeploying and deploying; infrastructure requirements and limitations; security requirements; traffic circulation; staging areas; distribution system capacity; quantity of supplies and materials to be redistributed; the amount and kinds of materials that require disposal agricultural inspections; and establishing accountability of retrograde cargo. The challenge is effective coordination and synchronization, vertically and horizontally, to ensure responsive simultaneous support to not only ongoing distributed operations, but also redeployment.

Chapter 7

The Advisor

Advisors are the most prominent group of U.S. personnel that serve with foreign security forces (FSF). They live, work, and fight with their FSF. The relationship between advisors and FSF is vital. Advisors are not liaison officers, nor do they command FSF units.

ROLES OF THE ADVISOR

7-1. The military advisor has three roles involving different responsibilities. First and foremost, advisors are members of a U.S. military organization with a well-defined chain of command and familiar responsibilities. Second, advisors embed themselves with their counterparts. Third, advisors are interpreters and communicators between U.S. forces and their foreign counterparts.

7-2. As members of military organizations, advisors receive and execute the orders of superiors. These orders may conflict with the orders their counterparts receive. Among other duties, advisors must act unobtrusively, but nonetheless positively, as inspector general—often observing, evaluating, and reporting on the performance of counterparts and their assigned unit.

7-3. Secondly, advisors live, eat, and work with the officers and men of their host units. Often, advisors soon regard themselves as one of them. The sharing of common hardships and dangers forges potent emotional ties. The success and good name of their units become matters of prime and personal importance to the advisor.

7-4. Finally, advisors are interpreters and communicators between U.S. superiors and foreign counterparts. Advisors must introduce and explain one to the other; they help resolve the myriad of problems, misunderstandings, and suspicions which arise in any human organization, particularly when people of starkly different cultures approach difficult tasks together. Advisors with quick and easy access to influential counterparts can sometimes be the best possible means of communicating.

7-5. To be effective, advisors obviously must gain their counterparts' trust and confidence. This relationship, however, is only a prelude to the advisor's major objective: inspiring and influencing a counterpart to effective action. In pursuing this goal—constantly, relentlessly, and forcefully, yet patiently, persuasively, and diplomatically—advisors must recognize conditions which can benefit or handicap their cause.

CONSIDERATIONS OF THE ADVISOR

7-6. An advisor follows ten considerations when serving with FSF.

BY, WITH, AND THROUGH

7-7. All planned operations of a combat advisory mission must be conducted by, with, and through the FSF. Not counting immediate action battle drill responses, the mark of an effective advisory effort is the amount of personal involvement the FSF take in their own operations. Civilians must see that they are secured by their own security forces to promote the legitimacy of the host nation and their capacity and to build trust and confidence. In a word, combat advisors are shadows whose presence is felt at all times.

Empathy Leads to Competency

7-8. Empathy can be defined as identification with and understanding of another's situation, feelings, and motives. This is tough for experienced U.S. leaders and often harder to explain, but it is the key to the success of an advisory mission. Truly understanding other human beings and where they come from allows for honest relationships to develop. In most cultures, the place to begin understanding another person's feelings and experiences is by understanding the other person's narrative. The narrative is a collective group's identity as an interpretation of both ancient history and recent collective experiences. Delving into the narrative, understanding it and how it affects people, is the beginning of empathy.

Success Is Built on Personal Relationships

7-9. Developing a sense of empathy allows advisors to begin nurturing relationships with those they are tasked to advise. No amount of resources and firepower can compensate for the lack of a relationship between advisors and their counterparts. It must be honest, genuine, and heartfelt. Mutual respect, trust, and understanding create success. Both parties rely on each other for mission accomplishment and often for survival. This relationship is likely to be challenged on numerous occasions; only one built on mutual trust can survive and ensure mission success. Furthermore, advisors learn to pick their battles. Camaraderie and rapport with FSF can vanish in an instant if one does not prioritize and define critical lines in the relationship.

Advisors Are Between U.S. and Foreign Security Forces

7-10. Living and fighting with FSF allows for the development of true bonds, empathy, and trust. These desired results have a price to pay. When addressing their own Army, advisors soon realize that they are also not one of them, increasing their level of frustration. Advisors are often alone, navigating between two military systems and two cultures, never quite fitting in with either one.

Advisors Will Never Win, Nor Should They

7-11. Often advisors find it difficult to understand what victory looks like in this environment. Progress tends to occur at a glacial pace and cannot usually be tracked on a day-to-day basis. The advisor attaining a tactical objective does not achieve success; success is achieved by the foreign security force achieving the objective. Advisors find it difficult to satisfy their own units, and they never fully satisfy the demands of their FSF. They are figuratively and literally caught in the middle. Advising is the art of striving to make a win-win situation for all parties. The advisor and foreign security force relationship is one in which the advisor removes the obstacles in the way. Only a plan by FSF will succeed—and it will only succeed if it is, in fact, their plan.

Advisors Are Not Commanders

7-12. Advisors do not command FSF personnel or units. Advisors provide advice, training, and access to coalition resources to FSF. They are not intended to lead FSF in combat; they are responsible for the command and control of their own team of advisors. However, they can and should influence the commander of the foreign unit.

Advisors Are Honest Brokers

7-13. Advisors are advocates for FSF with conventional forces. They transparently assess capabilities and reveal limitations of foreign units to the FSF's and advisors' higher command. They also provide access to U.S. resources—from close air support to medical evacuation—to FSF they advise.

Advisors Live with Shades of Gray

7-14. Black and white is not available to the advisor. Caught between two cultures, systems, and narratives, the advisor works within a gray area. Unlike operations in a conventional unit, many of the checks and balances are not present. Advisors often find themselves isolated with great autonomy and no supervision. These conditions present moral and ethical dilemmas daily. Advisors need to be comfortable in this environment. Often they lose sleep at night and question their actions and those of their FSF. Thus, this mission calls for Soldiers of enormous character, moral courage, and intellect.

Talent Is Everything, But Advisors Must Understand Rank

7-15. Military forces around the world approach rank and prestige in completely different ways. In some cultures, rank equals experience; in others, it is nobility or education. Often many militaries have rewarded loyalty with promotions, which results in rank-heavy armies. Some FSF recognize talent, while others recognize rank. Advisors likely advise counterparts much more senior in rank. They must understand that rank on the uniform is important to many armies.

Make Do

7-16. Advisors will never have everything they believe they need to succeed. However, their creativity often distinguishes between success and failure. Austerity and economy of force can fail to meet the prevalent expectations that FSF have of a U.S. unit in regards to resources. It is up to the advisor team to close the expectation gap by its creativity and imagination. Scrounging, bartering, and negotiating are daily activities of advisors. An enormous amount of energy must be devoted to these activities to achieve mission success and endear them to their counterparts.

PERSONALITY TRAITS OF THE ADVISOR

7-17. Not every Soldier is well suited to perform advisory functions; even those considered to be the best and most experienced have failed at being an advisor. Effective advisors are only the most capable individuals. Advisors are Soldiers known to take the initiative and who set the standards for others; however, they are also patient and personable enough to work effectively with FSF. Recognizing that not all Soldiers are capable of performing as advisors, leaders should immediately remove advisors who do not exhibit these qualities.

7-18. Because advisors operate in very subjective environments, it is difficult to establish objective criteria by which to assess potential advisors. However, research and experience indicate that several personality traits greatly enhance the advisor's ability to adapt and thrive in a foreign culture. These traits include—

- Tolerance for ambiguity.
- Realistic when setting goals and tasks.
- Open-mindedness.
- Ability to withhold judgment.
- Empathy.
- Communicativeness.
- Flexibility.
- Curiosity.
- Warmth in human relations.
- Motivation of self and others.
- Self-reliance.
- Strong sense of self.
- Tolerance for differences.
- Perceptiveness.
- Ability to accept and learn from failure.
- Sense of humor.

7-19. Of the traits listed above, no single trait is paramount. Advisor selection and training programs seek to develop an understanding of the contributions of each of them.

SKILLS OF THE ADVISOR

7-20. Advisor-specific skills fall into two subcategories: enabling skills and developing skills. Enabling skills build on the individual and collective skills of advisors. Developing skills are the main advisor tasks of teaching, coaching, and advising. Advisors use these skills to develop the capabilities of the foreign security force.

ENABLING SKILLS

7-21. Advisors use enabling skills to work individually and in a group. Enabling skills help advisors communicate across cultures, build rapport, influence, and negotiate. Successfully employing these enabling skills sets the conditions for advisors to move forward with their mission.

Individual Skills

7-22. Individual skills help advisors properly perform their mission. Initially advisors should be selected based on subject matter expertise and not just rank. However, tactical proficiency does not necessarily equate to proficiency as an advisor. Modern-day military advisors must be able to teach and advise FSF. To do this, they must be knowledgeable and proficient in procedures and able to impart this knowledge effectively. They may also be able to advise FSF in the areas of intelligence, communications, operations, and logistics. Some advisors may be required to call in U.S. supporting arms, principally artillery and air support, and coordinate both air and ground casualty evacuation. Finally, advisors should be proficient in basic combat skills. These common skills, warrior skills, and survival skills are often required by advisors. Since these skills are critical to an advisor due to the isolated and independent nature of the mission, advisors should refresh them during predeployment training.

Collective Skills

7-23. As a group of advisors form a team, they must practice their warrior tasks, organize to cover the functions associated with any team, and familiarize its members with their duties and responsibilities. As with individual advisor skills, some are refined before deploying and others are learned in country. Figure 7-1 depicts this graphically.

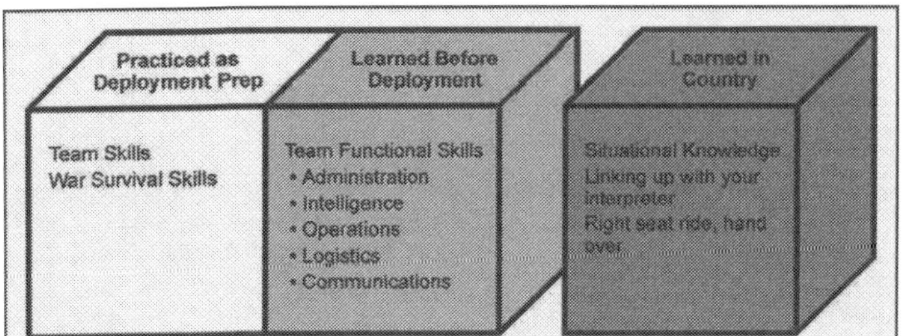

Figure 7-1. Collective advisor skills

DEVELOPING SKILLS

7-24. Developing skills—teaching, coaching, and advising—are required for every advisor. The advisor will make an initial assessment to determine the foreign unit's proficiency to determine the appropriate developing skill with which to start. Figure 7-2 illustrates the appropriate level of advice that corresponds to the foreign unit's capabilities.

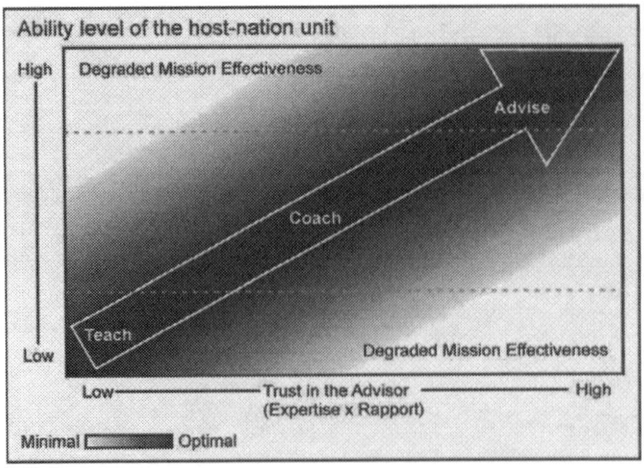

Figure 7-2. Appropriate advising level

7-25. Advisor teams have limited positional power and depend on personal influence to positively affect FSF efforts through teaching, coaching, and advising. Advisor team members often work with people of higher rank or grade than themselves. Advisors should remember the foreign unit's experience and capabilities and carefully choose opportunities to inject or impart knowledge. Foreign units are most receptive to advisor teams that teach unobtrusively. Foreign units most value those who are subtle in their teaching, coaching, and advising. Those advisors who master the ability to create a climate in which FSF personnel feel they are teaching themselves often prove the most effective.

Teaching

7-26. Teaching includes training and education. Methods of teaching can include classroom lectures, seminars, hands-on training, training exercises, and simulations.

Coaching

7-27. Coaching relies on guiding to bring out and enhance capabilities already present. Coaching refers to the function of helping someone through a set of tasks. Those being coached may or may not have appreciated their potential. The coach helps them understand their current level of performance and instructs them how to reach the next level of knowledge and skill. Coaching requires identifying short- and long-term goals and devising a plan to achieve those goals. The coach and the person being coached discuss strengths, weaknesses, and a course of action to sustain or improve the goals.

Advising

7-28. Advising is a combat multiplier that boosts supported unit capability. Advisors have experience in particular areas in which they are advising, but are not required to have similar backgrounds. Advisor relationships are not based on superior to subordinate relationships.

7-29. Advisors provide an expert opinion, advice, and counsel by focusing on both personal development (interpersonal and communication skills) and professional development (technical and tactical knowledge). Advising develops mutual trust and respect.

Advisor Versus Trainer

7-30. Advisors teach, coach, and advise work in a cycle. As FSF master one skill, the advisor can move on to other skills and initiate the process for the new skills. If FSF require additional teaching or coaching, the advisor can take them to that part of the cycle. Eventually, the teaching and coaching should decrease, and most of the advisor's time will be spent as an advisor providing an expert second opinion.

7-31. Ultimately, the question of what is the difference between being a trainer or an advisor will arise. Figure 7-3 shows that every advisor will have to be able to be a trainer, but all trainers are not expected to perform the more sophisticated task of advising. Hence, competent junior officers and enlisted Soldiers can be trainers, but only more experienced individuals can be advisors.

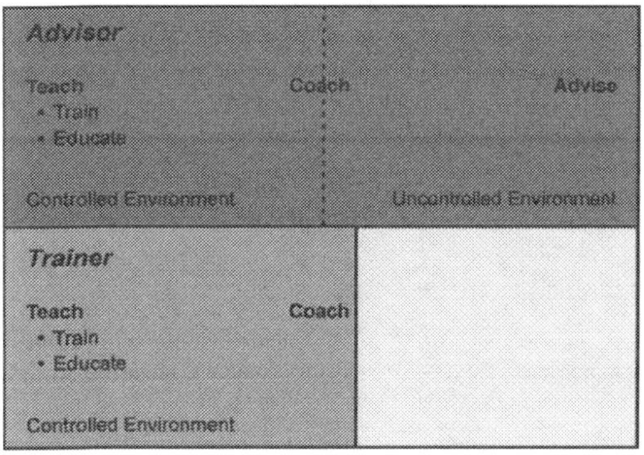

Figure 7-3. Advisor versus trainer

Chapter 8

Culture and Communication

As an advisor, understanding the host-nation population is a crucial element of premission planning and the development of foreign security forces (FSF). Prior knowledge of sociocultural differences aids in building effective relationships and prevents embarrassment, loss of rapport, and compromise of the mission. This chapter discusses three aspects of the sociocultural environment to provide a framework for analysis and comparison of foreign cultures for operational purposes. These aspects are society, culture, and communications.

SOCIETY

8-1. People who have a common culture, share a group identity, occupy a common territory, and are subject to the same political authority can be identified as a society. A society is not easily created or destroyed. Though they share the above factors, societies are not homogeneous nor static. A society usually has a dominant culture but can also have a vast number of secondary cultures, each influencing the other.

8-2. Advisors derive much of their effectiveness from their ability to understand and work with counterparts from a foreign society. Advisors must also consider societies outside their immediate operational environment whose actions, opinions, or political influence can affect the mission.

SOCIAL STRUCTURE

8-3. Society is composed of both social structure and culture. Social structure refers to the relations among groups of persons within a system of groups and is persistent over time. For the purposes of security force assistance, social structure includes groups, institutions, organizations, roles and statuses, and social norms. FM 3-24 contains a more in-depth discussion of social structure.

Groups

8-4. Within societies, groups are composed of two or more people that interact based on mutual interests and interrelated statuses and roles. These groups can be based on race, ethnicity, religion, or kinship. A racial group is defined by its members or by outside groups as different based on physical characteristics or descent. Race is a social category that has no biological basis. An ethnic group is a community whose learned cultural practices, language, history, ancestry, or religion distinguish them from others. Similar to racial groups, ethnic groups can be defined by their members or by outside groups. Religious groups may be subsets of larger ethnic groups and an ethnic group may contain members of different religions. Kinship groups are based on a link by blood relation. It may also be used to indicate a group of persons who live and work together to satisfy basic collective needs and goals. The role of a kinship group is generally seen to care for and educate offspring. This group provides for such needs as food, shelter, and clothing and prepares offspring to be functional members of the society in which they live.

Institutions

8-5. Institutions are the basic building blocks of societies. They are made up of groups that are organized to achieve a common task. Institutions are continuous through many generations and are built to continue to function even as individuals are replaced. An example of an institution from FM 3-24 is a military institution that brings together groups and individuals whose statuses and roles concern defense and security.

Organizations

8-6. Organizations are institutions that have a bounded membership with defined goals and established operations. They include fixed facilities from which they conduct business or operations and a set means of financial or logistic support. The four types of organizations are communicating, religious, economic, and social. Communicating organizations are designed to influence perceptions. Religious organizations are created to regulate societal norms, reaffirm beliefs and values, and provide social support. Economic organizations provide employment and assist in economic development. Social organizations provide support to groups in society, create social networks, and can influence ideologies. Organizations may belong to more than one category and exist to affect the behavior, values, and beliefs of society. As such, advisors must identify influential organizations both inside and outside of the FSF with which they are partnered.

Roles and Statuses

8-7. As members of a group, individuals interact based on assumed or given social positions. These social positions are referred to as statuses. Societies generally associate particular statuses with particular social groups (such as family, lineage, ethnicity, or religion). Each status inherently contains a set of expected behaviors known as a role, which includes how a person of that status should act, feel, and think. A status includes expectations about how others in a society should treat a person of that status.

Social Norms

8-8. Violation of a role derived from a given status results in social disapproval. The standard of conduct for social roles is a social norm. Norms may be either moral (incest prohibition, homicide prohibition) or customary (prayer before a meal, removing shoes before entering a house). Advisors must understand the roles, statuses, and social norms of groups they are advising. This knowledge will clarify the expected behavior of their FSF partners and their role as advisors.

LANGUAGE

8-9. Language is a learned element of culture that is a system of symbols that people use to communicate with one another. Successful communication within a society requires an understanding of the social setting, appropriate behaviors towards different statuses and roles, and nonverbal cues. The languages used in the advisors' operational environment have a major impact on the advising mission. Languages must be identified to facilitate language training during predeployment and employment, as well as in the requisitioning of translators. Translators are critical for communicating with FSF and interacting with local citizens and leaders.

CULTURE

8-10. Social structure comprises the relationships among groups, institutions, and individuals in a society; in contrast, culture provides meaning to individuals within the society. Social structure can be thought of as a skeleton with culture being the muscle on the bones. The two are mutually dependent and reinforcing. A change in one results in a change in the other.

8-11. Culture influences an individual's range of action and ideas, including what to do and not do, how to do or not do it, and whom to do it with or not to do it with. It is neither a random jumble of customs nor a laundry list of accepted practices and taboos that an advisor should know. It influences judgments about what is right and wrong and what is important and unimportant. Cultural rules are flexible in practice; the customs and actions that an individual practices are based largely on that particular environment.

CULTURAL CHARACTERISTICS

8-12. Culture is shaped by history, language, religion, and customs. Members of society use these shared beliefs, values, customs, and behaviors to cope with their world and with one another. It is learned behavior that is shared by members within a society. However, behaviors do not always define a culture.

Frequently, there are more telling cultural signs in the meaning attached to a behavior than in the actual behavior. Culture influences its members' identity, values, belief systems, and cultural forms. As such, it is arbitrary; advisors should make no assumptions regarding what a society considers right and wrong, good and bad. Culture is changeable through the social interactions between people and groups, both inside and outside of that society. TC 31-73 discusses cultural characteristics in great detail.

Identity

8-13. Each individual belongs to multiple groups through birth, assimilation, or achievement. Each group to which individuals belong influences their values and beliefs. Individuals consciously or unconsciously rank their identities into primary and secondary identities. Primary identities are those seen as core to their being; frequently these are national, racial, and religious. In contrast, secondary identities are more peripheral and may include such things as an individual's occupation or hobby.

Values

8-14. Values guide how individuals interact with the world around them. As deeply held beliefs about rightness, values shape how individuals think things should be. Values may never be articulated—in fact, they may not even be consciously held. At times, the beliefs of an individual may even contradict one another. Advisors should have some understanding of a counterpart's value system. Usually values are nonnegotiable. In working with a counterpart with conflicting values, advisors should persuade the individual that the advocated action or behavior really does not conflict with their values.

Belief Systems

8-15. Beliefs primarily take two forms: values (core) and opinions (peripheral). Values, or core beliefs, are not easily changed. Opinions, or peripheral beliefs, stem from thought and decisions. The totality of the identities, values, and beliefs that an individual holds compose that person's belief system.

Forms of Beliefs

8-16. Values are broad moral conclusions about the way to live life. These core beliefs are unstated, taken for granted, resistant to change, and not consciously considered. They reflect strong personal certainties drawn from faith, childhood environments, and life experiences. They lend shape and order to people's lives, helping them to decide if they settled on the right actions, friends, leaders, religions, and careers.

8-17. Opinions, or peripheral beliefs, are immediate conclusions about some aspect of the environment. They are usually quite specific and are open to debate, consciously considered, and easiest to change.

Types of Belief Systems

8-18. Belief systems involve cultural narratives and myths. The interpretation of these narratives and myths provide people with insight into how they should feel, think, and behave. The most prominent systems of beliefs tend to be those associated with religions and ideologies. However, any system of belief in which the interpretation of narratives impacts people's behavior (such as superstitions) can be a contributing component of a society's culture.

8-19. A belief system acts as a filter for new information. It is the lens through which people perceive the world. Beliefs of one group as to what is rational, normal, or true may appear strange, irrational, or illogical to outsiders. Understanding belief systems of various groups in an operational environment allows advisors to work with their foreign counterparts and influence civilians more effectively.

8-20. Belief systems play a fundamental role in the development of a society's values. Careful analysis of a region's belief systems is crucial to understanding the local value system. These systems influence religious doctrine, narratives, and myths and provide enormous insight into the culture of the community. Ultimately, the values which are most prized within a belief system are religious. These beliefs affect that culture's systems of social organization, learned behaviors, and relationship to the physical environment.

Religion as a Belief System

8-21. At its most basic level, religion is both the individual and communal expression of contact with the sacred. Notions of the sacred can vary widely in societies and may include deities and sacred items. In most cases, religions also involve symbols, rituals, and ethics or dogma. Religion can be both an individual practice and a group phenomenon.

8-22. Most societies have some variety of religion—typically a set of sacred beliefs and rituals that provide a common understanding of moral codes and proper conduct. Expressing a common religious understanding helps people to make sense of their world. Often religion provides explanations for human suffering, natural disasters, broken relationships, inequalities, and death. Some societies also have secular equivalents of religion. For instance, America is often remarked to have a civil or civic religion that can be seen in the images, symbols, ideas, and behavior of the political realm.

8-23. Because religion is such an integral part of culture, when advisors carefully prepare and analyze the mission, they must examine the religions and religious groups in the operational environment. Religious beliefs, leaders, and institutions are central to how many societies view the world. When planning any advising operation, planners consider how religion affects FSF and civilians.

Broad Impact of Belief Systems

8-24. In almost every society, followers of dominant religions have a strong societal impact. This influence may be exerted at varying degrees upon the individual citizen, sections of society, the economy, the military, and the political structure. TC 31-73 discusses these in detail.

THE IMPORTANCE OF UNDERSTANDING CULTURE

8-25. Advisors derive their effectiveness from their ability to understand and work with foreign counterparts from another culture. They must understand enough of their own culture and their counterpart's culture to accurately convey ideas, concepts, and purpose without causing counterproductive consequences.

8-26. Advisors must be aware of aspects of the local culture and history that influence behavior in their operational environment. Advisor team members must understand the reasons and motivations underlying personal interaction and must practice patience when working with their counterparts. Group norms guide individual behavior, and advisors must understand how individuals in a society tend to interact as members of a group, whether a race, ethnic, or kinship group.

8-27. Cultural understanding is not derived from demographic information provided to the military through country briefs prior to deployment. It is gained from studying, interacting, and understanding the people, religion, history, customs, and social and political structures within an area. For true understanding, it is necessary to live among the people, gradually understanding the subtleties and nuances of their culture.

CULTURAL EDUCATION

8-28. Culture influences a person's perception of the world. People from one society tend to view people from other societies through the prism of their own culture. Often, their perceptions are biased, are inaccurate, or involve assumptions of superiority. For example, Americans view their society and its culture very differently from the rest of the world. Likewise, as foreigners, advisors view local cultures through the prism of their own idealized culture. This creates a large gap between the culture the advisor observes and the host nation's real culture.

8-29. To narrow the gap of cultural understanding an advisor will encounter, a program of cultural learning should be undertaken. The end state of this cultural learning should be to gain the trust of foreign partners and civilians. Advisors will do so by demonstrating their understanding and sensitivity towards the host nation's culture and social structures.

8-30. An awareness of the cultural aspects of the operational environment can significantly enhance the effectiveness of an advisor. An extensive study program is best, but even a concise program is worthwhile if it provides advisors with the necessary cultural information. The best programs involve credible teachers who intimately understand the culture and have personal experience in the society. Training is dynamic and moves beyond the standard lecture and handout format. Small groups or a discussion panel work quite well. Cultural behavior and cultural familiarization handbooks can also be used. Since handbooks vary greatly in accuracy and quality, credible subject matter experts should be consulted to determine which documents to incorporate into the training. Table 8-1 discusses the training recommendations and tools.

Table 8-1. Cultural education recommendation

	Training Recommendations
Cultural Education	• Comparison of cultural values and social structures (United States compared to those of the operational area). • Local customs and traditions (for example, greetings do's and don'ts). • Geopolitical history (pre-colonial to contemporary and the orientation of each faction or party). • The role of religion in daily life.
	Training Tools
Resources	• Guest speakers native to the country of interest (for example, nongovernmental organization staff, foreign students, recent immigrants, or selected refugees). • Others who have worked in or studied the mission area (for example, special forces personnel, diplomats, and scholars). • Cultural familiarization handbooks.
Format	• Combination of briefings, small group discussions, and question and answer periods. • Handouts to augment—not replace—speakers. • Visual media, specifically slides and videos of the mission area.

8-31. Advisors with previous operational experience have addressed the importance of cultural education. Many civilian agencies have developed intercultural effectiveness programs designed for their personnel working overseas. Often these programs can be revised and tailored for military and paramilitary groups. If available, these programs are an excellent resource for predeployment preparation.

CULTURAL GAP

8-32. Each culture has its own societal rules regarding who a person may speak to, how and when the person may speak, and what topics the person may speak about. Many cultures rely on nonverbal signals to communicate. In such cultures, posture, expression, and actions convey more than spoken or written words. Advisors work to bridge the cultural gap by learning the language as well as nonverbal communication cues. To bridge the gap, advisors develop a sense of cultural awareness, recognize differences, learn to adapt, develop tolerance, and maintain personal contact.

Develop a Sense of Cultural Awareness

8-33. Advisors must recognize they are a product of their own culture. They must learn as much as possible about the culture of the people with whom they need to communicate. When communicating with people across cultures, advisors abandon any sentiments of ethnocentrism—the tendency of individuals to judge all other groups according to their own group's standards, behaviors, and customs. Such notions lead an individual to see other groups as inferior by comparison.

Recognize Differences

8-34. Each culture has its own way of accomplishing required daily tasks. Advisors must understand that each particular society may approach things differently; it does not mean that they are inefficient or less valid. Differences should not be seen as negative. Respect for counterparts must be maintained at all times.

Learn to Adapt

8-35. Advisors must be flexible and ready to adapt and adjust their behavior; however, they must be careful not to overdo their adjustment. Individuals who are overly flexible are often perceived as being insincere. The successful advisor must strive to act in a way that is appropriate to the target culture. Above all, advisors should be themselves and show sincerity.

Develop Tolerance

8-36. Advisors must develop a tolerance for deviations from accepted norms. Events or activities that may seem extraordinary to newcomers may be common practice in the culture. Advisors must be aware that members of the foreign culture may be astounded by that which is commonplace in the United States. Careful observation should be made before judgments are rendered about seemingly peculiar behavior.

Maintain Personal Contact

8-37. Personal contact is the most effective way to bridge organizational barriers. Interorganizational stumbling blocks are very real and the prejudices that arise from them are exacerbated by misunderstanding and ignorance. Often it is too easy to attribute negative attitudes and hostile motives to faceless groups. The advisor must employ superior interpersonal skills and work directly and closely with individual members of other organizations. This type of contact can effectively reinforce commonality and diminish the impact of disparity. These personal relationships are the key to effective interorganizational relationships.

Culture Shock and Adaptation

8-38. The term culture shock describes the anxiety experienced by persons in a completely new environment. It is not the result of a single event. Rather, the condition develops slowly from a series of minor events or conditions. Culture shock expresses the feelings of not having a set direction, not knowing what to do or how to do things in a new environment, and not knowing what is appropriate or inappropriate.

8-39. Culture shock occurs because the mind and body have to go through a period of psychological and physiological adjustment when individuals move from a familiar environment to an unfamiliar one. The cues received by all of the senses suddenly change. During the day the foreigner is bombarded with unfamiliar sights, sounds, smells, tastes, languages, gestures, rules, requirements, interactions, demands, systems, and expectations.

8-40. Some differences that advisors experience between their lives at their home station and their lives when deployed to a foreign location are obvious. These differences include language, climate, religion, food, educational system, and the absence of family and friends. Other differences may not be as obvious. These differences include how people make decisions, spend their leisure time, resolve conflicts, express their emotions, and use their hands, faces, and bodies to express meaning.

Stages of Shock

8-41. Five distinct stages occur in the culture shock process. These include enthusiasm, withdrawal, reemergence, achievement and reentry. Figure 8-1 depicts the stages of shock.

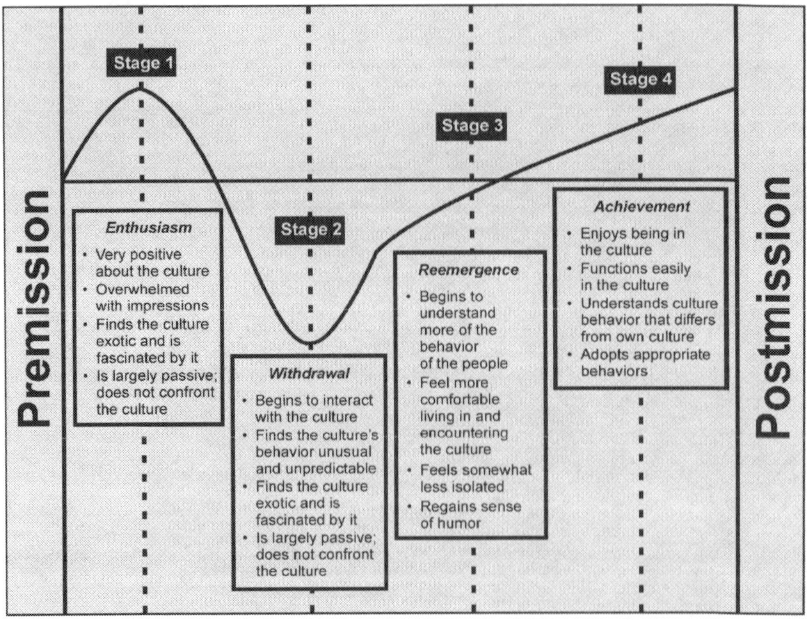

Figure 8-1. Stages of culture shock

Stage 1 – Enthusiasm

8-42. In this stage, a person may feel self-confident and pleasantly challenged. The person may be delighted by all of the new things encountered.

Stage 2 – Withdrawal

8-43. In this stage, a person begins to encounter difficulties and minor crises in daily life. It may be difficult to make oneself understood and foreign counterparts may prove more difficult to work with than anticipated. In this stage, one may have feelings of discontent, impatience, anger, sadness, and even incompetence.

Stage 3 – Reemergence

8-44. This stage is characterized by gaining some understanding of the new culture. A person may not feel as isolated, and a feeling of direction emerges. The individual is more familiar with the environment and is better able to adjust to the circumstances.

Stage 4 – Achievement

8-45. In this stage, the person realizes that the new culture has both good and bad things to offer. This stage is one of integration and the person is increasingly able to function in the new setting. A sense of accomplishment, a reduction of routine annoyances, and a more solid sense of belonging accompany this integration.

Stage 5 – Reentry

8-46. This stage occurs when people return to their own culture. They may find that things are not as they once were. Changes that occurred in their absence—their family, their friends, their communities, and themselves—combine to present a distorted image of home that differs greatly from the one imagined or remembered.

Culture Shock Symptoms

8-47. Being immersed in a society with extreme cultural differences tends to cause feelings of uncertainty and anxiety. The body and mind may react in unusual ways. Some persons may experience more pronounced physical symptoms of stress, such as chronic headaches or upset stomachs. Although uncomfortable, some degree of culture shock is a normal part of the adjustment process. Some common reactions include:

- Feeling irritable with (or even anger toward) one's own group or organization.
- Feeling isolated or alone.
- Tiring easily.
- Changing normal sleep patterns (either too much sleep or not enough).
- Suffering minor (but persistent) body pains, especially in the head, neck, back, and stomach.
- Experiencing feelings of hostility and contempt toward local people.
- Withdrawing from the local population (that is, spending excessive amounts of time alone reading or listening to music).

Overcoming Culture Shock

8-48. An important tool to overcome the obstacles of a new cultural environment is familiarity with the language. An ability to communicate in the new culture, even at the most basic level, reduces the effect and shortens the period needed for adjustment. Other means of overcoming culture shock can include having previous experience in the area (including a regular form of physical activity in personal routines) and acknowledging even slight progress in adjusting to the new culture.

8-49. TC 31-73 discusses culture shock in depth. It identifies the symptoms, cues, and treatment.

CROSS-CULTURAL COMMUNICATIONS

8-50. Communication is the transfer of messages from one person to another. These messages may be passed along verbally, in writing, or by signals (gestures, body language). The sender encodes the message and the receiver decodes it. The type and style of encoding used is based upon the sender's history, beliefs, values, opinions, and belief systems. Communication is determined by culture; as a result it can be either high context or low context. High context communication places an emphasis on clearly stating issues and problems and working towards resolving them in a straightforward manner. Low context communication places a premium on building relationships; as such, problem sets tend to be worked in a less formal matter in societies that utilize this type of communication. Quickly and effectively identifying whether a society uses high or low context communication is a key to effectively conveying material between the advisor and the host-nation population.

8-51. The receiver decodes messages based upon lifestyle, group membership, status and role, worldview, language, and social norms. Communication is a two-way process in which the encoding and decoding methods can affect both sending and receiving. Effective communication occurs when the message is perceived and responded to in the manner the sender intended. Ineffective communication can occur from poorly chosen words, flawed timing, a confused mixture of verbal and nonverbal signals, poor listening skills, and a failure to take culture into account. Communications can be broken into both levels and factors.

LEVELS OF COMMUNICATIONS

8-52. The cross-cultural communications capabilities required by an effective advisor to FSF can be described in three levels. At the lowest level is awareness followed by knowledge. The highest level is reached when these two are combined with well-trained and refined skills. TC 31-73 discusses communications in detail.

Awareness

8-53. Awareness is the basic level of cross-cultural capability. Awareness of cultural differences and their impact is the first prerequisite for successful work with a counterpart. Simply being sensitive to the fact that differences exist and carefully observing actions and reactions can assist the advisor in adjusting behavior and modifying actions to achieve greater influence with the counterpart. Awareness is not region specific and can be instilled in the advisor with relatively little training.

Knowledge

8-54. Knowledge of the details and nuances of a specific culture is the next level of cross-cultural capability. Advisors attain this level through academic study and immersion. Such knowledge is inherently area-specific and rarely transfers from one target area (or culture) to another. Developing in-depth regional knowledge necessary for effective cross-cultural communications requires extensive and time-consuming training. Appropriate personnel assignment policies must support this training.

Skills

8-55. Skills fundamental to effective cross-cultural communications, when combined with awareness and knowledge, form the highest level of cross-cultural capability. These skills include professional competence, language, nonverbal communication, negotiation, interpersonal skills, observation, problem solving, leadership, instruction, and fitness. Although some individuals show greater natural talent for these skills than others, all advisors require continual training in each to achieve and retain their full potential as advisors to foreign counterparts. These skills are discussed in more detail in TC 31-73.

FACTORS AFFECTING COMMUNICATION ACROSS CULTURES

8-56. There are four aspects that can impact communication across cultures. These factors are level of direct communication, perception of time, perception of the individual versus the group, and show of emotion.

Level of Direct Communication

8-57. Cultures fall along a spectrum of formality, with some cultures being direct and explicit in their communication and others less direct and more ambiguous. Cultures that rely on direct communication place more emphasis on verbal communications that are precise and explicit. Context and nonverbal cues are secondary to verbal communication. Cultures that rely on indirect communication rely less on verbal communication and more on context and nonverbal cues. Their verbal communication can seem vague and noncommittal to those from a culture that is more direct.

Perception of Time

8-58. Cultures can differ greatly in their perception of time. This difference divides into two cultural concepts of time: monochronic-time and polychronic-time. In monochronic-time cultures, members emphasize schedules, precise reckoning of time, and promptness. In such cultures, the schedule takes precedence over the interpersonal relation. In polychronic-time cultures, time is viewed as fluid. Members of polychronic-time societies do not observe strict schedules—agendas are subordinate to interpersonal relations.

Perception of the Individual Versus the Group

8-59. Cultures may be individualistic or collectivist in their orientation. An individualist culture is one in which the ties between individuals are loose—where people are expected to take care of themselves and their immediate families. In a collectivist culture, people are raised from birth into strong, cohesive groups. These groups offer a lifetime of protection in exchange for unquestioning loyalty. Advisors must understand from which culture their counterparts come to identify motivations accurately and communicate effectively.

Show of Emotion

8-60. Each culture has its own system of expressing emotion. Some cultures tend to express their emotions and show their feelings openly. Other cultures tend to repress those feelings more. Rather than showing their feelings openly, members of these cultures keep their emotions controlled and subdued. Those from the more expressive culture may view people from the repressive culture as cold or unfeeling. Similarly, those from the more repressive culture may view their more expressive colleagues as eccentric. To communicate effectively, advisors must be able to understand their foreign counterparts' emotional communication and display their own emotions carefully.

Chapter 9
Working with Counterparts

Rapport is a relationship marked by cooperation, conformity, harmony, or affinity. When people discuss good rapport, they describe a relationship founded on mutual trust, understanding, and respect. Relationships characterized by personal dislike, animosity, and other forms of friction often lack rapport. For the advisor, rapport describes the degree of effectiveness.

ESTABLISHMENT OF RAPPORT

9-1. Since advisors are in a unique military position, they establish rapport with their foreign counterparts. That position is one in which the advisor has no positional authority over the actions of their foreign counterparts. This lack of authority means that the doctrinal view of leadership is modified to emphasize interpersonal relationships and deemphasize authoritarian roles. Advisors instead use their people skills to influence the actions and decisions of their counterparts. Advisors use their people skills to building rapport. They have a genuine interest in other people, smile, remember and use people's names, encourage others to talk about themselves, listen to others, discuss what the other person is interested in, and make the other person feel important.

9-2. Advisors obtain certain knowledge before establishing effective rapport. First, they study FM 6-22 to gain basic leadership knowledge for understanding human nature and motivation. Advisors then incorporate information specific to the culture and society of their potential counterparts. This information may take the form of thorough area studies, operational area studies, and other research materials. Knowledge includes the components and techniques of rapport.

COMPONENTS OF RAPPORT

9-3. Three components compose rapport: understanding, respect, and trust. All the components of rapport are two-way streets; to have counterparts share about their culture, language, and experiences, advisors have to be willing and able to share also. Building this rapport may require the advisor to establish a personal level of understanding with their counterparts. Understanding is the first component of rapport. It begins before deployment and may include foreign cultural studies, language training, and equipment and doctrine familiarization. Once in country, advisors continue to broaden their understanding by observing and asking questions.

9-4. Respect is the next component of rapport. It is a reciprocal commodity. The foreign counterpart should grow to respect advisors, what they know, and how they perform. Advisors should look for those things that they respect in their counterparts. This may prove challenging. The counterpart may not fit the mold of U.S. or other coalition officers. Lacking formal training or education, the foreign counterpart may even be illiterate. The uniform standards of the host nation often may differ from the advisors' standards. Start with the basics and expand over time. Additional factors may affect the counterpart's willingness and motivation. For example, by accepting the duties, the counterparts (and their family) may be in mortal danger. The counterpart may have proven to be a fearless warrior, even without formal training. In short, traits deserving respect will exist even before the relationship matures. Mutual respect grows through shared experiences and shared dangers. Advisors should live, eat, and if authorized, fight with their counterparts. By sharing their hardships and dangers, advisors build respect.

9-5. Trust is the final critical component for building rapport. Trust grows gradually from understanding and respect. Building trust needs to begin on day one, but it will not mature until later in the relationship.

Advisors should begin by showing they are reliable and should do everything they say they will do. They should be in the right place at the right time. Two things can enhance this growth. First, the advisor should start out with confidence-building missions. Early success builds self-confidence and trust. Second, advisors should not promise any support they might not be able to deliver.

TECHNIQUES OF RAPPORT

9-6. Foreign security forces (FSF) respect advisors who actually show them the right way to do a task rather than just talk. Experience has shown that foreign units will not hesitate to go into danger if their advisors are right there with them. To further contribute to building rapport, advisors should maintain enthusiasm and a positive attitude; learn the language or how to use an interpreter; respect a counterpart's rank, age, status, and experience; develop negotiation skills; and improve interpersonal skills.

9-7. Other rapport-building techniques can include being confident, competent, and capable. Advisors never expect or demand their counterparts to do something that they are unable or unwilling to do. They demonstrate everything. Advisors learn their counterparts' names and spend time with them at meals and during holidays; they prepare mentally to interact with counterparts at all times. Advisors avoid creating an American-only enclave; they avoid giving the impression of favoring any one group. They frequently assess their counterparts' perceptions. Advisors recognize threats to discipline and enforce the chain of command.

ACCOMPLISHMENT OF RAPPORT

9-8. Advisors measure rapport by how well they can motivate their counterparts to act. The basic techniques of motivation (in the absence of authority) are advising, setting the example, seeking compromise, and coercing. TC 31-73 discusses rapport in detail.

ADVISING

9-9. Advising the counterpart to select a particular course of action is only effective if the counterpart trusts the advisor is professionally competent. If the counterpart does not perceive the proposed solution as realistic, the advisor's competence is questioned. Advisors carefully explain to their counterparts that recommended courses of action are realistic and will be effective.

SETTING THE EXAMPLE

9-10. Setting the example for the counterpart must be an ongoing effort for the advisor to avoid the appearance of a "do as I say, not as I do" attitude. In setting the example, advisors explain that what they are doing is the most effective action for the situation. This is particularly true when the counterpart really does not understand the behavior (or purpose).

SEEKING COMPROMISE

9-11. When seeking compromise with the foreign counterpart, advisors may create a situation in which the counterpart has a personal interest in successful execution. In some cultures, seeking a compromise may allow the counterpart to save face. Furthermore, in certain situations the counterpart—because of practical experience—may have a better solution to the problem at hand.

9-12. When seeking a compromise in certain cultures, the perceived competence of the advisors may suffer. This may be mitigated somewhat by approaching the compromise as two professionals (the advisor and the counterpart) reaching a mutual conclusion. To reach an effective compromise, advisors may have to conduct negotiations.

> ***Note***: Two areas of concern must never be compromised for the sake of maintaining rapport—operations security and human rights.

Coercing

9-13. Coercion is the least desirable method of motivation because it can cause irreparable damage to the relationship. Advisors should use coercion only in extreme circumstances. Advisors must avoid unintentionally forcing their counterparts into action. Instead, they establish and maintain good rapport by conveying they are sincerely interested in their counterparts, their nation, and its cause. Advisors will not belittle their counterparts' efforts. Advisors must demonstrate that they have come to help because they believe the counterparts' goals are just, fair, and deserving of success.

POLICIES AND AGREEMENTS

9-14. Advisors must remember their primary aim is to forward U.S. policy. Therefore, they establish relationships with foreign personnel, other U.S. forces, and other government agencies. Advisors promote U.S. policy through agreements, command relationships, and operations. TC 31-73 discusses policy in greater detail.

United States and Host-Nation Policies

9-15. Advisors not only support U.S. national policy in the areas where they operate, they forward the policies of the host-nation government (unless otherwise directed by higher authority). Advisors maintain communications with higher headquarters to ensure activities align with U.S. objectives.

9-16. Host-nation policies may dictate practices and procedures that appear inefficient or uneconomical. Advisors should avoid criticizing or condemning such practices and procedures until they thoroughly understand them. Advisors should make private recommendations critical or contrary to host-nation policy.

Agreements

9-17. The status of advisors is normally specified by a status-of-forces agreement or other arrangement between the U.S. and host-nation governments. These agreements vary widely; they may offer little political protection or may provide for full diplomatic immunity. If no such agreement exists, advisors are subject to local laws and customs. Even if a status-of-forces agreement or other agreement exists that provides immunity, advisors are still expected to observe local laws.

Command Relationships

9-18. Advisors must use the U.S. chain of command to obtain and disseminate guidance and assistance. They carefully distinguish between the U.S. and foreign security forces' chain of command. In particular, advisors prevent their counterparts from attempting to control subordinates through the U.S. chain of command. Advisors provide recommendations—not orders—to their counterparts. Only the counterpart should issue orders to subordinates.

9-19. Nongovernmental organizations and other government agencies may be co-located in the operational area. Advisors integrate the efforts of all organizations into their mission planning and should seek opportunities for integrated efforts with their counterparts. This type of unified action at the tactical level can be achieved through the planning process to integrate nonlethal actions. Holistically, nonlethal actions include development and diplomatic efforts.

Conduct of Operations

9-20. Advisors can expect to participate in stability operations. These operations use military and paramilitary forces, as part of the interdepartmental team, to support internal defense and development operations. U.S. assistance may include advice on military organization, training, operations, doctrine, and materiel. In addition, the United States may provide and control U.S. logistics and sustainment for FSF. The objective of this assistance is to increase the capability of foreign organizations to perform their missions and operate efficiently in the given operational environment.

LEVELS OF ADVISORS

9-21. Organizations and individuals possessing greater skills and resources assist by imparting their knowledge through assistance efforts. The success of assistance depends largely on effective interaction between advisors and their foreign counterparts. Advisors may operate in the national, operational, and tactical levels. The national level is considered the largest national subdivision. This may be a nation's combined military headquarters, Service components, national-level government officials, or a combination. Operational-level advisors work at a level below the national-level advisor. This may be with military corps or division officers, with regional government officials, or a combination. Depending upon the size of the nation and its military, operational-level advisors may be assigned as low as brigades. Tactical-level advisors normally advise military units at or below the brigade level and provide counsel and assistance across the spectrum of conflict. They advise and assist counterparts in developing unit combat effectiveness and serve as liaisons between FSF and U.S. combat, support, sustainment and protection forces. Chapter 3 of TC 31-73 discusses these levels in detail.

CONSIDERATIONS OF RAPPORT

9-22. When building and maintaining rapport, advisors consider intelligence, human rights, role shock, ambiguity, personal relationships, bureaucracy, responsiveness, and diversity. These considerations are discussed in greater detail in TC 31-73.

INTELLIGENCE

9-23. To promote national policy and attain overall objectives, advisors may need to plan and conduct intelligence training. The advisor should assist the counterpart in developing a local intelligence collection program, training intelligence personnel in their respective specialties, and utilizing trained intelligence personnel properly.

9-24. The advisor should be familiar with area studies and area assessments. Pertinent documents should be compared to detect trends or changes. The advisor also should evaluate the foreign intelligence staff, its standing operating procedures, its chain of command, intelligence projects, and reference material available from other intelligence agencies. Advisors should prepare and maintain a list of essential elements of information.

HUMAN RIGHTS

9-25. Advisors stress the consequences of mistreating suspects, prisoners, or other persons taken into custody. These persons must be treated per Article 3 of the Third Geneva Convention. Article 3 requires care for the sick and wounded, protection of prisoners and detainees of all types from abuse or other harm. Murder, mutilation, and torture are expressly forbidden, as is humiliating or degrading treatment. Advisors must never be active participants in the conduct of such punishment. Furthermore, they must explain to their counterparts that they are obliged to report any atrocities of which they have knowledge.

ROLE SHOCK

9-26. Role shock results from the discrepancy between the roles individuals expect to play and the roles they actually do play. Role shock may also be a product of the tension created by individuals trying to do jobs themselves versus advising others how to do the jobs. Role shock tends to increase until about the anticipated midpoint of the tour and seldom disappears completely until redeployment. Symptoms of role shock resemble those produced by other stressful life events.

9-27. Many advisors become entangled in complexities of bureaucracy—both foreign and American—and become increasingly frustrated when seemingly familiar things fail to respond in expected ways. This continuing frustration can be a key contributor to role shock.

9-28. Although most advisors undergo some degree of role shock, over time they learn to take the personal, living, and social conditions in stride. More challenging are the problems that arise in connection with their jobs—their professional roles, relationships, personal achievement, self-development, self-determination, and self-image. Often advisors find their duties and activities during deployments differ somewhat from those expected. The actual duties and responsibilities are greater in scope, involve technical work outside of their assigned specialties, and require honed administrative (rather than military) skills.

AMBIGUITY

9-29. Most advisory assignments are ambiguous. What an advisor is expected to achieve is not always made clear. Furthermore, assignments may conflict with what they have trained to do, what others expect of them, or what they want or expect to do.

9-30. U.S. and foreign security forces may spend months—even years—negotiating a project. Still, the overall objectives or definition of the mission to be accomplished might be expressed in general terms. Various interpretations of what is to be accomplished arise. Even more views surface regarding the most appropriate means of reaching those goals.

RELATIONSHIPS

9-31. Ideally, foreign counterparts provide a working context for advisors and help them to understand and adjust to the new culture. Historically, those with good working relationships with their counterparts make better adjustments than those who have no counterparts or bad relations with their counterpart. Good working relationships may be difficult to establish and maintain. Most advisors approach their assignments as teachers and mentors. Foreign counterparts may resist such a relationship, particularly if they were expecting someone to take over and do all or part of the job. In addition, many counterparts want their advisors to share in at least some of the blame if something goes wrong.

9-32. The United States has little control over the availability or capability of counterparts. The assigned individual may be technically incompetent, lack professional commitment, have other interests, or be unavailable when needed. In some cases, numerous counterparts are assigned and rotate in their workings with the advisor. A single counterpart may not be assigned until the advisor has been on the scene for some time. In some scenarios, a specific counterpart is never assigned.

BUREAUCRACY

9-33. Although some Americans might disagree, the United States maintains a system that is—by world standards—relatively permissive and nonbureaucratic. As such, advisors may become aggravated when confronted with or obstructed by host-nation bureaucracy, red tape, administrative centralization, and a cumbersome decisionmaking process. Cultures that emphasize consensus in the decisionmaking process often prove quite frustrating for advisors. To be successful, advisors strive to understand the bureaucracy in which they operate.

9-34. Advisors should recognize the value of being or knowing the middleman. This role is sometimes one of the key contributions an outsider can make. Middlemen can move in, around, and out of organizations somewhat independently of the local hierarchy, protocol, and customs. They can link persons and offices that might otherwise have difficulty communicating. Many advisors dislike this role and refuse to practice it; however, the middleman can be a significant catalyst in any advisory operation.

RESPONSIVENESS

9-35. As frustrating as foreign bureaucracies may be, advisors often get even more irritated with the responsiveness of U.S. forces. Advisors frequently fail to account for distances, lead times, diplomatic issues, and other challenges when conducting security force assistance. Moreover, few Soldiers in the unit conducting security force assistance have had substantive experience in working in an interagency environment. These issues are further complicated by the usual tensions between headquarters and field staffs and between administrators and technical specialists. Expectations, values, and styles of behavior vary from group to group.

9-36. Deploying personnel may not be familiar with the tasks of development and the vast range of problems typical of underdeveloped areas. They come in eager to contribute their knowledge and skill, and anxious to make maximum progress in the limited span of their assignment. In so doing, they may fail to recognize that the factors that frustrate them—such as the lack of professional skills and facilities—are actually their reason for being there.

DIVERSITY

9-37. Organizations attract individuals who differ in abilities, knowledge, skills, work ethics, social behaviors, values, and attitudes. Advisors must be able to recognize personality types and develop techniques to cooperate effectively with each type.

RECOMMENDED PRACTICES IN BUILDING RAPPORT

9-38. Advisors may adopt many practices to help work with counterparts. Practices that help advisors build rapport are discussed in paragraphs 9-39 through 9-51.

ENSURE THAT THE ADVISOR PRESENCE IS UNDERSTOOD

9-39. Advisors entering an operational area work under the sponsorship of a military commander, mayor, village head, or some other recognized local leader. Prior to this, the advisor's arrival may need to be coordinated through the district, sector, or other appropriate local office. Clearances from distant national, state, province, or sector governments cannot compensate for a clear, local explanation of why an American advisor is in the area. This particularly applies to small, isolated communities where it is unusual for a stranger to appear for even an hour without being acknowledged and accepted by local leaders. Without explanations from locally respected persons, civilians often arrive at their own explanations, often to the detriment of the advisor effort.

FIND A BASIS FOR COMMON INTEREST WITH THE LOCAL PEOPLE

9-40. If advisors show appreciation for civilians as individuals, cultural gaps and language barriers can be overcome and common ground can usually be found. Advisors should listen and show interest when the locals speak. Initial conversations usually center on universal matters, such as food, shelter, clothing, health, and education. Over time, discussions may naturally move to the matter the advisor wants them to consider. Advisors are better received if they know something about earlier contributions in such matters as agriculture, folk art, religion, and architecture. Naturally, advisors prove more effective and appreciated if they can speak the language.

TRY TO UNDERSTAND WHY THINGS ARE DONE THE WAY THEY ARE

9-41. Although some local practices may seem strange, they generally have good reasons behind them. Advisors can discover these reasons with careful observation and a creative imagination. Food habits, family traditions, folk cures, and festive celebrations almost always have a great deal of human experience at their root. The advisor also needs to be aware that many villages contain rival subgroups and factions. This tension needs to be accounted for when working with the local people.

START WITH WHERE THE PEOPLE ARE AND WHAT THEY WANT

9-42. Advisors find out what the local people really want and work with them to achieve this aim. The need felt by the local people often is the best starting point. Once advisors address this aim, people become more appreciative and cooperative—they begin to raise their expectations and become interested in working for other improvements. To address the initial desires of the people, the advisor may need to call in personnel with the specialized skills needed for the particular project. Although this can result in a delay, it will help the advisor to achieve better cooperation.

Work Within the Local, Cultural Framework

9-43. Advisors understand basic cultural matters such as ethnic backgrounds of the people, family relationships, leadership patterns, and value systems. They also need some knowledge of local services such as health, education, communications, and transportation.

Help People Believe They Can Improve Their Situation

9-44. Most traditional peoples of South America, Asia, and Africa live in a largely static environment. Through their experience with change, they are more fearful of losing status than they are hopeful of bettering their condition. Therefore, changes suggested by advisors are often viewed with fear. Concrete local projects that yield easily observed benefits help convince villagers that they can improve their situation and make them more willing to cooperate in other projects.

Be Content with Small Beginnings

9-45. Change tends to come slowly in areas where there have been few in recent times. Advisors remember that knowledge—technical or otherwise—is cumulative. Once a small beginning has been made, greater activity and additional changes soon follow. It is easier to achieve momentum than to maintain it. Regardless, advisors start within as promising a framework as possible and with the support needed to sustain the momentum achieved.

Utilize Local Organizations and Recognize Their Leaders

9-46. The surest way for an activity to be continued after the advisor leaves is for it to have been launched and carried forward within the local organizational and leadership framework. People respond best when their local organizations are recognized as important and useful. Advisors consult and encourage recognized local leaders to contribute as they can. A well-conceived technical activity reflects credit on the local leaders associated with it. Advisors give attention not only to officials and family heads of local groups, but also to the quiet, behind-the-scenes leaders.

Help the Government Get Organized To Serve The People

9-47. For advisors to be most effective, they must understand the organization of the local government and how their activities fit into the overall scheme. Usually a set of agreements exists between various local agencies and the national government (usually through some sort of interministerial council). This agreement provides for a coordinated effort in servicing the varied needs of civilians. The advisor may need to work with appropriate agencies to assist in getting such agreements made. If such agreements already exist, advisors should be careful to recognize and strengthen them.

Train and Use Subprofessional, Multipurpose Local Workers

9-48. Selected young people in the villages can be trained and used as subprofessional, multipurpose village workers. Otherwise, advisors' influence is restricted to where they are standing and the immediate vicinity. Furthermore, advisors may spend so much time establishing and maintaining enough rapport with the villagers that they may become incapable of rendering any real service at all. Often the gap between civilians and the advisor is formidable because of their educational and cultural differences. Often the advisor works with villagers who are poor, illiterate, and devoid of outside contacts. Volunteer or paid local workers have proven quite helpful as they serve as a liaison between villagers and advisors.

Expect Slow Progress

9-49. As civilians begin to see successes from their joint efforts and begin to have new hope, they naturally want a larger hand in their own matters. The advisor may sometimes feel they want to assume more responsibility than they can carry. These evidences of growing pains should be appreciated, for they are a necessary part of becoming able to assume responsibility. The advisor needs to adjust to the growing desires of the people to help themselves.

Transfer Controls Constructively

9-50. Advisors need to help civilians see how they can build the new—what they want—upon the foundations of the old—what they already have. From the beginning of a project, the advisor's team needs to talk with local leaders. Together they envision how to train local personnel, gather financial support, and transfer responsibilities before transferring control of activities. If operating responsibility is transferred too early, some breakage—usually of material things—will likely occur. If the team keeps control too long, the civilians who wanted to take over may grow disillusioned with them for not relinquishing control when they thought it should have been. Working with local leaders, the team determines when to phase out each technical activity.

Expect Little Gratitude from Those Helped

9-51. People who benefit from assistance sometimes feel defensive. In accepting assistance they are, in a sense, admitting their own insufficiency. The self-esteem of a person, community, or a nation as a whole is delicate. The advisor's team should not, therefore, expect thanks. Rather, the team should approach the people in a spirit of fraternity and humility, taking satisfaction in their forward progress. The team should do its job the best it can and accept work well done as is its own reward.

DYNAMICS OF ADVISING FOREIGN SECURITY FORCES

9-52. Advisors work in a complex environment that is heavily influenced by the unified effort of multiple organizations, relationships, compartmentalization, and organizational cultures. Paragraphs 9-53 through 9-57 discusses the probable dynamics that an advisor must take into account.

Advisors and Interdependent Operations

9-53. Advisors commonly operate in complex environments with interdependent operations. Interdependent operations require near-seamless integration of agencies representing the instruments of national power—diplomatic, informational, military, and economic. This same degree of integration must be extended to include external organizations and those agencies representing the instruments of power in partner states. The negotiation and cross-cultural skills that enhance their performance as advisors are also well suited to the complex interrelationships that characterize such operations.

Organizational Relationships

9-54. Often organizational relationships can often be misleading and must be clarified. The actual interrelationships among and within organizations seldom follow a line-and-block diagram. Instead, they are heavily influenced by circumstances, personalities, perceptions, and resources. All relationships and lines of authority are subject to negotiation. Advisors understand that clarifying the roles, functions, and responsibilities is a continuous process and that each statement or action sets a precedent for future interrelationships.

9-55. The advisor must become hypersensitive to the most minor organizational disparities. The apparent similarities—the common language, sociological backgrounds, and citizenship—can be disarming and lead the advisor to overlook very real differences. If ignored, these differences can quickly impede synchronization.

Compartmentalization

9-56. Advisors balance security and synchronization. Overcompartmentalizing must be avoided. Often distrust can be prevented simply by not being overly secretive. Although advisors must conduct a careful, continual risk assessment to avoid compromise, the sharing of critical information is necessary for harmonious operations.

Organizational Culture

9-57. Advisors should approach external organizations as they would foreign counterparts—as unique institutions with distinctive cultures. Organizational differences, biases, and approaches must be researched and—whenever possible—accommodated. Each Service in the Department of Defense has unique perspectives that lead to significant differences in objectives and approaches to various problems. Profound differences in perspective may even be found between the different branches of the Army. If not recognized and accommodated, these characteristics may create distrust, disharmony, and disruption during critical operations.

Chapter 10

Cross-Cultural Influencing and Negotiating

Leveraging foreign counterpart cooperation can prove a difficult and critical task. Although coercion may prompt surrogate forces to perform a specific action, it can never lead to established conduct and legitimacy. Lasting advisors can achieve long-term success only by patiently applying their influence.

Advisors analyze their counterparts and determine their motivations and needs. Basic needs, culture, and individual personalities influence people. Advisors determine what these three things are so the appropriate influence techniques can be applied. Then advisors negotiate. Negotiation is fundamental to problem solving using the study, development, and honing of negotiation skills critical to the advisor.

INFLUENCING

10-1. Advisors need to gain influence with their counterparts, leaders, civilians, coalition members, and others beyond their chain of command. Advisors are self-aware, nonjudgmental, and capable. It is not necessary to accept others' beliefs or values, but advisors need to understand others' ways of thinking and operating. Knowing their own culture and understanding why they believe and value what they do helps advisors find a common cause. A common cause unites a diverse set of decisionmakers with whom advisors interact and try to influence.

10-2. Once advisors gain a minimum amount of influence, they can make things happen through others. They can extend their influence beyond their chain of command. Influence is simply getting others to do what you want.

10-3. Advisors identify the types and principles of influencing. They determine the best techniques with which to influence and review vital considerations. Using measures of effectiveness, they check that their influences have the desired results.

TYPES

10-4. Commitment, compliance, and resistance clarify how advisors understand the impact of their attempts to influence their counterparts. These three types of influencing are determined by how committed the foreign security forces (FSF) are to the task.

Commitment

10-5. Commitment is a willing dedication to a requirement or a cause because one wants to. Commitment-focused influence generally produces longer lasting and broader effects. Whereas compliance only changes immediate behavior, commitment reaches deeper—changing attitudes, beliefs, and behavior. For example, when advisors build desire and responsibility for a task among FSF, they often demonstrate more initiative, personal involvement, and creativity.

Compliance

10-6. Compliance is conforming to a requirement or demand because they feel like they have to avoid punishment or to gain a desired reward. Compliance-focused influence is based primarily on the leader's authority. Giving a direct order to a follower is one approach to obtain compliance during a task.

Compliance is appropriate for short-term, immediate requirements and for situations where little risk can be tolerated. Compliance techniques are also appropriate for leaders to use with others who are relatively unfamiliar with their tasks or unwilling or unable to commit fully to the request. If something needs to be done with little time for delay, and there is not a great need for the target of influence to understand why the request is made, then compliance is an acceptable approach. Compliance-focused influence is not particularly effective when a leader's greatest aim is to create initiative and high esteem with a counterpart.

Resistance

10-7. Resistance is a reaction against change. Resistance generally takes one of four forms—reactance, distrust, scrutiny, and inertia. It is manifested behaviorally (I won't do it), affectively (I don't like it), and cognitively (I don't believe it). Recognizing and lowering potential sources of resistance are key components to gaining compliance and commitment.

MOTIVATIONS

10-8. Different motivations influence behavior. One type of motivation is the desire for something. Another is the fear of something undesirable. As an agent of influence, advisors determine how to increase approach motivations and how to decrease avoidance motivations. For instance, recruiting for a special mission can use special badges as incentive (approach). Advisors can assure Soldiers that the mission is relatively safe by providing extra training and better equipment (lowering risks or avoidance).

PRINCIPLES

10-9. Six influencing principles are relevant to virtually any culture and any group. Understanding and applying these principles can increase an advisor's effectiveness in persuading a group or individual:

- Principle of scarcity.
- Principle of authority.
- Principle of social proof.
- Principle of liking.
- Principle of reciprocity.
- Principle of consistency.

10-10. With the principle of scarcity, people value more of what they can have less. They typically associate greater value with things that are rare, dwindling in availability, or difficult to acquire.

10-11. In the principle of authority, people are more easily persuaded by individuals perceived to be legitimate authorities or experts. They defer to experts who provide shortcuts to decisions requiring specialized information.

10-12. With the principle of social proof, people often look to the behavior of those around them for direction about what choices to make. This action is heightened when those around them are similar in terms of age, education, social standing, and experience.

10-13. In the principle of liking, people prefer to say yes and to comply with the requests of those they like.

10-14. In the principle of reciprocity, if someone grants favors, invited or uninvited, an overpowering need to repay that favor immediately blooms within the recipient of that favor. This human trait transcends all cultures and races.

10-15. With the principle of consistency, the desire for consistency motivates behavior. The drive to look and be consistent is a highly potent weapon of social influence, often causing people to act in ways clearly contrary to their own best interests. People do not like to appear inconsistent to others. It is human nature that people strive to feel good about themselves, which includes behaving per their important values and beliefs. When the behavior is consistent with whom people are and what they value, they feel good. People align with their clear commitments.

TECHNIQUES

10-16. The art of influencing involves knowing what techniques to use based on the situation. Advisors have to know the limits to which can use their influence; their influence is finite. They need to know when to use their influencing skills to achieve the highest payoffs. Their counterparts will only accept so much influencing if they don't completely trust their advisors. Table 10-1 lists positive and negative techniques delineated in FM 6-22. Table 10-2 lists resistance-based influence techniques. These are used when the counterpart resists the advisor's influences.

Table 10-1. Influencing and rapport techniques

Technique	Description	Characteristics	Outcomes
Pressure	Explicit demands, persistent reminders, sometimes appropriate in high stakes, urgent situations	Frequent checking Nagging	Gains compliance Frequent use may cause resentment, undermine morale, and harm relationship
Legitimate Request	Source of authority is used	Source credibility is important	Gains compliance
Exchange	Quid pro quo Reward offered for task completion; punishment threatened for incomplete task	Offer of reward or punishment not tied directly to getting job done, just something that counterpart wants or doesn't want	Gains compliance
Personal Appeal	Request based on friendship or loyalty	Agent of influence highlights talents of counterpart that make target the right person for the job (this is why you are the right person for the job, will you do it for me?)	Gains commitment Builds relationship
Collaboration	Works with target of influence by providing resources of some type, or by removing barriers to task completion	Offer of assistance is directly tied to request	Gains commitment Builds relationship
Rational Persuasion	Using facts, evidence, or logic in a persuasive message	Relies on agent being viewed as expert	Gains commitment Builds relationship
Apprising	Explaining the consequences of a potential action or decision	Decision is good for unit, career, or mission accomplishment	Gains commitment Builds relationship
Inspiration	Arousing strong emotion to build conviction or commitment	Appealing to values of the counterpart	Gains commitment Builds relationship
Participation	Requests for counterpart to be a part of planning, active participation in development of solution	Promotes commitment Possibly more feasible and culturally appropriate course of action	Gains commitment Builds relationship
Relationship Building	Build positive rapport and trust	Provide the foundation for using influence or inspiring	Gains commitment

Table 10-2. Resistance-based techniques

Technique	Description	Characteristics	Outcomes
Sidestep Resistance	Agent defines own role as consultant or advisor, but not commander Pushes the consequence of choice into future Depersonalize task to gain commitment to general idea Focus on end state instead of expended effort	Does not sell, just advises to prevent immediate resistance Puts less focus on immediate impact, more focus on long-term effect Stating "leaders should" instead of "you should" Stress mission accomplishment	Gains compliance Potentially gains commitment if paired with other influence techniques like rational persuasion
Address Resistance Directly	Address reluctance by lowering costs associated with tasks	Provide some of own resources through collaboration or counterargue concerns	Builds commitment
Address Resistance Indirectly	Build confidence, esteem, and self-efficacy for the task	Acknowledge difficulties while showing support for counterpart's abilities	Builds commitment
Distract Resistance	Interfere with the development of counterarguments	Distracting attention	Gains compliance
Disrupt Resistance	Disrupt complacency to bring attention to message	Shake up to motivate	Gains compliance
Consume Resistance	Resistance is energy	Wear them down prior to opportunities to resist. Persist with requests until they concede	Gains compliance
Use Resistance to Promote Change	Reverse psychology	Issuing a challenge	Gains compliance; risky when you don't know target of influence really well

10-17. To succeed with any influence technique, the influence attempt should be perceived as authentic and sincere. Positive influence comes from leaders who do what is right for Army forces, the mission, the team, and each individual Soldier. Negative influence—real and perceived—emanates from leaders who focus more on personal gain and lack self-awareness. False perception may trigger unintended side effects such as resentment of the leader and the deterioration of unit cohesion. With each and every influence attempt, Army leaders should ensure that the influence objective aligns with Army Values, ethics, the Uniform Code of Military Justice, the Warrior Ethos, and the Civilian Creed.

Considerations when Choosing Techniques for Influencing

10-18. Advisors clearly understand what their counterparts value and fear. This information allows advisors to quickly relate the objective of the influence to something that their counterparts treasure. Also, a strong sense of self supports the advisor's recognition of which techniques they can successfully execute. Someone with an intimidating presence may achieve compliance with pressure more quickly than someone without that presence. Understanding the context of the request is important. For instance, using participation in the midst of flying bullets would not make sense. However, a long-term project that impacts many people might be a worthy situation for participation. Advisors also consider—

- The motivation of their counterparts.
- The relationship they have with their counterparts.
- What they do or not have to offer.
- An understanding of the second- and third-order effects their recommendations will have.
- What has worked in the past.
- Possession of an appropriate power based for use of a technique in particular context.
- Consistency of technique with prevailing social norms and role expectations.
- Level of resistance encountered or anticipated.
- Costs of using a technique in relation to likely benefits.

10-19. The cost of a technique refers to the amount of effort, favor, or resources that accompany the technique. A personal appeal that relies on friendship is quite effective; however, counterparts might expect the favor to be returned. If resources or time are plentiful, then the exchange might not seem expensive, but if they are not plentiful, then advisors must choose a different technique. Advisors balance the costs and benefits of each technique.

Moderating Factors

10-20. Sometimes moderating factors diminish the impact of any influence technique. These factors may include cultural misunderstanding, level of trust, or constraints on the counterpart like threats or corruption. Advisors analyze and learn about these factors. Advisors must be flexible when using influencing techniques. When one technique doesn't work, switching strategies or adding techniques may be appropriate.

CONSIDERATIONS

10-21. Seven considerations need to be addressed to influence behavior. These include determining the goal; who to influence; motives; beliefs, values, and attitudes; cultural beliefs, values, and attitudes; and susceptibility. These techniques are designed to help advisors achieve their goals. Just knowing that the counterpart wants something is not enough. Whether it is time off, new weapons, or money, advisors have a limited supply of stuff they can promise; they can only bend so much on rules, regulations, and laws. It is at this point that the advisor enters the realm of negotiations.

Determine the Goal

10-22. Advisors influence others to achieve some purpose. To be successful at exerting influence, advisors have an end state or goal in mind. Sometimes the goal will be very specific. Many goals are less distinct and measurable but are still valid and meaningful.

Determine Who to Influence

10-23. Advisors develop a mental list of those whom they might need to influence. Getting a FSF platoon to take action may be as simple as influencing the commander, but the advisor may have to influence the soldiers and commander of the platoon. When practical, advisors consider many variations of whom to influence. They look for circumstances in the immediate environment that significantly affect their counterparts. Individual or group characteristics to consider can include—

- Gender.
- Religion.
- Age.
- Socioeconomic status.
- Ethnicity.
- Political affiliation.

- Level of education (important in determining how to access and persuade).
- Occupation.

10-24. The challenge is to determine which characteristics impact emotional or behavioral responses and under what circumstances. Ethnicity may significantly affect behavior and emotions regarding voting and politics but may have almost no effect on the decision to enlist in the military. Individual and group characteristics may provide additional insight on motivations.

Determine and Prioritze Motives

10-25. Motives are anything that is required or desired (needs and wants). Motives are the reasons behind the decisions people make. Motives come from an inner desire to meet a need or want. Advisors use the knowledge of what motivates others to influence them. Knowing one's counterpart and others who may be influenced gives advisors insight into influencing their behavior. The desires to fulfill, alleviate, or eliminate a need or want provides the motivation to change behavior.

Basic Motives

10-26. Basic motives involve physiological needs—food, water, and air—and safety needs—security and shelter. When people feel hunger, they are motivated to eat; when people feel pain, they are motivated to reduce the source of the pain. Such basic motives are extremely powerful in driving behavior and overwhelm psychological needs and wants.

Social Motives

10-27. Social motives associate access to other tangibles (money, goods, education, infrastructure, and healthcare) or more complex psychological motives with basic motives. For example, people learn to want money because it can be exchanged for food and other desired goods. Power, achievement, reassurance, escape, justice, and acceptance illustrate types of learned social motives. Examples include:

- Wanting better educational opportunities for their children.
- Wanting better paying jobs.
- Wanting their interests represented by the government.
- Wanting revenge for perceived wrongs.
- Wanting self rule.

Prioritize Motives

10-28. Advisors prioritize motives by immediacy of the need or want. Critical motives are immediate needs. Short-term motives are currently being satisfied or active efforts are being made to satisfy them in the near future. Long-term motives are desired but are not immediately important; satisfaction may be delayed until some point in the future.

Determine Beliefs, Values, and Attitudes

10-29. Although often difficult to derive, beliefs, values, and attitudes can prove quite effective in persuasion. Beliefs are what is thought to be true. People do not believe in something if they think it is not true. People vary from being gullible to skeptical. Values prioritize what one believes. Values can be ethical, doctrinal, social, aesthetic, or economic in nature. Attitudes reflect whether one likes or dislikes something. Likes and dislikes can often be traced back to a person's beliefs and values. This includes loves, hates, frustrations, and fears. Frustrations occur when a want is not met. By examining causes of frustration, the advisor may find another motive to address.

Compare Beliefs, Values, and Attitudes

10-30. The predominant cultural beliefs, values, and attitudes; traditions; and norms may differ greatly from the group the advisor is trying to influence. Many Americans act differently in different cultures. This change in behavior relates directly to their predominant culture.

Determine Susceptibility

10-31. Susceptibility is the likelihood that the individual or group will be open to persuasion. Determining susceptibility aids in prioritizing efforts. If susceptibility is moderate to high, then advisors may not need to focus as much effort as they would if the susceptibility is low. Advisors determine susceptibility by assessing the strength of the underlying source of the motivation and the perception of the desired corresponding behavior. The strength of the motivation depends on the perceived risk and the perceived reward, while the perception of the behavior directly links directly to how consistent the behavior is with the individual or groups' values and beliefs.

Perceived Risks and Rewards

10-32. Risks may include the threat of physical harm, financial loss, or disapproval of peers. Rewards may include financial gain, political power, approval of peers, or increased self-esteem. Risks and rewards can vary from immediate- to long-term. Immediate risks and rewards are stronger than long-term risks and rewards. For example, few people will walk a tightrope (immediate risk), but many people smoke (long-term risk). The higher the perceived rewards and the lower the perceived risks, the more likely the behavior will occur.

Consistency with Values and Beliefs

10-33. Advisors determine if the desired behavior is consistent with values and beliefs. Generally, behavior is consistent with the relative moral or ethical, religious, political, and cultural beliefs, values, and norms. Behavior that is inconsistent or incongruent with values and beliefs often decreases over time. Alternately, if behavior continues to be incongruent with values and beliefs, then individuals are likely to justify this behavior by modifying their values and beliefs. The more consistent a desired behavior change is with the individual's values and beliefs, the more likely the behavior will occur.

10-34. The lower the susceptibility, the less likely influencing behavior will be successful, which does not mean that the person cannot still be influenced. The lower the susceptibility, the more conditions will need to be modified for behavior change to occur.

APPLICATION

10-35. Influencing is a dynamic process. Advisors make a plan before attempting to influence. If one technique fails, they have to be able to shift to other techniques. Advisors do not need to sit down and write a plan before attempting to influence people. However, even with apparently positive results, advisors review their tactics and techniques.

CHECKS

10-36. Advisors check measures of effectiveness, unintended consequences, and second- and third-order effects. Advisors assess if their influencing is achieving their immediate measures of effectiveness. However, with every action some unintended consequences may occur or the activity may not prepare counterparts for more complicated tasks later on.

CROSS-CULTURAL NEGOTIATING

10-37. The principal form of negotiation is likely to be cross-cultural negotiation. Cross-cultural negotiation brings with it a series of unique challenges and techniques. Culture fundamentally affects language and behavior. It also significantly impacts the way people handle conflict.

10-38. Negotiations among members of the same culture can be stressful; negotiating with members of other cultures can be exceedingly difficult. Logically, working with other cultures is a basic skill for the Soldier and an absolute requirement when acting in an advisory role. Understanding the components, perspectives, roles, and outcomes will help minimize the difficulty of cross-cultural negotiation. Chapter 4 of TC 31-73 discusses cross-cultural negotiations.

INFLUENCES ON NEGOTIATIONS

10-39. Cultural differences may lead to a conflict between what the parties expect and what their families and communities expect. This has an obvious impact on negotiating behavior. Negotiations are influenced by perceptions, individualism, punctuality and pace, relationship building, and language.

Perceptions

10-40. Most Americans have certain preconceived ideas about people from other cultures. These perceptions may not be factually based, but they exist and influence negotiations. Similarly, foreign negotiators have certain perceptions about Americans. Again, these perceptions may not be based on fact, but they exist. Advisors need to understand how other cultures perceive Americans to appropriately adjust their negotiating style. Successful advisors find ways to capitalize on the positive preconceptions of their counterparts and find ways to neutralize the negative preconceptions.

Individualism

10-41. The United States encourages a largely individualistic culture. As such, the typical American negotiator prefers to enter into negotiations alone or, if required, as part of a very small negotiating team (two or three people, at most). This behavior is largely a by-product of U.S. culture, which focuses on the individual, performance, initiative, and accomplishments. Most other cultures of the world place less emphasis on the individual and more on the group.

Punctuality and Pace

10-42. Americans place a great emphasis on punctuality, viewing it as an indication of an individual's basic ability and commitment. However, other cultures emphasize punctuality differently. Some emphasize punctuality more than Americans. In other cultures, time is considered relative or may not be important at all.

Relationship Building

10-43. Americans are very competitive during negotiations and are inclined to stress short-term results. Interpersonal relationships are of minimal importance to American negotiators. Building long-term relationships only occurs after the successful completion of negotiations.

Language

10-44. U.S. negotiators must be considerate of others' with respect to language. Even if the counterpart is fluent in English, they may expect negotiators to speak their language or use an interpreter. Members of some cultures do not like to speak English as they are very proud of their language. When U.S. negotiator counterparts do not speak English, they will need an interpreter. Counterparts may also expect American negotiators to present detailed written proposals in the local language. Many cultures appreciate foreigners who take the trouble to learn a few simple courtesy words. However, some prefer that foreigners who are not fluent refrain from using more than a few, basic courtesy words.

ELEMENTS FOR CONDUCTING CROSS-CULTURAL NEGOTIATIONS

10-45. The basic elements common to all negotiations must be applied in ways that allow for the differing cultures of the participants. Elements that must be considered include opening strategies, directness, strengthening behaviors, movement, power, face-saving, formal agreements, and mediation.

COMPONENTS OF NEGOTIATING

10-46. Two opposite types of negotiating exist: distributive and integrative. They reflect a philosophy towards negotiating in general. Distributive negotiating is win-lose while integrative is win-win. Distributive negotiating involves dividing a fixed amount of a thing to the parties involved. This negotiation normally involves people who have never had a relationship before. Integrative negotiating involves joining two parties to achieve something together. Successful negotiations can involve a mix of both strategies.

10-47. When using distributive negotiating, advisors know the point at which counterparts will walk away. Advisors push them almost to that point by manipulation and withholding information. The distributive negotiator believes—

- The disputants are adversaries and the goal is victory.
- Fixed resources are divided so one side receives more than the other after making concessions.
- One person's interests or position opposes the other's interests or position.
- The dominant concern is usually maximizing one's own interests.

10-48. Integrative strategies used by advisors include cooperation, sharing information, and mutual problem solving. The integrative negotiator believes—

- The disputants are joint problem solvers seeking a wise decision.
- An understanding of needs will allow resources to be divided and both sides can win.
- The relationship between the parties is what is critical.
- The dominant concern is to maximize joint outcomes.

Perspectives Affecting Negotiations

10-49. Four perspectives impact negotiations: time, environment, self, and nonverbal communication. Advisors consider how each party views these conditions when negotiating.

Time

10-50. Some cultures and individuals see time as a micro view. It is all about here and now. Other cultures and individuals see time in a macro view. Now is a point that bridges the past with the future. For example, imagine a person with the first scheduled event has a deadline to make an agreement now. The next scheduled event is 1500 so that person has two hours to come to an agreement. The second person might agree that they can come to an agreement now; but to the second person, now could be within the next month or two.

Environment

10-51. Some cultures believe they can impact, if not control, their environment. If something should go wrong, somebody is to blame. Every accident is preventable. Other cultures believe the environment impacts the person. Accidents just happen and no one could have done anything about it. One side may be quite ready to commit to something in the future *knowing* that nothing stands in the way, while the other side may be reluctant to talk in certainties of the future *knowing* that factors beyond their control may prevent them from fulfilling their commitments.

Self

10-52. In some cultures, when a person makes a mistake, blame can be isolated to that person. In other cultures, the individual is seen as an element of the group. So a mistake is made not by the individual but rather the whole group. In negotiations, one might accept a personal slight to get the goal achieved, whereas the counterpart represents the group and cannot allow shame to fall on them even though one might accept it personally.

Nonverbal Communication

10-53. Since advisors will often have to work through interpreters, nonverbal communication becomes more important. Some cultures respect people who mean just what they say and maintain their composure. Other cultures respect the eloquence in which something is said or the emotion used to convey it. Where words may be translated improperly, the advisor or counterpart can hear tone and see emotion.

Roles of Negotiation

10-54. When entering a negotiation, advisors must know what their role. The three major roles associated with negotiation are negotiator, mediator, and arbitrator. Each role has its own tactics or techniques and separate functions.

Negotiator

10-55. Advisors most often find themselves in the role of a negotiator. A negotiation is the process by which parties exchange commitments or promises to resolve disagreements and reach a settlement. Advisors prepare for each negotiation so they will not be at a disadvantage during the negotiation. To prepare for negotiations advisors examine their options, discover what motivates the parties, list all positive scenarios, and dispute resolutions.

10-56. Advisors examine their options. They question their goals, the importance of the goals, and the amount of leverage they have. They determine who is in power and what to do if no agreement is reached. Advisors uncover what motivates their counterparts. They delve into their counterparts' underlying needs.

10-57. By listing all the scenarios that satisfy their needs, advisors can maximize the gain for all parties. Advisors can add features to the scenarios that increase the opportunities for all parties to be satisfied. When advisors dispute resolutions, they often find some form of compromise. One of the best outcomes of compromise is when all parties leave the bargaining table satisfied but no single party feels like they have won or lost the argument. A skilled negotiator should strive to achieve a win-win result.

10-58. The negotiator represents one side and has a vested interest in the outcome of the negotiation. It is best to avoid situations where there is a conflict of interests. By blindly pushing the coalition agenda, the advisor could lose the trust of the counterpart. Considerations for a negotiator should include:

- If the foreign security forces ask the advisor to negotiate for them, they expect the advisor, as their negotiator, to have their best interests in mind. The coalition may try to use the advisor to push their agenda and influence rather than negotiate.
- If the foreign counterparts take illegal action, advisors cannot use commitment techniques. They need to use some form of compliant technique.
- If the foreign counterpart is not violating any laws, the advisor should focus on commitment techniques. Success will not be achieved all at once.

Mediator

10-59. Advisors may find themselves in the role of a mediator. The mediator has three general roles to perform: facilitator, formulator, and manipulator. Table 10-3 identifies the roles and tactics advisors use.

Table 10-3. Mediator roles and tactics

Roles	Tactics
Facilitator	• Make contact. • Arrange for interaction. • Clarify the situation. • Supply missing information. • Transmit messages between parties. • Ensure all interests are discussed.
Formulator	• Control the pace, formality, and physical environment of the meetings. • Highlight common interests. • Help parties save face. • Keep process focused on the issues. • Make substitutive proposals. • Suggest concessions parties could make.
Manipulator	• Keep parties at the table. • Change parties expectations. • Highlight costs of nonagreement. • Reward concessions. • Promise resources. • Threaten the withdrawal of resources. • Offer to verify compliance. • Threaten to withdraw mediation.

Arbitrator

10-60. The foreign security force may try to put the advisor in the position of an arbitrator. An arbitration is the process by which the parties to a dispute submit their differences to the judgment of an impartial person or group selected by mutual consent or statutory provision. They want the advisor to decide between two positions. This is a dangerous position for the advisor. When a win-win solution can be found, the advisor can improve rapport with both parties, but when it is a win-lose situation, the advisor could alienate one or both groups.

Expected Outcomes of Negotiation

10-61. The outcome of negotiations sometimes can determine the outcome. Advisors consider the desired outcome to determine which confrontation preference to use. Negotiation can focuses on physical outcomes of relationships. How advisors combine their preferences to their focus affects their perceptions of the outcome.

Confrontation Preferences

10-62. Negotiators will have one of three preferences for confrontation: avoidance, collaboration, or competition. Those that prefer avoidance will consider accommodation as a viable means to resolve an issue. Negotiators that prefer collaboration consider compromise as the preferred outcome. Those negotiators who use competition want to win. Intentionally losing in a negotiation to avoid confrontation seems counterintuitive; however, it tends to protect the current relationship. Collaboration requires a mutual respect of the existing relationship, while competition is generally indifferent. It would be fairly easy to take advantage of a negotiator who prefers to avoid confrontation. At the same time, expecting to be able to compromise with a highly competitive negotiator may become a point of frustration. For the competitor, winning may be more important than the deal itself.

Negotiation Focus

10-63. Defining whether the negotiation is a win-win situation is not as simple as which side gained the most from the negotiation. Advisors often focus on physical outcomes whereas diplomats focus on relationships. Every negotiation varies on which factor is more important. Just as car dealers may accept less profit on individual services, they gain overall if the customer returns.

Perception of the Outcome

10-64. Comparing confrontation preferences leads to nine potential combinations. If advisors can determine a participant's preference—avoidance, collaboration, and competition—and focus, they can determine how the participant might feel based on the outcome of the negotiation. If one participant is a collaborator and one is a competitor, then a win-win situation is possible, but neither side will be completely satisfied with the outcome. The competitors will not have gotten a clear win, and the collaborators will not have gotten what they considered fair. Subsequently, they will both have some minor negative feelings toward the other.

Appendix A

Army Special Operations Forces Imperatives

This appendix outlines the imperatives for Army special operations forces conducting security force assistance and foreign internal defense operations. Table A-1 lists these imperatives.

Table A-1. Army special operations forces imperatives

Imperatives	General Descriptors	Specific Actions
Understand the Operational Environment	• Knowing the U.S. command relationships (military and interagency) • Knowing the population (who, what, when, where, why, and how) • Knowing the enemy and other antigovernment forces (crime, militias, gangs) • Knowing the higher commander's intent • Knowing that the press is a part of the operational environment (news reports of events reach the U.S. population quickly)	• Define relationship as partnership or augmentation • Conduct analysis using ASCOPE (areas, structures, capabilities, organizations, people, events) • Analyze the enemy order of battle • Protect the population
Recognize Political Implications	• Identifying the political (and legal) restrictions and implications of actions by U.S. and foreign security forces • Identifying political implications of establishing and enforcing rules of engagement policy • Identifying short and long term political implications of security force assistance • Understanding the importance of establishing human rights (critical for tactical and strategic success) • Acknowledging the effects of public opinion on the mission locally, in the United States, and internationally • Understanding that insurgents conduct operations to have political effects among the local populace, external actors, and the U.S. and multinational civilian public	• Include rules of engagement in the planning and briefs of operations • Include local government leadership in operations • Reinforce human rights briefings and training to all foreign security forces • Prepare all U.S. and foreign security forces to talk with the press • Prepare information engagements for all operations • Anticipate and neutralize the methods enemy forces pass propaganda

Imperatives	General Descriptors	Specific Actions
Facilitate Interagency Activities	• Conducting security force assistance as both a military and an interagency effort; taking advantage of the greater impact of collective efforts • Coordinating the scope and limitations of each agency's influence and programs • (as opposed to a purely military program) • Obtaining the active cooperation of the foreign security forces • Including military and civilian agencies working together for unity of effort	• Include interagency members in planning and training as appropriate • Provide assistance to support limited agency resources, allowing agencies to prioritize their primary roles • Conduct regular coordination meetings with civil agencies to synchronize security efforts • Include host-nation agencies whenever including U.S. agencies
Engage the Threat Discriminately	• Avoiding the use of overwhelming firepower—to prevent alienating the population and turning them away from the host-nation government • Giving training and advice to foreign security forces based on the mission and threat • Targeting insurgents, criminal activities (lawlessness), or subversion—to avoid alienating the populace • Minimizing the use of force in populace and resources control measures—weighed for potential gain versus potential cost • Ensuring that rules of engagement and escalation of force are balanced between force protection and safeguarding civilians	• Take civilians into account when considering weapons systems and techniques for operations • Include civilians in risk assessment during mission planning • Train and rehearse rules of engagement and escalation of force to prepare Soldiers to react in accordance to policy • Publicize reasons for use of populace and resource control (for example, curfews protect civilians); avoid enemy propaganda or rumors dictating the psychological response of the populace • Update rules of engagement and escalation of force periodically • Rules of engagement or escalation of force reflects engaging targets that are visually verified and not simply suppressive fire in populated areas
Consider Long-Term Effects	• Building up the foreign security forces and alleviating the root causes of the current situation over time—not just treating the symptoms • Solving problems in a broader political, military, social and economic context • Making policies, plans, and operations consistent with national and theater priorities and the objectives they support	• Understand local grievances and underlying issues • Interact with all aspects of the population a government to identify root causes • Ensure the commander's intent remains focused on higher goals and policy related to addressing root causes, and protecting the local population • Monitor and update rules of engagement as local government and police gain control of rule of law. Conduct operation that adhere to local laws as possible • Understand legal and political constraints (rules of engagement) to avoid strategic failure while achieving tactical success

Imperatives	General Descriptors	Specific Actions
Ensure Legitimacy of Special Operations	• Reinforcing and enhancing the legitimacy and credibility of the foreign security forces and government • Gaining trust and credibility to ensure advisors and U.S. forces obtain legitimacy with the local population • Obtaining legitimacy in theater, in the United States, and internationally to ensure ongoing support	• Conduct joint and multinational patrols and operations. • Ensure information engagement is common in all operations and contributes to maintaining credibility • Engage the media with the host nation to inform as well as to justify actions • Focus on what the population considers the root causes, not the U.S. perspective • Develop long-term solutions to restore or improve basic services with host-nation leadership
Anticipate and Control Psychological Effects	• Considering the psychological effects of all operations (such as populace and resources control measures, combat operations, and civic action) • Avoiding negative psychological effects that can negate tactical victories • Appreciating the power of people's perceptions versus reality	• All operations are developed, analyzed and executed with information engagement • Continue to gauge the psychological effects with the local population, local government, and security forces as well as the enemy (propaganda and interrogation)
Apply Capabilities Indirectly	• Teaching host-nation forces to plan instead of giving them tasks • Supporting U.S. efforts enhancing the legitimacy and credibility of the host-nation government and foreign security forces • Balancing control and pressure on foreign security forces to ensure they conduct successful tactical operations and accomplish missions • Fulfilling primary role to advise, train, and help foreign security forces take the lead and stabilize the host nation • Influencing a foreign security force without commanding it; not accepting a lack of effort or a high level of corruption	• Conduct training of foreign security forces only to teach basics • Transition to mentor or advisor relationship during planning and operations to encourage the host-nation to take the lead • Use discretion when advising counterparts as not to embarrass foreign security forces leadership • Use host-nation methods whenever possible

Imperatives	General Descriptors	Specific Actions
Develop Multiple Options	• Considering the complex and irregular environment • Planning to use the full range of capabilities • Anticipating that an operational environment may dictate a change of rules of engagement or type of operations • Ensuring foreign security forces are capable of shifting from one option to another before and during mission execution	• Continue to refine and train on battle drills and contingencies • Develop an understanding of skills and abilities of the members of the unit and community • Elicit the input of foreign security forces on how to address situations to widen the options beyond conventional techniques • Include government agencies, special operations forces, nongovernmental organizations, local police, and the host-nation government to develop additional methods or options for solving problems
Ensure Long-Term Sustainment	• Avoiding advising and training foreign security forces in techniques and procedures beyond their capabilities to sustain • Modifying tactics, techniques, and procedures; training; operations; and sustainment to fit the culture, education, and technological capability of the foreign security forces • Developing host-nation sustainment before handing over economic, social, political, military, and security initiatives and projects • Avoiding host-nation dependency on U.S.-funded programs	• Understand and utilize the equipment and manpower present in the foreign security forces unit's inventory • Assess the capabilities of foreign security forces for leadership, training, sustainment (logistics, administration, and equipment), and ongoing development of a professional security force • Recognize the programs that are durable, consistent, and sustainable by the host nation • Do not begin programs beyond the economic or technological capacity of the host nation
Provide Sufficient Intelligence	• Ensuring detailed, near real-time, all-source tactical intelligence products support all activities and programs • Ensuring detailed and comprehensive intelligence on all aspects of the operational environment and its internal dynamics support operations • Ensuring human intelligence, urban recon, and internal security elements assess internal threats, warn the government, and take action to penetrate and neutralize instability • Considering aspects of the society not directly related to the tactical combat situation when assessing the enemy threat • Using information from intelligence assets to advise, train, and help foreign security forces counterparts and ease interagency efforts	• Ensure host nation forces are conducting operations using accurate, real-time intelligence • Establish priority of effort when identifying intelligence requirements • Assist foreign security forces to prioritize and focus counterintelligence on the commander's critical information requirement to receive timely and specific intelligence • Assess other organizations that impede the rule of law and government functions (such as gangs, criminals, smugglers, militias, mercenaries, drug lords, and narcoterrorists)

Imperatives	*General Descriptors*	*Specific Actions*
Balance Security and Synchronization	• Maintaining unity of effort and avoiding compartmentalizing key security forces from planning and intelligence contingencies even when security concerns exist • Avoiding either excessive or insufficient security • Anticipating reasons foreign security forces may not provide intelligence, such as lack of internal security or lack of U.S. intelligence support	• Encourage combined planning, preparation, and operations with foreign security forces' military, police, and governmental units • Resolve conflicting demands on mission planning and execution • Employ measures to mitigate risk of operations security instead of dismissing the host-nation government, police, or military needed for the overall effort • Strive to gain the trust of the local population to gain the most human intelligence and support from the foreign security forces

Appendix B
Legal Considerations

GENERAL LEGAL BASIS

B-1. Law and policy govern the actions of U.S. forces in all military operations, including security force assistance. For U.S. forces to conduct operations, a legal basis must exist. This legal basis profoundly influences many aspects of an operation. It affects the rules of engagement, how U.S. forces organize and train foreign forces, the authority to spend funds to benefit the host nation, and the authority of U.S. forces to detain and interrogate. Under the Constitution of the United States, the President is the Commander in Chief of U.S. forces. Therefore, orders issued by the President or the Secretary of Defense to a combatant commander provide the starting point in determining the legal basis. Laws are legislation passed by Congress and signed into law by the President and treaties to which the United States is party. Policies are executive orders, departmental directives and regulations, and other authoritative statements issued by government officials. This appendix summarizes some of the laws and policies that bear upon U.S. military operations in support of security force assistance. This summary does not replace a consultation with the unit's supporting staff judge advocate.

LEGAL AUTHORITY FOR SECURITY FORCE ASSISTANCE

B-2. If the Secretary of State requests and the Secretary of Defense approves, U.S. forces can participate in security force assistance. The request and approval can go through standing statutory authorities in Title 22, U.S. Code. Title 22 contains the Foreign Assistance Act of 1961, the Arms Export Control Act of 1976, and other laws. It authorizes security assistance, developmental assistance, security force assistance and other forms of bilateral aid.

B-3. The request and approval for security force assistance might also occur under various provisions in Title 10, U.S. Code. Title 10 authorizes certain types of military-to-military contacts, exchanges, exercises, and limited forms of humanitarian and civic assistance in coordination with the U.S. ambassador to the host nation. In such situations, U.S. forces may be granted status as administrative and technical personnel based on a status-of-forces agreement or an exchange of letters with the host nation. This cooperation and assistance is limited to liaison, contacts, training, equipping, and providing defense articles and services. It does not include direct involvement in operations. Assistance to police by U.S. forces is permitted but, generally, Department of Defense (DOD) does not serve as the lead government department. Without receiving a deployment or execution order from the President or Secretary of Defense, U.S. forces may be authorized to make only limited contributions during operations that involve security assistance.

EXISTING UNITED STATES LAW

B-4. The Constitution divides the power to wage war between the executive and legislative branches of government. Under Article I, Congress holds the power to declare war; to raise and support armies; to provide and maintain a navy; and to make all laws necessary and proper for carrying out those responsibilities. Balancing that legislative empowerment, Article II vests the executive power in the President and makes the President the Commander in Chief of the Armed Forces. This bifurcation of the war powers created an area in which the coordinate political branches of government exercise concurrent authority over decisions relating to the use of Armed Forces overseas as an instrument of U.S. foreign policy.

B-5. The Supremacy Clause of the Constitution (Article VI) states, in part, that all treaties made by the United States are the supreme law of the land. Therefore, ratified treaties, such as the United Nations (U.N.) Charter (see paragraphs B-7 through B-8) and the Geneva Conventions, create legal obligations on U.S. forces regarding their ability to perform various types of missions and functions.

B-6. The War Powers Resolution of 1973 ensures the collective judgment of both the executive and legislative branches to deploy U.S. forces by requiring consultation of and reports to Congress, in any of the following circumstances:

- Introduction of troops into actual hostilities.
- Introduction of troops, equipped for combat, into a foreign country.
- Greatly enlarging the number of troops, equipped for combat, in a foreign country.

B-7. The President is required to make such reports within 48 hours of the triggering event, detailing—

- The circumstances necessitating introduction or enlargement of troops.
- The constitutional or legislative authority upon which he bases his action.
- The estimated scope and duration of the deployment or combat action.

B-8. Since the War Powers Resolution was passed over the veto of President Nixon, no President has either conceded its constitutionality or complied fully with its mandates.

International Law and Treaties

B-9. The U.N. Charter became effective on 24 October 1945 after being ratified by the United States and a majority of other signatories. The U.N. Charter mandates that all member states resolve their international disputes peacefully. It requires that member states refrain in their international relations from the threat or use of force. The U.N. Charter also provides that all nations have the right to use self-defense to combat acts of aggression against them until such time as the Security Council takes action.

B-10. The U.N. Charter provides the essential framework of authority for the use of force, effectively defining the foundations for a modern *jus ad bellum.* Inherent in its principles are the requirements for necessity, proportionality, and timeliness. Necessity involves considering—

- The exhaustion or ineffectiveness of peaceful means of resolution.
- The nature of coercion applied by the aggressor State.
- The objectives of each party.
- The likelihood of effective community intervention.

B-11. Proportionality refers to limiting force in magnitude, scope, and duration to that which is reasonably necessary to counter a threat or attack. Timeliness refers to situations in which the delay of a response to an attack or the threat of attack attenuates the immediacy of the threat and the necessity to use force in self-defense.

B-12. U.S. forces obey the law of war. The law of war is a body of international treaties and customs, recognized by the United States as binding. It regulates the conduct of hostilities and protects noncombatants. The main law-of-war protections come from The Hague and Geneva Conventions. They apply at the tactical and operational levels and are summarized in ten rules shown in table B-1.

Table B-1. Summary of the law-of-war rules

• Fight only enemy combatants.
• Do not harm enemies who surrender; disarm them and turn them over to the chain of command.
• Do not kill or torture detainees.
• Collect and care for the wounded, whether friend or foe.
• Do not attack medical personnel, facilities, or equipment.
• Destroy no more than the mission requires.
• Treat all civilians humanely.
• Do not steal; respect private property and possessions.
• Do one's best to prevent violations of the law of war.
• Report all violations of the law of war to superiors.

B-13. During security force assistance operations, commanders must be aware of Common Article 3 of the Geneva Conventions and the status of insurgents under the laws of the host nation. One article contained in all four of the Geneva Conventions—Common Article 3—is specifically intended to apply to internal armed conflicts:

> *In the case of armed conflict not of an international character occurring in the territory of one of the High Contracting Parties, each Party to the conflict shall be bound to apply, as a minimum, the following provisions:*
>
> *(1) Persons taking no active part in the hostilities, including members of armed forces who have laid down their arms and those placed "hors de combat" by sickness, wounds, detention, or any other cause, shall in all circumstances be treated humanely, without any adverse distinction founded on race, colour, religion or faith, sex, birth or wealth, or any other similar criteria. To this end, the following acts are and shall remain prohibited at any time and in any place whatsoever with respect to the above-mentioned persons: (a) Violence to life and person, in particular murder of all kinds, mutilation, cruel treatment and torture; (b) Taking of hostages; (c) Outrages upon personal dignity, in particular humiliating and degrading treatment; (d) The passing of sentences and the carrying out of executions without previous judgment pronounced by a regularly constituted court, affording all the judicial guarantees which are recognized as indispensable by civilized peoples.*
>
> *(2) The wounded and sick shall be collected and cared for. An impartial humanitarian body, such as the International Committee of the Red Cross, may offer its services to the Parties to the conflict. The Parties to the conflict should further endeavor to bring into force, by means of special agreements, all or part of the other provisions of the present Convention. The application of the preceding provisions shall not affect the legal status of the Parties to the Conflict.*

Host-Nation Law and Status-of-Forces Agreements

B-14. After considering the type of baseline protections represented by fundamental human rights law, the military leader must be advised in regard to the other bodies of law that leader should integrate into planning and execution. This includes consideration of host-nation law. Because of the nature of most noninternational armed conflict missions, commanders and staffs must understand the technical and pragmatic significance of host-nation law within the area of operations. Although in theory, understanding the application of host-nation law during military operations is perhaps the simplest component, in practice it is perhaps the most difficult.

B-15. Judge advocates must recognize the difference between understanding the technical applicability of host-nation law and that law's application to control the conduct of U.S. forces during operations. In short, the significance of host-nation law declines in proportion to the movement of the operation toward violent conflict. Judge advocates should understand that U.S. forces enter other nations with a legal status that exists anywhere along a notional legal spectrum.

B-16. Judge advocates and planners must be familiar with any status-of-forces agreements or status-of-mission agreements that may be applicable. In any given country, there may be agreements short of status-of-forces agreements, such as diplomatic notes, on point. Relevant international documents affecting military operations may be difficult to locate. Several sources are available for locating applicable international agreements governing the status of U.S. forces or affecting military operations. Department of State (DOS) publications, such as *Treaties in Force,* contain unclassified international agreements. Both the relevant geographic combatant command's legal office and the defense attaché or military assistance group at the embassy normally have access to host-nation or international agreements impacting a military operation.

B-17. Status-of-forces and other forms of agreements frequently exist. They are essentially contractual agreements or treaties between two or more nations that establish the legal status of military personnel in foreign countries. Topics that are usually covered in a status-of-forces agreement include criminal and civil jurisdiction, taxation, and claims for damages and injuries. In the absence of an agreement or some other arrangement with the host nation, DOD personnel in foreign countries may be subject to its laws. Commanders ensure that all personnel understand status of U.S. forces in the area of operations and are trained accordingly.

LEGAL CONSTRAINTS ON THE SECURITY FORCE ASSISTANCE MISSIONS

B-18. U.S. law and regulation play a key role in establishing the parameters by which military forces may conduct security force assistance missions. These parameters tend to constitute constraints on the activities of military units. They range from the rules of engagement in combat situations to the ability to spend government funds in furtherance of a training or support mission.

GENERAL PROHIBITION ON ASSISTANCE TO POLICE

B-19. Historically, DOD is not the lead government department for assisting foreign governments. DOS is the lead when U.S. forces provide security assistance—military training, equipment, and defense articles and services—to host-nation governments. The Foreign Assistance Act of 1961 specifically prohibits assistance to foreign police forces except within specific exceptions and under a Presidential directive. When providing assistance to training, DOS provides the lead role in police assistance through its Bureau of International Narcotics and Law Enforcement Affairs. The President, however, may delegate this role to other agencies. This was done in 2004, when the President signed a decision directive granting the commander, United States Central Command, authority to train and equip Iraqi police.

TRAINING AND EQUIPPING FOREIGN SECURITY FORCES

B-20. All training and equipping of foreign security forces are specifically authorized. U.S. laws require Congress to authorize expenditures for training and equipping foreign forces. The laws of the United States also require DOS to verify that the host nation receiving the assistance is not in violation of human rights. The Secretary of Defense may authorize deployed U.S. forces to train or advise host-nation security forces as part of the mission. In this case, DOD personnel, operations, and maintenance appropriations provide an incidental benefit to those security forces. Numerous other programs to assist foreign forces are paid for with funds appropriated by Congress for that purpose. Consultation with a staff judge advocate or legal advisor early in the planning ensures the availability of funds for missions to train and equip foreign forces.

RULES OF ENGAGEMENT

B-21. *Rules of engagement* are directives issued by competent military authority that delineate the circumstances and limitations under which United States forces will initiate and/or continue combat engagement with other forces encountered (JP 1-02). Often these directives are specific to the operation. If there are no operation-specific rules of engagement, U.S. forces apply standing rules of engagement. When operating with a multinational force, commanders must coordinate the rules of engagement thoroughly.

FISCAL LAW CONSIDERATIONS

B-22. In security force assistance missions, like all operations, commanders require specific authority to expend funds normally found in the annual DOD appropriations act. As a general rule, operation and maintenance funds may not be used for security force assistance missions. In recent years, Congress appropriated additional funds to commanders for conducting more complex stability operations that were not typically covered by operation and maintenance funds. Examples include the Commander's Emergency Response Program (CERP), the Iraq Relief and Reconstruction Fund, Iraq Freedom Fund, and Commander's Humanitarian Relief and Reconstruction.

B-23. DOS has the primary responsibility, authority, and funding to conduct foreign assistance on behalf of the U.S. government. Foreign assistance encompasses any and all assistance to a foreign nation, including security assistance (assistance to the internal police forces and military forces of the foreign nation), development assistance (assistance to the foreign government in projects that will assist the development of the foreign economy or their political institutions), and humanitarian assistance (direct assistance to the population of a foreign nation). The legal authority for DOS to conduct foreign assistance is found in the Foreign Assistance Act of 1961.

B-24. There are two exceptions to the general rule requiring the use of Title 22 funds for foreign assistance. The first exception is for interoperability, safety, and familiarization training. DOD may fund the training (as opposed to goods and services) of foreign militaries with operation and maintenance dollars only when the purpose of the training is to enhance interoperability, familiarization, and safety training. Operation and maintenance funds may not be used for security force assistance training. This exception applies only to interoperability training. The second exception is congressional appropriation or authorization to conduct security foreign assistance. This exception allows DOD to fund foreign assistance operations if Congress has provided a specific appropriation or authorization to execute the mission.

Constraints on Funding

B-25. The Foreign Operations Appropriation Act (an annual DOS appropriations act) was enacted in 1997. Known as the Leahy Amendment, it prohibits the U.S. government from providing funds to a unit of the security forces of a foreign country if DOS has credible evidence that the unit has committed gross violations of human rights. The Leahy Amendment contains additional constraints on government funding of security force assistance missions. The law restricts funding until the Secretary of State determines and reports that the government of such country is taking effective measures to bring the responsible members of the security forces unit to justice.

B-26. Congress specifically appropriates funds for foreign assistance. The United States Agency for International Development (USAID) expends such funds under the legal authorities in Title 22. Provisions of Title 10 may also authorize amounts of money for these purposes. Standing funding authorities are narrowly defined and generally require advance coordination within DOD and DOS.

B-27. Effective foreign forces need training and equipment. U.S. laws require Congress to authorize such expenditures. U.S. laws also require DOS to verify that the host nation receiving the assistance is not in violation of human rights.

Funding Authorization for Security Force Assistance

B-28. All training and equipping of foreign security forces must be specifically authorized. Military and civilian personnel, operations, and maintenance appropriations should typically provide only an incidental benefit to those security forces. All other weapons, training, equipment, logistic support, supplies, and services provided to foreign forces must be paid for with funds appropriated by Congress for that purpose. Examples include the Iraq Security Forces Fund and the Afghan Security Forces Fund. Moreover, the President must give specific authority to DOD for its role in training and equipping foreign security forces. In May of 2004, the President signed National Security Presidential Directive (NSPD) 36 that made the commander, U.S. Central Command, under policy guidance from the chief of mission, responsible for coordinating all U.S. government efforts to organize, train, and equip Iraqi Security Forces, including police. Absent such a directive, DOD lacks authority to take the lead in assisting a host nation to train and equip its security forces.

Key Security Assistance and Foreign Assistance Funding Programs

B-29. Congress has appropriated funds to allow DOS to conduct its foreign assistance mission, including the following activities:

- Foreign Military Financing Program.
- International Military Education and Training Program.
- Economic Support Fund.
- Peacekeeping operations.
- Antiterrorism assistance.
- Global humanitarian demining.

- Refugee assistance.
- Personnel details.

B-30. DOS administers the following programs but no funds are appropriated to sustain them:

- Foreign Military Sales Program.
- Foreign Military Lease Program.
- Economy Act Security Assistance.
- U.S. Government Commodities and Services Program.
- Direct Commercial Sales Program.

B-31. Additional special support created by law for the foreign assistance mission includes—

- Excess defense articles.
- Presidential drawdowns.

B-32. DOS directly, or indirectly through USAID, finances numerous development assistance activities to address the following needs:

- Rule of law.
- Governance.
- Agriculture and nutrition.
- Population control.
- Health.
- Education.
- Energy.
- Environment improvement.

B-33. CERP (see paragraph B-19) is an example of a targeted humanitarian assistance fund program. CERP's primary purpose is to enable military commanders to respond to urgent humanitarian relief and reconstruction requirements within their fields of interest by carrying out programs that will immediately assist the host-nation populace. CERP was originally funded with seized Iraqi assets, but Congress later appropriated U.S. funds for the purpose. CERP is not applicable to missions outside Iraq and Afghanistan. Future missions, though, may have similar funding sources established to facilitate a humanitarian assistance mission. Commanders consult with the servicing judge advocate to determine the availability of funds.

Appendix C
Information Engagement

INFORMATION SUPERIORITY

C-1. Information shapes a commander's operational environment at every level. Twenty-first century conflicts occur in operational environments with instant communication. Because information systems are everywhere and exposure to news and opinion media is pervasive, an individual's words and actions can have immediate operational and strategic implications. Gaining and maintaining the trust of key publics has never been more important. Army forces conduct operations in the public eye and must maintain public confidence by being consistent in word and deed.

C-2. *Information superiority* is the operational advantage derived from the ability to collect, process, and disseminate an uninterrupted flow of information while exploiting or denying an adversary's ability to do the same (JP 3-13). Advisors build partner capacity in this mission area. Advisors ensure that host-nation counterparts integrate information superiority into operations as carefully as fires, maneuver, protection, and sustainment. Army information tasks contribute to information superiority. Information tasks are used to shape the operational environment (see FM 3-0). The information tasks are—

- Information engagement.
- Command and control warfare.
- Information protection.
- Operations security.
- Military deception.

C-3. Advisors prioritize the five information tasks based on the host nation's capabilities and strengths. This appendix discusses some information engagement capabilities that advisors can expect to employ while conducting security force assistance, as well as considerations when working with the media. For a further discussion of the other information tasks, see FM 3-0.

INFORMATION ENGAGEMENT CAPABILITIES

C-4. *Information engagement* is the integrated employment of public affairs to inform U.S. and friendly audiences; psychological operations, combat camera, and U.S. government strategic communications and defense support to public diplomacy, and other means necessary to influence foreign audiences; and leader and Soldier engagements to support both efforts (FM 3-0).

C-5. In general, advisors focus on information engagement because land operations occur among civilian populations. Advisors can have a significant impact by focusing on this task. Advisors work with host-nation counterparts to communicate information to the populace, build their trust and confidence in friendly forces, promote popular support for friendly operations, and positively influence public perceptions and behavior. Advisors integrate information engagement throughout the operations process.

C-6. Advisors ensure host-nation counterparts focus their information engagement activities on achieving desired effects locally. However, land operations always occur in a regional and even global context. Therefore, local information engagement activities must support and complement higher headquarters information engagement activities, host-nation and U.S. government strategic communications guidance (when available), and broader host-nation and U.S. government policy where applicable.

C-7. Three information engagement capabilities deserve particular attention as the advisor can directly influence them. These are public affairs, combat camera, and leader and Soldier engagements.

PUBLIC AFFAIRS

C-8. First, public affairs can have a profound and lasting effect on public perceptions of operational success or failure. This effect may have strategic consequences, even when related to tactical operations. Therefore, advisors ensure that public affairs efforts are integrated at tactical, operational, and strategic levels. (See paragraphs C-10 through C-20 for more information to help advisors working directly with the media.)

COMBAT CAMERA

C-9. Second, combat camera is used to document a wide range of tactical and operational successes and to counter enemy propaganda. If there is no dedicated host-nation combat-camera capability, the advisor recommends that a host-nation soldier be designated, equipped, and trained for this function. If that is not possible, the advisor should obtain a digital camera and, if necessary, photograph relevant host-nation activities personally. Positive photographs can have a tremendous benefit, and digital photos are easily taken and transmitted.

LEADER AND SOLDIER ENGAGEMENTS

C-10. Third, leader and Soldier engagements are a fundamental and highly effective method to influence people. They involve face-to-face interaction by leaders and Soldiers to influence the perceptions of the local populace. Military leaders engage key communicators and civilian leaders (often referred to as key leaders) through carefully planned meetings. Examples of key leaders include local tribal elders, elected or appointed officials, military and police commanders, and nongovernmental organization leaders.

WORKING WITH THE MEDIA

C-11. The appropriate public affairs officer normally manages media contacts. However, this is not always possible. A deployed advisor team seldom has a public affairs officer representative.

C-12. Occasions may arise when the host-nation press wants to question advisors. Even if an advisor team lacks a public affairs officer representative, advisors must avoid refusing to speak to the media. Apparent refusal to speak with accredited members of the media, for whatever reason, may inadvertently create strong negative impressions with strategic implications. Therefore, advisor teams designate a media spokesperson and keep this individual well informed. This individual coordinates media contacts for the advisor team. This individual ensures that official, published public affairs officer guidance and command policies regarding media relations, including anonymity and operations security, are observed. The following paragraphs provide general media relations guidance for advisor teams.

ENHANCING RELATIONSHIPS WITH THE MEDIA

C-13. Advisors strive to convey a positive impression and enhance relationships between the media and the military. If deemed acceptable within the public affairs guidance and command policies, the advisor team's designated media spokesperson—

- Maintains a list of reliable reporters and editors that cover operations in the team's operational area and keeps them informed of significant activities.
- Answers media inquiries promptly, accurately, and courteously.
- Finds out reporters' deadlines and uses them to the friendly force's advantage.
- Stresses the human aspects of a story that readers, viewers, or listeners can identify with, including the negative impact of opposing operations on the people.
- Points out how advisors and host-nation counterparts are working to address the needs of the unfortunate.
- Encourages the media to observe constructive work the host nation and advisors are doing and to interview them about their efforts.
- Identifies a location in advance for the press to take photos and videos.

Correcting Public Misconceptions

C-14. If information is withheld, journalists often fall back on speculation. Such speculation is usually inaccurate. However, it is often near enough to the truth that it is easily accepted by large sections of the public and possibly by established governments. In addition, partisan sources may find it advantageous to leak part of a story to the press to build up public support for their position. On occasion, such activities can grow into fully orchestrated press campaigns.

C-15. With modern satellites and communications technology, media are able to distribute reports and photographs extremely quickly. Incidents, sometimes fabricated or slanted toward a partisan viewpoint, may be aired in living rooms across the globe the same day.

C-16. However, advisors and host-nation counterparts frequently can correct public misconceptions and provide accuracy, adequate context, and proper characterization promptly. They can turn a potential liability into a major advantage by remaining prepared to coordinate and communicate with the media. They can take advantage of the media's powerful and far-reaching ability to communicate with critical audiences.

C-17. If the media report inaccurate or damaging information, advisors use caution in deciding on a course of action. Advisors (usually through the team's designated media spokesperson) inform their chain of command. Advisors consider the following questions:

- Is it important enough to correct or is it a detail that—in the long term—is not really important?
- How damaging is the charge, criticism, or error? Will a correction simply give greater visibility to an unfavorable point of view?
- Is a correction worth a restatement of the entire problem to new audiences who did not read or see it the first time?
- Is it possible to target only the audience originally exposed to the story?
- Can significant gain result from pointing out an error?

C-18. When incorrect or misleading information is released to the public through host-nation government or media sources, the advisor team determines a course of action appropriate to the situation. It must include a plan to ensure that relevant host-nation government officials and media recognize the inaccuracy of the information. This may prevent future incidents. Operational and combat experience since 2001 shows that in the struggle for information superiority, friendly forces must be candid and honest. Additionally, they must not allow lies disseminated by enemies to stand unchallenged, particularly in areas of great international interest.

Interviews With Advisor Team Members

C-19. Advisor teams sometimes arrange for media interviews with selected advisor team members. Properly conducted interviews can support the mission and even improve morale. When preparing for an interview, the team's designated spokesperson—

- Provides all team members with a single and simple theme to convey to the press, sometimes in the form of a mission narrative (a succinct account of the mission as a whole, explained consistently with the host nation's cultural view).
- Conducts rehearsals in front of video cameras, using mock interviews to practice communication of the theme.
- Selects interviewees based on their understanding of the theme covered, not their comfort level in front of a reporter. (Shyness is a normal trait that may add a human touch to a sensitive situation.)

C-20. If an advisor team member does not know the answer to a reporter's question, the individual should attempt to find out and get back to the reporter or refer the reporter to an appropriate source. This may establish the team as a helpful information source and develop a relationship for future balanced coverage.

C-21. Communicating with the media is the most difficult aspect of information engagement. Anything advisors say or do publicly can impact the mission. Therefore, advisors ensure consistency with national policy, professionalism, and clear communication. Advisors must know and follow the policies of the geographic combatant commander, the American Embassy, and the appropriate public affairs officer regarding media interviews. Additionally, advisors follow the specific guidelines shown in table C-1.

Table C-1. Specific guidelines for media interviews

- Maintain a friendly, yet professional attitude.
- Remain positive, greet interviewers, and welcome questions. (This is difficult under stressful or tragic conditions; however, a calm, mature appearance earns respect from audiences.)
- Use language that is clear and easy to understand.
- Avoid military jargon or terminology that others can misinterpret.
- Never enter an interview unprepared.
- Prepare notes and study them carefully; use facts to back up statements.
- Ensure the facts supporting their position are up to date and come to mind easily.
- Anticipate possible questions and think about various responses.
- Rehearse before interviews on location if possible and, if time permits, role-play tough questions.
- Learn and use reporters' names.
- Research interviewers' organizations, including—
  - The organizations' views or views they defend.
  - Their underlying motivations.
  - How members of the organizations previously conducted interviews.
- Keep the interests of the local nationals or other beneficiaries of the mission at the forefront.
- Avoid talking from the point of view of U.S. interests.
- Finish answering one question before answering another if two or more questions are asked at the same time.
- Avoid speculation and give only factual information that can be verified.
- Use caution in quoting statistics as such data may be easily disputed or reinterpreted.
- Avoid repeating a question that contains incorrect information or inflammatory language (this may result in a misquotation).
- Strive to tell the truth.
- Avoid saying *no comment*. (It gives the appearance the advisor has something to hide.)
- If unsure of something, admit it rather than mistakenly lie.
- If questioned about another agency's activities, refer to that agency for comment.
- Do not speak for other organizations.
- Avoid exaggeration or claims that cannot be backed up.
- Refuse all off-the-record discussions.
- Expect any statement they make to be quoted publicly.
- Avoid criticizing other groups or organizations, and avoid committing information fratricide.
- Remind reporters that photographing recognizable dead Soldiers, charts, maps, supply depots, or electronic warfare assets is off limits.
- Remind reporters that their personal security is not a primary military concern.
- Remain objective and calm.
- Demonstrate conduct befitting a representative the U.S. government
- Never give personal opinions and beliefs.
- Avoid reacting emotionally when interviewers appear skeptical or hostile.
- Maintain focus on the mission.

Glossary

SECTION I – ACRONYMS AND ABBREVIATIONS

ABCA	American, British, Canadian, Australian, and New Zealand Armies Program
AR	Army regulation
BCT	brigade combat team
CCIR	commander's critical information requirement
CERP	Commander's Emergency Response Program
COA	course of action
COIST	company intelligence support team
DA	Department of the Army
DOD	Department of Defense
DODD	Department of Defense directive
DOS	Department of State
DOTMLPF	doctrine, organization, training, materiel, leadership and education, personnel, facilities
FM	field manual
FSF	foreign security forces
IDAD	internal defense and development
ISR	intelligence, surveillance, reconnaissance
JP	joint publication
METT-TC	mission, enemy, terrain and weather, troops and support available, time available, civil considerations (mission variables)
MTT	military transition team
NSPD	national security Presidential directive
PIR	priority intelligence requirement
S-2	intelligence staff officer
S-3	operations staff officer
S-4	logistics staff officer
SFA	security force assistance
TC	training circular
UN	United Nations
US	United States
USAID	United States Agency for International Development

SECTION II – TERMS

foreign internal defense

(joint) Participation by civilian and military agencies of a government in any of the action programs taken by another government to free and protect its society from subversion, lawlessness, and insurgency. (JP 3-05)

foreign military sales

(joint) That portion of United States security assistance authorized by the Foreign Assistance Act of 1961, as amended, and the Arms Export Control Act of 1976, as amended. This assistance differs from the Military Assistance Program and the International Military Education and Training Program in that the recipient provides reimbursement for defense articles and services transferred. (JP 1-02)

***foreign security forces**

Forces, including but not limited to military, paramilitary, police, and intelligence forces; border police, coast guard, and customs officials; and prison guards and correctional personnel, that provide security for a host nation and its relevant population or support a regional security organization's mission.

host nation

(joint) A nation that receives the forces and/or supplies of allied nations, coalition partners, and/or NATO organizations to be located on, to operate in, or to transit through its territory. (JP 3-57)

humanitarian and civic assistance

(joint) Assistance to the local populace provided by predominantly US forces in conjunction with military operations and exercises. This assistance is specifically authorized by Title 10, United States Code, section 401, and funded under separate authorities. Assistance provided under these provisions is limited to (1) medical, dental, and veterinary care provided in rural areas of a country; (2) construction of rudimentary surface transportation systems; (3) well drilling and construction of basic sanitation facilities; and (4) rudimentary construction and repair of public facilities. Assistance must fulfill unit training requirements that incidentally create humanitarian benefit to the local populace. (JP 3-57)

internal defense and development

(joint) The full range of measures taken by a nation to promote its growth and to protect itself from subversion, lawlessness, and insurgency. It focuses on building viable institutions (political, economic, social, and military) that respond to the needs of society. (JP 3-07.1)

international military education and training

(joint) Formal or informal instruction provided to foreign military students, units, and forces on a nonreimbursable (grant) basis by offices or employees of the United States, contract technicians, and contractors. Instruction may include correspondence courses; technical, educational, or informational publications; and media of all kinds. (JP 1-02)

security assistance

(joint) Group of programs authorized by the Foreign Assistance Act of 1961, as amended, and the Arms Export Control Act of 1976, as amended, or other related statutes by which the United States provides defense articles, military training, and other defense-related services, by grant, loan, credit, or cash sales in furtherance of national policies and objectives. (JP 3-57)

security force assistance

The unified action to generate, employ, and sustain local, host-nation, or regional security forces in support of legitimate authority. (FM 3-07)

security sector reform

The set of policies, plans, programs, and activities that a government undertakes to improve the way it provides safety, security, and justice. (FM 3-07)

References

Field manuals and selected joint publications are listed by new number followed by old number.

REQUIRED PUBLICATIONS

These documents must be available to intended users of this publication.

FM 1-02 (101-5-1). *Operational Terms and Graphics.* 21 September 2004.

JP 1-02. *Department of Defense Dictionary of Military and Associated Terms.* 12 April 2001.

RELATED PUBLICATIONS

These documents contain relevant supplemental information.

JOINT AND DEPARTMENT OF DEFENSE PUBLICATIONS

Most joint publications are available online: <http://www.dtic.mil/doctrine/jpcapstonepubs.htm.>

DOD 5105.38-M. *Security Assistance Management Manual.* 3 October 2003. <http://www.dsca.osd.mil/samm/>

DODD 3000.05. *Military Support for Stability, Security, Transition, and Reconstruction (SSTR) Operations.* 28 November 2005. <http://www.dtic.mil/whs/directives/corres/html/300005.htm>

JP 3-0. *Joint Operations.* 17 September 2006.

JP 3-05. *Doctrine for Joint Special Operations.* 17 December 2003.

JP 3-07.1. *Joint Tactics, Techniques, and Procedures for Foreign Internal Defense (FID).* 30 April 2004.

JP 3-07.4. *Joint Counterdrug Operations.* 13 June 2007.

JP 3-57. *Civil-Military Operations.* 8 July 2008.

ARMY PUBLICATIONS

Most Army doctrinal publications are available online: <http://www.army.mil/usapa/doctrine/Active_FM.html>. Army regulations are produced only in electronic media. Most are available online: < http://www.army.mil/usapa/epubs/index.html>.

AR 11-31. *Army International Security Cooperation Policy.* 24 October 2007.

AR 12-1. *Security Assistance, International Logistics, Training, and Technical Assistance Support Policy and Responsibilities.* 24 January 2000.

FM 3-0. *Operations.* 27 February 2008.

FM 3-05.137. *Army Special Operations Forces Foreign Internal Defense.* 30 June 2008.

FM 3-05.202 (FM 31-20-3). *Special Forces Foreign Internal Defense Operations.* 2 February 2007.

FM 3-07. *Stability Operations.* 6 October 2008.

FM 3-24. *Counterinsurgency.* 15 December 2006.

FM 3-34.400 (FM 5-104). *General Engineering.* 9 December 2008.

FM 4-0 (100-10). *Combat Service Support.* 29 August 2003.

FM 4-02 (FM 8-10). *Force Health Protection in a Global Environment.* 13 February 2003.

FM 6-0. *Mission Command: Command and Control of Army Forces.* 11 August 2003.

FM 6-22 (22-100). *Army Leadership.* 12 October 2006.

FM 7-0. *Training for Full Spectrum Operations.* 12 December 2008.

FM 7-15. *The Army Universal Task List.* 27 February 2009.

TC 31-73. *Special Forces Advisor Guide.* 2 July 2008.

OTHER PUBLICATIONS

National Defense Strategy of 2008. Washington, DC: U.S. Government Printing Office, 2008. <www.defenselink.mil/news/2008 National Defense Strategy.pdf>

National Military Strategy of the United States of America. Washington, DC: U.S. Government Printing Office, 2004. <www.defenselink.mil/news/Mar2005/d20050318nms.pdf>

National Security Strategy of 2006. Washington, DC: U.S. Government Printing Office, 2006. <www.marforres.usmc.mil/docs/nss2006.pdf>

Quadrennial Defense Review of 2006. Washington, DC: U.S. Government Printing Office, 2006. <www.defenselink.mil/qdr/report/Report20060203.pdf>

Armed Forces. Title 10, U.S. Code. <http://www.loc.gov/law/help/guide/federal/uscode.php>

Arms Export Control Act of 1976. <http://www.loc.gov/law/help/guide.php>

The United States Constitution. <http://www.house.gov/house/Constitution/Constitution.html>

Foreign Relations and Intercourse. Title 22, U.S. Code. <http://www.loc.gov/law/help/guide/federal/uscode.php>

Foreign Operations Appropriation Act of 1997. <http://www.loc.gov/law/help/guide.php>

Foreign Assistance Act of 1961. <http://www.loc.gov/law/help/guide.php>

The Geneva Conventions of 1949. <http://www.loc.gov/rr/frd/Military_Law/MLR_site-map.html>

National Security Presidential Directive 36. "United States Government Operations in Iraq." 11 May 2004.

Treaties in Force 2007. <http://www.state.gov/s/l/treaty/treaties/2007>

Charter of the United Nations. <http://www.un.org/aboutun/charter/>

War Powers Act of 1973. <http://www.thecre.com/fedlaw/legal22/warpow.htm.>

OUTSIDE RECOMMENDED READING

von Steuben, Friedrick Wilhelm. *Regulations for the Order and Discipline of the Troops of the United States.* Philadelphia: Styner and Cist, 1779.

PRESCRIBED FORMS

None

REFERENCED FORMS

DA Form 2028. Recommended Changes to Publications and Blank Forms

Index

Entries are by paragraph number.

Entries are by paragraph number.

Entries are by paragraph number.

Entries are by paragraph number.

Made in the USA
Monee, IL
05 November 2026

67858278R00422